EQUILIBRIA, NONEQUILIBRIA, AND NATURAL WATERS

EQUILIBRIA, NONEQUILIBRIA, AND NATURAL WATERS

VOLUME I

RICARDO M. PYTKOWICZ
Oregon State University, Corvallis

A Wiley-Interscience Publication

JOHN WILEY & SONS

New York · Chichester · Brisbane · Toronto · Singapore

Library of Congress Cataloging in Publication Data:

Pytkowicz, Ricardo M. (Ricardo Marcos), 1929–
  Equilibria, nonequilibria, and natural waters.

  "A Wiley-Interscience publication."
  Bibliography: p.
  Includes index.
  1. Water chemistry.   2. Chemical oceanography.
3. Solution (Chemistry)   I. Title.
GB855.P97 1983   551.4   83-3598
ISBN 0-471-86192-8

Printed in the United States of America

10 9 8 7 6 5 4 3 2 1

PREFACE

This book has three purposes: (1) to guide the beginner from the fundamentals of chemistry, geology, and hydrology to the frontiers of aquatic chemistry; (2) to broaden the scope of the specialist in one of the water-related sciences; and (3) to stimulate the worker in any one field by presenting new concepts and speculations in his or her field.

The book consists of two volumes. Volume I, which may be called critical reviews and overviews, leads the reader from the fundamentals of chemistry, geology, oceanography, and thermodynamics through a rigorous study of electrolyte solutions. It is not merely a review of these fields—applications to aquatic chemistry and novel ideas are presented. Volume I provides the reader with a solid foundation for further work in the oceans, rivers, lakes, and geologic fluids.

Volume II also presents theoretical material for serious research but includes further applications. It covers acids, heterogeneous kinetics, new solid–solution equilibrium work, complexes, redox reactions, the geochemistry of trace metals, gas–water interactions, and a synoptic view of the main concepts presented in the book.

This work was supported in part by the Office of Naval Research and was completed while I was a guest at CSIRO-Oceanography in Australia. The formidable work of typing a work of this magnitude was achieved thanks to the dedication and talent of Mrs. Mary A. Stone. My wife, Joyce, did not only help in material ways but provided unflinching and priceless moral support.

RICARDO M. PYTKOWICZ

Corvallis, Oregon
April 1983

ACKNOWLEDGMENTS

I wish to express my deep gratitude to my wife Joyce, for her patient understanding and encouragement, as well as for her material help. Mrs. Mary Stone, who typed this work and took care of the logistics while I was abroad, showed uncommon ability and devotion to the job.

I benefited greatly through the years from discussions with colleagues and informal referees such as Werner Stumm, Wolfgang Nürnberg, Ross Heath, Arthur Chen, Geoffrey Skirrow, Bob Garrels, Dana Kester, Denis Mackey, Bob Collier, Charles Culberson, Sarah Ingle, Elliot Atlas, and John Hawley.

I thank Ocean Sciences–Office of Naval Research for partial support without which this book would not have been possible, and the National Science Foundation which sponsored part of the research presented in this work. My contact with the Thermodynamics Group, U.S. Bureau of Mines enlarged the scope of several chapters in this book.

It was a pleasure to work with the editorial and production staff of John Wiley & Sons because of their high standards of professionalism and dedication.

R.M.P.

CONTENTS

SYMBOLS

a_D	Debye distance of closest approach		k_B	Boltzmann constant
a_i	Activity of i		k_H	Henry's law constant
Cl‰	Chlorinity		m	Molality
c	Molarity, specific heat		N_A	Avogadro number
D_e	Dielectric constant		n	Number of moles
E	Internal energy		P	Pressure
Eh	Emf in the hydrogen scale		Q	Heat absorbed
F	Gibbs free energy		R	Gas constant
F	As a subscript refers to free quantity		S	Entropy
F'	The Faraday		S‰	Salinity
f	Number of degrees of freedom		s_i	Solubility of i
H	Enthalpy (heat content)		T	As subscript indicates total quantity
h	Moles of water of hydration per mole of salt		TA	Titration alkalinity
I	Ionic strength		TCO_2	Total dissolved inorganic carbon dioxide
J	Flux		W	Work done
K	Thermodynamic equilibrium constant; compressibility		X	Mole fraction
K'	Apparent equilibrium constant		β	Stability constant
K_{sO}	Thermodynamic solubility product		β_{pH}	Buffer capacity
K_s	Stoichiometric solubility product		γ_i	Practical activity coefficient
K^*	Association constant for ion pairs		Λ	Equivalent conductance
$K_{(c)}$	Step stability constant of complexes		ρ	Density
			τ	Residence or relaxation time
			Ω	Degree of saturation

INTRODUCTION

The purpose of this book is to present and illustrate key concepts in water chemistry. Its scope ranges from fundamental information to the frontier of this field and, at times, beyond it. This approach is used to provide readers of different backgrounds with an understanding of the topics and approaches used to solve problems in aquatic chemistry. Thus, this work serves as a foundation for further study and research. Care was taken to produce a book with a satisfactory balance between depth and breadth of coverage.

This work focuses on, but is not limited to, the concept of equilibrium because this is the state toward which most chemical processes occur, like the waters of rivers that flow toward the sea. Thus, equilibria provide us with a predictive capability for the direction and the extent of chemical reactions. The absence of equilibrium is itself of interest because it can shed light on the controlling rate processes.

The project is aimed at a wide audience as research in the aquatic sciences and is primarily a multidisciplinary endeavor. For this reason, at times introductory information is supplied, as a service to nonchemists and colleagues from outside the earth sciences. Still, concepts in this work also extend to the limits of our knowledge of aquatic chemistry.

The carbon dioxide–carbonate system is used often to illustrate concepts because of its intrinsic importance and because it has features in common with many other chemical systems.

The symbol (RMP) refers to thoughts by the author which may not yet have been submitted to the critical eye of the scientific community through journal referees or audiences at meetings. I apologize if I inadvertently claim credit for thoughts already presented by colleagues.

CHAPTER ONE

A CHEMICAL REVIEW AND THE EARTH

In this chapter, a brief review of the highlights of chemistry for nonchemists is presented. This will obviate the need for them to wade through thick introductory texts to refresh the hazy memories from freshman days.

1.1 DEFINITIONS ILLUSTRATED BY OUR PLANET

1.1.1 Chemistry

Chemistry is concerned with the forms in which matter is present and the transformations that it undergoes. It also seeks the reasons for these changes and their impact upon nature and humans. As an example of chemical transformations (reactions), carbon dioxide gas and water react with each other and are converted, in the presence of sunlight, into the green plant world upon which we depend so much. This reaction is an important component of the biogeochemical cycle of carbon dioxide in nature. Other examples are the formation of limestones by calcareous organisms, and the weathering of feldspars in rocks to soil clays. Remnants of calcareous organisms deposited in marine sediments are eventually uplifted by mountain building forces and form, for example, the Alps. Weathering dissolves limestones and rivers become an important source of calcium for plants, animals, and humans. The formation of clays from the mineral feldspar, present in rocks, is also vital to life, as it promotes the growth of plants. Furthermore, calcium carbonate and clays, as we shall see, play a major role in natural waters.

1.1.2 Elements

Elements are the simplest forms of matter if one does not include subatomic particles such as electrons, neutrons, positrons, and so on. Iron, for example, is an element because it is made only of iron when it is not associated to other elements, such as oxygen in the mineral magnetite. Water is not an element because it is composed of hydrogen and oxygen. There are a little over 100 elements from which the more complex forms of matter are built. Thus, elements are the building blocks of our universe. Some common ones are presented in Table 1.1.

Atoms are the smallest particles that still retain the properties of the elements. The atomic weights of the elements are the weights of their atoms relative to that of the carbon 12 isotope which is taken as 12.00000 awu (atomic weight units). Isotopes are forms of the same element which have different atomic weights due to different nuclear structures. Carbon, for example, has three isotopes; C^{12}, C^{13}, and C^{14}. Thus, C^{12} is the standard and the relative weights (actually, masses) are referred to it. By definition then, 1 awu is $\frac{1}{12}$ of the weight of the C^{12} atom.

Atomic weights were so designed that, when expressed in grams, the elements contain the same number of atoms, for example, 26.9815 g-Al and 79.909 g-Br (see Table 1.1). Atomic weights in grams are known as gram-atoms (g-atom) so that 26.9815 g-Al, for example, constitute 1 g-atom Al. The number of atoms contained in a g-atom is always $N_A = 6.02217 \times 10^{23}$, known as Avogadro's number. This number arose from the molecular theory of gases. It is convenient to have units such as g-atom which are related to a given number of particles because chemical reactions always involve fixed proportions of particle numbers. As an example, the formation of carbon dioxide always uses up carbon and oxygen atoms in a one to two proportion. By the way, 1 awu = 1 g/6.02217 × 10^{23},

TABLE 1.1   Some elements, their symbols, and atomic weights.

Name	Symbol	Atomic Weight	Name	Symbol	Atomic Weight
Aluminum	Al	26.9815	Lithium	Li	6.939
Argon	Ar	39.948	Magnesium	Mg	24.312
Barium	Ba	137.34	Manganese	Mn	54.9380
Boron	B	10.811	Mercury	Hg	200.59
Bromine	Br	79.909	Nitrogen	N	14.0067
Calcium	Ca	40.08	Oxygen	O	15.9994
Carbon	C	12.01115	Phosphorus	P	30.9738
Chlorine	Cl	35.453	Platinum	Pt	195.09
Copper	Cu	63.54	Potassium	K	39.102
Fluorine	F	18.9984	Silicon	Si	28.086
Gold	Au	196.967	Silver	Ag	107.870
Helium	He	4.0026	Sodium	Na	22.9898
Hydrogen	H	1.00797	Strontium	Sr	87.62
Iodine	I	126.9044	Sulfur	S	32.064
Iron	Fe	55.847	Uranium	U	238.03
Lead	Pb	207.19	Zinc	Zn	65.37

which is too small in practice; hence, the use of the g-atom.

Complex models have been proposed for the origin of the elements. Basically, they appear to result from nuclear reactions in severe energetic environments such as that of active red stars.

EXAMPLE 1.1.   What is the weight of $\frac{1}{2}$ g-atom of carbon and how many atoms does it contain?
weight $= 12.01115/2 = 6.0558$
number of atoms $= 6.02217 \times 10^{23}/2 = 3.01108$
$\times 10^{23}$

1.1.3 Compounds

Compounds are formed when elements combine in definite proportions to form new substances. As examples, each atom of carbon can combine with one atom of oxygen to form carbon monoxide,

$$C + \tfrac{1}{2}O_2 \rightarrow CO \qquad (1.1)$$

the toxic substance from car exhausts, or with two atoms of oxygen

$$C + O_2 \rightarrow CO_2 \qquad (1.2)$$

to form carbon dioxide, which is used in photosynthesis, for the carbonation of soft drinks, and so on. Molecules are the smallest particles that retain the properties of the compounds.

Note that in these two reactions O_2 is written instead of $2 \times O$. This is done because oxygen gas consists of oxygen molecules containing two atoms. Helium gas, on the other hand, is monoatomic and can be represented by He.

EXAMPLE 1.2.   CALCULATION OF THE MOLECULAR WEIGHT. The molecular weight of CO_2 is

$$MW_{CO_2} = AW_C + 2 \times AW_O = 44.01 \text{ awu} \text{ (i)}$$

where AW and MW represent atomic and molecular weights.

EXAMPLE 1.3.   CALCULATION OF G/L OF CONSTITUENTS OF CHEMICAL REACTIONS. In the reaction

$$CO_2 + H_2O + CaCO_{3(s)} = Ca^{2+} + 2HCO_3^- \text{ (i)}$$

how many g/L of CO_2 are needed to dissolve 10^6 g/L of solid $CaCO_{(s)}$?

Since 1 mole of CO_2 reacts with 1 mole of $CaCO_3$ and the respective MW's are 44.010 and 100.06,

$$\text{g-}CO_2/L = \frac{44.01}{100.09} \times 10^{-6} = 0.4397 \times 10^{-6} \text{ (ii)}$$

We could obviously have used moles, that is, 44.01 g and 100.09 g, with the same final result.

Ions are charged atoms or groups of atoms. The main constituent of kitchen salt is sodium chloride, NaCl, which dissociates in water to

$$NaCl \rightarrow Na^+ \; Cl^- \tag{1.3}$$

Sodium sulfate, on the other hand, dissociates according to

$$Na_2SO_4 \rightarrow 2Na^+ + SO_4^{2-} \tag{1.4}$$

Ions with positive charge are called cations and those with negative ones are the anions. They are known to exist because they can carry electric current in solution and are formed when, for energetic reasons, the cations donate electrons to the anions. Ions dissolve readily because their charges attract water molecules (hydration) and this process releases energy. The release of energy makes the process spontaneous as in the case of a falling object. More careful statements on this subject are made in Chapters 4 and 5. Ionic weights are essentially those of their atoms because the weight of electrons is slight.

Molecules and ions can coexist in solution. Acetic acid undergoes the partial dissociation

$$HAc \rightleftarrows H^+ + Ac^- \tag{1.5}$$

where Ac^- is a shorthand notation for CH_3COO^- The two-way arrows show that HAc molecules in solution exist at chemical equilibrium with H^+ and Ac^- ions, whereas arrows such as are shown in Equation (1.3) indicate a complete reaction with no NaCl molecules left in solution. We shall see later that the situation is not quite as simple as this.

1.1.4 Distribution of Substances in the Earth

This subject is introduced now and will be treated in greater detail in Chapters 2 and 13.

It is interesting to examine how certain elements are distributed in the earth as a whole, without considering variations in depth and compounds formed.

In the construction of tables such as Table 1.2, each element is entered after the following steps. Calcium may be present in limestones, mostly $CaCO_3$, dolomites, $CaMg(CO_3)_2$, Ca–feldspars, and so on. The weight of calcium in the earth is w_{CaCO_3}.

TABLE 1.2 Composition of the earth as a whole (Mason, 1966) in weight percent (wt%).

Element	wt%	Element	wt%
Fe	35	Na	0.14
Ni	28	K	0.01
Co	13	Ti	0.20
O	17	Cr	0.03
Si	2.7	P	0.07
Al	2.7	S	0.04
Mg	0.61	MN	0.09
Ca	0.44		

The form of $CaCO_3$ is then calculated as

$$w_{Ca} = \frac{AW_{Ca}}{MW_{CaCO_3}} + w_{CaCO_3} \tag{1.6}$$

where w_{CaCO_3} is the total weight of calcium carbonate in g, kg, or tons. The same operation is repeated for all the calcium compounds to obtain the total amount of calcium estimated to be in elemental form. This procedure is repeated for all the elements and weight percents are calculated. Table 1.2 yields an idea of the elemental composition of the earth.

Often another procedure is used, especially when dealing with the crust of the earth. Then, the substances are presented as if they were oxides, as is shown in Table 1.3. Thus

$$w_{CaO} = \frac{MW_{CaO}}{MW_{CaCO_3}} \times w_{CaCO_3} \tag{1.7}$$

This is a convention loosely based upon the fact that oxides are frequently found compounds in the earth. Thus, $CaCO_3$, for example, can be represented as $CaCO_3$ or as $CaO + CO_2$ and the silicate $Al_3Si_4O_8$ contains K_2O, SiO_2, and Al_2O_3.

The composition of the earth is not uniform and

TABLE 1.3 Relative composition of the crust (Mason, 1966).

SiO_2	55.2	CaO	8.8
Al_2O_3	15.3	Na_2O	2.9
FeO_3	2.8	K_2O	1.9
FeO	5.8	TiO_2	1.6
MnO	0.2	P_2O_5	0.3
MgO	5.2		

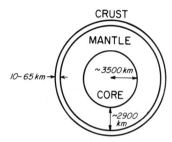

FIGURE 1.1   The structure of the earth.

6. Fe_2S (pyrite) produced in oxygen-depleted environments.

In general, igneous rocks (e.g., granites and basalts) are brought up from the mantle by magma, metamorphic rocks, such as marble, are formed when sedimentary rocks are buried and exposed to high temperatures and pressures, and sedimentary rocks (e.g., limestones and dolomites) are formed by the weathering, sedimentation, and postdepositional alteration of igneous rocks.
Schematically,

Of course, not all igneous rocks are weathered and not all sedimentary rocks are recycled because then we would not have a record of times past. (See Chapter 2 and Chapter 8 in Volume II for further details.)

It is interesting to observe that elemental iron, which was mentioned earlier, decreases from a large value in the core to 2.8%, on an oxide basis, in the crust. Still, it is an exceedingly important constituent because of its role in plant life, animal metabolism, industry, and the formation of rocks. The control of the mineral-forming reactions that occur results from the affinities (specificities) of the metals for species such as SiO_3^- (silicate group), OH^- (hydroxyl group) O_2, S^{2-} (sulfide group), Cl^-, and so on. The zonation of the earth is related to the relative densities of the substances present, which increase toward its center, and to their chemical reactions.

The composition of the crust is mainly that of igneous rocks (brought up by volcanic-type activity) because the sedimentary rocks plus the atmosphere and the hydrosphere constitute a small but very important fraction of the substances present. The crust contains primarily silicates of aluminum, calcium, magnesium, sodium, potassium, and iron. A fraction of these silicates, after chemical erosion (weathering), yields the above metals as well as the heavy ones such as lead, mercury, cadmium, and arsenic. It also supplies the clays required for fertile soils and, through sedimentary processes, carbonates, hydroxides, halides (Cl, I, Br, F), and sulfides.

reflects the importance of chemical reactions within and on the surface of our globe. The earth has three layers (see Figure 1.1); the thick inner core, which contains primarily elemental iron and nickel, the mantle, and the crust, which is the thin skin 3.5 km in depth, upon which we live. Mantle materials, when pushed by pressure toward the crust, undergo general reactions of the type

$$Fe–silicate + M = M–silicate + Fe \quad (1.8)$$

This reaction, among others, leads to the depth differentiation of the substances in the earth.

A simplified description of the composition of the three layers is shown below and in Table 1.3.

1. Core—Mainly elemental iron and nickel.
2. Mantle—Silicates, primarily of magnesium and iron and perhaps iron sulfides.
3. Crust—All the elements and compounds, including among others, silica, silicates, aluminum–silicates, oxides, carbonates, and halides.

The formulas and ultimate sources of some of the crustal compounds are:

1. H_2O (water) of volcanic origin as steam.
2. $KAlSi_3O_8$ (potassium feldspar), of magmatic origin, and present in igneous rocks.
3. SiO_2 (silica) present as quartz (crystalline), as amorphous silica (opal) in the shells of aquatic organisms, and present in siliceous sediments.
4. $CaCO_3$ (calcium carbonate) found in sedimentary rocks such as limestones and dolomites.
5. NaCl (halite) formed by evaporation.

One of the several possible reactions which illustrates the above paragraphs is that for the igneous mineral potassium feldspar:

$$2KAlSi_3O_8 + H_2CO_3 + nH_2O = K_2CO_3$$

potassium   carbonic   water   potassium
feldspar   acid   carbonate

$$+ Al_2(OH)_2Si_4O_{10} \cdot nH_2O + SiO_2 \qquad (1.9)$$

clay mineral   silica

H_2CO_3 results from the dissolution of atmospheric CO_2 in water, K_2CO_3 is soluble in natural waters as K^+, HCO_3^-, and CO_3^{2-}, and SiO_2 in general is present as a solute and as a finely divided hydrated silica in the aquatic environment.

The hydrated silica results from the reaction

$$SiO_2 + 2H_2O = Si(OH)_4 \qquad (1.10)$$

which produces dissolved $Si(OH)_4$ and which coats SiO_2 solid particles. SiO_2 is found, for example, in the tests (shells) of marine organisms such as diatoms. Silica in organisms is noncrystalline (amorphous silica or opal) but it can also be found as quartz, a crystalline, relatively unreactive form, present as a mineral in igneous rocks. Quartz is broken down physically during weathering and fragments are deposited in rivers, lakes, and oceans.

Carbon dioxide (CO_2), carbonic acid (H_2CO_3), and calcium carbonate ($CaCO_3$) formed during volcanic activity (CO_2) and during weathering play roles in metabolic activity as CO_2 is used in photosynthesis, $CaCO_3$ supplies calcium ions for the hard parts of organisms, while the system H_2CO_3–$CaCO_3$ buffers the pH's of body fluids and natural waters (rivers, lakes, and oceans) at values at which life can exist.

We then observe, in a preliminary way, the significance of chemical reactions upon biological, geological, and industrial processes.

1.1.5 Mixtures

Mixtures can have their constituents present in continuously varying proportions and the constituents can be separated by physical means. Examples are a mixture of iron filings and sulfur which can be separated by a magnet and a solution of kitchen salt in water which can be separated by evaporation.

1.2 CONTENTS OF NATURAL WATERS

In natural waters, one finds the following size distributions of substances:

1. *Particulate matter visible to the naked eye.* Examples include settling quartz particles resulting from the physical breakdown of rocks, dispersed organic particles present throughout the water and resulting from the breakdown of aquatic life, and floating pieces of wood.

2. *Colloids.* Visible in a microscope and dispersed in the water. They range in size, by definition, from $0.2\mu m$ (micrometer = 10^{-6} cm) to $0.002 \mu m$. Examples include small aggregates of organic molecules, proteins, and clays. From a chemical point of view, proteins and clays are single and well defined rather than aggregated entities. The bonds within them are regular and predictable.

3. *Solutes in true solution.* Examples include oxygen molecules, helium atoms, and ions of salts such as NaCl. The ions are charged atoms, such as Na^+ and Cl^-, which are present when salts are dissolved.

The size boundary between solutes and colloids is arbitrary and is based upon the limit of resolution of optical microscopes. Thus, in principle, true solutes are invisible. The size boundary between particulate and "dissolved" (dissolved plus colloidal) matter is also conventional and is often based upon the pore size of the filter used for their separation (for example, a $0.450 \mu m$ Millipore filter).

Solutions play a role in the laboratory, in our bodies, and in our environment. Our body fluids are salt solutions that control a variety of processes such as intercellular diffusion, the pH and its buffering, and the osmotic regulation. Rivers bring breakdown products of erosion to lakes and oceans. This leads to the sedimentation of the products, often in a manner controlled by processes in solution, such as the uptake by organisms, and the sediments eventually return to our environment as rocks by seafloor spreading, partial metamorphosis, uplift, and evaporation. An example of dissolution is that of oxygen, which permits marine, riverine, and lake fishes to survive. Many more examples will appear later in this book.

In the case of true solutions, we can distinguish the following categories:

1. Nonelectrolytes, which consist of uncharged particles. Examples include gases, such as oxygen, and sugar molecules.
2. Strong electrolytes, which, except for coulombic interactions between oppositely charged ions, dissociate completely. For example,

$$HCl \rightarrow H^+ + Cl^-$$
$$\text{(hydrochloric acid)} \quad (1.11)$$

3. Weak electrolytes, which only dissociate in part as in the case of Equation (1.5), where HAc represents acetic acid and the equality represents the equilibrium $\rightleftarrows$ in which the rate of dissociation of HAc equals the rate of recombination of $H^+ + Ac^-$.

1.3 CONCENTRATION UNITS AND CONVERSION EQUATIONS

In this section the concentration units and the equations for their interconversion which are most frequently used in oceanography and solution chemistry in general are presented. Units and conversion equations of special interest to oceanography will be covered in Chapter 3.

It is necessary, when we study chemical reactions, to define the volume or weight in which they occur. It would not do in Equation (1.5), for example, to react x g-H^+/L with y g-Ac^-/2L because these ions interact in the same space. Units of amounts usually are g, kg, cm^3, and L.

Reactants and products are expressed in μg (10^{-6} g), g, moles, equiv, g-equiv, and so on. Thus, the units of concentration, in principle, can be expressed as shown below. Of course, many other units can be devised.

We add the following to the units mentioned earlier.

1.3.1 Primary Quantities

Atomic Weight (AW) is the weight of an element relative to the C^{12} isotope, expressed in awu.
Molecular Weight (MW) is the weight of a molecule of a compound, expressed in awu.
Gram-Atom (g-atom) is the atomic weight expressed in g. Mole is the molecular weight expressed in g.
Gram-Equivalent (g-equiv) is the g-atom or mole divided by the charge of the ion. For example, 1 g-atom of Ca^{2+} weighs 40.0 g while 1 g-equiv weighs 20.04 g. The reason for this definition is that, when $CaCl_2$ is formed there is 1 g-atom Ca^{2+} for 2 g-atom Cl, but there are 2 g-equiv Ca^{2+} for 2 g-equiv of Cl^-. Thus, we can predict that each ion of Ca^{2+} (2 equiv) will react with two atoms of Cl^- for electrical neutrality.
Equivalent (equiv) is the maximum number of positive or negative charges that may be produced from an atom (Al has 3 equiv) or molecule ($CaCl_2$ has 2 equiv).

It is useful to observe that 1 g-atom of an element or 1 mole of a compound contains 6.02217×10^{23} atoms or molecules. Many properties of solutions depend upon the number of particles present per unit weight or volume of the solution.

1.3.2 Concentration Units

In these definitions, the symbol soln (solution) can be replaced by SW (seawater). The most frequently used units are:

Molarity (*M*), moles/L-soln.
Normality (*N*), g-equiv/L-soln.
Molality (*m*), moles/kg-H_2O, per kg of water in the solution.

	μg	g	moles	equiv	g-equiv	g-atom
g	μg/g	g/g	moles/g	equiv/g	g-equiv/g	g-atom/g
kg	μg/kg	g/kg	moles/kg	equiv/kg	g-equiv/kg	g-atom/kg
cm^3	μg/cm^3	g/cm^3	moles/cm^3	equiv/cm^3	g-equiv/cm^3	g-atom/cm^3
L	μg/L	g/L	moles/L	equiv/L	g-equiv/L	g-atom/L

Parts per thousand (‰), g/kg-soln, at STP (standard temperature and pressure). The standard temperature is 0°C and the standard pressure is 1 standard atmosphere, which equals the average sea level pressure of 76.0 cm Hg.

Weight per weight units are recommended because the concentrations are independent of the temperature and pressure. One kg-H_2O or 1 kg-soln contains the same amount of salt before and after compression or thermal expansion but 1 L does not. The symbol (‰) refers primarily to dissolved gases although it can be used for solutions.

1.3.3 Conversion Equations

In general,

$$weight/kg\text{-}H_2O = (weight/L\text{-}soln)$$

$$\frac{1}{\rho_{soln} - 10^{-3}M_{soln}(MW)_{solute}} \quad (1.12)$$

The weight units may be g, kg, g-atom, moles, and so on. In moles,

$$m_{soln} = M_{soln}\frac{1}{\rho_{soln} - 10^{-3}M_{soln}(MW)_{solute}}$$

$$(1.12.1)$$

For an electrolyte in seawater

$$m_{SW} = M_{SW}\frac{1}{\rho_{SW} - 10^{-3}\Sigma_T} \quad (1.12.2)$$

where Σ_T is the sum in g of the electrolytes present in 1 L.

$$weight/kg\text{-}H_2O = (weight/kg\text{-}soln)$$

$$(\rho_{soln} - 10^{-3}g/kg\text{-}soln) \quad (1.12.3)$$

$$weight/L\text{-}soln = (weight/kg\text{-}H_2O)$$

$$\frac{\rho_{soln}}{1 + 10^{-3}M_{soln}(MW)_{solute}} \quad (1.13)$$

$$M_{soln} = m_{soln}\frac{\rho_{soln}}{1 + 10^{-3}M_{soln}(MW)_{solute}} \quad (1.13.1)$$

$$M_{SW} = m_{SW}\frac{\rho_{SW}}{1 + 10^{-3}\Sigma_T} \quad (1.13.2)$$

$$weight/kg\text{-}soln = (weight/L\text{-}soln)\frac{1}{\rho_{soln}} \quad (1.14)$$

$$weight/L\text{-}SW = (weight/kg\text{-}SW)\frac{1}{\rho_{SW}} \quad (1.14.1)$$

$$weight/kg\text{-}H_2O = (weight/kg\text{-}SW)$$

$$(\rho_{SW} - 10^{-3}\Sigma_T) \quad (1.14.2)$$

$10^{-3}M_{soln}(MW)_{soln}$ represents the number of g of solute per mL of solution so that $\rho_{soln} - 10^{-3}M_{soln}(SW)_{soln}$ is the density compensated for the contribution of the solute.

One mL = 10^{-3} L where the liter is the volume occupied by 1.000 g-H_2O at 3.96°C and at a pressure of 76.0 cm Hg. The cm^3, however, is based upon the standard meter, 1 m = 100 cm and is, therefore, based upon the length of the standard platinum rod. Due to the use of two types of standards, the volume and the length ones, the mL and the cm^3 do not correspond exactly but 1 mL = 1.000027 cm^3. This difference is negligible for our purposes. Let us consider some examples involving additional units next.

EXAMPLE 1.4. μg-atom ILLUSTRATED. One often uses μg-atom/kg-SW or μg-atom/L-SW (10^{-6} g = 1 μg and 10^{-6} g-atoms = 1 μg-atom). As an example, $4 \times 10^{-6}/(AW)_{Ca} = 4 \times 10^{-6}/40 = 10^{-7} = 0.1$ μg-atom.

EXAMPLE 1.5. μg-atom ILLUSTRATED FOR A GAS. Gases are usually reported in mL/L (STP). Upon occasion, μg-atom/L are used instead. One mole of any gas contains 22,400 mL (STP). Because O_2 has two atoms, 1 g-atom is contained in 11,200 mL/STP. Therefore, 1 μg-atom corresponds to 0.0112 mL (STP).

EXAMPLE 1.6. A solution of density 1.020 at 25°C and 1 atm pressure contains 2.00 g-$CaCl_2$/L-soln. What are the molarity, the normality, and the molality at 25°C and 1 atm?

molarity $= 2/MW = 0.00450$ M
normality $= 2 \times$ molarity $= 0.00900$ N

moles/kg-soln $= \dfrac{0.00450}{\rho} = 0.00442$

molality $= (1 - 10^{-3} \times \dfrac{2}{1.020}) \times 0.00442$

$= 0.00441$ m

1.3.4 Density and Specific Gravity, Overturn in Natural Waters

For the most accurate work we must distinguish between the density ρ and the specific gravity, although the term density will be used throughout this book.

> *Density*—Mass per unit volume of a substance at a given temperature, solute content in the case of a solution, and pressure. It is usually expressed in g/cm^3 in the cgs (centimeter-gram-second) system of units.
>
> *Specific Gravity*—Ratio of the density of a substance relative to that of pure water at 4°C.
>
> *Specific Volume*
>
> $$\frac{1}{\rho}, \quad \text{usually in} \quad cm^3/g \qquad (1.15)$$
>
> *Mass*—Measure of the quantity of matter, usually expressed in g, kg, lb, and so on, while weight is the attractive force on the body (the mass) toward the earth. The unit of weight in the cgs system is the dyne, 1 dyne $= cm \cdot g/sec^2$. Weight and mass are almost equal numerically. The weight, however, changes with distance from the center of the earth. These terms will be used interchangeably although it should be remembered that the weights vary slightly with latitude and altitude.

The density of most natural waters ranges roughly from 1–1.025 although brines and laboratory solutions can have values above 1.025. The densities of pure water and laboratory solutions can be found in volumes such as the *Handbook of Chemistry and Physics* (CRC Press) and sources for seawater will be mentioned in Chapter 3.

The densities of lakes with low salinities (salt contents) are nearly unity.

The specific gravity measured in the past did not take into account the isotopic composition of the standard pure water used at 4°C and, therefore, numerically it deviates slightly from the density (Pytkowicz and Kester, 1971).

The density or the specific volume is an important quantity in the study of motion in water bodies such as lakes and oceans because denser waters tend to sink. The numbers below apply to pure water at atmospheric pressure.

Form	Temperature	Specific Volume (cm^3/g)	
Ice	−40°C	1.085	Density increases
Melt	0°C	1.091	
Water	4°C (actually 3.96°C)	1.0001	
Water	<4°C		decreases

Water columns are stratified when they are not overly stirred by winds. Thus, less dense waters tend to remain on top and denser ones sink.

Normal lake waters, excluding hypersaline ones, behave like pure water. Thus, surface waters are gradually cooled in winter and sink to the bottom at 4°C. This brings oxygen and nutrients needed by organisms to the deep waters. Oxygen is often depleted in summer and fall during the decomposition (oxidation) of organic matter. After the whole lake is filled with 4°C water, further surface cooling produces freezing. The ice remains on the surface, because it is less dense than the water, and prevents the freezing of the whole water column which would occur if it were denser than the 4°C water. This permits life to continue beneath the surface, a result of the anomalous thermal behavior of water.

The anomalous density of water does not occur for high salt contents, such as those that occur for seawater of salinity (see Chapter 3) above 24.7‰ (Gross, 1972). The oceanic water column remains stratified but not in a static manner as cold and/or salty dense waters sink to their proper density levels. Sea ice, formed in winter, has a low salt content because most of the solutes are expelled during freezing. It has, therefore, a low density relative to those of saline waters, and floats on the sea surface. Saline oceanic surface waters have a density of roughly 25 g/cm^3, which is higher than that of ice.

1.4 EQUILIBRIUM AND STEADY-STATE CONCEPTS

In this section I appeal primarily to your intuition to contrast chemical equilibria and steady states. A more rigorous definition of chemical equilibrium is presented in the chapter on thermodynamics (Chapter 4), which some readers may wish to skip.

Equilibrium and steady state are two examples of time-invariant (stationary) conditions. Equilibrium is essentially a case of macroscopic rest with a cancellation of forces, whereas the steady state is a case in which there are net fluxes of matter and/or energy. No net work is done.

Those two conditions occur in physical, chemical, biological, and geochemical systems and provide different controls for the states of living, laboratory, and mineral systems.

An insulated wire at a uniform temperature is in thermal equilibrium because there is no net flow of heat (see Figure 1.2). The equilibrium is macroscopic in that individual atoms still undergo thermal vibrations. The same is true of a gas in a cylinder with a piston, which is at equilibrium with the external temperature and pressure. The molecules undergo thermal motions but no net energy is used (flows) because the piston is at rest.

On the other hand, if a time-invariant thermal gradient is imposed upon the wire, there is a stationary flow of heat, that is, the heat flux is constant. In this case, energy must be supplied to the system (the wire plus the source at T_1 and the sink at T_2) at T_1 to hold this temperature constant.

Further examples and an analysis of equilibrium conditions are presented in the following pages.

The chemical equilibrium state is very important in nature because chemical reactions tend toward it. A knowledge of this state helps us, therefore, predict the direction and the maximum extent of chemical processes. It is essential to distinguish between equilibrium and steady states because, although both represent stationary (time-invariant) conditions, they pertain to altogether different control mechanisms. Furthermore, they are described by different systems of equations in, for example, the study of geochemical cycles and their perturbations (see Chapter 2).

1.4.1 Mechanical Equilibrium and Steady State

Let us ease into the concept of chemical equilibrium by first looking at mechanical equilibrium, which is illustrated in Figure 1.3.

Balls released on the high slopes of the well in Figure 1.3a will come to rest at point A or B, depending upon the height from which they started. When at rest, at A or B, they are said to be in mechanical equilibrium. A chemical counterpart of this occurs when calcium carbonate precipitates from a solution. It may form the less stable (more soluble) aragonite, equivalent to point B, or the more stable calcite, analogous to point A. A is more stable than B because more energy is required to lift the ball from A than from B to a common level higher than A and B. The chemical states are more complex than those of objects and the detailed argument on relative stabilities is more involved. In essence, however, aragonite releases chemical energy when it is converted to calcite. This is analogous to the release of potential energy if the ball at B was moved to the more stable position A.

The springs shown in Figure 1.3b, when stationary, are in equilibrium states that depend on the weights 1 and 2. Again, the chemical equilibrium state shows an analogous feature to the mechanical

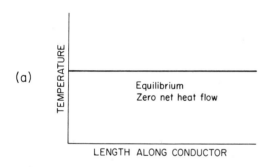

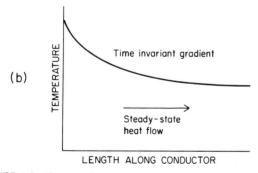

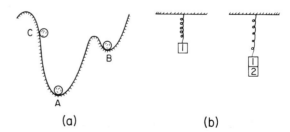

FIGURE 1.2 Contrast between thermal equilibrium and steady state.

FIGURE 1.3 Examples of mechanical equilibrium: (a) a ball at the bottom of a well and (b) springs at rest.

one in that it depends upon the amount of reacting substances (reactants) present at the beginning of the process. As an example, an increase in CO_2 in the atmosphere, due to the burning of fossil fuels, leads to a larger concentration of CO_2 in the water. This displaces the equilibrium state of the hydration reaction $H_2O + CO_2 = H_2CO_3$ (carbonic acid) to the right.

Mechanical equilibrium has the following properties:

1. It is a stationary (time-invariant) state.
2. It is a state of minimum potential energy compatible with the constraints imposed upon the system (the height of the bottom of the well and the weight hung from the spring in Figure 1.3).
3. A corollary of (2) is that the system will return spontaneously to equilibrium after a forced temporary and slight perturbation upon it, because it will seek anew the state of minimum potential energy. This will happen, for example, if we raise the ball up the wall of the well.
4. No energy is spent to keep the system at equilibrium and no work is done to displace it temporarily from equilibrium by an infinitesimal amount. This occurs because the forces upon the system are balanced, that is, $\mathbf{F}_{net} = 0$. Therefore, $dW = \mathbf{F}_{net}\, d\mathbf{x} = 0$ where $d\mathbf{x}$ represents the displacement.

These properties will be shown later to have their counterparts in chemical equilibria.

Steady states and equilibria are not only described by different equations, but also obey different energy conditions. An example is the case of a ball held at point C in Figure 1.3. It is stationary but muscular energy is required to keep it there and, therefore, a steady state occurs. Another example of a mechanical steady state is the height of water in a reservoir with an inflow equal to the outflow. The driving force is that of gravity and work is done at the expense of the potential (gravitational) energy of the water.

For readers familiar with thermodynamics, equilibrium corresponds to a zero entropy change, whereas a minimum rate of entropy production occurs in steady states.

1.4.2 Chemical Equilibrium and Steady State

The topics of equilibria and steady states as controls of geochemical processes will be developed in Chapter 2. In this subsection chemical and geochemical examples will be used only to illustrate concepts.

An example of chemical equilibrium is given by Equation (1.5). A steady state can be the concentration of KCl at a point P in a river, taken over a time period short enough to hold the KCl constant. In this case, as much KCl is brought to P as is removed. A further example would be a series of one-way chemical reactions $A + B \rightarrow C + D \rightarrow E + F$ in which $C + D$ is held constant.

Consider the dissociation of acetic acid represented by Equation (1.5). If we add HAc to water, it will start to dissociate into $H^+ + Ac^-$. These two ions will then recombine to form molecules of HAc at a rate that becomes faster as the ionic concentrations increase because the number of collisions of H^+ with Ac^- goes up. Eventually the rates of association and of dissociation will become equal and a stationary condition will have been reached. We remind you that the term stationary is used here to describe the macroscopic state of the system when the numbers of ions and molecules present in solution become time invariant. Still, association and dissociation reactions are occurring constantly so that one has a dynamic stationary condition in contrast to that of the ball at the bottom of the well.

We may now ask whether the acetic acid solution reached an equilibrium or a steady state and we must seek the answer in the properties of these conditions. We find that:

1. The system is stationary (time invariant).
2. Thermodynamics teaches us that the free energy (capacity to do useful work) of the solution is at a minimum, compatible with the temperature, pressure, and amount of HAc added, when the stationary condition is reached (see Chapter 4).
3. If we perturb the solution momentarily by, for example, raising the pressure and then releasing it, the distribution of HAc, H^+, and Ac^- will return to the original values.
4. While stationary, the solution does not absorb or release a net amount of energy.

These properties, by analogy with the case of mechanical equilibrium, show us that we are deal-

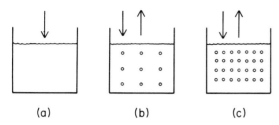

FIGURE 1.4 Entry and eventual equilibrium of dissolved oxygen.

ing with a chemical equilibrium rather than with a steady state. It is interesting to note also that, as in the case of the spring, we can load the system by changing the pressure, temperature, or amount of HAc added to the water, and bring it to a new equilibrium configuration. Vinegar is an aqueous solution of HAc which can be used by us only because there is not much H^+, the hydrogen ion that causes acidity, at equilibrium.

Another example of chemical equilibrium is shown in Figure 1.4, which presents the uptake of oxygen. This gaseous element is involved in photosynthesis, respiration, and the oxidation of organic matter and inorganic rocks. The container originally has degassed seawater, which is then placed in contact with the atmosphere. Oxygen will enter the water at a given rate and will leave it at a rate that increases as more gas is dissolved. Eventually, the influx and exit rates become equal and the seawater is at equilibrium with the atmosphere (it is saturated with oxygen). The thermal energy of the oxygen molecules maintains one-way fluxes into and out of the water but the net flux is zero.

Let us consider and illustrate some aspects of steady states starting with one of their thermodynamic properties. These properties will be considered in greater detail in Chapter 4. First, we already know that energy is used. Another important property of systems is their entropy, which measures the loss of available energy (degradation of energy). If, for example, we place a hot body in touch with a cold one, they will reach a common temperature. Engines operate between a warm source and a cold sink (its surroundings). Therefore, the loss of thermal contrast between the two rocks makes their heat content unavailable to do useful work, even though the total energy did not change. Entropy is a measure of the degradation of energy during chemical, biological, physical, and geological processes.

The equilibrium state is under a condition of maximum entropy compatible with the constraints imposed upon the system, as in the case of the two rocks mentioned above. Equilibrium is achieved when they reach the same temperature.

The steady state, if not far from equilibrium, is one in which entropy is produced (energy is degraded) but this occurs at a minimum rate (Prigogine, 1955).

In general, systems through which there are net flows of matter and energy are not at equilibrium. A case in point is that of organisms in general, which, to remain alive, must absorb matter and produce energy from it. If relatively invariant, such systems approach steady states (Katchalsky and Curran, 1965; Morowitz, 1968). An example is that of our bodies and their biochemistry, which can best be described as systems that approach steady states, at least during short periods when we are not changing much. The reason for this is that we must constantly burn energy to maintain life in contrast to the fourth property of the equilibrium state, namely, that no net energy flow is needed at equilibrium. The composition of the oceans, in an average sense, approaches a steady state (Pytkowicz and Kester, 1971).

There are also transient systems, in which the concentrations of components change with time, building up toward equilibria or steady states. As an example, the sedimentary reservoir of $CaCO_3$ may be growing (Pytkowicz, 1980). Other examples are the concentrations of heavy metals, such as mercury and lead, which are increasing in lakes and oceans due to the activities of humans, when considered on a time scale of months or years.

1.4.3 The Law of Mass Action

Many reactions do not fully consume the reactants but reach an equilibrium at which reactants and products coexist, as shown in the case of Equation (1.5). Let us represent a general reaction that reaches equilibrium by

$$A + B = C + D \qquad (1.16)$$

At equilibrium, the reaction rate of A with B to form C plus D is the same as that of the reverse one. We also know that, in the simplest cases, rates of reactions are proportional to the concentrations, which reflect the number of particles of the reactants and products in a given volume.

Thus, at equilibrium,

$$r_f = k_f[A][B] = k_r[C][D] = r_r \quad (1.17)$$

r_f and r_r are the rates of the forward and reverse reactions, k_f and k_r are the corresponding proportionality constants, and brackets represent concentrations. From Equation (1.15) one obtains

$$K' = \frac{k_f}{k_r} \frac{[C][D]}{[A][B]} \quad \text{or} \quad \frac{[C][D]}{[A][B]} \quad (1.18)$$

if the constant ratio k_f, k_r is incorporated into K'. Equation (1.16) represents the law of mass action and K' is the stoichiometric equilibrium constant, that is, the constant in terms of concentrations. We shall see later that terms called activities often replace concentrations.

If a mixture of A, B, C, and D does not meet the condition of Equation (1.16) then the reaction will proceed until Equation (1.16) is obeyed.

The system will remain stationary after this and will thus meet the first condition for equilibrium. The stationary state is necessary but not sufficient to ensure equilibrium and the second condition must be added, namely, that no work is done while Equation (1.16) is obeyed.

Broad geochemical cycles, when not in transient conditions such as those of the heavy metals, appear to be nearly at steady states in the sense of average amounts in natural reservoirs and fluxes. The reasons for this are the fluxes of matter and energy when substances are weathered, brought to the oceans, sediment, and are eventually returned to the earth's surface. Sodium and its compounds appear to be at steady states with a cycling time of 10^8 years (see Chapter 2).

the entropy is at an unchanging maximum. Examples include fast reactions such as $H_2CO_3 = H^+ + HCO_3^-$ in open systems, and fast or slow reactions in closed containers.

2. *Steady state.* Concentrations do not change, energy is utilized, and entropy is generated at a minimum rate. An example includes the amount of sodium in the oceans, a step in its geochemical cycle.

3. *Transient state.* Concentrations change, energy is utilized or generated, and the entropy increases. Examples include the marine concentration of lead, the height of a falling object (energy produced), and the height of a body in the process of being raised (energy utilized).

A specific case of the law of mass action is shown in Example 1.7.

EXAMPLE 1.7.   0.001 mole of HAc is added to 1 L of distilled water at 25°C.

$$K' = 1.8 \times 10^{-5} = \frac{[H^+][Ac^-]}{[HAc]} = \frac{x^2}{0.001 - x}$$
$$(i)$$

Then x, the number of moles of HAc which dissociate, is 1.25×10^{-4} moles/L. Next, 0.01 mole of HAc is added to 1 L of water (in analogy to the spring, one can load the system by adding extra HAc).

Then $x = 4.15 \times 10^{-4}$ moles/L and

	Initial [HAc]	Final [HAc]	Final [H$^+$]	Final [Ac$^-$]
Case I	0.001	8.75×10^{-4}	1.25×10^{-4}	1.25×10^{-4}
Case II	0.01	9.58×10^{-3}	4.15×10^{-4}	4.15×10^{-4}

Note that there can be local equilibria in systems that are in a steady state or in a transient state. Thus, the fast reaction $H_2CO_3 = H^+ + HCO_3^-$ can reach equilibrium in our body fluids. Fast or slow reactions in isolated containers reach equilibrium because no flows of matter or energy can occur.

In summary, for chemical and mechanical states:

1. *Chemical equilibrium.* Concentrations are invariant, no energy is used or released, and

The equilibrium distributions are quite different, in analogy with the spring in Figure 1.2b, but K' does not change.

An example of equilibria that are important in freshwaters, body fluids, and the oceans is the case of the hydration and dissociation reactions of carbon dioxide in solution. When carbon dioxide is dissolved in water, the following steps occur:

$$CO_2 + H_2O \rightleftarrows H_2CO_3 \quad \text{(hydration step)}$$
$$(1.19)$$

$$H_2CO_3 \rightleftarrows H^+ + HCO_3^- \quad \text{(first dissociation)}$$
$$(1.20)$$

$$HCO_3^- \rightleftarrows H^+ + CO_3^{2-} \quad \text{(second dissociation)}$$
$$(1.21)$$

The gaseous CO_2, carbon dioxide, hydrates to carbonic acid, H_2CO_3, which in turn dissociates to H^+, bicarbonate, HCO_3^-, and to carbonate ions, CO_3^{2-}. All these species coexist and play a role in the formation of calcareous sediments, in life (photosynthesis and respiration), in the control of the pH of seawater, and in the impact of CO_2 produced by the burning of fossil fuel on our environment. This system will be examined in detail later in this book.

1.5 ACTIVITY AND ACTIVITY COEFFICIENTS: A FIRST VISIT

Experience has shown that many properties of gases and of dilute solutions depend on the number of particles (ions, atoms, or molecules) present per unit volume or weight (the concentration). As examples of this we have the law of mass action and the pressure of ideal gases.

1.5.1 Ideal Gas Law

Ideal gases, which in effect are most gases at pressures below a few atmospheres, obey the law

$$PV = nRT \quad \text{or} \quad P = \frac{RTn}{V} \quad (1.22)$$

P is the pressure, V the volume, n the number of moles in V, R the gas constant or the proportionality constant between PV and nRT, and T the absolute temperature in degrees Kelvin (°K). This ideality implies that the gas molecules are neither attracted nor repelled by each other and only interact through elastic collisions.

At a given temperature, if one multiplies both sides of Equation (1.22) by $N_A V$, where $N_A = 6.02217 \times 10^{23}$ is Avogadro's number, one obtains

$$P = \text{constant} \times \frac{nN_A}{V} \quad (1.23)$$

Because N_A is the number of molecules per mole, the pressure of an ideal gas is proportional to the number of particles nN_A in V.

1.5.2 Interacting Particles

Solute particles in aqueous solutions that are not very dilute and atoms or molecules in compressed gases can be close enough to interact with each other. Such interactions can be electrostatic (coulombic) in the case of ions or result from other types of bonds for nonelectrolytes.

Two processes should be kept in mind because they help explain and qualify the above paragraph. First, particles are in motion in all directions due to their thermal energy and, therefore, at any one time a fraction of them will be close together. This enhances the effectiveness of attractive or repulsion energies. Second, the ideal state in solutions differs from that in gases because, even at very high dilutions, solute particles interact with water molecules. The state of ideality in solution is defined, therefore, in terms of the lack of interactions between solute particles but hydration is accepted as part of this state. Their pressure, which results from collisions with the walls of the container, is less than expected from Equation (1.23).

1.5.3 Activities and Activity Coefficients

When interactions occur in aqueous solutions, one uses an effective concentration

$$a_i = \gamma_i m_i \quad (1.24)$$

known as the activity a_i of the chemical species i. The symbol γ_i, the activity coefficient, converts the actual concentration m_i into a_i. The use of a_i permits properties, such as the law of mass action, to still be valid even when they cannot be expressed in terms of concentrations of noninteracting particles. An important aspect of a_i and γ_i is that

$$\lim \gamma_i = 1 \quad \lim a_i = m_i \quad (1.25)$$

when m_i tends to zero, that is, when the interactions reflected by γ_i disappear.

An analogy would be the political opinions of married couples. The husband and wife cannot be considered to be separate entities, if one wishes to analyze political trends, because usually there are interactions and changes in their premarital points of view. Thus, the actual political leanings of a society must be interpreted in terms of single, relatively isolated, individuals, married couples, and members of various groups.

Equation (1.18) now becomes

$$K_O = \frac{a_C a_D}{a_A a_B} = \frac{\gamma_C \gamma_D}{\gamma_A \gamma_B} K' \qquad (1.26)$$

In the case of the dissociation of HCO_3^-,

$$K_{aO} = \frac{a_H\, a_{CO_3}}{a_{HCO_3}} \qquad (1.27)$$

$K^{(t)}$ is known as the thermodynamic equilibrium constant. Equation (1.26) is valid at any concentration for a given pressure and temperature.

Certain conventions, which will be explained in Chapter 4, later are used for activities and activity coefficients based on the reference state and the standard state, as shown below:

1. *Definitions.* Reference state $a_i = m_i$ ∴ $\gamma_i = 1$, that is, the system behaves ideally.

Standard state $a_i = 1$, which will be shown later to correspond to the standard state for the chemical potential that drives reactions.

2. *Reference and standard states.*

	Reference State	Standard State
Gas	Zero pressure	1 atm
Solid	Pure solid	Pure solid
Liquid	Pure liquid	Pure liquid
Solvent	Pure solvent	Pure solvent
Solute	Infinite dilution	Unit concentration plus ideal behavior, that is, $\gamma_i = 1$

The standards are arbitrary, just as sea level is arbitrarily chosen as the standard for the potential energies of falling objects.

Thus, in the important case of Urey's simplified reaction,

$$CaCO_3 + SiO_2 = CaSiO_3 + CO_2 \qquad (1.28)$$

we have

$$K_O = a_{CO_2} \cong pCO_2$$

as the other participants in the reaction are solids or pure fluids at high temperatures and pressures. This reaction tells us that perhaps the atmospheric pCO_2, before the industrial revolution, was controlled by reaction (1.28) in the magma. We tend to accept this hypothesis only in part (see Chapter 2). The relationship $a \cong pCO_2$ is obtained by definition although we shall see later that the use of a term called the fugacity is exact while the use of partial pressure is approximate.

As another example, the activity coefficient of a 0.1 m NaCl solution is 0.778 and its activity is 0.0778. For most practical purposes, the properties of the solution are those of a 0.0778 m solution instead of a 0.1 m solution.

1.6 PERIODIC TABLE AND ELECTRONIC STRUCTURE

1.6.1 Electronic Structure, Periodic Table, and Properties of Atoms

At this point it is worthwhile to briefly review the Periodic Table in relation to the electronic structure of atoms for those readers who are not chemists. These topics provide insight into the solution properties of the elements.

Chemists realized in the nineteenth century that certain chemical properties are similar for elements present in the vertical columns of the periodic table, which is shown in a contemporary form in Table 1.4. The understanding of the Periodic Table, which was first arrived at empirically, followed the elucidation of the electronic structure of atoms during the present century.

We now know that atoms are constituted of nuclei surrounded by electrons present in shells. These shells, starting from the one nearest to the nucleus, are indicated by the letters K, L, M, N, and O. Within each shell a maximum of two electrons are present in orbitals labeled s, p_x, p_y, and p_z. The distribution of electrons around the nuclei of the first 18 elements in the order of increasing atomic weights are shown in Table 1.5.

The orbitals can be represented by the surface types on which electrons are most probably located (Figure 1.5). The electrons in the orbitals have spectral properties which may be explained by electron spins, with electrons in the same orbital having opposite (paired) spins as the most probable configuration.

The chemical properties of the elements depend in great part upon the number of electrons in the outer shell and primarily upon the unpaired ones.

TABLE 1.4 Periodic Table of the elements.

	1a	2a	3b	4b	5b	6b	7b	8b	9b	10b	1b	2b	3a	4a	5a	6a	7a	0
Period																		
I	H																H	He
II	Li	Be											B	C	N	O	F	Ne
III	Na	Mg											Al	Si	P	S	Cl	Ar
IV	K	Ca	Sc	Ti	V	Cr	Mn	Fe	Co	Ni	Cu	Zn	Ga	Ge	Ar	Se	Br	Kr
V	Rb	Sr	Y	Zr	Nb	Mo	Tc	Ru	Rh	Pd	Ag	Cd	In	Sn	Sb	Te	I	Xe
VI	Cs	Ba[a]		Hf	Ta	W	Re	Os	Ir	Pt	Au	Hg	Tl	Pb	Bi	Po	At	Rn
VII	Fr	Ra[b]																

[a] Lanthanide series.
[b] Actinide series.

Thus, note that lithium and sodium have a single *s* electron in their outer shell. They are both metals, form singly charged ions by the reactions

$$Li \rightarrow Li^+ + e^- \qquad Na \rightarrow Na^+ + e^- \qquad (1.29)$$

and have relatively low ionization energies (the energy required to ionize the atom).

The noble gases such as helium, neon, and argon have filled shells and, at the same time, have the highest ionization energies. This indicates that the electronic configuration of noble gases is the most stable one. Note, furthermore, that elements in their reactions tend to seek a noble gas con-figuration, that is, a most stable configuration. An example in point is that given by reactions (1.29) in which Li^+ and Na^+ have the electron configuration of helium and neon, respectively. Noble gases tend to be unreactive because of their stability.

Nonmetals have high ionization energies and tend to reach noble gas configurations by gaining electrons. Thus, an atom of chlorine ionizes according to

$$Cl + e^- \rightarrow Cl^- \qquad (1.30)$$

In a crystal of NaCl, one finds the ions Na^+ and Cl^-, as is to be expected from reactions. These ions

TABLE 1.5 Electronic structure of the first 18 elements. MNE is the maximum number of electrons that can fit in a shell.

Shell	K						M		
MNE	2			L 8				18	
Orbital	$1s$	$2s$	$2p_x$	$2p_y$	$2p_z$	$3s$	$3p_x$	$3p_y$	$3p_z \cdots$
H	1								
He	2								
Li	2	1							
Be	2	2							
B	2	2	1						
C	2	2	1	1					
N	2	2	1	1	1				
O	2	2	2	1	1				
F	2	2	2	2	1				
Ne	2	2	2	2	2				
Na	2	2	2	2	2	1			
Mg	2	2	2	2	2	2			
Al	2	2	2	2	2	2	1		
Si	2	2	2	2	2	2	1	1	
P	2	2	2	2	2	2	1	1	1
S	2	2	2	2	2	2	2	1	1
Cl	2	2	2	2	2	2	2	2	1
Ar	2	2	2	2	2	2	2	2	2

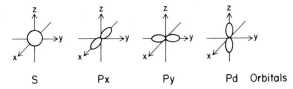

FIGURE 1.5 Orbitals as surfaces on which electrons are most probably found.

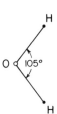

FIGURE 1.6 The water molecule. The bond length of O–H is 0.9572 Å, the angle is actually 104.52°, and the dipole moment is 1.834 D.

then separate when NaCl is dissolved. Not all elements are ionic in the solid state [e.g., the mineral brucite, $Mg(OH)_2$] because the ionic character depends on the ease to gain or lose electrons.

1.6.2 Types of Chemical Bonds

The five major types of chemical bonds that occur in nature are: ionic, covalent, hydrogen, Van der Waal's, and metallic.

The ionic bonding occurs when one atom gives up electrons to another so that both can reach a noble gas configuration. They are then held by electrostatic forces, because the two ions have opposite charges. Ionic bonding occurs, as we just saw, in crystals of NaCl.

Covalent bonding is distinct from ionic bonding in that the noble gas configuration is attained by sharing rather than by transferring electrons. The simplest case of this type of bond occurs in hydrogen molecules, which can be represented by

$$H:H \quad \text{or} \quad H\text{—}H \qquad (1.31)$$

where the dots and the dash pairs of electrons represent electrons in the outer shells that form the bond.

Hydrogen bonds will be discussed in conjunction with the structure of water. Van der Waal's bonds are short range, resulting from quantum mechanical forces. These forces are the product of induced dipole moments.

1.7 THE STRUCTURE OF WATER

1.7.1 Water

Water is by far the major solvent of crustal materials and is of fundamental importance in fields as diverse as water and solution chemistry, geology, physiology, agriculture, industry, and so on. This importance led to considerable work toward an

understanding of its structure and the relationships between structure and properties. Among properties of interest one finds solvent characteristics, viscosity, boiling and freezing points, and vapor pressure. Water as a solvent will be considered in this chapter and further in Chapter 5.

The individual water molecules have the structure shown in a simplified manner in Figure 1.6. If, however, the molecules existed as individual entities [see, for example, Horne (1969) for more advanced representations], then the properties of water would be similar to those found in low molecular weight liquids. The MW of water is 18.

That, however, is not the case and water behaves as if it were made of large molecules. For instance, the vapor pressure of water is lower and its boiling point is higher than those of methane (CH_4 with MW = 16). This result led chemists to realize that water molecules are associated to each other and the association is the result of hydrogen bonds (Figure 1.7).

Still, the problem of the structure of liquid water has not been completely cleared. We wish to know the detailed intermolecular structure and how it affects the properties of water and solutions. Several models are under consideration at present.

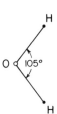

FIGURE 1.7 Hydrogen bonds.

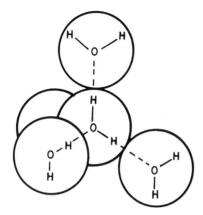

FIGURE 1.8   The tetrahedral structure of water (Horne, 1969).

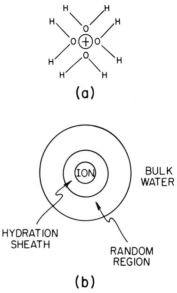

FIGURE 1.9   Ionic hydration.

These models must explain the vapor pressure, the boiling and freezing points, the thermal conductivity, the viscosity, and other properties of water. They can be divided into two classes; continuum and mixture ones. There is essentially one continuum model; that of Pople (1951). It represents bonds in liquid water as having continuous distributions of H-bond energies, angles, and interatomic distances, centered around the average values for the water as a whole. The structure is that of ice—I_h with four H bonds per molecule and a tetrahedral configuration (see Figure 1.8).

Mixture models are of several types:

1. *Interstitial.*   Some of the molecules leave their lattice sites and meander in interstices found among regularly arranged molecules.

2. *Vacant lattice point.*   An icelike structure is present, with vacant sites and interstitial molecules. The lattice-type order is short range due to the disturbances in the orderly tetragonal bonding. This accounts for the fluidity of water because there is no fixed geometry among distant molecules.

3. *Flickering cluster.*   Part of the molecule is monomeric while others form flickering clusters (Frank and Wen, 1957; Nemethy and Scheraga, 1962). Individual molecules jump from one configuration to the other and, just as in (2), there is no long-range order.

1.7.2   Electrolyte Solutions

The solubility of electrolytes in water is due in good part to the high dielectric constant which weakens interionic forces (see Chapter 5). Ionic hydration occurs when ions are added to water (see Figure 1.9). The negative end of the water molecule, that is, the oxygen atom, is attracted to the cation. Outside of the hydration sheath there is a random transition region in which water molecules are not oriented either by the ion or by the configurational energy of the bulk water. Energy is released during hydration when water–water bonds are replaced by water–ion bonds. This leads to the dissolution of electrolytes such as HCl, NaCl, and K_2SO_4.

1.8   TYPES OF EQUILIBRIA

There are several general classes of equilibria as is shown below.

1.8.1   Dissociation of Acids and Bases

For a generic weak acid HA or base BOH (this B should not be confused with the symbol for boron)

$$HA = H^+ + A^- \qquad BOH = B^+ + OH^- \qquad (1.32)$$

The second reaction may be written, as we shall see later, as

$$B^+ + H_2O = BOH + H^+$$

$$BOH + H^+ = B^+ + H_2O \qquad (1.33)$$

In their simplest definitions acids are proton (H^+) donors and bases are proton acceptors (see Chapter 1, Volume II).

1.8.2 Ion Pairing

Results from electrostatic (coulombic) attraction between oppositely charged ions

$$C^+ + A^- = CA^0 \qquad (1.34)$$

with zero net charge if the cation C^+ and the anion A^- have equal absolute charges. An example is $NaCl^0$ in solution. For $NaSO_4^-$ there is a net negative charge.

1.8.3 Complexing

Complexing occurs when strong electron sharing is present

$$M^+ + A^- = MA^0 \qquad (1.35)$$

as, for example, in $PbCl_2^0$.

1.8.4 Solubility

In this case a solid or gas dissolves in water until saturation occurs and

$$CA_{(s)} = C^+ + A^- \qquad (1.36)$$

Examples are the dissolutions of O_2 and $CaCO_3$.

1.8.5 Redox Reactions

These reactions involve electron transfer. When a zinc is dipped into a solution of sulfuric acid, the following reactions occur:

$$Zn = Zn^{2+} + 2e^- \qquad (1.37)$$

$$2e^- + 2H^+ \; H_2 \qquad (1.38)$$

The net reaction is

$$Zn + 2H^+ = Zn^{2+} + H_2 \uparrow \qquad (1.39)$$

where e^- represents an electron. Zinc is dissolved and hydrogen is evolved. The equilibrium constant is

$$K = \frac{[Zn^{2+}] \, p(H_2)}{[H^+]^2} \qquad (1.40)$$

where $p(H_2)$ is the partial pressure of hydrogen. The concentration of pure metallic zinc is 1 by definition. Activity coefficients are set as unity for simplicity.

Redox reactions provide chemical energy when electrodes made of two metals dipped in a solution are connected by a wire (see Chapter 5).

Redox reactions can be made to convert chemical into electrical energy. Consider two rods, one of zinc and one of copper, dipped in a $ZnSO_4$–$CuSO_4$ solution, and connected by an external wire. The reactions will be

$$Zn = Zn^{2+} + 2e^- \qquad \text{(dissolution)} \quad (1.41)$$

$$2e^- + Cu^{2+} = Cu \qquad \text{(deposition)} \quad (1.42)$$

Electrons will flow along the wire from the zinc to the copper rod and there will be an electrical current (see Chapter 5 for more details). There will, therefore, be chemical energy. The cell may be written as

$$Zn \mid ZnSO_{4(s)}; \qquad CuSO_{4(s)} \mid Cu \qquad (1.43)$$

1.9 A CLARIFYING NOTE

Thermodynamic constants are given, for $A + B = C + D$ by

$$K = \frac{a_C a_D}{a_A a_B} = \frac{\{C\} \{D\}}{\{A\} \{B\}} \qquad (1.44)$$

where $\{ \}$ is an alternative notation for activities.

At infinite dilutions or in the standard states,

$$K = \frac{[C] [D]}{[A] [B]} \qquad (1.45)$$

where brackets represent concentrations because the activity coefficients are unity so that, as an example, $\{A\} = [A]$.

At finite concentrations stoichiometric constants are given by

$$K' = \frac{[C] [D]}{[A] [B]} \qquad (1.46)$$

Under certain conditions that will be examined, K' is indeed a constant which obviates the use of activity coefficients.

In the first three chapters activity coefficients

will often, but not always, be set as unity for simplicity and the equations

$$K = \frac{[C] [D]}{[A] [B]} \qquad (1.47)$$

and

$$K_{sO} = [C] [A] \qquad (1.48)$$

will be used, where K_{sO} is the thermodynamic solubility product.

SUMMARY

Chemistry is concerned with forms of matter and the transformations they undergo. Such transformations are the core of the chemistry, geochemistry, and biological chemistry of natural waters.

Chemical concepts include atoms and molecules, as well as their weights, mixtures, and compounds, and the ions which are charged atoms or groups of atoms usually present in electrolyte solutions.

The distribution of elements as rocks, minerals, liquids, and gases in the earth is related to the history of our planet and elemental properties.

Aqueous solutions may contain gases, nonelectrolytes, or electrolytes (acids, salts, etc.), and are vital for body fluids, oceans, lakes, rivers, and so on.

Concentrations and interconversion equations used in aqueous chemistry are described.

The equilibrium and steady-state concepts are contrasted. The former is a balanced state that does not require energy, whereas the latter one (e.g., the constant height of water in a reservoir) involves energy flow. Both states are present in nature. Equilibrium is described by the law of mass action.

Properties of dilute gases and solutions depend on the number of particles per unit volume (concentration) but, when particles interact, the activity must replace the concentration.

The Periodic Table, the electronic structure of atoms, and the properties of the elements are related. The properties considered are mainly the nature of bonds that are formed and the consequent behavior of the elements in nature.

Water, as the major crustal component, is of fundamental importance. Its properties are intimately related to hydrogen bonding.

The major types of aqueous equilibria, namely, the dissociation of acids and bases, ion pairing, true complexing, solubility, and redox are described in an introductory way.

REFERENCES

Frank, H. S., and M. W. Wen (1957). *Discuss. Faraday Soc.* **24**, 133.

Gross, M. G. (1972). *Oceanography: A View of the Earth*, Prentice-Hall, Englewood Cliffs, N.J.

Horne, R. A. (1969). *Marine Chemistry: The Structure of Water and the Chemistry of the Hydrosphere*, Interscience, New York.

Katchalsky, A., and P. F. Curran (1965). Nonequilibrium Thermodynamics in Biophysics, Harvard University Press, Cambridge.

Mason, B. (1966). *Principles of Geochemistry*, Wiley, New York.

Morowitz, H. J. (1968). *Energy Flow in Biology*, Academic Press, New York.

Nemethy, G., and H. W. Scheraga (1962). *J. Chem. Phys.* **36**, 3382.

Pople, J. A. (1951). *Proc. R. Soc. Lond. Ser. A.* **205**, 163.

Prigogine, I. (1955). *Thermodynamics of Irreversible Processes*, Interscience, New York.

Pytkowicz, R. M. (1980). *Geochim. J.* **14**, 47.

Pytkowicz, R. M., and D. R. Kester (1971). In *Oceanography, Marine Biology Annual Review*, Vol. 9, H. Barnes, Ed., Allen and Unwin, London, p. 11.

CHAPTER TWO

GEOCHEMISTRY

Geochemists study those chemical processes which have in the past and are in the present shaping up the structure and the composition of the earth. Those aspects of geochemistry that deal with the interactions between natural waters (the hydrosphere) and the solid crust (part of the lithosphere) are emphasized. Oceans are emphasized because their size enhances their geochemical importance.

The major constituents are treated here. Minor constituents, whose study requires a knowledge of complexation and redox processes, will be examined in Chapter 8, Volume II. A substantial section on the theories of cycles is developed in the present chapter. These theories, which already provide insights on the effects of control mechanisms, eventually lead to a quantitative understanding and prediction of unperturbed and perturbed cycles. The concepts of equilibrium, steady states, and transients appear again in these studies.

To start this chapter, we expand our understanding of the role of the oceans as reservoirs that link the continents with the submarine sediments. This understanding has helped us elucidate the reactions and the pathways of many minerals and rocks throughout nature's reservoirs (RMP).

2.1 DEVELOPMENT OF OUR UNDERSTANDING, CYCLES, AND THE HISTORY OF THE OCEANS

2.1.1 The Oceans as Accumulators

The first human contact with oceanic chemistry and geochemistry was practical in that there was interest in sea salt and chemicals concentrated by molluscs as sources of food. Still, a purely scientific interest can already be found in the writings of the early Greeks and Arabs who wondered about the source of the saltiness of the oceans. These early thinkers concluded that the marine salt resulted from the leaching of land products by rivers and that these products accumulated in the oceans. This was the first view of the oceans as accumulators.

The practical and the scientific outlooks have become more intimately related through the years due to the broader foundation of knowledge required by modern technology. Some examples are the knowledge needed to extract magnesium and bromine from seawater, clarify properties of seawater, such as the attenuation of sound, for military applications, and understand the solution behavior of minerals which sediment and are then involved in lithification and orogenesis, and the behavior of toxic metals. Lithification simply means rock formation, whereas orogenesis pertains to mountain building activity.

Robert Boyle, in the seventeenth century, tested the accumulator concept of the oceans by adding silver nitrate to river water and seawater. The reaction was

$$AgNO_3 + Cl^- \rightarrow AgCl + NO_3^- \qquad (2.1)$$

A slight white precipitate of AgCl was observed in rivers while a larger one occurred in seawater. This justified the concept of a slight constant leaching of soils by rivers with accumulation of the river-borne leachates in the oceans.

The accumulator concept led to a seventeenth-century attempt to calculate the age of the oceans by means of the relation

$$\text{age} = \frac{B}{J_{AB}} \qquad (2.2)$$

B is the oceanic salt content, A is the content of the sedimentary reservoir, and J_{AB} is the annual flux of solutes from the land to the sea (Halley, 1715). In terms of contemporary data, this leads to an oceanic age of about 25×10^6 yr (see Example 2.1) a figure well below that of 3.5×10^9 yr established by isotopic studies of early marine rocks. The reasoning of Halley is of historical interest because it represents the scientific frame of mind during the early eighteenth century. It is reproduced in part next.

A short Account of the Cause of the Saltness of the Ocean, and of the several Lakes that emit no Rivers; with a Proposal, by help thereof, to discover the Age of the World. Produced before the Royal-Society by Edmund Halley, R.S. Secr.

There have been many attempts made and proposals offered, to ascertain from the appearances of nature, what may have been the antiquity of this globe of Earth; on which, by the evidence of Sacred Writ, mankind has dwelt about 6000 years; or according to the Septuagint above 7000. But whereas we are there told that the formation of man was the last act of the Creator, 'tis no where revealed in Scripture how long the Earth had existed before this last creation, nor how long those five days that preceeded it may be to be accounted; since we are elsewhere told, that in respect of the Almighty a thousand years is as one day, being equally no part of eternity; nor can it well be conceived how those days should be to be understood of natural days, since they are mentioned as measures of time before the creation of the sun, which was not till the fourth day. And 'tis certain Adam found the Earth, at his first production, fully replentified with all sorts of other animals. This enquiry seeming to me well to deserve consideration, and worthy the thoughts of the Royal Society, I shall take leave to propose an expedient for determining the age of the world by a medium, as I take it, wholly new, and which in my opinion seems to promise success, though the event cannot be judged of till after a long period of time; submitting the same to their better judgment. What suggested this notion was an observation I had made, that all the lakes in the world, properly so called, are found to be salt, some more some less than the ocean sea, which in the present case may also be esteemed a lake; since by that term I mean such standing waters as perpetually receive rivers running into them, and have no exite or evacuation.

The number of these lakes, in the known parts of the world is exceeding small, and indeed upon enquiry I cannot be certain there are in all any more than four or five, viz. first, the Caspian Sea; secondly, the Mare Mortuum or Lacus Asphaltites; thirdly, the lake on which stands the City of Mexico, and fourthly, the Lake of Titicaca in Peru, which by a Channel of about fifty leagues communicates with a fifth and smaller, call'd the Lake of Paria, neither of which have any other exite. Of these the Caspian, which is by much the greatest, is reported to be somewhat less salt than the ocean. The Lacus Asphaltites is so exceedingly salt, that its waters seem fully sated, or scarce capable to dissolve any more; whence in summertime its banks are incrusted with great quantities of dry salt, of somewhat a more pungent nature than the marine, as having a relish of Sal Armoniac; as I was informed by a curious gentleman that was upon the place.

The Lake of Mexico properly speaking is two lakes, divided by Causways that lead to the city, which is built in islands in the midst of the lake, undoubtedly for its security; after the idea, tis probable, its first founders borrowed from their beavers, who build their houses on damms they make in the rivers after that manner. Now that part of the lake which is to the northwards of the town and causways, receives a river of a considerable magnitude, which being somewhat higher than the other, does with a small fall exonerate it self in the southern part, which is lower. Of these the lower is found to be salt, but to what degree I cannot yet learn; though the upper be almost fresh.

And the Lake of Titicaca, being near eighty leagues in circumference, and receiving several considerable fresh rivers, has its waters, by the testimony of Herrera and Acofta, so brackish as not to be potable, though not fully to salt as that of the ocean; and the like they affirm of that of Paria, into which the Lake of Titicaca does in part exonerate it self, and which I doubt will not be found much salter than it, if it were enquired into.

Not I conceive that as all these lakes do receive rivers and have no exite or discharge, so 'twill be necessary that their waters rise and cover the land, until such time as their surfaces are sufficiently extended, so as to exhale in vapour that water that is poured in by the rivers; and consequently that lakes must be bigger or lesser according to the quantity of the fresh they receive. But the vapours thus exhaled are perfectly fresh, so that the saline particles that are brought in by the rivers remain behind, whilst the fresh evaporates; and hence 'tis evident that the salt in the lakes will be continually augmented, and the water grow salter and salter. But in lakes that have an exite, as the Lake of Genesaret, other wise call'd that of Tiberias, and the upper Lake of Mexico, and indeed in most others, the water being continually running off, is supply'd by new fresh river water, in which the saline particles are so few as by no means to be perceived.

Now if this be the true reason of the saltness of these lakes, tis not improbable but that the ocean it self is become salt from the same cause, and we are thereby furnished with an argument for estimating the duration of all things, from an observation of the increment of saltness in their waters. For if it be observed what quantity of salt is at present contained in a certain weight of the water of the Caspian Sea, for example, taken at a certain place, in the dryest weather; and after some centurys of years the same weight of water, taken in the same place and under the same circumstances be found to contain a sensibly greater quantity of salt than at the time of the first experiment, we may by the rule of proportion, take an estimate of the whole time wherein the water would acquire the degree of saltness we at present find in it.

And this argument would be more conclusive, if by a like experiment a simular encrease in the saltness of the ocean should be observed: for that, after the same manner as aforesaid, receives innumerable rivers, all which deposite their saline particles therein; and are again supplyed, as I have elsewhere shown, by the vapours of the ocean, which rise therefore in atoms of pure water, without the least admixture of salt. But the rivers in their long passage over the earth do imbibe some of the saline particles thereof, though in so small a quantity as not to be perceived, unless in these their depositories after a long tract of time. And if upon repeating the experiment, after another equal number of ages, it shall be found that the saltness is further encreased with the same increment as before, then what is now proposed as hypotheticall would appear little less than demonstrative. But since this argument can be of no use to ourselves, it requiring very great intervals of time to come to our conclusion, it were to be wished that the ancient Greek and Latin authors had delivered down to us the degree of the saltness of the sea, as it was about 2000 years ago: for then it cannot be doubted but that the difference between what is now found and what then was, would become very sensible I recommend it therefore to the Society, as opportunity shall offer, to procure the experiments to be made of the present degree of saltness of the ocean, and of as many of these lakes as can be come at, that they may stand upon record for the benefit of future ages.

If it be objected that the water of the ocean, and perhaps of some of these lakes, might be at the first beginning of things, in some measure contain salt, so as to disturb the proportionality of the encrease of saltness in them, I will not dispute it: but shall observe that such a supposition would by so much contract the age of the world, within the date to be derived from the aforegoing argument, which is chiefly intended to refute the ancient notion, some have of late entertained, of the eternity of all things; though perhaps by it the world may be found much older than many have hitherto imagined.

EXAMPLE 2.1. AGE OF THE OCEANS BASED ON THE ARGUMENT OF HALLEY (1715)

age of the oceans (Halley) = total oceanic salt

$$\frac{\text{content}}{\text{annual river flux}} \qquad (i)$$

total oceanic salt content = oceanic mass (in kg)

$\times$ average salt content/kg-SW

$= 14.13 \times 10^{20}$ kg $\times$ 34.44 g

salt/kg-SW $= 4.866 \times 10^{22}$ g-salts   (ii)

Annual river flux $= 1.914 \times 10^{15}$ g-salt/yr (Clark, 1959)

age of the oceans $= 4.866 \times 10^{22}/1.914 \times 10^{15}$ yr

$= 2.542 \times 10^7$ yr

2.1.2 Sedimentary View

The case for the above discrepancy is simply that the earlier workers did not take into account marine sedimentation. In reality, Equation (2.2), for the buildup of the oceanic salt content, should be written as

$$\text{age} - \frac{B}{J_{AB} - J_{BC}} \qquad (2.3)$$

J_{BC} is the flux into the submarine sediments. This is, of course, a simplified equation, because J_{AB} and J_{BC}, as well as B, were functions of time, at least in the early history of the oceans. Now the age becomes significantly larger because the denominator decreases.

2.1.3 The General Geochemical Cycle

The sedimentary view succeeded the accumulator view but was not yet sufficient to explain geochemical fluxes and reservoir sizes. The awareness that some mountain chains, such as the Alps and the Dolomites, consisted of lithified calcareous residues of marine sediments, which were in turn partially eroded and returned to the oceans, indicated the existence of broad geochemical cycles. This view was further supported by the fact that the present sedimentary mass, consisting of unconsolidated sediments and sedimentary rocks, should be much greater than it is if rivers had always deposited brand new loads in the oceans. Instead, the river load contains in part material that had already been sedimented and then was returned to the weathering environment by tectonic forces such as uplift and folding. The general cycle is shown in Figure 2.1. A more complete diagram of the geochemical cycle, which does not center upon the oceans, can be found in Mason (1966). Thus, we finally reached a view of geochemistry in which substances undergo cyclic pathways. The pathways are slow and the cycling time is on the order of 10^8 yr.

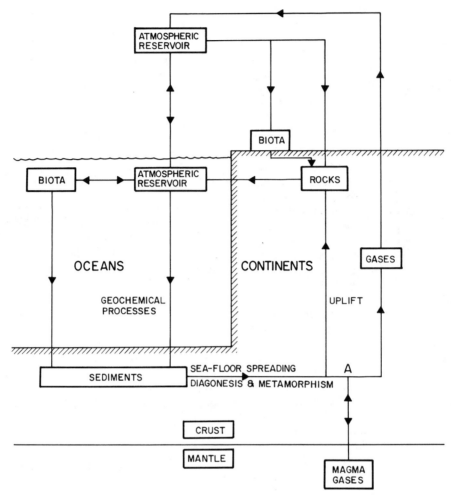

FIGURE 2.1   The geochemical cycle.

2.1.4   History of the Oceans

Another aspect of pathways in geochemistry is the chemical development of the oceans. This topic is of great interest to many geochemists but it transcends the scope of this work in that we are mostly concerned with the present. It will, therefore, be treated only briefly.

The primitive atmosphere was lost early, as noble gases above the earth were depleted relative to their cosmic abundances (Rubey, 1951; Brown, 1953). They were replaced by present volatiles such as H_2O, CO_2, CH_4, H_2S, SO_2, N_2, and HCl. This replacement occurred either by a large initial injection of gases into the atmosphere or by a gradual release from the earth. This process may have started soon after the creation of the earth, believed to have occurred about 5×10^9 yr ago. It is important to determine how the volatiles were released,

whether suddenly or gradually, in order to understand the early chemistry of the oceans and the sediments because the acid volatiles weathered the minerals that were added to the incipient oceans. Therefore, the composition of the newly formed and later oceans, as well as that of the developing sediments, depend on the manner in which gases were added to the atmosphere. The term acid volatiles arose because CO_2, HCl, SO_2, and H_2S have an acidic reaction in seawater and induce the chemical weathering of rocks and minerals. This is still a subject of controversy. Harbauch and Bonham-Carter (1970), Garrels and Mackenzie (1971a), and Li (1972) made computations with constant mass (early formation of the sedimentary mass) and accumulation models (gradual release of volatiles and buildup of sediments). There is not enough contrast in the results of the two types of models because cycling has destroyed some of the characteristics of ancient sediments and, therefore, a choice of a model is not feasible at this time.

Garrels and Mackenzie (1971a) summarized our knowledge of the history of the oceans and sediments by outlining three periods of development. The early stage, which lasted until 3×10^9 yr ago, is characterized by a cooling crust that reacted with an acid atmosphere and hydrosphere and led to the weathering of basaltic and andesitic rocks (see Section 1.3). Conditions were reducing and minerals, such as compounds of ferrous ions, played an important role in sedimentary processes (Engel and Engel, 1964; Ronov, 1968). This was followed by a transition period, lasting until about 1.5 to 2×10^9 yr ago, during which oxygen was gradually built up by atmospheric processes and photosynthesis, with life starting roughly 3×10^9 yr ago. During this period sulfate, resulting from the oxidation of sulfur gases, became important in the sedimentary record. Then, around 1.5×10^9 yr ago, sediments acquired modern characteristics with sandstones, carbonates, and shales differing little from their Paleozoic counterparts.

Another valuable source on the history of the oceans is Holland (1972).

Before proceeding further, I shall present some background material on dissolved carbonates (RMP). In recent oceans the upper waters are heavily supersaturated with calcium carbonate, whereas the deeper waters are undersaturated. One may expect inorganic precipitation where there is supersaturation. This, however, does not occur primarily because magnesium inhibits the crystallization (Pytkowicz, 1965a, 1973a) and most of the $CaCO_3$ removed from the oceans is biogenic through the tests of calcareous organisms. Also, organic coatings on nuclei may prevent crystal growth (Chave, 1965). Partial dissolution occurs at depths when the tests sink although a significant fraction of the biogenic $CaCO_3$ reaches the sediments. Inorganic precipitation occurs to a slight extent in regions such as the Bahama Banks, where carbonate particles are stirred into the shallow water column and serve as nuclei for crystallization.

The existence of pre-Paleozoic carbonates over 1.5×10^9 yr old is intriguing. The reason for this puzzle is that calcareous tests first appeared during the Cambrian, 6×10^8 yr ago. This suggests that pre-Cambrian carbonates were settled down inorganically, a process that is only occasionally found today.

A couple of possibilities occur to me to explain the above problem. First, perhaps the magnesium/calcium ratio was smaller in earlier time. This hypothesis is improbable because, as I show later [see also Pytkowicz (1979)], the amount of calcium in the oceans probably has been increasing, whereas that of magnesium may have stayed relatively constant, at least for long periods. Thus, the magnesium/calcium ratio may actually have decreased with time. An alternative hypothesis is that old carbonate minerals were precipitated inorganically from shallow seas. The precipitation may have resulted from concentration by evaporation or by the stirring of shallow sediments. The stirring could have provided suspended particles which served as sites for the nucleation and precipitation of further carbonates.

The second alternative receives some support from the fact that many older carbonate rocks appear to have precipitated from shallow waters (Garrels and Mackenzie, 1971a). Thus, since the Cambrian there may have been a shift from a predominantly shallow inorganic precipitation of carbonates to the settling of calcareous organisms over a range of depths (RMP).

Supersaturation of dissolved $CaCO_3$ in lakes and rivers should seldom occur because of the low magnesium content of the waters. Still, protective organic coatings resulting from the decay of organisms can coat nuclei and inhibit precipitation.

The geochemistry of rivers will be treated later in this chapter and the formation of lakes will be surveyed in Chapter 3.

2.1.5 The Sedimentary Mass

The development of sedimentary rocks is, of course, another important aspect of the history of the ocean–sediment system. With regard to the overall masses involved in weathering processes, Goldschmidt (1933, 1954) concluded that 8.1×10^{23} g of igneous rocks have been weathered by the action of 14×10^{23} g of primary acid volatiles. The volatiles were shown by Rubey (1951) to have been supplied by the degassing of the interior of the earth because the amounts present in sediments, oceans, and atmosphere exceed those which could have been released by the weathering of igneous rocks.

Li (1972) estimated from oxygen 18 mass balance computations that the total sedimentary mass is 2.4×10^{24} g. He concluded further that each 100 g of

TABLE 2.1   Total released magmatic volatiles according to Li (1972).

Volatile	Mass Released 10^{20} g	Mole %	Mole % in Oceans
HCl	11.9	33 NaCl	67 in oceanic Cl^-
CO_2	46.4	80 in carbonate	Minor
		20 in organic carbon	Minor
H_2S	4.2	55 in FeS_2	10 in SO_4^{2-}
		35 in $CaSO_4$	
N	2.8	~100 in the atmosphere	Minor
H_2O	953	2 in clay minerals	98 in the hydrosphere

In addition, a net 9.3×10^{20} moles (298×10^{20} g) of O_2 have been produced by photosynthesis.

igneous rocks produced 113 g of sediments with the remainder being supplied by volatiles and that 2.7 $\times 10^{24}$ g of igneous rocks have been weathered. His results yield a larger extent of weathering than those calculated by Goldschmidt (1954). The relative weight abundance of sedimentary rock types obtained from Li's model was evaporite:carbonate:sandstone:shale + clay = 2:15:12:71 (see Section 1.3 on rocks and minerals). These results agree reasonably well with those calculated by Garrels and Mackenzie (1969). See Table 2.1.

Data in the preceding paragraphs imply that, if the river composition has remained relatively constant, the total sedimentary mass can pass through the oceans in 1.2×10^8 yr. The distribution of the sedimentary mass shows little trend in composition and depositional rates may have averaged about constant since the Cambrian, 6×10^8 yr ago (Garrels and Mackenzie, 1971a). This suggests that, as a rough average, river loads have indeed remained invariant. Therefore, all sediments may have been recycled about five times during this period. The cyclic aspect of sediments has been recognized, for example, by Siever (1968a) and by Garrels and Mackenzie (1971a) who point out that it affects the interpretation of the geological record through the destruction of most pre-Cambrian sediments. Cycling mechanisms may consist of shallow sedimentation, with subsequent evaporation or uplift, and deep sedimentation with seafloor spreading, metamorphism, and eventual uplift. The 1.2×10^8 yr time pertains to the latter.

It is interesting to consider that perturbations of geochemical fluxes by the activities of humans will not be buried in submarine sediments but will propagate and distribute themselves through the various natural reservoirs and fluxes because of the cyclic nature of the processes that occur (Pytkowicz,

1971, 1972, 1973a). This is somewhat academic with regard to the deep sedimentation because of the long cycling times (1000 yr within the oceans and 1.2×10^8 yr within the earth) required for the redistribution. It can become important, however, within the smaller subcycles that occur in the hydrosphere, atmosphere, biosphere, and shallow sediments.

As an aside (RMP), it is interesting to observe how submicroscopic properties of the elements, which play an enormous role in the distribution of their compounds, are distributed in nature. HCl, for example, is a strong acid because of the highly polar nature of the H:Cl bond. Thus, it is highly soluble and forms H^+ and Cl^- ions in solution. Cl^- tends to stay in the oceans, except when Cl^- is removed as NaCl during evaporation of seawater in shallow isolated waters. H^+, on the other hand, reacts with (titrates) igneous and sedimentary rocks while it is in the hydrosphere and yields sediments and sedimentary rocks. These titrations occur because HCl is an acid that donates protons, H^+, to basic (proton acceptor) minerals, according to the simplest definitions of acids and bases. A typical titration is the weathering of feldspars in igneous rocks to clays. The bond between Ca^{2+} and CO_3^{2-} in solid carbonates is much less polar than in HCl and is, consequently, more stable. Therefore, the solubilities of minerals such as calcite and aragonite are slight. Although seawater contains Ca^{2+} and CO_3^{2-} ions, the bulk of the carbonates is present in nature as solid phases (limestones, dolomites, unconsolidated sediments, etc.). Water, as we saw earlier, is a liquid at average earth surface conditions due to its association by hydrogen bonding. For this reason, it is one of the key substances in our environment. Life would not exist without it and the same is true of this book.

2.2 MINERALS, INCLUDING CLAYS, ROCKS, AND GEOLOGIC TIME

In this section introductory material for nongeologists is presented, which will be useful in the following sections.

2.2.1 Minerals

Minerals are homogeneous substances with individual physical and chemical properties. Examples are quartz (SiO_2), pure calcite ($CaCO_3$), halite ($NaCl$), and gypsum ($CaSO_4.2H_2O$).

The chemical composition alone does not define a mineral because the crystal structure also affects its properties such as the melting point, solubility, hardness, specific gravity, luster and cleavage, that is, the surfaces along which crystals break. Examples of crystals versus composition are shown below:

SiO_2 (quartz, hard):   Single compound

SiO_2 (cristobalite):   Different crystal form from quartz. Single compound.

$SiO_2.nH_2O$ (opal):   Biogenic, soft, amorphous. Single compound.

$CaCO_3$ (aragonite):   Single compound.

$CaCO_3$ (calcite):   Different crystal form from aragonite. Single compound.

$Ca_xMg_{1-x}CO_3$ (magnesian calcite): $0 < x \le 0.28$. Solid solution of $CaCO_3$ (calcite) and $MgCO_3$ (magnesite). Two compounds with the crystal form of calcite.

$CaMg(CO_3)_2$ (dolomite):   Single compound with characteristic properties and crystal form.

Minerals made of layers of oxides, such as those of silicon and of aluminum, can be scalelike. They may undergo substitutions that lead to varieties that differ enough in properties to warrant separate names. Thus, one finds muscovite, $KAl_3Si_3O_{10}(OH)_2$, and biotite, $K(Mg, Fe)_3AlSi_3O_{10}(OH)_2$, in which there is a partial replacement of Al by Mg and Fe. The former is known as white mica, in general colorless, whereas the latter, black mica, is dark brown to black.

Rocks are consolidated (lithified) aggregates of minerals, that is, they are solid mixtures. Some minerals are presented below, classified primarily in terms of the negative groups, since these groups tend to be associated to the mineral sources. Some notes accompany the chemical compositions of the minerals. More detailed comments and crystal form coverages can be found in geological and geochemical texts, for example, Gilluly et al. (1959), Gass et al. (1971), Krauskopf (1967), and Mason (1966). A brief outline of crystal forms is presented in Section 2.3.3 of this chapter.

Carbonates

(See this chapter and Chapter 8, Volume II.)

Aragonite ($CaCO_3$):   Orthorhombic crystals. Tends to be converted into the more stable calcite at low pressures. Present in calcareous organisms, in sediments, and to a relatively small extent, in sedimentary rocks.

Calcite ($CaCO_3$):   Rhombohedral form. Present in calcareous organisms, sediments, in sedimentary rocks (limestones and dolomites), and as a vein mineral in many rocks. Solid solutions of $MgCO_2$ in $CaCO_3$, indicated by $Ca_xMg_{1-x}CO_3$ with $0 < x \le 0.28$, are often found in nature.

Dolomite ($CaMg(CO_3)_2$):   Rhombohedral. A definite compound found in dolomitic sedimentary rocks. See Chapter 8, Volume II.

Magnesite ($MgCO_3$):   Difficult to precipitate and rare. Possibly formed, in a very simplified representation, by $MgSiO_3 + CO_2 = MgCO_3 + SiO_2$.

Nesquehonite ($MgCO_3 \cdot 3H_2O$):   Formed at low (Ca^{2+}/Mg^{2+}) ratios and high pCO_2 values in natural waters.

Siderite ($FeCO_3$):   Hexagonal. Sedimentary by the action of H_2CO_3 upon other iron minerals. Also present in igneous rocks as ore veins.

Chlorides

Chloride minerals are present primarily in evaporites, resulting from the evaporation of salt waters in shallow water bodies.

Halite ($NaCl$):   Cubic. Formed in ancient times by evaporation. Interstratified with other sediments.

Sylvite KCl: Cubic.

Hydroxides

Brucite (Mg(OH)$_2$): Hexagonal. Formed by the action of hot waters upon serpentine, $Mg_3Si_2O_5(OH)_4$.

Oxides of Aluminum

Corundum (Al$_2$O$_3$): Formed at high temperatures and pressures. Of metamorphic origin and found in schists and metamorphosed limestones.

Diaspore (AlOOH): A constituent of bauxite, the main aluminum ore, which is produced by the leaching of silica from clays.

Gibbsite (Al$_2$O$_3 \cdot$ 3H$_2$O): A possible product of the breakdown of feldspars that are present in igneous rocks. Its structure appears in the octahedral layer of clays.

Oxides of Iron

Goethite (FeOOH): An anhydrous compound of iron formed in the weathering of iron-bearing minerals, for example, the oxidation of FeS_2. (See Vol. II, Chapter 7.)

Hematite (Fe$_2$O$_3$): An important iron ore.

Limonite (FeOOH $\cdot$ nH$_2$O): Hydrated goethite.

Magnetite (Fe$_3$O$_4$): A mixture of ferric and ferrous oxides.

Oxides of Silica

Chalcedony (SiO$_2$): Very fine quartz crystals.

Cristobalite (SiO$_2$): Tetragonal sheets.

Opal (SiO$_2$ $\cdot$ nH$_2$O): Amorphous hydrous silica found in tests of siliceous organisms and in sedimentary rocks.

Quartz (SiO$_2$): Trigonal hard component of igneous rocks.

Sulfates

Often present in evaporites.

Anhydrite (CaSO$_4$): Orthorhombic. Found in sedimentary beds but rare because it is easily hydrated.

Baryte (BaSO$_4$): It is more common in sedimentary rocks although it is found with minerals of sulfide ores.

Gypsum (CaSO$_4 \cdot$ 2H$_2$O): Forms massive deposits from early evaporation.

Sulfides

Formed in reducing (low oxygen) environments and often used as ores; are found in sulfide veins.

Chalcocite (Cu$_2$S)

Galena (PbS)

Pyrite (FeS$_2$)

Pyrrhotite (FeS)

Silicates are listed separately because of the complexity of their structures. They can be present as minerals in rocks, for example, feldspars, or as individual minerals in the case of soil clays.

Silicates can be present in massive and grainlike crystalline forms (feldspars, pyroxine, amphibole, etc.) or they may appear as scales. In the latter case, the silicates are made of parallel sheets as in the case of micas and clays.

Note that layered silicate formulas can be presented in several ways. This depends on the extent to which possible substitutions are shown and on the manner in which oxygens bound to silica and oxygens or hydroxides bound to aluminum or its replacements are presented. Thus, chlorite may be represented by $Mg_5Al_2Si_3O_{10}(OH)_8$ or as $(Mg, Al)_6[(Si, Al)_4O_{10}](OH)_8$ which shows interchanges between Mg and Al as well as those between Al and Si. The simpler formula is for a roughly average case. A uniform formulation throughout the book will not be attempted, therefore the simplicity or complexity required by different problems can be preserved. As another example, montmorillonite, $C_x(Al_{2-x}, Mg_x)(Si_4O_{10})(OH)_2$, is often simplified by the use of $C \equiv Na$ and $x = 0.5$.

In the tabulation that follows, a few silicates with their most variable formulas, for example, illite, are presented to show how the compositions can vary. The others are shown in their most common compositions.

Silicates

Amphibole: Similar to pyroxene but containing hydroxyls.

Albite: $NaAlSi_3O_8$

Anorthite: $CaAl_2Si_2O_8$

Biotite: $K(Mg, Fe)AlSi_3O_{10}(OH)_2$

Chlorite: $(Mg, Al)_6[(Si, Al)_4O_{10}](OH)_2$

Epidote: $Ca_2(Al, Fe)_3(SiO_4)_3(OH)$

Garnet: Silicates of calcium, iron, and aluminum silicates. Six-sided crystals.

Illite: $K_{1-x}Al_2[(Al_{1-x}, Si_{3+x})O_{10}](OH)_2$ or, for $x = 0$, $KAl_3Si_3O_{10}(OH)_2$

Kaolinite: $Al_2Si_2O_5(OH)_4$

Kyanite: Al_2SiO_5

Microline: $KAlSi_3O_8$

Montmorillonite: $C_x(Al_{2-x}Mg_x)(Si_4O_{10})(OH)_2$ or $CAl_2Mg(Si_4O_{10}) \cdot (OH)_2$ for $x = 1$. The symbol C represents a cation.

Muscovite: $KAl_2[Al, Si_3)O_{10}](OH)_2$

Olivine: $(Fe, Mg)_2SiO_4$

Phillipsite: $KAlSi_2O_6 \cdot 2H_2O$

Plagioclase feldspar: $NaAlSi_3O_8$ to $CaAl_2Si_2O_8$. Albite to anorthite.

Potassium feldspar or orthoclase: $KAlSi_3O_8$

Pyrophyllite: $Al_2Si_4(OH)_2$

Pyroxene: Silicates of Ca, Mg, and Fe with divalent iron.

Sepiolite: $Mg_2Si_3O_8 \cdot 2H_2O$

Amphibole, garnet, and pyroxene are solid solutions of the various silicates.

2.2.2 The Structure and Importance of Clays

Clays are hydrous aluminum silicates which consist primarily of alternating sheets. One type of sheet contains aluminum at the center of octahedrous made of oxygen and hydroxyl ions. The other type has silica at the center of a tetrahedron with oxygens and hydroxyl ions at the corners as is shown in Figure 2.2.

Clays play two important roles in nature. First, they are a link between igneous rocks and natural waters, and they are involved in the marine geochemistry of SiO_2 and some cations, such as

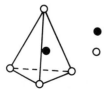

FIGURE 2.2 Unit cell in tetragonal sheet.

Mg^{2+} and K^+. Igneous rocks are altered chemically and structurally yielding clays, cations, and silica (SiO_2). These three products then are transported by rivers to oceans and lakes where clays sediment and undergo diagenesis (postdepositional alteration). At some stage in early diagenesis there may be at least a partial uptake of silica and cations by clays leading to the reverse of the initial weathering process.

The second role of the clays already hinted at in the previous paragraph is due to their capacity to adsorb and exchange ions (major constituents as well as metals) on their surface and absorb them in interlayer positions as well as by lattice substitutions. In this way clays help control the compositions of natural water. There is some debate on whether clays are effective in the water column or primarily in the pore waters of sediments (Pytkowicz, 1975a).

From the standpoint of human activities, clays are present in many topsoils as very finely powdered solids and exchange ions needed by plants when in contact with roots. Thus, they play a significant role in agriculture. Clay loams, however, form heavy soils in which root systems cannot develop easily. Clays are used in industry for pottery, bricks, earthenware, tiles, stoneware, and porcelain.

EXAMPLE 2.2. CONTROL OF CATIONS BY CLAYS. An example of the effect of interclay reactions on the partial control of cations is the conversion of albite to kaolinite

$$2NaAlSi_3O_{8(s)} + 2H^+ + 9H_2O = 2Na^+ \\ + 2H_4SiO_4 + Al_2Si_2O_5(OH)_{4(s)} \qquad (i)$$

with the equilibrium

$$K = \frac{\{Na^+\}\{H_4SiO_4\}}{\{H^+\}} \qquad (ii)$$

H_4SiO_4 is the hydrous form of SiO_2 present in solution and generated by

$$SiO_2 + 2H_2O = H_4SiO_4 \qquad (iii)$$

Equilibrium (2.4) shows the effect of clay equilibration and two other factors that enter into the control of cations such as Na^+ in natural open and pore waters. These are the effect of siliceous organisms on SiO_2 used in their test and the effect of the pH.

The equilibrium constant for the above reaction is $K = 10^{-1.9}$ according to Stumm and Morgan (1970). Thus if the pH is 10^{-8} and $\{H_4SiO_4\} = 10^{-6}$, then $\{Na^+\} = 1.9 \times 10^{-2}$ in the molar scale that was used for K.

I remind you that, in addition to clay–clay interactions, there also is the conversion of rocks into clays as a supplier of silica and cations. The processes will be discussed later on in this chapter.

The properties of clays depend more upon the structures than on the elements present in them (RMP). As an example, kaolinite, as well as illite, contains primarily Si, Al, O, and H, although illite usually but not always adsorbs K between lattice layers (see clay structures below). In the case of $x = 0$ one obtains for illite $Al_3Si_3O_{10}(OH)_2$, whereas kaolinite has the formula $Al_2Si_2O_5(OH)_4$.

The key property of clays, from the standpoint of major constituents as well as trace metals in solutions, is the uptake of the cations by adsorption, ion exchange, interlayer absorption, and structural replacement.

The structural replacement, for example of Mg^{2+} for Al^{3+}, leaves a net charge which, in conjunction with the ionization of silenol groups Si—OH to Si—O$^-$, provides adsorption sites. Different ions are more or less attracted to the available sites and ion exchange ensues. Interstitial uptake, such as that of K$^+$ in the case of illite, is fairly specific. However, it cannot be distinguished from simple ion exchange by elution alone. Thus, if a MgCl$_2$ of a CsCl$_2$ solution is passed through a column containing the clay, exchanged and interstitial ions will be removed roughly at the same time.

Structural uptake can be characterized by the change in the composition of the solution in contact with the clay (e.g., Fe^{3+} and glauconite) and by the lag to retrieve the ion.

Next, the structures and properties of clays based in part upon Berner (1971) are presented. The symbol C refers to an unspecified cation.

Structure	Clay Class	Formula	Cation Exchange Capacity (meq)/100 g
	Kaolinite	$Al_2Si_2O_5(OH)_4$	0–10
H$_2$OC$_i$H$_2$O	Montmorillonite	$C_x(Al_{2-x}Mg_x)(Si_4O_{10})(OH)_2$ with $0 \le x \le 2$	80–140
K$^+$	Illite	$K_{1-x}Al_2[(Al_{1-x}, Si_{3+x})O_{10}](OH)_2$ with $0 \le x \le 1$	10–40
	Chlorite	$(Mg,Al)_3[(Si,Al)_2O_5](OH)_4$	5–30
H$_2$OCH$_2$O	Vermiculite	$C(Mg)_3[(Al_x, Si_{4-x})O_{10}](OH)_2$	100–180

Kaolinite always has approximately the ideal composition $Al_2Si_2O_5(OH)_4$.

Montmorillonite never has the ideal composition $Al_2Si_2O_5(OH)_2$ due to extensive isomorphous substitution. The most common replacement in the tetrahedral sheet of montmorillonite is Al^{3+} for Si^{4+} (up to 15%). In the octahedral sheet the usual replacement is that of Mg^{2+} and Fe^{3+} for Al^{3+}. The ions Zn^{2+}, Ni^{2+}, Li^+, and Ca^{2+} also act occasionally as replacements and there may be complete substitution of Al^{3+}.

Illite represents the transition between montmorillonite and the nonclay mineral mica [muscovite with the ideal formula $KAl_2[(Si_3Al)O_{10}](OH)_2$) with less K^+ than muscovite]. The structure is dominated by the strong adsorption of K^+ in the interlayer positions.

In muscovite Al^{3+} is replaced by Mg^{2+} and some Fe^{2+} in the octahedral sheet.

Chlorite is somewhat similar in structure to montmorillonite but it has a stronger negative charge due to the replacement of Al for Si in the tetrahedral sheet and of Mg^{2+} and Fe^{2+} for Al^{3-} in the octahedral one.

Glauconite is a variety of illite with Al^{3+} substituted extensively by Fe^{3+} and the interlayer K^+ replaced by Ca^{2+} and Na^+.

EXAMPLE 2.3. A ROUGH EXAMPLE OF THE MEAN ION EXCHANGE OF CLAYS. Garrels and Mackenzie (1971a) present the following values for the wt% of recent clays in shales 6.5×10^8 yr old:

chlorite $\cong$ 15   illite $\cong$ 20
kaolinite $\cong$ 15   montmorillonite and others $\cong$ 60

Let us assume for simplicity that these are the proportions in which the clays passed through the oceans and the sediments. Do keep in mind that this is a very rough assumption made for illustrative purposes only. Then, we obtain, from mean values of the ion exchange capacities (IEC) of individual clays presented earlier, the average capacity

$$0.05 \times 18 + 0.15 \times 5 + 0.20 \times 25 + 0.60 \times 120 = 78 \text{ meq}/100 \text{ g}$$

Thus, 100 g of mixed clays can exchange 78 meq of ions. This IEC is smaller than that of organic matter, 100–200 meq/100 g (Babcock, manuscript). This does not mean that organic matter is more effective in removing ions from water bodies and/or

pore waters, because the overall capacities depend on the relative amounts of clays and organic substances.

Values for the total clay flux in rivers have not been found but, once it is known, a simple multiplication will yield the flux of exchange capacity into the oceans. Then, a knowledge of the exchange equilibria of the various ions in seawater will produce their rate of removal by clays. Of course, a direct knowledge of the proportions of clays in rivers is better than the indirect approach through shales.

The organic matter problem is attacked differently because the quantity of interest is the amount of such matter that is buried in the sediments. Although we know it in terms of grams of carbon, it is necessary to convert it into grams of average organic matter, a nontrivial problem.

2.2.3 A Brief Survey of Crystals and Their Role in Nature

Crystal systems refer to the axes of macroscopic crystals, whereas forms concern the arrangements of faces, such as those of a cube or a rhomb. On an atomic, ionic, or molecular scale the unit cell concept can be illustrated by that of copper, with one atom at each of the eight corners of a cube.

The crystal of halite, NaCl, is cubic in general and so is its unit cell. If, however, NaCl is precipitated from an aqueous solution containing 10% area, then the crystal form is octahedral although the unit cell is still cubic. This can be demonstrated by x rays. Another example of a mineral that has a different unit cell geometry from that of the crystal is the case of clays. The unit cells are tetrahedral and octahedral but the crystal is made of sheets that are scalelike. Therefore, the unit cell geometry cannot always be inferred from the crystal form.

The major crystal systems and forms are shown in Figure 2.3 and their properties are presented below.

1. *Cubic or isometric.* The axes are at right angles to each other and have the same lengths. In addition to cubic crystals this system includes tetrahedral and octahedral crystals. Examples include NaCl, ZnS, diamond, and Pb.

2. *Tetragonal.* Three axes are at right angles with two of the same length. Examples include Sn, SnO_2, and TiO_2.

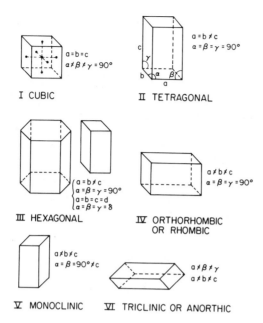

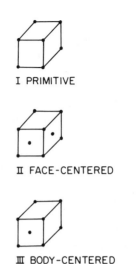

I PRIMITIVE

II FACE-CENTERED

III BODY-CENTERED

FIGURE 2.4 Cubic unit cells.

FIGURE 2.3 Crystal systems and forms. The axes are depicted in 2.3I while the angles and axes lengths are shown in 2.3II.

1. *Primitive cubic unit cell.* Examples include Cu, Ag, and Pb.
2. *Face-centered cubic.* Examples include NaCl and KCl.
3. *Body-centered cubic.* C is at the center of the cube. Examples include Na, K, and Li.

3. *Hexagonal.* Horizontal axes are of equal length with equal angles between them. Vertical axes are perpendicular to the plane of the other axes and are longer. Examples include ice, HgS, Zn, and graphite.

4. *Orthorhombic or rhombic.* Three axes are at right angles to each other and are of unequal lengths. Examples include: $PbCO_3$, $BaSO_4$, $MgSiO_3$, and K_2SO_4.

5. *Monoclinic.* One axis is inclined relative to the plane of the other two. The lengths of the axes are all different. An example is $CaSO_4 \cdot 2H_2O$.

6. *Triclinic or anorthic.* All axes are at different angles to each other and have different lengths. An example is $CuSO_4 \cdot 5H_2O$.

7. *Rhombohedral or trigonal.* Some authors accept this crystal habit, whereas other ones include it into the triclinic habit. All angles differ but the lengths of the axes are the same. Examples include calcite, $MgCO_3$, quartz, and As.

The term lattice is used to describe unit cells. Thus, various types of cubic unit cells are shown in Figure 2.4.

Lattices play an important role in nature because they stabilize minerals through the crystal bonds that are formed (RMP). Let us look at the face of the NaCl crystal (see Figure 2.5). This is an ionic crystal, held together primarily by coulombic (elec-

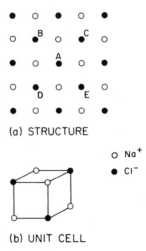

(a) STRUCTURE

○ Na$^+$
● Cl$^-$

(b) UNIT CELL

FIGURE 2.5 A fragment of the face of an NaCl crystal. The ions represented by *A, B, C, D,* and *E* reveal that the basic structure is face-centered although the unit crystal appears to be a simple cube. The reason for this apparent discrepancy is that the same type of particle must be used to define the structure crystallographically.

trostatic) forces, whereas other crystals may be held by covalent, metallic, or hydrogen bonds. Still, the principle is the same.

A single NaCl molecule, actually Na^+-Cl^-, would be held by the electrical energy

$$E = -\frac{ze^2}{D_e r} + \frac{b}{r^n} \qquad (2.4)$$

z, the valence, and D_e, the dielectric constant in a vacuum, are unity in this case. The term e is the elementary (electronic) charge, 4.80325×10^{-10} esu, and r is the interionic mean distance. The last term is the Born short-range repulsion energy. This is a quantum mechanical term.

In the lattice, however, there are the additional long-range electrostatic attractive and repulsion forces (not to be confused with the B force) due to all of the ions. Madelung (1918) [see also Kittel (1959)] showed that the attractive energy is increased by a factor A_M which is 1.7476 for NaCl. Thus, Equation (2.4) becomes (for NaCl),

$$E = -A_M \frac{e^2}{r} + \frac{b}{r^n} \qquad (2.5)$$

Examples of Madelung's constants for several crystalline substances are:

	A_M		A_M
NaCl	1.7476	ZnS (zinc blende)	1.6381
CsCl	1.7627	ZnS (wurtzite)	1.641

The excess attractive energy due to the formation of lattices decreases the solubilities of crystals relative to those of individual metal atoms, individual molecules, and pairs of ions because more energy is required to separate the particles in the crystal. An example is amorphous opal which is more soluble than quartz.

Therefore, without lattices, the amounts of solids on earth would decrease while the concentrations of solutes in natural waters would go up. The formation of bones and teeth would be more difficult because apatites (calcium phosphates) would dissolve to a greater extent in body fluids.

Lattice structures are subject to isomorphous substitutions in which the crystal forms are not changed. An example is the replacement of part of the calcium by magnesium in Mg calcites. Properties do change, however, as in the case of the increased solubilities of Mg calcites with an increasing magnesium content. Polymorphism refers to changes in form as in the case of aragonite and calcite.

These types of alterations alter the lattice energies of the isomorphs and polymorphs.

EXAMPLE 2.4. THE SOLUBILITIES OF CALCITE AND ARAGONITE. The thermodynamic solubility products of calcite and aragonite at 25°C and 1 atm total pressure are $10^{-8.35}$ and $10^{-8.22}$ even though

	K_{sO}	Crystal Form
Calcite	$10^{-8.35}$	Hexagonal
Aragonite	$10^{-8.22}$	Orthorhombic

Thus, calcite is less soluble and aragonite is converted to calcite in pure water under standard (STP) conditions.

2.2.4 Rocks

Rocks are constituted of single or mixtures of minerals, where the latter roughly represent compounds. There are three main categories of rocks:

1. *Sedimentary.* These are formed at the surface of the earth by sedimentation from water bodies or by accumulation on the surface followed by consolidation. Examples include: sandstone made of sand grains cemented by calcite; limestone made of calcareous (calcitic or aragonitic) tests cemented by $CaCO_3$; shale containing mainly clay minerals; chert constituted of opal from siliceous tests; dolomite mainly with the mineral dolomite; and evaporites with halite, anhydrite, and gypsum. The term lutites is used for fine-grained siliceous sedimentary rocks containing slate, shale, schist, argillite, and so on.

2. *Igneous.* These are brought from the interior of the earth as magma. In eruptions, magma is released as lava or volcanic ash. Lava forms volcanic rocks such as basalt when it cools on the surface. When it crystallizes deep below the surface, it forms plutonic igneous rocks (e.g., granite). Plutonic rocks cool slowly and form relatively large grains. The mineral composition of rocks depends on the parent mineral, the rate of cooling of the

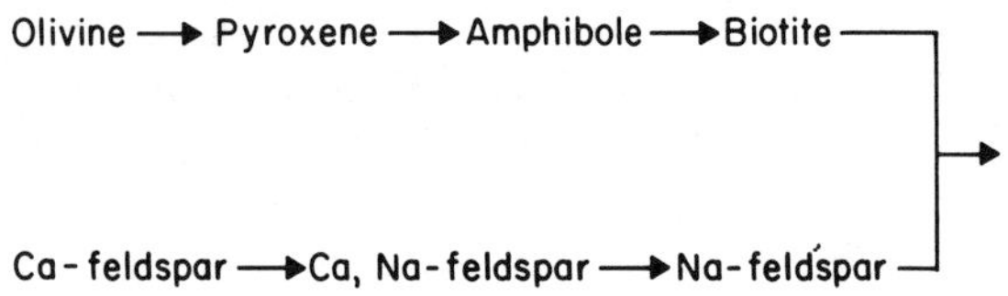

Olivine → Pyroxene → Amphibole → Biotite

Ca - feldspar → Ca, Na - feldspar → Na - feldspar

FIGURE 2.6 The Bowen reaction series.

magma, and of the segregation. Some common igneous rocks are

| Basalt | Andesite (fast cooling) | Rhyolite |
| Gabbro | Diorite (slow cooling) | Granite |

Composition

Plagioclase feldspar   Alkali feldspar
Ferromagnesian minerals   Quartz

of solidified fractions. The sequences of minerals formed during the crystallization of magma are given by the Bowen reaction series (Figure 2.6).

In the upper branch of the reaction series olivine first precipitates from the melt when it cools. This occurs when the magma is at equilibrium with this mineral. If olivine then settles, the remaining magma changes composition and is poorer in iron. Eventually equilibration occurs with pyroxene which crystallizes. Note that the overall system is not an equilibrium system, that is, there is not a common melt composition stable relative to all the minerals formed.

3. *Metamorphic.* These are formed at high temperatures and pressures from other rocks. Examples include marble from the metamorphism of limestone and slate from the metamorphism of shale.

2.2.5 Equilibrium, Nonequilibrium, and Mineral Assemblages (RMP)

Next, mineral reactions that may or may not reach equilibrium are examined because this will provide us with insights into control mechanisms in geochemistry. This section starts with a simple hypothetical example to illustrate principles.

EXAMPLE 2.5. MINERAL EQUILIBRIA AT DEPTH AND ON THE SURFACE OF THE CRUST. Let us assume that $CaCO_{3(s)}$ and $BaCO_{3(s)}$ can be in contact and that they can equilibrate deep in the crust and on the surface of the earth, to illustrate principles. Fur-

thermore, let us assume that water vapor is essentially absent at depth, for simplicity.

1. At depth the thermodynamic equilibrium constant is

$$K = \frac{\{CaCO_{3(s)}\}}{\{BaCO_{3(s)}\}} = 1 \qquad (i)$$

if CO_2 is not present, because the activities of pure solids are unity. The subscript (s) indicates a solid phase. There is no control on the relative amounts of $CaCO_{3(s)}$ and $BaCO_{(s)}$ at equilibrium.

2. The thermodynamic solubility products with water present are

$$CaCO_{3(s)} \qquad K_{sO} = 10^{-8.8} \qquad (ii)$$

$$BaCO_{3(s)} \qquad K_{sO} = 10^{-9.3} \qquad (iii)$$

Then, at equilibrium

$$K = \frac{Ca^{2+}}{Ba^{2+}} = 10^{0.5} \qquad (iv)$$

3. The phase rule: Let us anticipate the phase rule, which will be discussed later on but which is already known by many readers. The rule states that $f = c - p + 2$, where f is the number of degrees of freedom, c the number of the components, and p is the number of phases.

In our system at depth $c = 3$ because the components are CaO, BaO, and CO_2 if CO_2 is present as a separate substone. Also, $p = 3$, since the phases are $CaCO_{3(s)}$, $BaCO_{3(s)}$, and $CO_{2(g)}$, and, therefore, $f = 2$. For each temperature and pCO_2, there is an equilibrium state of the system. No further conditions, such as the ratio of the amounts of $CaCO_{3(s)}$ and $BaCO_{3(s)}$, can be imposed on the system because the two degrees of freedom have been used. All we learn is that a thermodynamic equilibrium can exist for the system. If we assume that a vapor phase $CO_{2(g)}$ can exist then the equilibrium constant is

$$K = pCO_2 \qquad (v)$$

instead of (i). The reactions are

$$CaCO_{3(s)} = CaO_{(s)} + CO_2 \qquad (vi)$$

$$BaCO_{3(s)} = BaO_{(s)} + CO_2 \qquad (vii)$$

We added one phase and one component so f does not change.

T and the total pressure P or T and pCO_2 can be used to define the system when pCO_2 is the equilibrium value $pCO_2 = K$. The total P is the pressure produced by the weight of overlaying rocks which may equal pCO_2 if CO_2 is the only gas present. This is not a necessary condition for the interchangeability of P and pCO_2 or phase rule parameters because we do have a degree of freedom (the choice of the type of pressure used) after T is fixed.

In the CaO–BaO–CO_2 system all that we learn is that there is an equilibrium state for any given T and pCO_2 or T and P, and that the densities of the solids are fixed by T and P. We learn nothing about the relative amounts of $CaCO_{3(s)}$ and $BaCO_{3(s)}$ at equilibrium.

Note that, if a hypothetical melt of the solids exists at equilibrium with them, then p becomes 4 (CaO, MgO, CO_2, melt) while $c = 3$ and $f = 1$. Then, for each temperature there is only one pCO_2 at equilibrium with the system and vice versa.

Next, let us consider the aqueous system. Now $c = 4$ and $p = 4$ because we added the component water and the aqueous solution phase. Then, $f = 2$ and one equilibrium state, which defines $[Ca^{2+}]/[Ba^{2+}]$, exists for every temperature and pCO_2 or total pressure. Actually, we shall see later that $f = 2$ is correct in terms of mole fractions but that $f_m = f + 1 = 3$ is the value in terms of molalities or molarities. Thus, the aqueous system has an additional degree of freedom relative to the solid one. When fixed it determines the individual concentrations.

Therefore, the aqueous system differs from the solid or the solid–melt system in additional freedom, and in that it specifies a ratio of $CaCO_3$ and $BaCO_3$ in solution. Furthermore, the aqueous concentrations $[Ca^{2+}]$, $[Ba^{2+}]$, and $[CO_3^{2-}]$ are also known at equilibrium if one concentration-related parameter is fixed. This parameter may be $[Ca^{2+}]$ or one of many others as we shall see later.

Next, let us consider to what extent rates of reactions and equilibria can affect the masses of various types of rocks and their interconversions. Emphasis shall be placed upon environments, especially those in which liquid water is present.

Let us consider further the processes that may occur, using the metamorphic reaction $CaCO_3 + SiO_2 = CaSiO_3 + CO_2$ as an example. Now $c = 3$

(CaO, SiO_2 CO_2), $p = 4$ (the reactants and products), and $f = 1$ in the absence of water. Thus, at any temperature there is only one value of pCO_2 for which the system can be at equilibrium. The minerals are $CaSiO_3$ (wollastonite), $CaCO_3$ (calcite), and SiO_2 (silica or quartz). The equilibrium constant is $K = pCO_2$.

If $CO_{2(g)}$ is present in fissures that do not reach the surface of the earth and is not compressed by the overlaying rocks, then the equilibrium condition only requires that enough $CaCO_3$ react with SiO_2 to generate an amount of CO_2 such that $pCO_2 = K$. The equilibrium condition, however, plays no role in the relative amounts of $CaCO_3$, SiO_2, and $CaSiO_3$ which are, therefore, controlled by transport rates, that is, by fluxes. Actually, the solid concentrations change a little with P and T if they are expressed in weight/volume units, because their densities change. This is a minor effect. Wollastonite is often found at boundaries between granite and limestones. The system is assumed to be closed, so that the pCO_2 is controlled by the extent of the reaction. Wollastonite, however, is also formed when the solids are under a pressure caused by the weight of overlaying rocks but CO_2 is in fissure contact with the atmosphere, with $pCO_2 \cong 1$ atm. Actually, pCO_2 is somewhat larger than 1 atm due to the high temperature.

In contrast to the wollastonite reaction, which may reach equilibrium, the cooling of volcanic and plutonic masses leads to nonequilibrium differential crystallization. The relative amounts of minerals formed during the Bowen reaction series depend on the rates of crystallization and segregation of the minerals. Still basalts can react with acid volatiles and equilibrate with aqueous solutions and clays during cooling, and reach aqueous equilibria. Thus, no general type of control mechanism occurs when solid phases coexist.

Another example of mineral interactions is that of the pure calcite–aragonite transformation. This is an interesting case because both minerals have the same composition, $CaCO_3$. Thus, $c = 1$, $p = 2$, and $f = 1$ at equilibrium. Therefore, at any given temperature there is only one pressure at which the two minerals can coexist. If the pressure is not the right one then there will be, at least in principle, a complete conversion of one mineral into the other if the reaction is fast enough.

The equilibrium condition only refers to the P and T which ensure their coexistence but do not yield their relative amounts.

Other examples of solid–solid reactions are presented below (Garrels and Christ, 1965).

$$PbCO_3 + CaSO_4 = PbSO_4 + CaCO_3$$
$$\text{cerussite} \quad \text{anhydrite} \quad \text{anglesite} \quad \text{calcite}$$
$$(2.6)$$

Cerussite and anhydrite are stable relative to anglesite and calcite at 25°C and 1 atm. Therefore, the latter two minerals are eventually completely converted to those on the left. There are temperatures and pressures, however, at which all four minerals can coexist at equilibrium, as $f = 2$.

Another reaction with a gas phase is $FeO + CO_2 = FeCO_3$ and its equilibrium constant is $K = 1/pCO_2 = 10^{6.1}$. FeO is unstable relative to siderite, $FeCO_3$, at the earth's surface, since pCO_2 is $10^{-3.5}$ atm. Then, $(pCO_2)_{equil} = 10^{-6.1} < pCO_2 = 10^{-3.5}$ and CO_2 is consumed while the reaction proceeds to the right. This occurs because the concentrations of the two polymorphs do not vary whether they be at equilibrium or not. The concentration of $CaCO_3$ (calcite) is always 10 moles/kg-calcite because the MW is 100 and the mole is 100 g. The same is true of aragonite. Therefore, $K' = $ (calcite)/(aragonite) $= 1$. This, however, is also true until one form is completely converted into the other stable one away from equilibrium.

Still, the equilibrium state is real because, as stated earlier, for each pressure there is only one temperature at which the two minerals can coexist (see also Chapter 5, Volume II) regardless of their amounts. At other temperatures only one of the two forms can survive when the system is given enough time to reach its final state.

Not all minerals can reach solid–solid equilibrium on the surface of the earth or even at depths. If two minerals are at the proper temperature and pressure then equilibrium can exist. Motion, however, is the rule in our planet and the minerals may and probably will shift to a (P, T) set at which they are not at equilibrium. Furthermore, we saw that the Bowen series does not yield equilibrium assemblages. Thus, we have seen so far that there is no general rule regarding the nature of stationary (time-invariant) states.

Let us look further at the earth surface, mentioned in Example 2.5, where new forces appear. Solid–solid disequilibria can continue for long times in the absence of water because equilibration is exceedingly slow. It requires solid diffusion and/or recrystallization and is sluggish even if the minerals are in solid solution or in an intimate mixture of grains.

There is, however, a major difference between the surface of the earth and the depths because there are forces on the surface which are not as active deep in the crust. They are derived primarily from solar energy and gravity, which can regulate the amounts of the different minerals present, be they at equilibrium or not, and accelerate equilibria and steady states. These forces act primarily through life, a nonequilibrium system, and through running water. Thus, water is now introduced as a new component.

Let us first consider the mass control. The rate of weathering of the solid crust depends on the reactivity of its minerals to the erosional power of water plus acid volatiles.

The weathered products are brought to oceans and lakes where inorganic and biological processes lead to sedimentation at a rate that depends on the hydrological cycle. The rate of sedimentation is a function of the rates of chemical reactions and life processes.

Finally, the cycle is closed by subduction, metamorphism, and uplift. The subdivision of substances among the various stages and between primary and sedimentary rocks, once a steady state is achieved, is a function of the relative rates of the processes described above.

If the sedimentation is fast, then the oceanic content will be lower than if it is slow, if the other fluxes (weathering, river transport, and uplift) are considered fixed (Pytkowicz, 1971, 1972, 1973b). The sedimentary process appears, therefore, to be controlled by steady states.

Next, let us consider the problem of equilibration in solution. Pairs of minerals can coexist at equilibrium under specific conditions. For those readers familiar with pH–pe diagrams, $FeCO_3$ and Fe^{2+} are at equilibrium if the pH is about 7 and the pe is -10 (see Chapter 7, Volume II). If, however, the pH is changed, then only one of these forms is stable. Thus, pairs or groups of minerals have large ranges of acidity and oxidation conditions in which they are not at equilibrium.

Furthermore, even single minerals may not reach equilibration for kinetic (rate-controlling) reasons. As an example, calcite is supersaturated in upper oceanic waters due to the inhibitory effects of Mg^{2+} and organic matter on precipitation. In the case of opal, seawater is undersaturated due to the high rate of uptake by siliceous organisms.

Disequilibrium may also occur because minerals may not be exposed long enough to each other to equilibrate. Thus, glacial lakes in volcanic mountains contain primarily silicates, whereas those with drainage basins in limestone terrains are calcitic. The silicates and the calcites from the two types of lakes do not interact.

The oceans, which receive the runoff from all types of terrains, may be the sites of a tendency toward multimineral equilibration to the extent that life and ion adsorbing on settling particles permit it.

In summary, fast local equilibrations between minerals may or may not occur, depending on conditions. The oceans may tend toward general biologically controlled steady states coupled with some local equilibrations.

Thus, some processes that lead to disequilibrium on the surface are:

1. The slow dissolution of calcareous tests and the inhibition of $CaCO_3$ precipitation by Mg^{2+} keep the oceans away from equilibrium with dissolved carbonates.

2. CO_2 is taken up by organisms and this leads to undersaturation in some near-surface oceanic regions.

3. The high solubility of $CaSO_4 \cdot 2H_2O$ in seawater (higher than in fresh waters) prevents a solution equilibration of $CaCO_3$ and gypsum in the submarine sediments.

4. CO_2 may be removed from natural waters fast enough to prevent solution equilibration in the simplified reactive $CaCO_3 + SiO_2 = CaSiO_3 + CO_2$.

5. Drainage basins from igneous rocks and limestones may feed separate lakes so that silicates and carbonates cannot interact.

6. Microbial action in the absence of oxygen can convert sulfates into hydrogen sulfide and heavy metal sulfides.

2.2.6 Minerals and Equilibria

Let us summarize the results so far regarding the states of mineral assemblages (RMP).

First, the number of degrees of freedom helps us determine whether equilibria can and do exist deep in the crust and mantle. If $f = 2$, as in the hypothetical $CaCO_{3(s)}$–$BaCO_{3(s)}$ or $CaO_{(s)}$–$BaO_{(s)}$–$CO_{2(g)}$ system, then equilibrium can occur at any (P, T) or (pCO_2, T) set. It will indeed occur if the minerals

and the CO_2 remain long enough at a given set for the equilibrium $K = pCO_2$ to be reached. Other examples of $f = 2$ and possible equilibrium were presented in this section. One must recognize, however, that motion is the rule within the earth because of thermal and pressure forces (the term force is used in the general sense of driving energies rather than its physical definition). Therefore, minerals may not remain at a given site long enough to equilibrate.

In the cases of the wollastonite and the aragonite–calcite reactions, however, $f = 1$ and T or P fix the equilibrium state. It is doubtful, therefore, that more than a fraction of the systems are at the correct (P, T) set for equilibration. The same is true if a melt is present.

On or near the surface of the earth the Bowen reaction series yields minerals that are not at solid–solid equilibrium with each other. The presence of water containing CO_2 tends to accelerate processes between minerals by weathering, dissolution, and aqueous reactions. This can yield fast local equilibria such as the dissociations of carbonic acid in natural water bodies. Life intervenes, however, so that the total fluxes of the components of the CO_2 system, as well as the degree of saturation of $CaCO_{3(s)}$, are controlled by rate processes (Pytkowicz, 1975a). The same is true, for example, of metal ions, which may undergo complexation and redox equilibria, but which are introduced into and removed from water bodies by fluxes. Thus, no generalization can be made about local equilibria versus steady states because the control mechanism depends upon specific reactions and locations. A generalization can be made, however, in global terms.

In effect, we saw earlier that systems that undergo fluxes of matter and energy are the domain of irreversible thermodynamics and are either in transient or in steady states. Thus, equilibria in nature are local states that occur only for fast reactions but the overall controls are due to steady states in stationary (time-invariant) systems. I shall return to this topic later because of its importance in control mechanisms and quantitative descriptions of geochemical cycles.

My general conclusions are that local, relatively fast equilibria may occur at times deep in the crust and on the surface of the earth. The existence of fluxes of matter and energy and the irreversibility of life, however, leads me to believe that stationary (time-invariant) systems are controlled by steady states and other systems by transients. This is true on the surface of the earth because of the effects of water and life. Disequilibrium at depth probably occurs when $f = 1$ and the (T, P) set is not the equilibrium one for the minerals under consideration.

An example of the two types of controls in the CO_2–carbonate system in natural waters is given next. The dissociation reactions of H_2CO_3 reach fast equilibration but the total CO_2 present is controlled by fluxes (rate processes) and $CaCO_3$ may not be at saturation.

2.2.7 The Geological Time Scale

The geologic time scale is presented in Table 2.2.

2.3 EXAMPLES OF GENERAL CYCLES AND REVERSE WEATHERING

In this section reactions that lead to reverse weathering and selected geochemical cycles of substances are examined. The number of cycles studied, unperturbed cycles as well as those perturbed by humans, is quite large even though most results are still tentative and incomplete, and transcends the scope of this book. Therefore, references will be provided at

the end of this chapter for material that will not be covered in this book.

2.3.1 Hydrological Cycle (RMP)

Let us first examine the hydrological cycle because it is vital for weathering. Consider some relevant numbers. The average depth of the oceans is about 3800 m and the annual rate of evaporation of oceanic waters is on the order of a column 1 m deep. Thus, in 3800 yr all the oceanic waters pass through the atmosphere, since the mixing time of the oceans is about 1000 yr.

About 90% of the evaporated seawater reprecipitates directly on the oceans and only 10% falls on the oceans. It requires, therefore, 38,000 yr for the total water content of the oceans to precipitate on land, do its erosion work, and return to the oceans as rivers, groundwaters, and ice (see Figure 2.7). This length of time is but an instant in geologic time since the age of the oceans is roughly 3.5×10^9 yr. This gives an idea of the erosional power of water. A more quantitative view of the hydrological cycle is presented in Chapter 3.

The erosional power of water is such that the sedimentary mass at present is on the order of $24,000 \times 10^{20}$ g (Ronov, 1968) or $32,000 \times 10^{20}$ g

TABLE 2.2 Geologic time scale (10^6 yr).

Era	Period	Epoch	Starting Years Ago $\times 10^6$
Cenozoic	Quaternary	Recent	2×10^{-2}
		Pleistocene	2
	Tertiary	Pliocene	12
		Miocene	30
		Oligocene	40
		Eocene	55
		Paleocene	60
Mesozoic	Cretaceous		130
	Jurassic		168
	Triassic		200
Paleozoic	Permian		235
	Carboniferous		315
	Devonian		350
	Silurian		375
	Ordovician		445
	Cambrian		550
Proterozoic			800
Archeozoic			2000
Azoic			3000

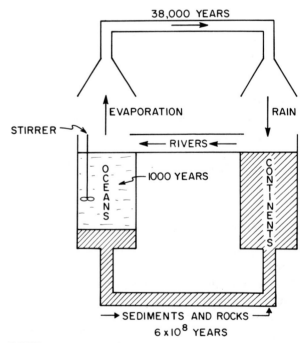

FIGURE 2.7 An analog diagram of the geochemical and the hydrological cycles (RMP).

(Garrels and Mackenzie, 1971a). According to the last two authors, a rough distribution of the sedimentary mass is 3500×10^{20} g of limestone, $17,300 \times 10^{20}$ g of shale, 2600×10^{20} g of sandstone, and 8600×10^{20} g of volcanogenic altered or unaltered debris, with 2100×10^{20} g of neutralized CO_2 and 450×10^{20} g of HCl used during the titration of igneous rocks. It should be remembered that the actual amount of chemical erosion is larger than that expressed above since the sedimentary rocks have undergone several cycles during the history of the earth.

It is not my intention to support one type of estimate over another because the estimates of weathering and sedimentary mass are rough estimates. Let us, however, use some figures presented earlier and above to estimate the erosional power of water.

EXAMPLE 2.6. THE EROSIONAL POWER OF WATER (RMP). Ronov (1968) and Li (1972) set the sedimentary mass at 2.4×10^{14} g. We saw earlier that this mass is generated during one geochemical cycle which lasts 1.2×10^8 yr.

We shall look at the first cycle in which only igneous rocks produced sedimentary rocks, rather than at more recent cycles in which igneous plus sedimentary rocks are weathered. The 1.2×10^8 yr, which applies to more recent cycles, is used only as an illustration since we do not know its early value.

According to Li, 100 g of igneous rock plus acid volatiles generate 113 g of sedimentary rocks. Thus $\frac{100}{113} \times 2.4 \times 10^{14}$ g $= 2.12 \times 10^{14}$ g of igneous rocks have been weathered.

During 1.2×10^8 yr there have been $(1.2 \times 10^8)/(3.8 \times 10^4) = 3.2 \times 10^3$ cyclings of the oceans or $3.2 \times 10^3 \times 1.37 \times 10^{21}$ L $= 4.38 \times 10^{24}$ L where 1.37×12^{21} is the volume of the oceans.

Thus, the grams of igneous rocks weathered by each liter of water are $(2.12 \times 10^{14}$ g$)/(4.38 \times 10^{24}$ L$) = 4.84 \times 10^{-11}$ g/L. It is quite surprising that such a small amount of rocks has been dissolved or suspended in the cycling water.

One should not think of the oceans as a simple pipeline that connects the continents to the submarine sediments, because the relative compositions of rivers and oceans are quite different. This is shown in Table 2.3. Biological, chemical, and geochemical processes alter the chemical nature of the river input and it is only after postdepositional reactions (low-grade diagenesis and metamorphism) that the original composition of eroded rocks may

TABLE 2.3 Relative riverine and oceanic compositions.

Constituent	Relative River Composition	Relative Oceanic Composition
Cl^-	1.00	1.00
Na^+	0.81	0.55
Mg^{2+}	0.53	0.07
SO_4^{2-}	1.44	0.14
K^+	0.29	0.02
Ca^{2+}	1.92	0.02

perhaps be reestablished, if the overall geochemical cycle is in a rough steady state. This point will be examined in greater detail later.

2.3.2 Main Weathering and Sedimentation Pathways

Two of the main steps in the geochemical cycle, sedimentation and weathering, are discussed next, after a brief digression on inputs to the weathering environment. Juvenile material (magma and gases) is brought up for the first time as the result of the high temperatures and pressures in the interior of the earth. Not all the minerals and gases brought to the surface of the earth are juvenile because a considerable fraction already underwent one or more cycles within the crust or in the crust–mantle exchange.

The oceanic sediments can be returned to the weathering environment through several pathways. If they are shallow deposits, for example, gypsum or halite, in evaporites, and some carbonates, they may be exposed to air as solids by the evaporation of shallow seas, by changes in sea level, or by the uplift of submarine sediments. Deep deposits may be transported toward the continents simply by seafloor spreading, subducted (sunk), and later uplifted as sedimentary rocks. Examples are many limestone, dolomite, and apatite formations. Finally, sedimentary and detrital rocks may undergo metamorphism at high pressures and temperature, as in the case of limestones that are converted to marble. The term detrital refers to unreactive fragments that settle after being brought by rivers and winds.

The weathering process consists in general of the action of acid volatiles dissolved in water which react with the rocks. This action does not always result from a direct dissolution of the atmospheric volatiles into the water. It is thought, in the case of

CO_2, to occur through the uptake of this gas by plants and the subsequent release to groundwaters during decay. This mechanism explains why groundwaters can be manifold supersaturated with regard to the atmospheric pCO_2. Acid volatiles, such as CO_2 and HCl, are gases that have an acid reaction when dissolved in water.

The main classes of weathering processes are illustrated in a simplified manner by:

1. *Igneous rocks*

$$2KAlSi_3O_8 + 2CO_2 + 3H_2O = 2K^+ + 2HCO_3^-$$
potassium
feldspar

$$+ Al_2Si_2O_5(OH)_4 + 4SiO_2 \qquad (2.7)$$
kaolinite

Equation (2.7) can be written in the equivalent way

$$6KAlSi_3O_8 + 6H^+ + 3H_2O = 6Na^+$$
$$+ 3Al_2Si_2O_5(OH)_4 + 6K^+ \qquad (2.8)$$

Equation (2.8) is more general than (2.7) in that it includes H^+ from $CO_2 + H_2O = H_2CO_3 = H^+ + HCO_3^-$ as well as from other acid volatiles such as HCl. Note that the weathering consumes H^+ (or CO_2) and that cations plus SiO_2 are released. The acid volatile constituents appear as the anions, such as HCO_3^- and Cl^-. Reactions similar to (2.7) and (2.8) may be written for sodium and calcium feldspars.

Siever (1968) split Equation (2.7) into steps represented by:

(a) *Primary weathering:*

$$5KAlSi_3O_8 + 4H^+ \rightleftarrows KAl_5Si_7O_{20}(OH)_4$$
potassium   illite
feldspar

$$+ 8SiO_2 + 4K^+ \qquad (2.9)$$

(b) *Secondary weathering:*

$$2KAl_5Si_7O_{20}(OH)_4 + 2H^+ + 5H_2O$$
illite

$$\rightleftarrows 5Al_2Si_2O_5(OH)_4 + 4SiO_2 + 2K^+ \quad (2.10)$$
kaolinite

Equation (2.9) represents the weathering of igneous rocks to clays while (2.10) shows the further weathering of clays to less siliceous ones in soils.

These reactions are examples of net overall processes but do not apply in detail to individual soils, which can be either acid or alkaline. The separation of weathering into steps is important when reverse weathering is considered, as we shall soon see. The counterparts of Equation (2.7) for sodium and calcium are

$$2NaAlSi_3O_8 + 11H_2O = 2Na^+ + 2OH^-$$
albite

$$+ 4H_4SiO_4 + Al_2Si_2O_5(OH)_4 \qquad (2.11)$$
kaolinite

$$CaAl_2Si_2O_8 + 3H_2O = Ca^{2+} + 2OH^-$$
anorthite

$$+ Al_2Si_2O_5(OH)_4 \qquad (2.12)$$
kaolinite

Kaolinite can be further degraded in silica and gibbsite

$$Al_2Si_2O_5(OH)_4 + 5H_2O = 2H_4SiO_4$$
kaolinite

$$+ Al_2O_3 \cdot 3H_2O \qquad (2.13)$$
gibbsite

2. *Carbonate rocks.* Limestone weathering can be expressed by the generic reaction

$$Ca_xMg_{1-x}CO_3 + CO_2 + H_2O \rightleftarrows xCa^{2+}$$
Mg calcite

$$+ (1 - x)Mg^{2+} + 2HCO_3^- \qquad (2.14)$$

which applies to magnesian calcites ($x \leq 0.28$), aragonite, and vaterite. During the weathering, half of the carbon in HCO_3^- originates in the atmosphere and half comes from the limestones.

Other reactions are

$$CaMg(CO_3)_2 + 2CO_2 + 2H_2O = Ca^{2+}$$
dolomite

$$+ Mg^{2+} + 4HCO_3^- \qquad (2.15)$$

$$MgCO_3 + 2H_2O = Mg(OH)_2 + H^+ \quad (2.16)$$

Of course a fraction of the HCO_3^- in reactions (2.14) and (2.15) is CO_3^{2-} due to the dissociation of bicarbonate (see Vol. II, Chapter 4).

3. *Phosphate rocks*

$$Ca(PO_4)_3(OH) + 3H_2O = 5Ca^{2+}$$
hydroxyapatite

$$+ 3HPO_4^{2-} + 4OH^- \qquad (2.17)$$

$$Ca_5(PO_4)_3F + H_2O = Ca_5(PO_4)_3OH$$
fluorapatite   hydroxyapatite

$$+ H^+ + F^- \qquad (2.18)$$

4. *Evaporites.* Equation (2.15) represents the weathering of halite which, as in

$$NaCl \rightarrow Na^+ + Cl^- \qquad (2.19)$$
halite

$$CaSO_4 \cdot 2H_2O = Ca^{2+}$$
gypsum

$$+ SO_4^{2-} + 2H_2O \qquad (2.20)$$

the case of gypsum, is a simple dissolution.

5. The weathering of quartz

$$SiO_2 + 2H_2O = H_4SiO_4 \qquad (2.21)$$
quartz   silica or
               silicic acid

is also a dissolution.

6. Amorphous hydroxides, oxides, and sulfides can be sources of metal ions.

Thus, some of the main products of weathering, which often consumes hydrogen ions, are cations, silica, and bicarbonate. In a roughly stationary (time-invariant composition) ocean, these substances must be removed by sedimentation and diagenetic processes. We shall see later that there is fair evidence for such an ocean that is roughly invariant. The fluxes of calcium bicarbonate resulting from the weathering of limestones and of calcium feldspars are easily disposed of because organisms such as foraminifera use them to produce tests and the resulting calcium carbonate eventually forms sediments. The overall cycle is closed by the lithification of limestones and by metamorphism of part of the carbonates back to calcium feldspars. CO_2 is released during test formation when bicarbonate is converted to carbonate and carbon dioxide, and by metamorphism (Urey, 1952; Pytkowicz,

1973b). We shall see later that perhaps the calcium cycle actually is not quite invariant and that new $CaCO_3$ is being formed in excess to that required by a stationary ocean.

Note that the weathering reactions have been represented as equilibrium reactions because the symbol $=$, which is a simplification of $\rightleftarrows$, was used instead of $\rightarrow$. The latter represents a one-way reaction. The reason for this is that both symbols apply and either one can be used if we keep in mind what we mean by it. As an example, $CaCO_{3(s)}$ may saturate running waters that cause weathering so that $CaCO_{3(s)} = Ca^{2+} + CO_3^{2-}$ and $K'_{SP} = [Ca^{2+}][CO_3^{2-}]$ pertains to this aspect of erosion. On the other hand, the $CaCO_3$ reservoir is not at equilibrium with its environment but is being eroded by a one-way process, that is, weathering. The proper notation from this point of view is $CaCO_{3(s)} \rightarrow Ca^{2+} + CO_3^{2-}$. The $CO_2 + H_2O$ term is neglected in this paragraph for the sake of simplicity although in reality the products of weathering are Ca^{2+}, HCO_3^-, and CO_3^{2-}.

The cycles of sodium and potassium feldspars have been challenging, as is shown in the next subsection.

There are other approaches to the weathering problem which deserve attention. Siever and Woodford (1979) examined the weathering kinetics of igneous rocks and found an acceleration of incongruent dissolution at low pH's and in the presence of oxygen. They interpreted their results as being due to the precipitation of $Fe(OH)_3$ which formed protective coatings on the igneous minerals. Holdren and Berner (1979) scanned mineral surfaces under earth surface conditions and concluded that the control mechanism for the dissolution was a surface-reaction one rather than protective coatings. Gregor (1980) examined weathering rates using linear models, that is, rates proportional to the abundance of rock types. He estimated the extents of weathering of sedimentary and igneous rocks.

2.3.3 Reverse Weathering

Sillen (1961) proposed that clay reactions may control the major cations of seawater and that reaction (2.10) may be considered as an example of such processes. He thought in terms of actual equilibria within seawater fixing, for instance, K^+ and H^+. Garrels (1965) recognized that reverse weathering [reactions (2.9) and (2.10) proceeding to the left] must be total, that is, must yield not only re-generated clays but also feldspars, to prevent the oceans from becoming soda lakes. Such lakes would occur if the HCO_3^- added by rivers was not eventually removed. The concept of Garrels is a generalization of that of Sillen in that he considered weathering reactions on land and diagenesis in sediments instead of only considering reactions in seawater. Siever (1968a) then proposed that reverse secondary weathering occurs in sediments, mainly in coastal subsiding sedimentation belts, and that reverse primary weathering occurs later and results from metamorphism. Subsidence refers to the sinking of sediments under the continents due to seafloor spreading. Siever and Woodford (1973) and Wollast (1974) also concluded that secondary reverse weathering is the result of high-grade post-depositional diagenesis rather than processes in open waters. The term high grade simply refers to an extensive change in contrast to the early diagenesis that occurs in the upper layer of aquatic

sediments. The change in magnesium content of a Mg calcite illustrates early diagenesis.

It is not necessary, in these developments of Sillen's (1961) original idea, to resort to a set of chemical equilibria within the oceans to obtain a fairly constant composition. A stationary (time-invariant) ocean may result simply from steady states (see Chapter 1) in which weathering and reverse weathering rates have eventually become about equal. A sketch of the weathering–reverse weathering cycle is shown in Figure 2.8.

Reverse weathering, as was mentioned earlier, consists of reactions such as (2.9), (2.10), and (2.14) proceeding to the left. These three reactions are only illustrations of the many processes that can occur and additional reactions are shown below in the following pages.

Mackenzie and Garrels (1966) presented a budget for the removal of all the major constituents of seawater plus clays, as a working hypothesis to explain the balance between river input and sedimentation,

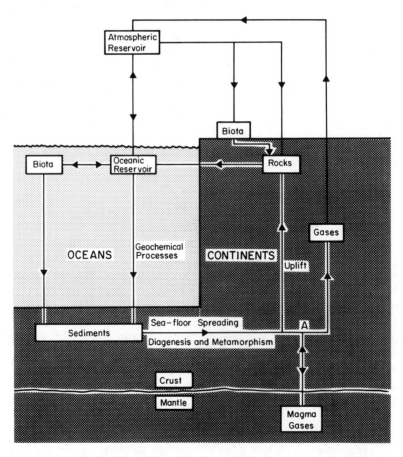

FIGURE 2.8 A sketch of the weathering–reverse weathering cycles (Pytkowicz, 1975a).

with reverse weathering included. The results are based on what is known about weathering on land and on a fragmentary knowledge of quantitative marine sedimentation.

The quantitative model of Mackenzie and Garrels (1966) must remain as a useful hypothesis because the rates of formation of the authigenic minerals proposed by them are poorly known. There are divided opinions on the evidence for the formation of authigenic minerals they predicted (e.g., Drever, 1974; Holland, 1978). It is to be expected that reverse weathering does occur if the oceans are at steady states and the main question, if this hypothesis is accepted, is at what stage in the geochemical cycle the diagenetic processes occur. It appears that the reverse weathering cannot be well documented in the upper part of the sedimentary column (early diagenesis) but occurs during high-grade diagenetic alteration. These problems are more serious for aluminum silicates than for calcareous sediments and evaporites for which reverse weathering is more easily documented. Still, reverse weathering must occur during high-grade diagenesis if the oceans are indeed at compositional steady states.

The term authigenic reflects the local formation of a mineral, in contrast to detrital, which means the sedimentation of minerals in the form in which they were brought by rivers.

Two examples of the series of reactions proposed by Mackenzie and Garrels (1966) are shown below because it may be that their hypothesis will be accepted more than semiquantitatively in time.

$$95.5\text{FeAl}_6\text{Si}_6\text{O}_{20}(\text{OH})_4 + 191\text{SO}_4 + 47.8\text{CO}_2$$

$$+ 55.7\text{C}_6\text{H}_{12}\text{O}_6 + 238.8\text{H}_2\text{O} = 286.5\text{Al}_2\text{Si}_2\text{O}_3(\text{OH})_4$$

$$+ 95.5\text{FeS}_2 + 382\text{HCO}_3^- \qquad (2.21)$$

This reaction consumes 191×10^{21} mmoles of SO_4^{2-} out of each 382×10^{21} mmoles brought by rivers.

$$191\text{Ca}^{2+} + 191\text{SO}_4^{2-} = 191\text{CaSO}_4 \quad (2.22)$$

In this step the remaining SO_4^{2-} is sedimented and 191×10^{21} mmoles of Ca^{2+} out of each 1220 mmoles brought into the oceans.

Examples of potential reverse weathering reactions are listed below. Note that the reverse of these reactions represents weathering processes on land.

Reverse weathering can occur in principle not only in marine sediments but also in sediments of lakes if the proper ranges of values of Eh, pH, cation, and silica concentrations exist in the pore waters. Of course, the proper clays must also be present.

1. *Clays.* Based on Siever (1968a),

$$5\text{Al}_2\text{Si}_2\text{O}_5(\text{OH})_4 + 4\text{SiO}_2 + 2\text{K}^+$$
kaolinite

$$= 2\text{KAl}_5\text{Si}_7\text{O}_{20}(\text{OH})_4 + 2\text{H}^+ + 5\text{H}_2\text{O}$$
illite

$$(2.23)$$

From Drever (1974)

$$6\text{Al(OH)}_3 + 9\text{Mg}^{2+} + 5\text{SiO}_2 + 8\text{H}_2\text{O}$$
gibbsite

$$= \text{Mg}_9\text{Al}_6\text{Si}_5\text{O}_{20}(\text{OH})_{16} + 18\text{H}^+ \qquad (2.24)$$
chlorite

$$\text{Al}_2\text{Si}_2\text{O}_5(\text{OH})_4 + 0.8\text{K}^+ + 0.5\text{Mg}^{2+} + 1.2\text{SiO}_2$$
kaolinite

$$= \text{K}_{0.8}\text{Mg}_{0.5}\text{Al}_{2.2}\text{Si}_{3.4}\text{O}_{10}(\text{OH})_2 + 18\text{H}^+ + 0.3\text{H}_2\text{O}$$
illite

$$(2.25)$$

A comparison of 2.10 and 2.12 shows that there is some flexibility in structure exchange in clays.

From Stumm and Morgan (1970)

$$3\text{Na}_{0.33}\text{Al}_{2.33}\text{Si}_{3.67}\text{O}_{10}(\text{OH})_2 + 6\text{Na}^+ + 10\text{H}_4\text{SiO}_4$$
sodium montmorillonite

$$\text{NaAlSiO}_8 + 6\text{H}^+ + 2\text{H}_2\text{O} \qquad (2.26)$$
albite

From Berner (1971)

$$\text{Al}_2\text{Si}_2\text{O}_5(\text{OH})_4 + 2\text{Mg}^2 + 2\text{Na}^+ + 6\text{HCO}_3^-$$
kaolinite

$$+ 10\text{H}_4\text{SiO}_4 = 4\text{Na}_{0.5}\text{Al}_{1.5}\text{Mg}_{0.5}\text{Si}_4\text{O}_{10}$$
montmorillonite

$$+ 6\text{H}_2\text{CO}_3 + 19\text{H}_2\text{O} \qquad (2.27)$$

Laboratory work on clay reactions with seawater was made by a number of investigators (e.g., Mackenzie and Garrels, 1965; Siever, 1968b). The main conclusion was that clays will combine with alkalies and with silica at high pH's and high silica concentrations, but that they will dissolve in part if the pH and silica concentrations are low. These results indicate that clays can follow a reverse weathering behavior under the proper conditions characteristic of deep sediments. In the measurements by Siever, the final forms of the clays could not be established in terms of known ones. Still, the convergence of the dissolved SiO_2 concentration from the high and low initial values suggests the possibility of at least a partial equilibration under laboratory conditions.

2. *Carbonates.* The reverse weathering of carbonates is simply their biogenic and occasional inorganic sedimentation in the sea, with a larger relative extent of inorganic precipitation in fresh waters. The main reaction is

$$x Ca^{2+} + (1 - x) Mg^{2+} + 2HCO_3^-$$
$$= Ca_x Mg_{(1-x)} CO_3 + CO_2 + H_2O \quad (2.28)$$

In this equation, $Ca_x Mg_{1-x} CO_3$ represents the formation of aragonite ($x = 1$) and magnesium calcites ($x \leqslant 0.28$). For dolomite

$$2CaCO_3 + Mg^{2+} = CaMg(CO_3)_2 + Ca^{2+}$$
$$(2.29)$$

Equations (2.28) and (2.29) are balanced stoichiometrically and represent the equilibrium states. They do not, however, necessarily represent the course of the incongruent reactions that may take place (see Chapter 8, Volume II).

3. *Amorphous silica.* The reverse weathering consists of the removal of H_4SiO_4 into the tests of siliceous organisms such as diatoms and radiolaria. Amorphous silica ($SiO_2 \cdot 2H_2O$, that is, opal) is sedimented.

4. *Evaporite formation*

$$\underset{\text{halite}}{Na^+ + Cl^- = NaCl} \quad (2.30)$$

$$Ca^{2+} + SO_4^{2-} + 2H_2O$$
$$= \underset{\text{gypsum}}{CaSO_4 \cdot 2H_2O} \quad (2.31)$$

5. *Redox reactions (anoxic conditions).* From Stumm and Morgan (1970)

$$Fe(OH)_3 + 4H^+ + 2SO_4^{2-}$$
$$= \underset{\text{pyrite}}{FeS_2} + 2.33O_2 + 1.5H_2O \quad (2.32)$$

6. *Redox and complexation reactions (oxygenated waters).* Oxides and hydroxides such as ferromanganese nodules.

7. *Detritus.* Quartz (crystalline SiO_2) derived from land.

The nature and the extent of reverse weathering can differ considerably in the sediments present in various locations in the oceans. Sediments beneath regions of low fertility (red muds) have little organic content. Thus, there is no extensive oxidation of organic matter and the reactions of anoxic conditions do not occur. In predominantly biogenic sediments ($Ca_x Mg_{1-x} CO_3$ or $SiO_2 \cdot 2H_2O$), clay reactions play a minor role.

Let us consider lakes. Alpine lakes near glacial cirques are not fertile and anoxic conditions do not occur, in contrast to what may happen to those lakes in rich agricultural lands. If mountain lakes result from an igneous drainage then silica may be liberated inorganically or by siliceous organisms, leading to the formation of clays. Alpine lakes in limestone terrains contain weathered carbonates. It is doubtful that a significant extent of reverse weathering occurs in such high altitude waters although some may occur in alkaline conditions.

More extensive reverse weathering can occur in lakes located in fertile soils rather than in rocky terrains, although the major process occurs later on, in marine sediments. The net process on land is one of direct weathering. In soils one may have redox conditions which lead to the removal of metal ions, clay–clay interactions with silica released inorganically or biogenically, apatite removal in the hard parts of organisms, evaporites, and so on.

It is interesting to examine the work of Drever (1974) because he took into consideration some processes that are not included above, even though his main concern was restricted to the reverse weathering of Mg^{2+} magnesium balance. Drever considered:

1. The uptake of Mg^{2+} by calcite and during the following dolomitization.

2. Ion exchange of the Ca^{2+} present on exchange sites of clays in rivers by Mg^{2+}, Na^+, and K^+ once the waters reach the oceans, that is, the uptake and reverse weathering of these three ions.

3. The kaolinite–illite and the kaolinite–chlorite transformations. Drever at the time concluded that there was no proof of the presence of authigenic illite and chlorite in submarine sediments.

Holland (1978) concluded that now there is ample evidence for the formation of illites, as sinks for K^+ and Mg^{2+}, in the oceans. Chlorite may be formed but is masked by large amounts of this detrital clay. These reactions, however, are not yet generally accepted.

4. Uptake of brucite, $Mg(OH)_2$, between the silicate layers of montmorillonites.

5. Magnesium ion exchange with iron in anaerobic sediments according to

$$2Fe \text{ (clay)} + 3Mg \text{ (soln)} + 4S$$
$$= 3Mg \text{ (clay)} + 2FeS_2 \qquad (2.33)$$

6. Gibbsite–chlorite transformation.

7. Precipitation of minerals relative to which seawater is supersaturated (an example would be sepiolite, $Mg_8Si_{12}O_{30}(OH)_4(OH_2)_4 \cdot 8H_2O$).

8. Burial of interstitial water and subsequent reactions. Drever estimated that the main mechanisms generated the fluxes shown in Table 2.5 and that he could only account for the removal of 50% of the river-borne Mg^{2+}, which amounts to 1.3×10^{14} g/yr. This is probably an underestimate because the formation of illite was not included. The residue may be explained in part by hydrothermal activity, as we shall see later.

Another possibility for part of the unbalance resides in the fact that Drever used data obtained in the laboratory and uncorrected for the in situ pressure and temperature. Shifts in equilibria or in the approach to equilibria could occur when the samples were decompressed and warmed. Bischoff et al. (1970) and Sayles et al. (1973) obtained data in which these shifts were avoided. In any event, the reactions considered by Drever (1974) and his general conclusions are valid even if the actual rate of Mg^{2+} uptake may be somewhat uncertain.

Sayles et al. (1973), Sayles and Manheim (1975), and Manheim (1976) took the temperature effect into account when they examined pore waters obtained during the Deep-Sea Drilling Project. In situ squeezing of sediments was used to avoid shifts in mineral–seawater equilibria during decompression. The following conclusions were reached with regard to those pore waters that are deep in the sediments, where diagenesis is expected to occur:

1. The composition of seawater changes considerably during deep burial. As an example it becomes depleted in Na^+ (1%), K^+ (23%), Mg^{2+} (16%), and SO_4^{2-} (62%) in terrigenous sediments (sediments brought from land).

2. Pelagic clays and biogenic particles may be almost at equilibrium with the pore solution, which differs primarily from seawater by its $Si(OH)_4$ content. I have reservations about this conclusion because of the complexity of the phases present.

3. K^+ and Mg^{2+} are only enriched where there is diffusion from evaporitic deposits. Elsewhere, they are taken up by the sediments.

4. Ca^{2+} is greatly enriched in regions of carbonate deposition, because of dissolution and dolomitization. This suggests to me that Ca^{2+} diffuses from the sediments to the oceans. Still, this return flux is only part of the amount settled as $Ca_xMg_{1-x}CO_3$ so that there is a net removal of calcium from the sea.

5. Diffusion is the main mechanism for seawater–pore water exchange. This process may lead the oceans toward equilibrium through the trend toward equilibration with minerals which occurs in the sediment–pore water system. Still, settling and solid–pore water interactions of particulate $Ca_xMg_{1-x}CO_3$ and SiO_2 play an important seawater–pore water transfer role (RMP). Note that a trend toward equilibrium does not mean that this state is reached. If it were, one would not observe gradients in concentrations between the oceans and the pore waters.

6. Dissolved SiO_2, as H_4SiO_4, remains roughly constant because of the abundance of biogenic SiO_2. This presupposes equilibrium (saturation) of dissolved silica with opal in the sediments and contradicts the control of SiO_2 by clay equilibria (RMP).

7. The alkalies are removed by the reverse weathering reactions of the type of Equation (2.23).

TABLE 2.5 Mg^{2+} removed from the oceans according to Drever (1974).

Process[a]	Mg^{2+} Removal[a]	Percentage of River Input
Carbonate formation	0.075	6
Ion exchange	0.097	8
Glauconite	0.039	3
Mg–Fe exchange	0.29	24
Interstitial water burial	0.11	9
	0.61	50

[a]Should also include illite formation (RMP).

An interesting point is brought up by Sayles and Manheim (1975) who concluded that dissolved SiO_2 is about constant down the sediment column and that this results from the large amount of solid SiO_2. This implies saturation of the dissolved silica relative to opal. Jones and Pytkowicz (1973), who measured the solubility of SiO_2 at high pressures in seawater, concluded tentatively that most pore waters are undersaturated. Furthermore, as was mentioned earlier, clays tend to fix the levels of dissolved silica. These two mechanisms, interactions with opal and clays, need not be incompatible if amorphous silica, where predominant, fixes the SiO_2 in solution at kinetically controlled levels. Clays may cause a tendency toward clay–seawater equilibrium where they predominate in the sediments.

Reactions that represent facets of reverse weathering, as detected and/or inferred by Drever (1974), Sayles and Manheim (1975), and Manheim (1976), are presented in right column.

In general terms, the conclusions of Sayles et al. (1975) and Manheim (1976), as well as some of those of Drever (1974), are in accord with the accepted reactions for reverse weathering. These reactions require the uptake of cations, such as K^+ and Mg^{2+}, of SiO_2, and of alkalinity as the release of H^+. Calcium ions are removed from the oceans by sedimentation of carbonates minus the upward diffusion of this ion.

2.3.4 Hydrothermal Activity

Questions have been raised about the adequacy of the usual reverse weathering reactions to balance the weathering ones. Examples are some uncertainty regarding the formation of authigenic clays and a sufficient uptake of Mg^{2+} to compensate for the river input.

Recently, it has become clear that submarine hydrothermal activity, the contact of seawater with hot basalts, plays an important role in the geochemical balance of the oceans (Holland, 1978; Edmond et al., manuscript). Edmond et al. based their conclusions on measurements made by Corliss et al. (1979) at the submarine thermal springs of the Galapagos Rift.

Edmond et al. (manuscript) observed that the warm seawater, released at the springs after contact with the basalt, was enriched in H_4SiO_4, K^+, Ca^{2+}, Li^+, Mn^{2+}, and CO_2, but impoverished in Mg^{2+}.

Ion or Molecule in Pore Water	Depleted[a]	Released	Reaction
K^+	Yes	—	Kaolinite to illite
K^+	Yes	—	Kaolinite to montmorillonite
Na^+	Yes	—	Calcite to clinoptilotite
Na^+	Yes	—	Anorthite to anacalcite
Mg^{2+}	Yes	—	Kaolinite to chlorite
Mg^{2+}	Yes	—	Kaolinite to illite
Mg^{2+}	Yes	—	Calcite to dolomite
Mg^{2+}	Yes	—	Kaolinite to montmorillonite
Mg^{2+}	Yes	—	Ion exchange with calcium
Mg^{2+}	Yes	—	Iron exchange by magnesium
Ca^{2+}	—	Yes	Calcite to dolomite
Ca^{2+}	Yes	—	Calcite to clinoptilotite
Ca^{2+}	—	Yes	Anorthite to anacalcite
SiO_2	Yes	—	Most above reactions
H^+	—	Yes	Most above reactions

[a]Removed from pore waters by sediments

Alkalinity was removed in the sense that CO_2 (or H^+) was released. The release of H^+ was attributed to proton-freeing reactions. I remind you at this point that CO_2 is consumed in many direct weathering reactions and that this leads to the formation of HCO_3^-, that is to the generation of alkalinity.

The springs may be a source of SO_4^{2-}, which is reduced to elemental sulfur and H_2S, but which may be reoxidized when the waters in contact with the basalt are returned to the oceans (Corliss, personal communication). Edmond et al. concluded that the above results confirm the reverse weathering mechanism conceived by Sillen (1961) and Garrels (1965). I find this hard to accept. Granted, the springs are a sink for alkalinity through the release of H^+ or CO_2 used up in direct weathering. At the same time, however, they are a source for H_4SiO_4 and several cations, all of which should be removed in reverse weathering.

We shall see later that dissolution or removal of CO_2 does not affect the alkalinity of a solution such as seawater. This point may already concern some readers because of the previous paragraph. The difference between weathering, in which case a solid coexists with the solution and a solution alone, is that, in the former case, CO_2 can react with the solid to generate HCO_3^- and CO_3^{2-}, whereas in solution it can only change two equivalents of CO_3^{2-} to the same number of equivalents of HCO_3^-. The latter process does not alter the alkalinity, which is expressed in equiv/kg-SW.

2.3.5 Pathways of the CO_2 System Before Humans

The system perturbed by fossil fuel burning is discussed in Chapter 2, Volume II.

The CO_2 system is exceedingly important because it participates in photosynthesis and respiration, the weathering and the formation of limestones and dolomites, the weathering of igneous rocks, the formation of marine shells, the control of the pH of natural waters, and the potential hazards of fossil fuel burning.

The fluxes of that system are presented in a simplified manner, for the sake of clarity, in Figure 2.9. This diagram corresponds to the fluxes before the advent of the Industrial Revolution (ca. 1860) since the production of CO_2 by the accelerated burning of fossil fuels such as petroleum, coal, and natural gas is not shown.

The main pathways are:

1. The release of juvenile CO_2 from the mantle to the atmosphere.
2. The weathering of carbonate and igneous rocks by CO_2 and water, followed by the

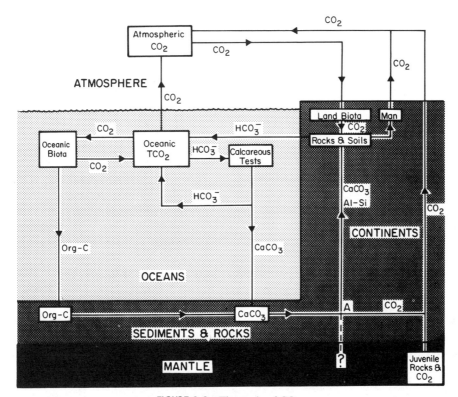

FIGURE 2.9 The cycle of CO_2.

river transport of the resulting HCO_3^- to the oceans.

3. Photosynthesis and oxidation on land and at sea.

4. Marine sedimentation of refractory (nonoxidized) organic matter.

5. Formation of marine calcareous tests with partial redissolution and sedimentation of the remainder.

6. Transport of calcareous and organic sediments by seafloor spreading, subduction, and uplifting, with low-grade diagenesis and metamorphic processes converting the carbonate sediments back to the limestones, dolomite, and plagioclase feldspars present in the weathering environments.

7. The hydrothermal injections of CO_2 and removal of alkalinity are not shown at this time because results are still limited to the Galapagos Rift.

2.3.6   Estimates of Fluxes and Reservoirs in the CO_2 System

These estimates are aimed at those readers who are specifically interested in the types of calculations involved in the determination of geochemical budgets, because the procedures are tedious. Still, such estimates are of value not only because of the data they present, but also because they reveal uncertainties and gaps in our knowledge. A steady state is assumed and an excessive number of significant figures is used for mass balance purposes. Hydrothermal inputs are not entered since their global values are too uncertain at this time.

Now, interested readers are referred to Figures 2.10 and 2.11.

EXAMPLE 2.7.   QUALITATIVE FLUXES AND RESERVOIRS OF THE CO_2 SYSTEM.   The first step in obtaining a model of the CO_2 system is to select the minimum numbers of reservoirs and fluxes needed to represent it. The selection of reservoirs is not unique and is biased by the problem on hand. We seek a general geochemical description and, to prevent the diagram from becoming unmanageable, will not subdivide the oceans into water masses, look at individual mineral phases of carbonates, or consider biological speciation or trophic levels.

The diagram of the CO_2 system is presented in Figure 2.10 in more detail than it was presented

earlier in this book and is a key to Figure 2.8. The primary input into the atmosphere is the juvenile CO_2 which is produced by the degassing of the mantle. It is taken up by biological and geological processes on and in the crust and by the oceans. The crustal uptake is represented by flux J_{1-A-2}, the flux from reservoir 1 to 2 via photosynthesis at A. The resulting organic carbon (org-C) may be lithified (J_{2-6}), transported to the oceans as dissolved and particulate organic matter (J_{2-C-8}), or may be oxidized back to CO_2.

The soil solution is usually supersaturated with CO_2 and part of its CO_2 content returns to the atmosphere (J_{3-1}). The remainder is used to weather aluminum silicates at D and sedimentary carbonates at E. The weathering reactions may be illustrated by the generic Equations (2.9), (2.10), and (2.14) plus (2.34),

$$CaAl_2Si_2O_8 \cdot 2NaAlSi_3O_8 + 4H_2CO_3 + 2(nH_2O)$$
$$\text{plagioclase feldspar}$$

$$= Ca^{2+} + 2Na^+ + 3HCO_3^-$$
$$+ 2Al_2(OH)_2Si_4O_{10} \cdot nH_2O \qquad (2.34)$$

The HCO_3^- formed by these reactions is transported to the oceans by rivers (J_{D-7} and J_{E-7}) and becomes part of the total dissolved inorganic carbon dioxide of seawater

$$TCO_2 = (CO_2) + (H_2CO_3) + (HCO_3^-) + (CO_3^{2-})$$
$$(2.35)$$

The HCO^- that results from the weathering of sedimentary carbonates is removed as $Ca_xMg_{1-x}CO_3$ by calcareous tests at H. The CO_2 generated, which corresponds to the atmospheric CO_2 used in the weathering, is returned to reservoir 7 by J_{H-7} and eventually reenters the atmosphere. A stationary (time-invariant) condition exists if biological utilization by tests equals the weathering rate. Inorganic precipitation of $CaCO_3$ was not mentioned because Pytkowicz (1965a) showed that the removal of carbonates from the open oceans is strictly biogenic. More than half of the settling or settled tests are redissolved at I by reacting with the CO_2 that results from the oxidation of organic matter. At least part of the remaining $CaCO_3$, which is incorporated into the submarine sediments (J_{9-J}), is eventually returned to the weathering environments on the continents by flux J_{J-5} but some of it

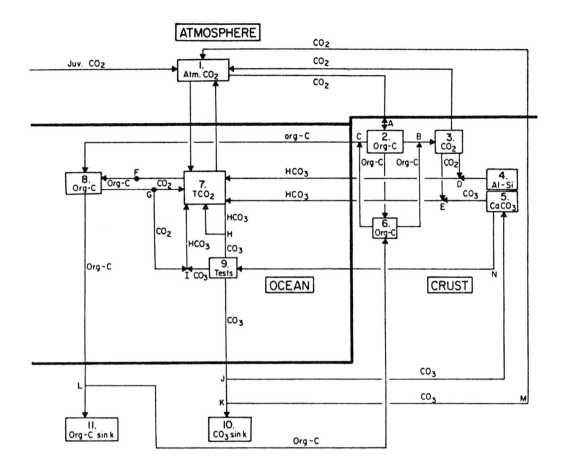

The symbols have the following meaning:

A Photosynthesis on land
B Oxidation on land
C Physical weathering of organic carbon
D Weathering of aluminium silicates
E Weathering of carbonates
F Photosynthesis in the oceans
G Oxidation in the oceans
H Formation of calcareous tests
I Dissolution of tests
J Junction
K Junction
L Junction
M Metamorphism
N Physical weathering of $CaCO_3$

1 Atmospheric CO_2
2 Organic carbon on land
3 Dissolved inorganic carbon on land
4 CO_2 content of aluminium silicates
5 Sedimentary carbonates
6 Lithified organic carbon
7 Dissolved inorganic carbon in the oceans
8 Organic carbon in the oceans
9 Calcareous tests plus detrital carbonates
10 Carbonate sink
11 Organic carbon sink

FIGURE 2.10 A block diagram of the CO_2 system (Pytkowicz, 1973b).

may be permanently incorporated into reservoir 10. This point as well as others left unanswered in this section will be discussed shortly.

Calcium-bearing igneous rocks will be represented for simplicity by CaIg in the remainder of this text. The $Ca(HCO_3)_2$ generated by reaction (2.34) is utilized by calcareous organisms in reac-

tion (2.28) and part of the resulting $CaCO_3$ may form new carbonate rocks. The weathering of CaIg probably was the original source of most carbonate rocks. The remaining $CaCO_3$ may eventually be brought by seafloor spreading to metamorphic environments where it can undergo reactions of the general type $CaCO_3 + SiO_2 = CaSiO_3 + CO_2$ (at M in

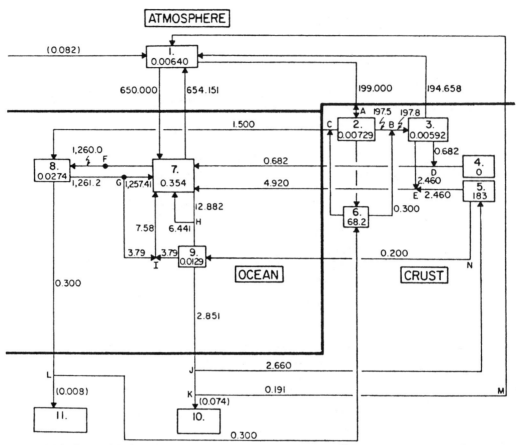

FIGURE 2.11 A quantitative model of the CO_2 system. The reservoir sizes are in 10^{20} g-C and the fluxes in 10^{14} g-C/yr (Pytkowicz, 1973b).

Figure 2.8) and be brought by tectonic activity back to reservoir 5. The CO_2 released during metamorphism is returned to the atmosphere (J_{M-1}).

The weathering of NaIg and KIg yields soluble $NaHCO_3$ and $KHCO_3$ which must undergo reverse weathering, if the oceans are not to become soda lakes. The CO_2 released by reverse weathering is represented as joining J_{7-1}.

The organic carbon cycle will be examined next. Photosynthesis exceeds oxidation on land and the surplus organic carbon (org-C) is transported to the oceans by rivers (J_{2-C-8}). The primary supply of org-C to reservoir 8 comes from photosynthesis at F. Most of it is oxidized and returns to 7 by J_{8-G-7}. Some of the CO_2 produced by oxidation is used in the dissolution of calcareous tests in deep waters (J_{G-1}), as was mentioned earlier. That part of the org-C which is not oxidized settles to the sediments, is eventually lithified, and either joins the weathering reservoir (flux J_{L-6}) or is removed into sink 11. This alternative depends on whether the system is entirely cyclic or not. The material in reservoir 6

can be physically weathered (fJ_{6-C}) or it can be oxidized (J_{6-B}).

J_{1-7} and J_{7-1}, the exchange fluxes between the atmosphere and the oceans, are not equal since the excess of photosynthesis over oxidation on land is compensated by net outflows of CO_2 into the atmosphere and, as org-C, into the sediments. Thus, surface seawaters are, on the average, slightly supersaturated with CO_2 resulting from oxidation in and upwelling of deep waters. This supersaturation occurs primarily at low latitudes, possibly because the saturation level is further enhanced by the warming of the waters and the decrease in pH following biogenic removal of $CaCO_3$.

It has been shown in this section that the CO_2 system in nature, although extremely complex, can be represented for geochemical purposes by 11 reservoirs and 30 fluxes. This large number of interconnected reservoirs and fluxes must be kept in mind in the selection of those that are relevant to the problem under study. Thus, in the case of fossil fuel CO_2, fluxes of CO_2 to the deep marine sediments are too slow to play a role in terms of decades.

A numerical model, corresponding to the block diagram in Figure 2.8 and shown in Figure 2.12, will be obtained next.

EXAMPLE 2.8. QUANTITATIVE FLUXES AND RESERVOIR SIZES OF THE CO_2 SYSTEM. The reservoir sizes in Figure 2.7 are in 10^{20} g-C (grams of carbon) and the fluxes in 10^{14} g-C/yr. The number of significant figures is well in excess of the accuracy of the data but is needed for mass balance computations. The uncertainties in the data as well as the sources were presented in an earlier model of Pytkowicz (1967) but are not reproduced here in order to keep the diagram as simple as possible. A stationary ocean and a residence time for deep oceanic waters of 1000 yr were used in the calculations.

The sizes of reservoirs 1, 2, 5, 6, 7, and 8 were obtained from the compilation by Revelle and Fairbridge (1957). The size of reservoir 3 was estimated from the average pCO_2 of groundwaters (Garrels and Mackenzie, 1971b), the volume of groundwaters (Clarke, 1959) which is roughly $\frac{1}{10}$ of that of the oceans, and the solubility of CO_2 in seawater (Murray and Riley, 1971). A very rough idea of the size of reservoir 9 was obtained by assuming that only the upper 10 cm of the submarine sediments are in exchange contact with seawater and by using the average $CaCO_3$ content of the sediments from Sverdrup et al. (1942). Estimates of the amount of carbonate present in calcareous tests in the water column were not available.

The flux of juvenile CO_2 is 0.082×10^{14} g-C/yr and $J_{L-11} + J_{K-10} = 0.082 \times 10^{14}$ g-C/yr. J_{L-11} and J_{K-10} were taken as proportional to the values of J_{8-L} and J_{9-J}, which will be calculated later. These three fluxes are small compared to the cyclic ones and the apportionment of juvenile CO_2 through the system is cumbersome and would detract from the clarity of what follows. For these reasons, J (juvenile CO_2), J_{K-10}, and J_{L-11} are shown in parentheses and are not included in the mass balance computations.

$J_{5-E} = 2.460 \times 10^{14}$ g-C/yr. $J_{E-7} = 2 \times J_{5-E} = 4.920 \times 10^{14}$ g-C/yr [see reaction (2.8)]. This is twice the value from Clarke (1959) since his data were obtained from incinerated river bicarbonate samples with a consequent loss of half of the CO_2. Also from reaction (2.14), $J_{3-E} = J_{5-E}$. $J_{5-N-9} = 0.200 \times 10^{14}$ g-C/yr. $J_{J-5} = J_{5-E} + J_{5-9} =$

2.660×10^{14} g-C/yr according to the stationary hypothesis.

$J_{3-D} = J_{D-7} = 0.682 \times 10^{14}$ g-C/yr. Part of J_{D-7}, $J_{D-7}^{(Ca)}$, results from the weathering of CaIg. The carbon in the CO_2 used in this weathering follows paths 1–A–2–B–3–D–7–H–9–J–K–M–1 and 1–A–2–B–3–D–7–H–7–1. This means that half of the CO_2 is returned to the atmosphere during the biogenic conversion of $Ca(HCO_3)_2$ to $CaCO_3$ and the other half is returned following metamorphism of $CaCO_3$ to CaIg at M. $J_{D-7}^{(Ca)}$ was calculated from J_{D-7} and the average composition of igneous rocks, and was found to be 0.382×10^{14} g-C/yr. Half of this flux appears as $CaCO_3$ at H and the rest, as CO_2, returns to reservoir 7. Thus, $J_{K-M-1} = 0.5 \times J_{D-7}^{(Ca)} = 0.191 \times 10^{14}$ g-C/yr.

J_{9-I}, the rate at which tests dissolve, is 3.79×10^{14} g-C/yr, and J_{G-I}, the rate of CO_2 utilization in the dissolution, is also 3.79×10^{14} g-C/yr, as can be seen from the reverse of reaction (2.28). The returning flux of CO_2 in the form of HCO_3^- to reservoir 7 is $J_{I-7} = J_{9-I} J_{G-I}$. J_{H-7}, the rate of utilization of HCO_3^- by tests, is $J_{I-7} + J_{E-7} + J_{D-7}^{(Ca)} = 12.882 \times 10^{14}$ g-C/yr. Thus, tests utilize bicarbonate brought by rivers plus that recycled from deep waters. Half of this bicarbonate becomes test carbonate and half becomes CO_2. Thus, $J_{H-9} = J_{H-7} = 6.441 \times 10^{14}$ g-C/yr.

Next, I examine the org-C cycle. I found no information on J_{2-6}, the rate of lithification of org-C on land, and J_{6-C}, the subsequent rate of physical weathering of the lithified org-C. J_{8-L-6}, the flux of org-C into submarine rocks, and J_{6-B}, the flux of chemical weathering of this org-C by oxidation, are 0.3×10^{14} g-C/yr.

J_{1-A-2}, the gross rate of photosynthesis on land, is 199×10^{14} g-C/yr (Hutchinson, 1954) and 1.5×10^{14} g-C/yr of the resulting org-C are transported to the oceans as dissolved and particulate organics. J_{2-B} is $J_{1-A-2} - J_{2-C-8} = 197.5 \times 10^{14}$ g-C/yr. J_{B-3} is $J_{2-B} + J_{6-B} = 197.8 \times 10^{14}$ g-C/yr. $J_{3-1} = J_{B-3} - (J_{3-D} + J_{3-E}) = 194.658 \times 10^{14}$ g-C/yr. Thus, there is a net uptake of CO_2 by land and, as will be seen later, there is a net release of CO_2 by the oceans. It is difficult to calculate this net release from pressure heads across the sea surface, since these pressure differentials are masked by seasonal and geographical variations that are strong enough to cause spacial and temporal reversals in the local directions of the net gas fluxes.

J_{7-F-8}, the rate of gross photosynthesis in the

oceans, is 220×10^{14} g-C/yr. This quantity was estimated from the total net productivity of 200×10^{14} g-C/yr (Ryther, 1963) and from the fact that the net productivity is roughly 90% of the total productivity. The total productivity does not include the effect of respiration. The rate of oxidation is $J_{7-F-8} + J_{2-C-8} - J_{8-L} = 221.2$ g-C/yr since $J_{8-L} = 0.3 \times 10^{14}$ g-C/yr. Pytkowicz (1968) estimated from the oxygen utilization that org-C equivalent to 23.8×10^{14} g-C/yr is oxidized in deep oceanic waters. Thus, 89.3% of the oxidation of org-C in seawater occurs in the near-surface layers. $J_{G-7} = J_{8-G} - J_{G-I} = 217.41 \times 10^{14}$ g-C/yr. $J_{1-7} = 650 \times 10^{14}$ g-C/yr. J_{7-I} is 654.151×10^{14} g-C/yr. Again, it should be emphasized that the number of significant figures used in this work is that required for mass balance and does not represent how well the fluxes are known.

When hydrothermal processes become better known, the flux of titration alkalinity,

$$TA = [HCO_3^-] + 2[CO_3^{2-}] + [B(OH)_4^-]$$
$$+ [OH^-] - [H^+] + \Sigma (A) - \Sigma (C) \quad (2.36)$$

will include a term for its removal from the oceans. TA is the titration alkalinity, while $\Sigma (C)$ and $\Sigma (A)$ represent the sums of the concentrations of pH-dependent cations and anions, assuming single-charged species (see Chapter 2, Volume II). It is not known at present what the global hydrothermal TA flux is but the release of H^+ in proton formation reactions withdraws TA from the oceans by the titration of bases. This should be compensated, at least in part, by weathering reactions that generate Ca^{2+} since HCO_3^- should be formed at the same time (RMP).

2.3.7 Elemental Cycles

An example of estimated elemental cycles is presented in Figure 2.12. It has the attractive feature that the elemental contents of old rocks are distinguished from those of more recent ones.

The term "elemental" is used in the sense that the fluxes and reservoirs of a number of substances are presented, and this is done in terms of the amounts, or fluxes, of the elements of interest rather than their compounds. An example of such cycles is presented in Figure 2.12 which concerns only sedimentary rocks. It has the attractive feature

that the contents of young and old rocks are considered separately. Garrels and Mackenzie (1972) proceeded on the assumptions that young rocks (Mesozoic and Cenozoic) are cycled every 1.5×10^8 yr whereas old ones are more resistant and require 6.0×10^8 yr (Garrels and Mackenzie, 1971b). The model was built on the present-day dissolved and suspended load of streams, cycling through the atmosphere, the average compositions of sediments and oceans, and so on.

Unfortunately, in this first approach, the interconversions of sedimentary and igneous rocks were not taken into consideration for the sake of simplicity. Therefore, the fluxes of metals from igneous rocks and the sizes of igneous reservoirs were not made explicit. This means that the early evolution of geochemical systems, when new sedimentary rocks were being generated from juvenile igneous ones, cannot be studied by this approach. The method can only be used for the contemporary steady-state sedimentary mass.

Old rocks contain, among others, quartz, chlorite, illite, and limestones, which, of course, are also present, but in different proportions, in young rocks. The old rocks tend to survive longer than young ones because they are at the bottom of basins or infolded with intrusive or basement rocks, and in other sheltered positions.

A more detailed examination of sedimentary cycles, in which residence times of individual elements were taken into consideration, was made by Mackenzie and Wollast (1977).

Studies of cycles, assessing human influences, were derived by Garrels et al. (1973), although very little was shown about how the fluxes were estimated.

2.3.8 Oxygen

The main source of atmospheric oxygen appears to have been photosynthesis, with a smaller contribution from the photodissociation of water in the upper atmosphere. The total amount of oxygen formed by the first mechanism corresponds to the total amount of plant life ever produced. The net amount, on the other hand, is that due to the photosynthesis minus the amount required for respiration and to oxidize the organic matter present in soils, rocks, and sediments.

The above argument can best be seen in terms of the simplified photosynthetic-oxidative reaction

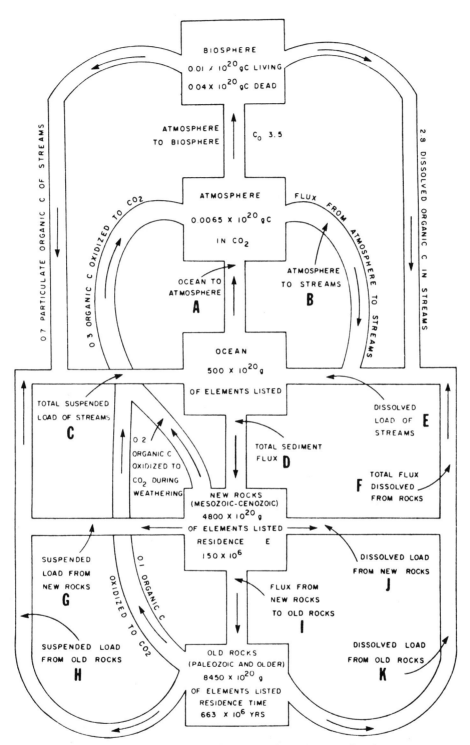

FIGURE 2.12 The geochemical cycle of the elements (Garrels and Mackenzie, 1972).

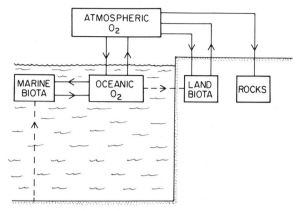

FIGURE 2.13 The oxygen cycle. Broken lines point in the reverse direction of excess organic fluxes such as the organic matter brought to the oceans by rivers.

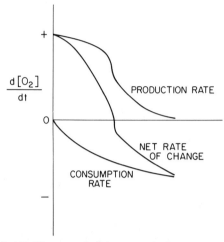

FIGURE 2.14 The control of the atmospheric oxygen [based on Holland (1978)].

$CO_2 + H_2O = CH_2O + O_2$. The reaction to the left, which is the oxidative one, lagged behind the photosynthetic formation of carbohydrates, CH_2O, and oxygen built up in the atmosphere. This was the transient accumulation process.

It is believed that by now a steady state has been reached for the oxygen cycle, which is sketched in Figure 2.13. The rate of photosynthesis is still slightly faster than the oxidative rate in the oceans since a slight residue of organic matter reaches the marine sediments. This organic matter is not oxidized until it undergoes the complete geochemical cycle and does not, in the short run, utilize the oxygen that was released during its production. At the same time, however, the above surplus added to the atmosphere is believed to be utilized in the oxidation of carbon, iron, and sulfur compounds on the continents. Thus, the overall cycle of oxygen is in a steady state.

The simple reaction presented above for photosynthesis can be used in conjunction with the carbon fluxes shown in Figure 2.11 to calculate the fluxes of oxygen. This is done by means of the equation

$$J_{O_2} = \frac{MW_{O_2}}{AW_C} J_C \qquad (2.37)$$

where J is the flux. A more rigorous conversion would require the use of a realistic equation such as (Odum, 1961)

$$106CO_2 + 16NO_3 + HPO_4^{2-} + 122H_2O + 18H^+$$
$$= C_{106}H_{263}O_{110}N_{16}P + 138O_2 \qquad (2.38)$$

We must also keep in mind the effects of human activities on the pathways of oxygen. Humans act in two ways; first through the enhanced combustion of fossil fuels, which burns oxygen and, second, by decreasing the biomass through net deforestation, which decreases the production of this gas. Both mechanisms reduce the atmospheric oxygen although this decrease is still small.

Holland (1978) views the present control of the atmospheric oxygen as follows (see Figure 2.14). If the atmospheric oxygen increases then the dissolved content in seawater also goes up. This leads to an increase in the amount of organic matter so that less of it is buried. Therefore, the back flux of oxygen to the atmosphere decreases. The opposite happens for a decrease in atmospheric oxygen, that is, a smaller amount is returned by the oceans.

2.3.9 Silica

The important cycle of silica has been treated by Heath (1974), Wollast (1974), and others and the results were summarized by Holland (1978).

The broad geochemical cycle of clays is probably controlled mainly by the release of SiO_2 during weathering on land and its uptake during reverse weathering in high-grade diagenesis.

Within the oceans H_4SiO_4, the soluble form of SiO_2, is taken up by siliceous organisms and sedimented as tests (shells) after the organisms die. The silica is then lithified to chert or, after uptake by clays, winds up in shales (lithified clays). These rocks can have a cycle (rock → seawater → siliceous tests → rock) of their own or may undergo reverse weathering back to feldspars. The relative importance of shales (lithified clays) and cherts should depend upon the nature of the sediments (clays versus tests of siliceous organisms).

Heath (1974) pointed out that most of the silica in the oceans comes from subaerial weathering and that silica-secreting diatoms and radiolaria extract enough opal to strip the dissolved silica from the oceans in 250 years if further silica were not being added. This is the residence time. Because of the rapid dissolution of tests, only about 4% of the opal survives long enough to be buried in the sediments while the remainder is recycled in the sea. These 4% correspond to the river flux in a steady-state situation. The copresence of volcanic materials with opal in the geologic record is thought by Heath not to correspond to geochemical reactions. This is in contrast to those views in which the ultimate amount of SiO_2 results from aluminum–silicate interactions with silica.

2.3.10 A Note on the Impact of Humans

Humankind, in the words of E. D. Goldberg, has become a geologic (weathering) agent since a significant fraction of the river and wind-born flux of substances is due to its activities.

Some natural reservoirs show only a small effect from perturbations so far, as in the case of CO_2, whereas others reflect their impact. Examples of the latter are lead and, in some bays and estuaries, mercury. In the same vein, some fluxes have been altered significantly, whereas others have not. The reservoirs of CO_2 in nature are so large that they have not yet changed much but the fluxes have increased greatly due to fossil fuel burning.

The important consideration is that the production of industrial, agricultural, and household pollution tends to increase exponentially so that the impact of humans will be enhanced rapidly. From the standpoint of geochemical cycling this means that stationary systems will become simply baselines upon which the perturbations (transients) will have to be imposed.

Some examples of pollutants are (Garrels et al., 1973; MIT Press, 1970; Goldberg, 1971):

1. *Carbon monoxide (CO).* Human input is only a local problem so far. The natural CO level is

0.1 ppm but concentrations of up to 100 ppm have been measured in areas of heavy automobile traffic. This gas is potentially harmful due to the replacement of oxyhemoglobin by carboxyhemoglobin during respiration but toxic levels have not been established.

2. *Carbon dioxide (CO_2).* Fossil fuel burning was already adding 388×10^{12} moles-CO_2/yr in 1971. This is about 95 times the natural one, flux, and the anthropogenic production of fossil fuel is increasing at about 4.2% per year (Pytkowicz, 1978). The total CO_2 in the atmosphere, oceans, and biosphere appears to have increased by over 0.2%. The associated problem, namely, the warming of the earth by infrared absorption, will be discussed in a later chapter.

3. *Particles.* Industrial and agricultural particles in the atmosphere (aerosols), which operate in conjunction with volcanic and desert particles, have increased the turbidity of the atmosphere by about 4% since 1940. This effect decreases the solar infrared radiations that reach sea level. The particulates consist mainly of

Sulfur oxides	8×10^{13} g/yr
Nitrogen oxides	5×10^{13} g/yr
Smoke particles	2×10^{13} g/yr

The anthropogenic sulfur is produced by fossil fuel burning. Other substances present are lead, pesticides, and so on.

4. *Nutrients.* Phosphates and nitrates from agriculture can cause eutrophication in water bodies, that is, an excess of life which utilizes all the dissolved oxygen and leads to anoxic conditions and death.

5. *Trace metals.* Trace metals resulting from mining enter the atmosphere, rivers, and the oceans. Several of them, such as Pb, As, Cd, and Hg, are toxic in sufficient amounts. Examples of fluxes, in 10^{12} g/yr, are:

	Pb	Cd	Hg
Emission	0.40	0.004	0.01
Rainout	0.31	0.004	0.001
Streams	0.42	—	0.005

As an example of the acceleration of metal fluxes by humans, the estimated prehuman flux of mercury to the oceans was 0.0013×10^{12} g/yr so that a substantial increase is indicated by the above tabulation.

2.4 THE HYPOTHESIS OF A QUASI-INVARIANT OCEAN

2.4.1 A Stationary Ocean?

The term stationary refers to a time-invariant state in the sense that its long-term time average is roughly constant. In other words, there are short- and long-term fluctuations around the average but little trend.

The stationary hypothesis, as a rough description of the time behavior of the chemical composition of seawater, is reasonably well accepted at this time although at least one exception to be discussed is supported by several considerations.

The times required to bring the solutes in the oceans to their present levels range from years to about 10^8 yr, when estimated from present river inputs and the sedimentary records. Actually, we shall see that these times correspond to a buildup to $1-1/e$ of the present values. They are short when compared to 3.5×10^9 yr, the minimum age of the oceans (Barth, 1952; Goldberg, 1965). This suggests that a permanent one-way unbalance, a trend due to nonstop addition, would have led to much more concentrated oceans than the present ones. The reason for this is that the buildup process would have lasted much longer than 10^8 yr.

The residence time is one of the two concepts used in the above argument. It was defined by Barth as $\tau = B/J_{AB} = B/J_{BC}$, where B is the oceanic content of the element, J_{AB} is the river influx, and J_{BC} is the sedimentation flux. τ actually does not lend itself directly to the stationary argument because it applies to the ratio of the oceanic content and the fluxes only under stationary (time-invariant) conditions. This leads us to the relaxation time introduced below (RMP).

First, however, I discuss an important aspect of the Barth approach and my approach. Note that linear rate laws are intrinsic to the Barth concept because the fluxes can be written as $J_{AB} = k_{AB}B$ and $J_{BC} = k_{BC}B$ with the rate constants k_{AB} and k_{BC} equal to $1/\tau$. This linear assumption is quite an oversimplification for real processes but is a useful approach in conceptual studies of classes of rate behavior. Linear laws have the key property required for this, a property shared with nonlinear models, that fluxes increase monotonically with increasing reservoir sizes. I shall discuss this in greater detail later in this chapter (Pytkowicz, 1971, 1972, 1973b).

Let us consider an ocean controlled by linear processes and let A, B, and C be the amounts of an element in the continents, oceans, and submarine sediments.

Let us abandon the residence time for oceanic buildup periods and replace it by relaxation times for the time spans when the composition of the oceans was increasing. Linear laws in terms of masses lead to (Pytkowicz, 1972, 1973b) yield

$$\frac{dB}{dt} = k_{AB}A - k_{BC}B \qquad (2.39)$$

where $J_{AB} = k_{AB}A$ and $J_{BC} = k_{BC}B$ are the fluxes in and out of the oceans and k_{AB} and k_{BC} are the rate constants for river input and for sedimentation of the element in all of its forms as we saw above. Total reservoir contents A, B, and C are used because it is not convenient to use concentrations such as those of $CaCO_3$ in a limestone or of SiO_2 in a chert, as they may be nearly invariant.

For simplicity, $A = A_0$ is assumed to be roughly constant because of uplift or cyclic replacement. A_0 is the invariant value when A is constant. This leads, by integration, to

$$B = \frac{k_{AB}}{k_{BC}}A_0(1 - e^{-kBCt}) \qquad (2.40)$$

The residence time τ for a stationary ocean is obviously $1/k_{BC}$, since B/J_{BC} yields this quantity. For a transient ocean, however, the term $B/J_{BC} = 1/k_{BC} = \tau$ is simply the relaxation time for the oceanic composition to reach $1-1/e$ of its eventual stationary value. Thus, if τ is short relative to the age of seawater, it shows that if present river inputs are characteristic of the past, then the oceans have had ample time to reach a stationary condition.

There are other lines of evidence for a stationary ocean. Holland (1972) concluded that wide excursions in oceanic composition are not compatible with the sedimentary record. Weyl (1966) suggested that life as we know it could only withstand limited changes in the composition of seawater. Garrels and Mackenzie (1971a) summarized sedimentary evidence for a fairly constant ocean at least in the last 6×10^8 yr. Finally, Goldberg and Arrhenius (1958) found that residence times based on river influx and on sedimentation were roughly similar.

Let us consider an example of the use of relaxation times.

EXAMPLE 2.9. AN APPLICATION OF RELAXATION TIMES. Estimate how long it has taken for (Na$^+$) and (Mg^{2+}) to have reached 80% of their present oceanic values. Assume that $\tau_{Mg} = 4.5 \times 10^7$ yr and $\tau_{Na} = 2.6 \times 10^8$ yr.

The equation for the oceanic content is

$$B = B_\infty(1 - e^{-t/\tau}) \rightarrow e^{-t/\tau} = \frac{B_\infty - B}{B_\infty} \qquad \text{(i)}$$

and, therefore,

$$e^{-t/\tau} = \frac{B_\infty - B}{B} \qquad \text{(ii)}$$

Equation (ii) yields

$$e^{-t/\tau} = 0.2 \quad \text{and} \quad \frac{t}{\tau} = 1.6 \qquad \text{(iii)}$$

Thus, $t_{Na} = 4.16 \times 10^8$ yr and $t_{Mg} = 7.2 \times 10^7$ yr.

I accept the stationary or, rather, quasistationary hypothesis as being valid for oceanic concentrations and geochemical cycles of most elements before the advent of the Industrial Revolution. This hypothesis is quite useful for the study of geochemical control mechanisms. Let us consider two types of exceptions: the potentially very slow undetected ones and the documented ones.

It is conceivable, if there has been a gradual evolution of volatiles, that part of the NaCl resulting from the continuous weathering of feldspars by HCl does not find its way into evaporites but still accumulates at a very slow rate in the oceans. If, on the other hand, paleosalinity studies in pore waters of deep sediments show no gradual increase in the salt content of seawater, then the argument can be reversed and used to support an early massive degassing of the mantle during which most of the sedimentary mass was formed. If this is the case, a large fraction of the volatiles released by present volcanoes may be products of metamorphism and recycling of earlier sediments and do not weather additional igneous rocks to produce further NaCl. Then, a closed cycle prevails.

Another disturbing aspect is that most sedimentary carbonates must have resulted from the weathering of calcium feldspars, which releases soluble $CaHCO_3$ to the oceans. The $CaHCO_3$ is converted by organisms into the $Ca_xMg_{1-x}CO_3$ of calcareous tests, a process that continues today (Pytkowicz, 1972, 1973b). If truly juvenile CO_2, formed from the degassing of primary magmatic material, is being released then new sedimentary carbonates may be formed and added to the cycling component. These carbonates can then serve as sinks for the juvenile CO_2. It is well known in chemistry that an increase in the amounts of reactants accelerates chemical reactions and that part of the excess reactants shows up as products. Similarly, in reservoir kinetics, I have shown (Pytkowicz, 1971, 1972, 1973b) that any one reservoir in nature can only remain stationary if all the other linked reservoirs do the same. Thus, if the sedimentary carbonate reservoir is increasing, then part of this increase will be transmitted to the oceans, since they still have a capacity for carbonates before reaching saturation (Pytkowicz, 1965b, 1968).

A puzzling feature of the oceans is the large variation in the calcite compensation depth during the last 50 million yr (van Andel and More, 1974). This is the depth below which carbonates are completely dissolved. This variation may perhaps indicate changes in the carbonate content of the oceans and in the CO_2 system fluxes. It is of interest, therefore, to establish whether the change represents simply fluctuations around a mean or if they indicate a trend.

An ocean that, on the average, has remained very roughly time invariant for a while leads one to seek mass balances, which were discussed earlier, and general control mechanisms, the subject of a later section. This topic is relevant not only to a description of past and present oceans but also to the fate of perturbations induced by the activities of man and, thus, to the geochemistry of the future.

In summary, indications are that the composition of seawater and the geochemical fluxes have not changed drastically for a long but undefined time period. Fluctuations may have been imposed on a roughly constant average for some elements and those slow trends may still be occurring for other elements. The quasi-invariant hypothesis leads one to seek mass balances and control mechanisms such as equilibria or steady states.

I should add at this point that the quasistationary hypothesis receives credence for several elements, in addition to the reasons stated earlier, from the work of Ronov and Yaroshevsky (1969). They found a fairly constant ratio of the major sedimentary components of the crust to their amounts in igneous rocks. It is hard to visualize elements which undergo different processes behaving in such a manner in a transient situation.

Calcium, and consequently the CO_2 system, appear to be an important exception to the stationary rule except for relatively short geologic time spans. The trends in calcium and the CO_2 system are related to the conversion of plagioclase feldspars into sedimentary carbonates that were mentioned earlier. These trends are natural ones that occur even in the absence of the fossil fuel CO_2 contributed by humans, and will be discussed next.

2.4.2 The Excess Crustal Calcium Problem (Pytkowicz, 1979)

1. *Background.* Ronov and Yaroshevsky (1969) estimated a mass balance for crustal materials and found an excess of calcium. They considered this excess to be one of the most significant problems in the geochemistry of the outer crust, since it appears to be a single exception for crustal elements.

This problem is examined in five steps: the background information, the constant ratios for the elemental oxides besides calcium, the excess calcium question, types of calcium changes in relation to a steady-state ocean, and an example of how the amount of sedimentary carbonates may change with time.

Symbols Used for the Excess Calcium Problem
(Amounts in Terms of Oxides)

A, B, D	Relative amounts of different rock types in the crust.
I	Crustal igneous rocks content.
C	Crustal contents.
S	Contents of sedimentary rocks, sediments, and oceans.
O	Oceanic contents.
I_1, I_2	Minerals in hypothetical rocks.
f_j	Fraction of rock type j in I.
x_{ji}	Fraction of mineral i in j.
$M_{i_{(m)}}$	Mass of oxides of element i in the mantle.
I_i	Fraction of crustal I_i which becomes S_i.
S_i	Mass of sedimentary rock, sediment, and oceanic content originating in igneous mineral i.
k	Rate constant.

Loewengart (1975) interpreted this excess as the result of an original excess of magnesium, produced when the early mantle was titrated by acid volatiles, and followed by ion exchange of the magnesium with mantle calcium via the oceans. The hypothesis that will be presented is complementary to that of Loewengart and does not require an initial excess of magnesium. In other words, magnesium may be removed from seawater in part in hydrothermal sites but this is just a step in a steady-state cycle in which it is supplied by weathering reactions on land. Thus, the hydrothermal uptake of magnesium and release of calcium (Edmond et al., manuscript) in different amounts does not necessarily support the conclusion of Loewengart (1975).

Consider the data presented in Table 2.6, a modification of that of Loewengart. One observes that all elements, calculated as their oxides, are present in $(S + O)$ as 5.7–6.0% of their amounts in the crust, except for calcium. It is obvious, therefore, that the composition of $S + O$, except for CaO, is about the same as that of igneous plus metamorphic crustal rocks, I, from which the sediments and the oceanic salt content were derived. Therefore, there has been a unique buildup of calcium. I shall use the term igneous and the symbol I for igneous plus metamorphic rocks in the crust and S for $S + O$ for simplicity. $C = I + S + O$ will indicate the crust.

2. *The constant ratios.* Before examining the calcium problem let us consider the significance of the constant proportion of elemental oxides in $(S + O)$. Garrels and Mackenzie (1971a) concluded that the crust is quasiconservative in that material brought up from the mantle may have had a fairly

TABLE 2.6 Distribution of major elements, expressed as 10^{24} g of their oxides, between the sedimentary rocks plus sediments (S) and oceans in relation to their amounts in the crust (C).

	C	S	$(S)/C$
SiO_2	16.403	0.962	0.0586
Al_2O_3	4.412	0.253	0.0573
Fe_2O_3	2.037	0.117	0.0574
MgO	1.104	0.064	0.0580
CaO	2.002	0.240	0.120
Na_2O	0.821	0.049	0.0597
K_2O	0.683	0.041	0.0600
	27.462	1.726	

constant composition over the last 3.5×10^9 yr. They then explained the 6% ratio by assuming that igneous rocks were weathered as if they acted as a single volume-average rock.

Let us illustrate the conclusion of Garrels and Mackenzie and extend it (RMP).

EXAMPLE 2.10. THE WEATHERING OF THE AVERAGE IGNEOUS ROCK ILLUSTRATED BY A HYPOTHETIC EXAMPLE. The assumed composition of average crustal igneous rocks is

$$A : B : D = 0.2 : 0.5 : 0.3 \qquad (i)$$

Capital letters represent reservoir labels as well as their contents. The mineral compositions of the rocks are, in terms of elemental oxides,

A	$40\%I_1$	$60\%I_2$
B	$30\%I_1$	$70\%I_2$
C	$60\%I_1$	$40\%I_2$

I am assuming that only two minerals, I_1 and I_2, are present in the system. The proportions in which elements (represented as oxides) are present in the crustal igneous rocks are

$$(I_1) \quad 0.4 \times 0.2 + 0.3 \times 0.5 + 0.6 \times 0.3 = 0.410 \qquad (ii)$$

$$(I_2) \quad 0.6 \times 0.2 + 0.7 \times 0.5 + 0.4 \times 0.3 = 0.590 \qquad (iii)$$

This means that if the rock types are weathered in the proportions in which they are present, that is, as an average igneous rock, then the elements will weather in the proportions in which they are present in this average rock (41% of I_1 and 59% of I_2). Then the sedimentary lithosphere plus the oceanic solutes (S) will contain the elements in the same proportions and with all of them in constant ratios to their amount in I, the igneous plus metamorphic content of the crust.

In Example 2.4, if 6.38% of I has been weathered, then $S = 0.06C = 0.06 (I + S)$ where C is the content of the crust. Furthermore, $I_1 = 41\%$ and $I_2 = 59\%$ are the proportions of I_1 and I_2 in S, while I_1 and I_2 in the sediments divided by their amounts in the crust will be 0.06.

We conclude, therefore, that the weathering of igneous rocks in the proportions in which they are present in the crust is a sufficient condition for constant ratios of elements in S to their amounts in I.

Notation is changed in what follows. Let each rock type be j and each mineral be i. Then, algebraically, each igneous rock type contributes mineral flux proportional to $f_j x_{ji}$ to the transfer from I to S. The term f_j is the fraction of j in I and x_{ji} is the fraction of i in j. This leads us to an important question that has not yet been answered, namely, why should the different rock types (e.g., basalt and granite) weather in proportion to their amounts. It should be expected that the specific weathering (per unit mass) would vary substantially with the rock type.

I shall not be able to explain the constant 0.06 but shall propose a reason why all the values of S_i/C_i are almost the same, with the exception of calcium.

Consider a steady-state sedimentary composition and mass, resulting from the fluxes shown in Figure 2.15. The symbols M_i, $I_i^{(n)}$, and S_i will be used to represent the mass of the oxide of element i, present in the mantle, in that fraction of I_i (igneous plus metamorphic rocks of the crust) which goes to S_i and in S_i (sedimentary rocks, sediments, and oceanic solutes). Note that oxides are used simply to normalize the elements rather than represent specific minerals since i may in reality be present as carbonates, silicates, sulfides, oxides, and so on. The steady state will be, if again we use the linear approximation,

$$\frac{dS_i}{dt} = (k_{IS})_i I_i^{(n)} = (k_{SM})_i S_i = 0 \qquad (2.41)$$

We cannot think about the crust in terms of chunks of igneous and other rocks just sitting there, because there are the above net fluxes connecting all the reservoirs. The steady state is used as an ap-

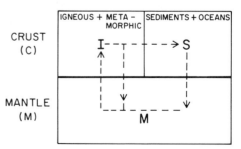

FIGURE 2.15 A diagram of the fluxes among the mantle and crustal reservoirs.

proximate condition of the system which is related to the quasi-invariance of the oceans.

From Equation (2.41) we obtain

$$\frac{I_i^{(n)}}{S_i} = \frac{(k_{SM})_i}{(k_{IS})_i} \qquad (2.42)$$

or

$$\frac{I_i^{(n)} + S_i}{S_i} = 1 + \frac{(k_{SM})_i}{(k_{IS})_i} = \frac{(k_{IS})_i + (k_{SM})_i}{(k_{IS})_i}$$

$$= 1 + \frac{(k_{SM})_i}{(k_{IS})_i} \qquad (2.43)$$

From Equation (2.30),

$$\frac{S_i}{C_i} = \frac{S_i}{I_{i(n)} + S_i} = \frac{1}{1 + (k_{SM})_i/(k_{IS})_i} \cong 0.06$$

$$(2.44)$$

The subscript i refers to $I^{(n)}$. This result means that $(k_{SM})_i/k_{IS})_i \cong 16.6$.

Note that $k = 1/\tau$, where τ may be a residence of a relaxation time. At a steady state, the fluxes J are related by

$$(J_{IS})_i = (k_{IS})_i I_i^{(n)} = (k_{SM})_i S_i = (J_{SM})_i$$

$$(2.45)$$

The ratio $(k_{SM})_i/(k_{IS})_i = 16.6$ shows, according to Equation (2.45), that $I_i^{(m)}/S_i = 16.6$. This helps us understand why the igneous crustal reservoir is much larger than the sedimentary one because the rate constant for $S \rightarrow M$ is faster than that for $I \rightarrow S$.

These results also show that the $I_i^{(n)} \rightarrow S_i$ step is quite slow relative to $S_i \rightarrow M_i$ so that the former one, the weathering by water and CO_2 and/or sedimentation, is the rate-controlling one. The existence of a rate-controlling reaction that is common in general terms for all i's may provide a lead for the constant ratio $(S_i/C_i) = 0.06$. The exit rate, which involves subduction, may also be similar for all i's.

3. *The Excess Calcium.* Let us see how the buildup of the $CaCO_3$ reservoir and its relatively large size can be rationalized. Originally, no sedimentary carbonates were present. They were gradually formed primarily from the weathering of plagioclase feldspars, a process that still occurs today.

The weathering and sedimentary reactions are of the types shown by Equations (2.28) and (2.34).

These processes were primarily inorganic before the advent of calcareous organisms, approximately 6×10^8 yr ago. Now it is mostly biogenic (Pytkowicz, 1965a).

Reactions (2.28) and (2.34) represent the conversion of feldspars to calcium carbonate but the process is not entirely unidirectional. Carbonates settled in shallow waters probably remain as $CaCO_3$. However, at least part of the carbonates deposited in deep oceanic waters, above the carbonate compensation depth, may undergo metamorphosis back to feldspars after burial following seafloor spreading and subduction (Pytkowicz, 1975a). Note that the weathering of carbonates is not included, since it is part of a cyclic process that does not increase the $CaCO_3$ (or $Ca_xMg_{1-x}CO_3$) reservoir.

Thus, the present hypothesis (RMP) is that the calcium excess, at least in part, is due to the conversion of calcium feldspars into calcium carbonate. This does not exclude the mechanism proposed by Loewengart but is attractive because it does not require the postulate of an initial magnesium excess.

The 0.12 ratio for CaO in sediments versus crust may indicate continuing production of $CaCO_3$, faster weathering, faster biologically mediated sedimentation, or less reverse weathering (part of the $I_i^{(n)}$ to S_i path).

4. *Calcium and a Steady-State Ocean.* The picture that emerges is one in which CO_2 (or H^+) is consumed and ions plus molecules such as Na^+, K^+, Ca^{2+}, HCO_3^-, and SiO_2 are released during weathering.

These species are brought to the oceans, eventually become part of the sediments, and undergo reverse weathering, yielding a mostly invariant system.

Calcium may cause a problem, if indeed the generation of $Ca_xMg_{1-x}CO_3$ from plagioclase feldspars has not yet reached a steady state. This would introduce transient states for calcium, magnesium, the titration alkalinity, silica, and plagioclase feldspars. Granted, these changes may be so slow that they do not introduce observable trends over long time periods.

The alternative is, as we saw above, that the 12% ratio of CaO in the sediments to CaO in the crust results from a fast J_{IS} flux due to the biogenic production of $Ca_xMg_{1-x}CO_3$. In this case, a steady state may have been reached for the cycle feldspar–

carbonate–feldspar and there is no contradiction with the general stationary hypothesis.

5. Next, a rough estimate of the rates of $CaCO_3$ formation from Ca–feldspars before and after the onset of calcareous tests during the Cambrian, roughly 6×10^8 yr ago, is made, with the assumption that the plagioclase transformation has gone on until now. The numerical inputs come primarily from the data of Ronov and Yaroshevsky (1969) in the form presented by Pytkowicz (1979). Example 2.10 illustrates the effect of the onset of calcareous organisms on the rate of formation of new carbonates.

EXAMPLE 2.11. PRE- AND POST-CAMBRIAN RATES OF FORMATION OF SEDIMENTARY CARBONATES. (All masses are expressed in terms of oxides.)

Igneous mass of the crust (actually igneous + metamorphic), 25.736×10^{24} g.

Sedimentary mass of the crust (sediments, sedimentary rocks, and oceanic solutes), 1.726×10^{24}.

CaO mass in crustal igneous rocks 1.762×10^{24}.

CaO mass in sedimentary rocks (assumed to be present as limestones), 0.240×10^{24}.

Age of the oceans, 3.5×10^9 yr.

Pre-Cambrian period during the ocean history, 2.9×10^9 yr.

Post-Cambrian period, 6.0×10^8 yr.

Sedimentary rock mass excluding CaO, 1.486×10^{24} g.

Average rate of weathering of igneous rocks excluding CaO, VIII/V = $(1.486 \times 10^{24}) / (3.5 \times 10^9)$ = 4.246×10^{14} g/yr.

Average rate of formation of CaO from igneous rocks (Pre-Cambrian), (III/I) × IX = $[(1.762 \times 10^{24}) / (25.736 \times 10^{24})] \times 4.264 \times 10^{14}$ = 0.2907×10^{14} g/yr.

Amount of sedimentary CaO formed before the Cambrian, VI × X = 0.0843×10^{24} g.

Amount of sedimentary CaO formed since the Cambrian, IV − XI = $(0.240 - 0.0843) \times 10^{24}$ = 0.156 g.

Average rate of formation of sedimentary CaO formed since the Cambrian, XII/VII = $(0.156 \times 10^{24}) / (0.6 \times 10^9)$ = 2.600×10^{14} g/yr.

We see from Example 2.11, that the rate of post-Cambrian sedimentary $CaCO_3$ formation, ex-pressed as CaO, could be 2.600×10^{14} g/yr versus 0.2907×10^{14} g during the pre-Cambrian. This sharp increase would be due to the appearance of calcareous organisms roughly 6×10^8 yr ago. Furthermore, Example 2.11 may indicate that the rate-limiting step is the rate at which calcareous organisms deposit into the sediments that part of the test carbonates which originates in the weathering of Ca–feldspars.

The rate of formation of new carbonates at an average of 2.600×10^{14} g-CaO may have decreased because the igneous mass in the weathering environment gradually changed and became enriched with $CaCO_3$ at the expense of plagioclase feldspars. Thus, in time, an increasing fraction of the $Ca(HCO_3)_2$ dissolved in rivers could be derived from the weathering (recycling) of sedimentary carbonates.

The whole picture is quite speculative and incomplete at this time. One aspect that I find especially intriguing is the fate of new (juvenile) mantle material that is added. Can such material be part of a stationary two-way exchange between the crust and the mantle or is it in part enriching the continental, oceanic, and sedimentary reservoirs? I do not believe that this question has been answered at this time.

2.4.3 An Equilibrium Ocean?

The proper choice of control mechanisms for the composition of the oceans is exceedingly important to explain the geochemical history and predict the response of future oceans to human perturbations. Stationary (time-invariant) conditions can arise from steady states and from equilibria (see Chapter 1). These alternatives lead to different mathematical representations for calculations of the chemical behavior of the oceans during the buildup phase and in response to perturbations.

In closed chemical systems, such as insulated test tubes, reactions reach eventual equilibria with no net transformations of matter and with a thermodynamic condition of minimum free energy (capacity to do work; see Chapter 4) compatible with the constraints (T, P, etc.) imposed upon the system. This is similar in mechanical terms to a ball that falls into a well and reaches the minimum potential energy compatible with the topography (Chapter 1).

Chemical systems that are open can, however, operate indefinitely away from equilibrium. Net fluxes of matter occur and are maintained at steady states by the flow of energy through the system (e.g., Glansdorff and Prigogine, 1971). The thermodynamic condition for a nonequilibrium steady state is a minimum generation of entropy, which implies that as little structure as possible is dissipated.

Steady-state systems, when looked at in detail, are shown to contain negative feedback loops. This means that outputs are coupled to and limit variations of inputs. If the input A starts to increase, then the output C acts in such a manner as to tend to bring A back to its original value. A simple example is that of the oxygen cycle.

I should remind some readers that entropy, which will be studied in Chapter 4, expresses the loss of useful energy to a system. If a hot and a cold body are placed together, they will eventually reach a common temperature. When there is thermal contrast, an engine can operate by drawing heat from the hot body and releasing it to the cold one. This is no longer possible after thermal equilibrium is reached. The total energy of the system does change but the available energy goes to zero. Entropy is a measure of the loss of available energy. It also meaures the loss of contrast in nature and the change of an ordered system toward randomization. The theorem of Prigogine (1955) shows that systems which are near but not at equilibrium become stable when they generate a minimum of entropy. Biochemical cycles are often at steady states; they utilize solar energy to maintain structure with a minimum gain in entropy. Perhaps geochemical cycles follow a similar law but the thermodynamics of such cycles have not yet been established.

Most geochemical work has been done in terms of equilibria, the linear model, and simple feedback loops where the last two mechanisms lead to steady states. The linear steady-state model presented earlier in this book does not include a feedback since the input is not affected directly by the output. Still, there is an element of stability because the outflux J_{BC} increases if the influx J_{AB} is accelerated so that variations of B are limited.

Let us next examine arguments in favor of an equilibrium control of the composition of seawater. This will be followed by a discussion of steady states.

Sillen (1961, 1967) was the main proponent of the equilibrium view for the oceans and presented two types of arguments to support his views. First, he suggested that clays interact with the ions in the water column according to reactions of the type

$$clay_1 + H^+ = clay_2 + SiO_2 + cation$$

$$(2.46)$$

He proposed that the equilibria for these reactions is reached and has the form

$$K = \frac{a_{cation}}{a_H} \qquad (2.47)$$

The activities are unity for the solids. Enough equations, such as (2.47), for a number of clays would presumably fix ionic ratios in solution. SiO_2 was assumed to be at equilibrium with quartz and, therefore, to have a fixed activity.

I cannot accept the use of Equation (2.47) because in reality silica is present in seawater as reactive H_4SiO_4 and quartz does equilibrate with it or participate in equilibria at low temperatures (0–25°C), at least within the time of its exposure to the oceans. Therefore, the equilibrium for reaction (2.46) should be

$$K = \frac{a_{cation} \, a_{H_4SiO_4}}{a_H} \qquad (2.48)$$

$a_{H_4SiO_4}$ varies by orders of magnitude within the oceans because its distribution is controlled by hydrographic processes and by the uptake and release of SiO_2 by siliceous organisms (Heath, 1974; Wollast, 1974). Therefore, if equilibrium (2.48) is operative, it would lead to large changes in the concentrations of the major ions with depth because a_H is nearly constant. The pH ranges from 7.8 to 8.2 (Pytkowicz, 1975a). Such changes are not observed and the slight vertical gradients of most major elements, except for calcium and bicarbonate, are due simply to the density stratification of seawater. This indicates that Equation (2.48) is not effective in open oceanic waters.

The effect of reaction (2.48) on the major ions in the oceans may occur indirectly in the following way. Equilibrium may be approached in part or completely in pore waters, and then diffusional exchange between seawater and pore waters limits the buildup of the major ions and silica in the oceans. This view is plausible but is still speculative because different types of equilibria are reached in carbonate, siliceous, terrigenous, and anoxic sediments and because the extents of equilibrations are not known. Also, the equilibrium constants for clay reactions in seawater at low temperatures and high pressures encountered in the pore waters have not yet been determined.

Holland (1965) extrapolated clay equilibrium data obtained by Hemley (1959), Hemley et al. (1961), and Hemley and Jones (1964) at 300–600°C, and concluded that the oceans may be approximately at equilibrium with illite, montmorillonite, and kaolinite. He based his conclusion upon oceanic ionic ratios. However, Holland also set the activity of silica as being the value at equilibrium with quartz. Furthermore, extrapolations from 300–25°C are uncertain and I have shown that if equilibrium exist at all, it occurs in pore waters rather than in open seawater. Thus, Holland's work does not establish an ocean with a composition fixed by equilibrium processes. Still, equilibrium values may play a role by entering into rate laws of systems tending toward equilibrium if partial equilibration occurs in the water column or if diffusion into the sediments is regulated by partial or total equilibria in pore waters.

Subsequent workers attributed to Sillen a partially improved argument by not assuming solid–water equilibration in the water column (e.g., Drever, 1974). They stated that equilibration occurred in the pore waters of sediments and that the oceans actually reached equilibrium by diffusional exchange of solutes with the pore waters. The very diffusion of seawater ions into the sediments indicates, however, that the former are not at equilibrium with the bottom solids.

Drever presented the reaction

$$Mg_3(OH)_6Mg_3Si_4O_{10}(OH)_2 + 12H^+$$
chlorite

$$= 4SiO_2 + 6Mg^{2+} + 10H_2O \quad (2.49)$$
quartz

as controlling the ratio

$$K = \frac{(a_{Mg})^6}{(a_H)^{12}} \quad (2.50)$$

in pore waters.

Let us consider the formation of quartz in Equation (2.49). At 25°C, the dissolution or precipitation rate of quartz is negligible for kinetic reasons, even though it is the most stable form of silica. Indeed,

the equilibrium constant $K = a_{H_4SiO_4}$ is 1.7×10^{-4} for the reaction

$$SiO_2 + 2H_2O = H_4SiO_{4(soln)} \quad (2.51)$$
quartz

The solubility of amorphous silica, on the other hand, is $K \cong 2 \times 10^{-3}$ (Krauskopf, 1956), in terms of

$$SiO_2 + 2H_2O = H_4SiO_4 \quad (2.52)$$
opal

if a_{SiO_2} and $a_{H_4SiO_4}$ are assumed to be unity.

In spite of its greater stability, the slow kinetics of quartz–solution interactions and the possibility of the formation of H_4SiO_4 coatings on quartz particles (Berner, 1971) leads one to question the possibility that the SiO_2 in Equation (2.49) is quartz. It is more likely that it represents opal in the following equation (RMP):

$$K = \frac{a_{H_4SiO_4(soln)}}{a_{SiO_2(s)}a_{H_2O}^2} \quad (2.53)$$

The correct equilibrium for Equation (2.49) then becomes

$$K = \frac{(a_{Mg})^6(a_{H_4SiO_4})^4}{(a_H)^{12}} \quad (2.54)$$

This brings to pore waters of clay sediments the same question asked earlier with regard to the supernatant waters, namely, to what extent $a_{H_4SiO_4}$ varies and prevents the equilibrium (2.54). (H_4SiO_4) ranges from 19 to 60 ppm in pore waters with the high value being from special environments (Gulf of California and Bering Sea) according to Heath (1974). Again, the existence of other sediment types with which seawater is in contact should be called to the attention of the reader.

The results of Jones and Pytkowicz (1973) and the conclusions of Berner (1971) indicate that the H_4SiO_4 in pore waters are supersaturated with respect to quartz and undersaturated relative to opal in further support of nonequilibrium.

Even if an equilibration of opal, cations, and H^+ with assemblages of clays did occur in some types of sediments, this does not mean that the oceans would equilibrate for the reasons mentioned earlier; diffusion of cations would have ceased and cationic concentrations in pore waters would have to vary

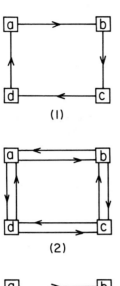

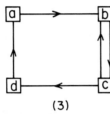

FIGURE 2.16 Reservoirs with equilibria and steady states: (*a*) weathering environment; (*b*) oceans; (*c*) submarine sediments; (*d*) metamorphic environment.

greatly to compensate for changes in [H_4SiO_4] and keep K constant.

In summary, clay equilibrations with silica, H^+, and cations do not exist in open oceanic waters. Such equilibria may be approached but are not generally reached in clay sediments. Diffusion of ions from seawater to pore waters may reflect the equilibrium tendency of pore waters. The compositional control in pore waters, as well as in seawater, is kinetic but driven ultimately by a tendency toward equilibria with various sediment types. Life, acting through opal and $CaCO_{3(s)}$ among others, plays an important role in the extent of disequilibrium of some seawater constituents.

The diffusional argument is based on the theory of cycles illustrated in Figure 2.16 (RMP). The qualitative statement is sufficient to disprove the equilibrium hypothesis for the oceans. However, the theory of cycles provides a broader insight into the controls of the oceanic composition. The symbol *a* represents the weathering environment, *b* the oceans, *c* the submarine sediments, and *d* the metamorphism sites. Capital letters refer to the reservoir contents. Figure 2.16(1) represents the case

of a steady-state cycle with a net flux $A \rightarrow B \rightarrow C \rightarrow D \rightarrow A$. In Figure 2.16(2) there are inter-reservoir equilibria for all the reservoirs, that is, $A \rightleftarrows B \rightleftarrows C \rightleftarrows D \rightleftarrows A$, and there are no net fluxes.

Figure 2.16(3) presents an important case. In this figure an equilibrium between the oceans and the sediments ($B \rightleftarrows C$) is mixed with steady-state fluxes elsewhere. This cannot happen, as I shall show next, and means that either only equilibria or only steady states can exist in a cyclic stationary system.

If the system is stationary then $dA/dt = dB/dt = dC/dt = dT/dt = 0$ and, consequently

$$k_{DA}D - k_{AB}A = 0 \qquad (2.55)$$

$$k_{AB}A + k_{CB}C - k_{BC}B = 0 \qquad (2.56)$$

$$k_{BC}B - k_{CA}C - k_{BC}C = 0 \qquad (2.57)$$

$$k_{CD}C - k_{DA}D = 0 \qquad (2.58)$$

If we introduce the equilibrium

$$k_{BC}B = k_{CB}C \qquad (2.59)$$

we find that $k_{CA}C = 0$. In other words, the system comes to a stop and there are no net fluxes. Thus, case (3) really is the same as (2). As further confirmation, note that, if there were net fluxes and $C/B = K$ remained constant, then $A + D$ would decrease relative to $B + C$ and the system would not remain stationary. Thus, equilibrium between oceanic waters and the sediments, via the pore waters, cannot exist because there are continent–seawater and pore waters–metamorphic environment net fluxes (RMP).

The absence of equilibria between reservoirs does not preclude equilibrations within reservoirs, as long as these equilibria are fast relative to the inter-reservoir fluxes. An example in point is that of the carbonic acid dissociation reactions.

I would like to point out, pursuing further the topic that led to these kinetic considerations, that equilibrium within pore waters does not necessarily imply an ocean at equilibrium. Again, consider a linear model:

$$\frac{dB}{dt} = k_{AB}A - k_e(B - B_e) \qquad (2.60)$$

The last term expresses diffusional exchange between the oceans, in which the total amount of the substance is B. The term B_e is the value if the oceans were at equilibrium with the sediments and k_e is the ocean–sediment diffusion constant. Under stationary conditions $dB/dt = 0$ and

$$B = \frac{k_{AB}}{k_e} A + B_e \qquad (2.61)$$

B will only become equal to B_e if $k_{AB}A$, the river flux of the substance, is negligible relative to k_e. Then $k_{AB}A \cong 0$, $B = B_e$, and $dB/dt = 0$. The flux $k_e(B - B_e) = J_{BC}$ becomes zero and there is no diffusion from seawater to the sediments. If k_{AB} is not negligible and the system is stationary then the ocean is at a steady-state B different from B_e. This point deserves further investigation for the major elements, although evidence points to the existence of diffusion, but first B_e has to be established at the pressures and temperatures encountered in deep sediments. The work of Sayles and Manheim (1975) may indicate a nearness to but not an achievement of equilibrium in pore waters.

Let us next consider the phase rule equilibrium approach to the control of the chemical composition of seawater. This may require readers to review phase rule and triangular diagram material from Chapter 4. If we have two minerals, C_1A_1 and C_2A_2, which do not form solid solutions and which are in contact with water, the relevant triangular diagram is shown in Figure 2.17. At point D, and only at this point, can A_1B_1 and A_2B_2 be simultaneously at equilibrium with their aqueous solutions, at the temperature and pressure implicit to the diagram. Let us prove that this is the case. From the standpoint of the phase rule, $f = c - p + 2 = 2$. The number of degrees of freedom is $f = 2$ because there are three components ($c = 3$) which are C_1A_1, C_2A_2, and H_2O, and three phases ($p = 3$) which are C_1A_1, C_2A_2, and their aqueous solution. If P and T are fixed, the number of degrees of freedom decreases by 2 and that of free compositional variables is $f = c - p = 0$. Therefore, no such variables can be set arbitrarily at equilibrium and the equilibrium mole fractions are fixed.

If there had been $c = 4$ (C_1A_1–C_2A_2–water–alcohol) and $p = 3$ (C_1A_1–C_2A_2–solution), then f would have been 3. This would mean that one compositional variable was free in addition to P and T, and that there would be a range of compositions instead of a point for which the three phases could be at equilibrium. For each arbitrarily selected (P, T, 1 mole fraction) set there would be one equilibrium composition of the system, that is, one set

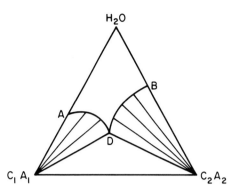

FIGURE 2.17 Minerals C_1A_1 and C_2A_2 in contact with water.

$X_{C_1A_1}$, $X_{C_2A_2}$, X_{H_2O}. This argument is developed in terms of mole fractions, to illustrate the effect of changes in c and f. It should not be confused with the example in terms of molalities which follows.

$$K_{sO}^{(1)} = \frac{[C_1^+][A_1^-]}{[C_1A_1]} \qquad (2.62)$$

$$K_{sO}^{(2)} = \frac{[C_2^+][A_2^-]}{[C_2A_2]} \qquad (2.63)$$

$$[C_1^+] + [C_2^+] = [A_1^-] + [A_2^-] \qquad (2.64)$$

where the last equation is the condition of electrical neutrality. Activity coefficients of ions were set as unity for simplicity.

The number of unknowns, if we replace mole fractions by molalities or molarities, is four: $[C_1^+]$, $[A_1^-]$, $[C_2^+]$, and $[A_2^-]$. The terms $[C_1A_1]$ and $[C_2A_2]$ are constant because the solid phases are pure. Since there are three equations, $f = 4 - 3 + 2 = 3$ and there is one free compositional variable at a given (P, T). This appears to contradict the phase rule for which the number of free compositional variables was found to be zero earlier. The apparent discrepancy arises because the phase rule is derived in terms of mole fractions. To obtain molalities it is necessary to add one further compositional parameter, namely, the total number of moles of substances in 1 kg of water Σn_i. Then, the molality of a given substance is $m_i = X_i \times \Sigma n_i$.

Thus, in summary, the phase rule sets $f = c - p + 2$ if the concentrations of all the components of the aqueous phase and solid solutions are presented in mole fractions, that is, as $X_i = n_i/\Sigma_i n_i$, while those of vapors are expressed in partial pressures. When the phase rule is converted to molar, molal, or ppt units, then

$$f_m = c - p + 3 \qquad (2.65)$$

The number of compositional degrees of freedom c_m (with P and T fixed) is given by

$$c_m = c - p + 1 \qquad (2.66)$$

Sillen (1967) performed an exercise with an equilibrium ocean controlled by the phase rule but clearly stated that the real ocean was not quite as simple. His model is simply an illustration of the type of state which the oceans may attempt to reach.

Sillen selected nine components and then, arbitrarily, nine phases for his examle, so that $f = 9 - 9 + 2 = 2$. The phase rule could, therefore, be said to control the oceanic composition in terms of mole fractions or of the ratios of the components to any 'one of them. To show that mole fractions used in the phase rule are equivalent to such ratios I set, for the case of four components,

$$X_2 = \frac{n_2}{n_1 + n_2 + n_3 + n_4}$$

$$= \frac{n_2/n_1}{1 + n_2/n_1 + n_3/n_1 + n_4/n_1} \qquad (2.67)$$

The components used by Sillen were HCl–H_2O–SiO_2–$Al(OH)_3$–$NaOH$–KOH–MgO–CaO–CO_2 and the phases the aqueous solution–quartz (SiO_2)–kaolinite ($Al_2Si_2O_5(OH)_4$)–potassium mica (KAl_3 $Si_3O_{10}(OH)_2$)–chlorite ($Al_2Si_4O_{10}(OH))_2$ which is the idealized composition without magnesium–montmorillonite ($Al_2Si_4O_{10}(OH)_2$)–phillipsite (Mg_3 $Al_3Si_5O_{16}(H_2O)_6$)–$CaCO_3$–CO_2. The numbers of components and phases used by Sillen are not sufficient, however, to describe all mineral–seawater equilibria. Furthermore, one set of components may constitute more than one phase (igneous rock and clay; opal and quartz) so that metastable minerals may exist which do not yield an equilibrium state for the overall system.

In conclusion, the Sillen approach may turn to be useful to indicate reactions which may tend to control chemicals in pore waters and, by diffusion, in overlaying oceanic waters. One should not, however, consider only the minerals selected by Sillen because of the presence of biological processes, redox reactions, oxides, hydroxides, sulfides, phosphates, and evaporites. The phase rule approach was also used by Helgelson and Mackenzie (1970), who considered a larger number of components.

The whole argument of steady states versus equilibria in nature is in a way a problem in semantics because one may say that equilibria limit compositional excursions due to life or that life causes changes from equilibria. Which statement is closest to the truth depends on the relative rates of biogenic (steady-state inducing) and mineral–seawater (equilibrium-inducing) processes. Still, there is a clear conceptual line between the two types of processes in terms of energetics and entropy as well as in their analytical representations

2.4.4 Steady States and the Control of the pH

Let us first consider some generalities about steady states. It was shown by Prigogine (1978) and Glansdorf and Prigogine (1971) that there are two types of stable processes that are not at equilibrium. First, there are those that remain near-equilibrium and obey the condition of a minimum rate of generation of entropy. These systems are the subject of conventional irreversible thermodynamics, which include the Onsager (1931) reciprocity relations. There is another set of stable systems in which these systems remain far from equilibrium and the conditions for this to occur are lucidly explained by Prigogine (1978).

In any event, no system through which energy flows can be an equilibrium one because the entropy is generated. Local equilibria can occur within the system but the whole is not at equilibrium. Since energy flows through the oceans, their stability must be of the steady-state type.

Biogeochemical processes in nature are extremely complex and of the nonequilibrium type and we hardly understand their mechanisms and feedback loops. Still, the elementary mathematical modeling of reservoir transfers presented later in this chapter provides useful guidelines for thought.

An equilibrium view of the oceans requires specific equilibria that fix the composition of the seawater (Sillen, 1961, 1967; Holland, 1965). The steady-state view (Pytkowicz, 1971, 1972, 1973b) simply states that chemicals build up in the oceans; this accelerates removal processes because increasing concentrations are well known to accelerate reactions, and that eventually input and output rates become roughly equal.

Weyl (1966) first proposed oceanic steady states and based his argument on the occurrence of surface coatings on mineral grains, which do not reach equilibrium with seawater. This has not yet been established to be the case since surface coatings, which may have different compositions than the bulk solids, conceivably can reach equilibrium with seawater (see Chapter 8, Volume II).

Broecker (1971) and Pytkowicz (1971, 1972) pointed out that the concentrations of many constituents of seawater are controlled by biological reactions and, therefore, are not subject to mineral–seawater equilibria. Pytkowicz further showed that, under certain conditions, even for those constituents for which there is a tendency toward mineral–seawater equilibria, equilibrium is not achieved if the rate constants k_{AB} and k_{BC} for input and output of B from the oceans are of comparable size. Equilibria and steady states can coexist in the oceans but may apply to different chemicals or to different reactions of the same substances. A case in point would be the CO_2 system in which carbon enters the soft parts of organisms and also calcareous tests which are subject to saturation equilibrium at least as a driving force, even though saturation is not achieved.

We have already seen that the oceanic CO_2 system is probably not stationary, due to the production of excess $CaCO_3$, and we will find out in the next section that, when one reservoir in a cycle changes, so do all the other ones.

Next, let us look at a specific problem, the control of the pH of seawater. In a general sense, the pH is constrained by weathering and reverse weathering since the consumption of acids by the former is balanced by their release during the second. These processes, and the biogenic removal of calcium carbonate by calcareous tests (Pytkowicz, 1972, 1973a), control the bicarbonate content (actually the alkalinity) of seawater. This carbonate, rather than clay reactions, then acts as the buffer agent since it was shown (Pytkowicz, 1967) that the bicarbonate already present in the oceans provides enough buffer capacity for at least 1000 yr, even if reverse weathering is neglected. This capacity is enormously stretched in the presence of reverse weathering and because the bicarbonate is constantly being renewed by weathered products from rivers.

The actual pH within this buffer capacity is controlled by any two relevant quantities because the CO_2 system is described by four equations in six unknowns (Pytkowicz, 1968). At this point we can invoke a steady-state argument to fix the pH (Pytkowicz, 1972), which goes as follows. pCO_2, the partial pressure of carbon dioxide in the atmosphere

before 1860, was maintained constant because sedimentation of organic carbon and carbonates originating in the weathering of plagioclase feldspars removes juvenile CO_2. Thus, the atmospheric pCO_2 is controlled by biological activity. The surface layers of the oceans are roughly at equilibrium with the atmosphere, whereas the deeper ones, below the wind-mixed layer, reflect the results of organic oxidation. Therefore, the oceanic pCO_2 is also biologically controlled. The dissolved carbonates are not at equilibrium with solid calcium carbonate (Pytkowicz, 1968), and the removal of titration alkalinity (a measure of dissolved carbonate and bicarbonate) is primarily biogenic (Pytkowicz, 1965a, 1973a). Thus, the pH is biologically controlled and no equilibrium is called for.

Furthermore, since the pCO_2 and the dissolved carbonate are fixed by life, then the remaining four equations in four variables imply that all components of the oceanic CO_2 system are at approximate short-term steady states which are biogenically controlled, or so were before the Industrial Revolution. In this argument I neglect the slow drift caused by the weathering of plagioclase feldspars. In Pytkowicz (1975a) I also examine the equilibrium argument based upon the Urey reaction (1952).

Little is known about the effectiveness of feedback loops in weakening or reinforcing controls of geochemical cycles but an example is of interest. A kind of negative feedback is provided, for example, by the buffer factor. The essence of this process is that, when waters absorb CO_2, they become acidified and a smaller proportion of the CO_2 is converted into HCO_3^- and CO_3^{2-}. Thus, there is negative feedback because CO_2 becomes less and less soluble in the water.

2.5 THEORY OF RESERVOIR TRANSFERS AND CYCLES

2.5.1 Linear Models

Pytkowicz (1972) developed a simple linear theory of reservoir transfers in geochemical cycles. The time behavior of the models that will be studied is shown in Figure 2.18. The linear approximation was adopted because of its simplicity and because it does not detract from the conceptual validity of the results since the resulting fluxes increase with increasing reservoir contents. This is the key property required for my modeling. The value of theories for open or cyclic reservoir transfers is that the

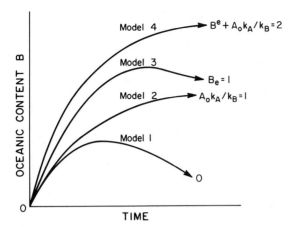

FIGURE 2.18 The time behavior of the various models.

resulting models and equations provide insights into the controlling features of the transfers and, furthermore, that they improve our capability to predict the response of systems to perturbations. One important consequence of such models is that they show that changes in any one reservoir are transmitted eventually to all the other ones. This means that we can only select part of a system for study if the remainder has long relaxation times relative to those of interest.

Open Models. In my simple models, I considered five cases. The symbols A, B, and C represent the total contents of a given chemical in three linked reservoirs a, b, and c with the paths shown below:

$$a \rightarrow b \rightarrow c \qquad (2.68)$$

where a might be the weathering environment on land, b the oceans, and c the submarine sediments. Total contents are used to avoid the problem of expressing the rate of transfer from a solid reservoir as a function of the concentration (always unity for a single-component reservoir regardless of its size).

Model 1. In this model it is assumed that equilibrium does not play a crucial role in the intra- and inter-reservoir transfer of the chemical under consideration. This occurs when B is controlled by the biota, settles as detrital material, or is physically adsorbed by particulate matter

$$\frac{kA}{dt} = -k_{AB}A \qquad (2.69)$$

$$\frac{dB}{dt} = k_{AB}A - k_{BC}B \qquad (2.70)$$

$$\frac{dC}{dt} = k_{BC}B \qquad (2.71)$$

This model reflects exit rates that are simply proportional to the amounts of an element which are present in the reservoirs. The rate of weathering of sedimentary rocks containing calcium, for example, should increase with an increasing amount of such rocks. Also, the rates of the reactions that remove a given ion from seawater should increase if the amount of this ion goes up. This model does not imply, however, that concentrated ions will necessarily be removed from the oceans faster than dilute ones since the rate constants for different entities may not be the same. The rate constants represent in a simplified manner not only complex chemical reactions but also the possible effect of environmental parameters, such as light and temperature, on the reaction rates.

The solutions of Equations (2.69) and (2.70), with the initial conditions $A = A_0$ and $B = 0$ at $t = 0$, are

$$A = A_0 e^{-k_A t} \qquad (2.72)$$

$$B = A_0 \frac{k_{AB}}{k_{BC} - k_{AB}} (e^{-k_{AB} t} - e^{-k_{BC} t}) \qquad (2.73)$$

A is gradually depleted while B, the oceanic content of the given element, will increase to a maximum value and will then gradually decrease to zero. Eventually all of A_0, the amount initially present in the weathering environment, will be present as C_∞ in the submarine sediments.

This model does not yield a nonzero stationary state for B. There may, however, be an apparent steady state in the region around the maximum value of B_1 since this region may last millions of years. Thus, if the rates of solute removal increase with the amount of solute present, then the chemical composition of the oceans may appear to be stationary for long periods. A heavily mined and scarce mineral may tend to behave as is idealized in this model. The third step may, however, involve a tendency toward equilibrium if, for example, the element is removed from pore waters and is precipitated as a sulfide. On the other hand, the metal may be scavenged and settled on clays or in organisms, in which case Equation (2.71) applies.

Model 2. This model is a special case of model 1 in which it is assumed that $A = A_0$ is essentially constant because of the large amount of A relative to outflux or replenishment (exposure of new rocks in or uplift into the weathering environment). This type of situation occurs more frequently than that of model 1 since natural reservoir sizes on land tend to change little over geologically short time spans, unless perturbed by human activities. Then, the solution of Equation (2.70), with the initial condition $B = 0$ when $t = 0$, is

$$B = A_0 \frac{k_{AB}}{k_{BC}} (1 - e^{-k_{BC}t}) \qquad (2.74)$$

Equation (2.74) tends toward the steady state $B_0 = A_0 k_{AB}/k_{BC}$, which makes $dB/dt = 0$ in Equation (2.70). This is an interesting result because it implies that the oceans may reach a steady state if two conditions that may be reasonable are satisfied; the rates of removal increase with increasing amounts of reactants in the oceans; and the amounts in the continental weathering reservoir are not seriously depleted either because of the reservoir size or because of replenishment. This model corresponds to an ocean in which the amounts of solutes gradually increase. This increase causes a corresponding increase in the exit rates until eventually the exit and entry rates become equal and a steady state is reached. It should be pointed out that, for stationary systems, a rate constant is the inverse of the residence time, which is defined as the ratio of the amount in the reservoir to the exit rate. If perturbations cause transients, then the rate constant is the inverse of the relaxation time for the perturbation.

Note that major constituents have large values of the steady state B_0 by definition and long τ's while minor ones reach smaller steady-state values but do it faster.

The steady-state system has some stability to perturbations because if B, for example, should increase then the flux from B to C, $k_{BC}B$, would increase and would counteract at least in part the increase in B. As an example, if the amount of dissolved calcium carbonate in the oceans should increase then two factors could tend to enhance the rate of calcium carbonate removal from the oceans. The rate of production of calcareous organisms could increase and the rate of dissolution of settling calcareous tests would decrease because the deep oceans would be less undersaturated. Thus, more calcareous tests would be incorporated into the sediments. This example goes beyond the scope of Equation (2.74) because it touches on the effect of intrareservoir equilibration when the degree of undersaturation is considered.

Other cases, which do not involve equilibrium, would be the enhanced rate of deposition of organic carbon if carbonic acid, nitrates, and phosphates were added to the oceans at accelerated rates, and the possible deposition of excess trace metals by adsorption on clays or organic matter. The treatment of cases in which there may be equilibration within reservoirs will be treated next.

Let us take a look at mineral–seawater equilibrium within the water column. Models 1 and 2 corresponded to processes for which possible equilibrium states within reservoirs did not have to be considered explicitly. They may apply, as I stated earlier, to biological utilization and to physical adsorption on organic and clay matter but not to mineral–seawater reactions. The rates of mineral–aqueous solution depend on departures from equilibrium as will be shown below, and this will lead to models 3 and 4.

In effect, let us consider a solubility reaction which is schematically represented by

$$D_s(\text{solid}) \underset{k_r}{\overset{k_f}{\rightleftharpoons}} D_a(\text{solution}) \qquad (2.75)$$

For a first-order reaction

$$\frac{dD_a}{dt} = k_f D_s - k_r D_a \qquad (2.76)$$

k_f and k_r are the reaction rates for the forward and reverse steps in (2.75). This is quite an oversimplification for heterogeneous kinetics as we shall see in Chapter 6, Volume II. In concentration units, D_s is constant by definition, whereas in terms of amounts D_s is held invariant in the oceans because of the constant input. Therefore, Equation (2.76) may be written as

$$\frac{dD_a}{dt} = K - k_r A \qquad (2.77)$$

At equilibrium $dD_a/dt = 0$ and, therefore, $K = K_r D_a^e$ where the superscript e refers to the value at equilibrium. Thus,

$$\frac{dD_a}{dt} = k_r (D_a^e - D_a) \qquad (2.78)$$

It can be seen from Equation (2.78) that solubility reactions have rates which, for first-order reactions, are proportional to the departure from equilibrium. For higher order reactions the rates would still be functions of the departures from equilibrium. This result will be included in the next model.

EXAMPLE 2.12. SEAWATER–SILICA INTERACTIONS (D. HURD, PERSONAL COMMUNICATION). Let us consider the case of deep waters through which opal falls. If the molarity c is used instead of the total content B,

$$\frac{dc}{dt} = kS(c_e - c) \qquad \text{(i)}$$

Integrating (i)

$$\ln \frac{c_e - c}{c_e} = kS\tau \qquad \text{(ii)}$$

S is the surface area of the solid per unit mass and τ is the residence time of deep waters.

The right value for c in deep waters was found for the proper values of k, S, τ, and c_e introduced in Equation (ii). The term k was estimated for a dissolution-controlled reaction rather than the diffusion-controlled reaction treated, but formally the resulting equations are the same.

At the present time the relative importance of processes described by models such as 1 and 2 and by seawater–mineral equilibria has not been entirely ascertained. Although geologists tend to assume that the mineral equilibria play a predominant role in the control of the oceanic composition, processes illustrated by the first two models must be considered especially in the case of biogenic reactions. Even the removal of the mineral calcium carbonate is not the result of mineral equilibria alone but rather of both types of processes. The removal of calcium carbonate from the near-surface oceanic waters is essentially biogenic (Pytkowicz, 1965a) and may depend only weakly, if at all, on the degree of supersaturation of the seawater in the photic zone.

As the calcareous tests settle through the oceans, however, they are not completely incorporated into the sediments because there is a partial dissolution (Pytkowicz, 1967, 1968). The rate of dissolution should depend on the degree of undersaturation of the deep oceanic waters. Another example of the possible coexistence of two types of mechanisms is the removal of trace metals which may occur by specific reactions with clays, physical adsorption on clays and organic matter, incorporation into the refractory parts of dead organisms that are not oxidized in the water column, and adsorption and formation of oxides and sulfides.

Model 3

$$\frac{dA}{dt} = -k_{AB}A \qquad (2.79)$$

$$\frac{dB}{dt} = k_{AB}A = k_{BC}(B - B_e) \qquad (2.80)$$

This model implies that the solids, such as clays that are brought by rivers and settle through the oceans, will remove dissolved salts and dB/dt will gradually change to zero when eventually B becomes equal to B_e. From this time on all the relevant salts will leave the oceans at the entry rate. With regard to the addition by rivers of ions of an insoluble salt, once B grows to B_e, solids will deposit in the sediments at the same rate at which its constituent ions enter the oceans and B_e will be constant.

Models 3 and 4 are a simplified formal representation of, among others, the processes visualized by Sillen (1961, 1967), namely, the chemical interactions of seawater ions with settling or recently settled clays. A represents solid material and, therefore, there is no term A_e because A will tend eventually to zero rather than to an equilibrium if it is depleted.

The solution of Equation (2.80), with the initial conditions $A = A_0$ and $B = 0$ when $t = 0$, is

$$B = B_e(1 - e^{-k_{BC}t})$$

$$+ A_0 \frac{k_{AB}}{k_{BC} - k_{AB}} (e^{-k_{AB}t} - e^{-k_{BC}t}) \qquad (2.81)$$

The system tends toward

$$A_0 = 0 \qquad B_0 - B_e \qquad C_0 = A_0 - B_e$$

$$(2.82)$$

There may be an apparent steady state when B reaches its maximum value, as in model 1, but in time the oceans will reach equilibrium with the minerals settling through them. It is conceivable though that if competing reactions such as biological production and oxidation as well as mixing are fast, relative to the mineral–seawater interactions, then equilibrium cannot be reached within the oceans. As was mentioned earlier, mixing may be the reason why the distribution of the major ions in the oceans does not reflect regional variations in the nature of the clays.

Model 4. In this model, the counterpart of model 2 and a special case of model 3, A is assumed to be essentially constant. It will be shown that, as a consequence of $A = A_0$, the constant influx from reservoir a may prevent equilibration and lead to a steady state in the oceans. Then, Equation (2.80), with $A = A_0$ and the initial condition $B = 0$ at $t = 0$, leads to

$$B = B_e(1 - e^{-k_{BC}t}) + A_0 \frac{k_{AB}}{k_{BC}} (1 - e^{-k_{BC}t})$$

$$(2.83)$$

B tends toward $B_0 = B_e + A_0 k_{AB}/k_{BC}$ for which dB/dt in Equation (2.80) is zero. B will only tend toward its equilibrium value B_e if $A_0 k_{AB}/k_{BC}$ is negligible relative to B_e. Thus, the oceans may tend toward a steady state if exit rates increase with increasing amounts of solutes and if the continental reservoir is not seriously depleted. This model again can be applied to the settling clay–seawater system.

Let us elaborate further on the models and examples. We see that the processes that control the time behavior of the various models are quite different, the main processes being the steady state, and the secondary processes, the equilibrium. Biological processes lean toward the former whereas geochemical processes tend toward the latter and require a knowledge of the equilibrium state even after more complex and realistic equations are developed. The condition B_e is crucial when we seek to understand quantitatively the past and even more crucial when we wish to predict the long-term response of a system perturbed by humans, since $B - B_e$ directs this response. Transients are not controlling but temporary conditions.

Often it is still not quite clear to what extent substances in geochemical cycles are controlled by steady states versus equilibria. Let us consider a few examples starting with steady states. Organisms are not equilibrium systems although local equilibria can occur within their body fluids, as, for example, the dissociation equilibria of carbonic acid. Energy and matter flow through organisms leading, during short periods in which a population changes little, to a steady state. Keeping this in mind, that part of the oceanic CO_2 system which undergoes photosynthesis, oxidation, and sedimentation of organic carbon was roughly at a steady state before the advent of the Industrial Revolution.

Calcium carbonate uptake in tests of organisms in the upper oceanic waters occurs from within the cells and is, therefore, another example of a steady-state process from the standpoint of a long-term average. Another example of such a process is the diffusion of oceanic cations such as K^+ into the deep sediments. In the deep sediments, of course, the K^+ and clay interaction introduces a term C_e but in the oceanic waters and in the upper layers of the sediments there appears to be little reaction (Pytkowicz, 1975a; Drever, 1977).

Cyclic Linear Model. Next, the general properties of cyclic systems are examined because many chemical constituents of seawater may cycle from the continents to the oceans and then back to the continents via the submarine sediments.

Model 5. A three-reservoir model is used for simplicity even though in some cases there may also be a metamorphic stage. The system will be considered cyclic with A going to B, to C, and then back to A. The differential equations are

$$\frac{dA}{dt} = k_{CA}C - K_{AB}A \qquad (2.84)$$

$$\frac{dB}{dt} = k_{AB}A - k_{BC}B \qquad (2.85)$$

$$\frac{dC}{dt} = k_{BC}B - k_{CA}C \qquad (2.86)$$

The open models 1 and 2 are special cases of model 5 for negligible values of $k_{CA}C$. The cyclic counterparts of models 3 and 4 will not be treated, since model 5 is sufficient to illustrate the desired concepts.

Cyclic systems are of special interest because they occur often in nature. The carbon dioxide system, in addition to its broad cycle from the continents to the oceans and back to the continents, undergoes several subcycles within the oceans (Pytkowicz, 1967) and cycles through the atmosphere. Nutrients such as phosphate and nitrate, as well as carbon dioxide and oxygen, are cycled through the biosphere. Silicate minerals are weathered on the continents and the resulting aluminum silicates that are brought to the oceans undergo eventual reconstitution to silicate minerals (Garrels, 1965; Siever, 1968a).

Stability of Cyclic Models (Longuet-Higgins). The first question that arises from Equations (2.84) through (2.86) is whether such a system tends toward a stationary value. This is shown in the appendix by Professor M. S. Longuet-Higgins to Pytkowicz (1971) to be indeed the case. Of course, fluctuations around the stationary values will occur if changes in the environment cause variations of the rate constants.

The general solution of Equations (2.84) through (2.86) is of the type

$$A = C_1\alpha_1 + C_2\alpha_2 e^{p_2 t} + C_3\alpha_3 e^{p_3 t} \quad (2.87)$$

$$B = C_1\beta_1 + C_2\beta_2 e^{p_2 t} + C_3\beta_3 e^{p_3 t} \quad (2.88)$$

$$C = C_1\gamma_1 + C_2\gamma_2 e^{p_2 t} + C_c\gamma_3 e^{p_3 t} \quad (2.89)$$

If the system starts from some nonstationary state, then as time goes on, the terms in p_2 and p_3 will vanish and a steady state of the type $A = C_1\alpha_1$, $B = C_1\beta_1$, and $C = C_1\gamma_1$ will be reached.

The steady-state solution can easily be shown to be

$$A_0 = \frac{T}{k_{AB}r} \quad (2.90)$$

$$B_0 = \frac{T}{k_{BC}r} \quad (2.91)$$

$$C_0 = \frac{T}{k_{CA}r} \quad (2.92)$$

with

$$r = \frac{1}{k_{AB}} + \frac{1}{k_{BC}} + \frac{1}{k_{CA}} \quad (2.93)$$

and

$$T = A_0 + B_0 + C_0 \quad (2.94)$$

These equations show that a steady-state amount in any one reservoir is a function of the total amount and the rate constants in all the reservoirs. Therefore, it is not possible to explain the steady state in any one reservoir only by processes within that reservoir but it is necessary to consider the whole system.

Two types of perturbations are possible. First, T may change in which case A, B, and C will gradu-ally approach new values which will still maintain the proportionality

$$A_0 : B_0 : C_0 = \frac{1}{k_{AB}} : \frac{1}{k_{BC}} : \frac{1}{k_{CA}} \quad (2.95)$$

Second, a rate constant, for instance k_{AB}, may change to a new value. Again, a steady state will be reached. The system is not buffered in the usual sense but B and C do not change in proportion to changes in k_{AB} because $1/k_{AB}$ is only one of the terms in r. It can also be seen easily that a given change in the content of a large reservoir will have a greater effect on the values of A, B, and C than the same change in a small reservoir. This occurs because large reservoirs at a steady state have relatively small values of k_i and, therefore, $1/k_i$ in the large reservoir becomes an important term in r. The subscript i may be AB, BC, or CA.

A chemical stability of the oceans which results from a steady state requires that the values of T and of the rate constants remain fairly constant. The constancy of T, the total amount of an element present in a geochemical cycle, may be easy to accept but it is harder to visualize the relative constancy of k_i, the rate constants, because they depend on a complex of environmental factors.

Thus, for example, the rate of removal of calcium carbonate from the oceans depends not only on the amount of calcium carbonate dissolved in seawater but also on the efficiency with which calcareous organisms can extract the carbonate. This efficiency may be a function of many factors, such as the water temperature, its pH, the amount of light, the availability of nutrients, and the species present.

Variations in k_i for ions adsorbed by clays can also occur, for example, during periods of peneplanation (low relief) when the particulate load of rivers decreases. Research is needed on the effects of parameters such as these on the rate constants and possible control mechanisms that may prevent large excursions in the values of k_i, to ascertain whether the steady-state hypothesis can be justified. It was shown earlier that the equilibrium hypothesis requires further work before it can be proven. It is seen now that further work is also needed before the alternative, the steady state, can be accepted.

General Problem of n Reservoirs. The general problem of n reservoirs, in which each reservoir may interact with other $n - 1$ reservoirs, is mentioned next. This is the type of problem that will occur when the fluxes of some chemicals through and within the oceans are studied. Such fluxes may include branching and subcycles as in the case of the carbon dioxide system (Pytkowicz, 1967, 1968).

The general system of differential equations for n reservoirs is of the type

$$\frac{dX_i}{dt} = C_{i1}X_1 + \cdots + C_{in}X_n + D_i$$

(2.96)

with

$$i = 1, \ldots, n$$

where X_i represents the amount of the component in reservoir i. The terms C_{ij} may represent sums or differences of rate constants and, therefore, Equation (2.96) includes the case of possible equilibration between reservoirs. It may correspond to open systems if the appropriate rate constants are negligible. D_i may be the sum of terms of the type $k_{ij}X_i^e$ for internal equilibrium in individual reservoirs.

The system with $dX_i/dt = 0$ in general has a unique solution

$$X_1 = X_1^e, \ldots, X_n = X_n^e$$

(2.97)

We then define a new variable $Y_i = X_i - X_i^e$, which represents the departure from equilibrium, and which converts system (2.95) into a homogeneous one. It is easy to see then that the solution of (2.96) will be of the type

$$X_i = X_i^e + K_1\alpha_{i1}e^{p_1t} + \cdots + K_n\alpha_{in}e^{p_nt}$$

(2.98)

This solution may tend toward internal equilibrium if the exponential terms tend to vanish. If, on the other hand, one or more roots are zero, then the solution will tend toward a steady-state value $X_{i,e} + K\alpha$.

These simple models were presented to illustrate the general conditions under which one may expect a steady-state or an equilibrium ocean, from a chemical point of view. It was shown that, if chemical reservoirs on land are held fairly constant either because of their size or replenishment, then the steady state will occur. In the steady-state model, the amounts of solutes in the oceans grow or decay

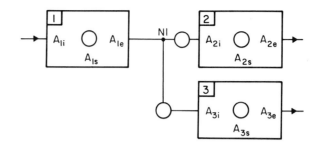

A_{ji}	Amount of chemical in reservoir j in entry form
e	Exit
s	Storage
N	Node

In circle function
  = Equilibrium
  − Steady state
  ~ Transient

FIGURE 2.19 More detailed representation of reservoirs.

from initial values and this causes corresponding growths or decays in their exit rates until the exit rates become equal to the entry rates.

The equations derived in this work may provide a basis for the quantitative description of oceanic trends and transients caused by perturbations such as those resulting from human intervention in nature. An example will be presented when phosphates are examined.

Internal Behavior of Reservoirs. I would like to suggest in conclusion that, as our knowledge of reservoir kinetics increases, more complex models of the type shown in Figure 2.19 be taken into consideration. The term reservoir kinetics is reserved for laws obtained for a whole reservoir such as, for example, the exponentially increasing input of CO_2 into the atmosphere, and does not refer to the next stage which is the detailed study of rate processes within reservoirs. The advantage of reservoirs is that the number of reactions beyond our capacity to handle can be lumped together for a first-cut approach to the problem on hand.

Models such as those presented above were used by Lerman et al. (1975) to describe and study the phosphate system as perturbed by mining.

2.5.2 Nonlinear Models

Holland (1978) also presented equations for reservoir transfers and, although he did not consider the equilibrium situation, the possibility of feedback was included. In his approach if a component i moves from, let us say, reservoir 1 to reservoir 2, then the amount of i in reservoir 2 at time t is

$$A_{2i} = A_{2i}^0 + \int_0^t J_{1,2,i} dt \qquad (2.99)$$

if 2 is purely an accumulator. A_{2i}^0 is the initial amount of i in 2. This is the case for the rare gases in the atmosphere except for He and Xe.

If reservoir 2 is not accumulative then

$$A_{2i} = A_{2i}^0 + \int_0^t (J_{1,2,i} - J_{2,3,i}) dt \qquad (2.100)$$

but $J_{1,2,i} = J_{2,3,i}$, and, therefore, $A_{2i} = \text{constant}$. Such conditions may only occur in terms of very long-term averages.

Holland then considers nonsteady-state reservoirs without feedback, which he describes by means of the equation

$$\frac{dA_{ji}}{dt} = J_{ji}(T) - k_{ji} g(A_{ji}) \qquad (2.101)$$

This equation pertains to the case in which the influx of i into the reservoir j, given by $J_{ji}(t)$, is independent of the contents of j. The term k_{ij} can be a complication function of the contents of j. The term $g(A_{ji})$ expresses g as a function of A_{ji}. Equation (2.101), as used by Holland, includes among others, a step function and one in which a steady-state value of A_{ji} starts to increase asymptotically toward a new value. The work is then extended to inputs that depend on the state of the reservoir, and to feedback situations, such as that of atmospheric oxygen.

Lasaga (1980) used matrix algebra for the kinetic treatment of geochemical cycles to ease the solution of the relevant differential equations. Only first-order rate laws were considered. Lasaga applied the method to a part of the CO_2 system.

At present, kinetic laws are of conceptual value but limited practical use due to the number of equations used, possible variations of k_i, uncertainties in the order of fluxes, and uncertainties in the values of the fluxes and the reservoirs. Still, they provide insights into geochemical cycles and point out needed data.

SUMMARY

Aquatic geochemistry deals with the influence of oceans, rivers, lakes, and so on, on the composition and distribution of substances in the earth's crust. Emphasis in this chapter is placed on the major constituents of water bodies.

Early work led to a model of the oceans as accumulators of leached matter. Later it was realized that sedimentation prevented oceans and lakes from being pure accumulators. Now we understand that even the sediments are not final stops for sediments, due to broad geochemical cycling.

There are several hypotheses for the origin and development of the oceans and some features have not yet been explained. One important development was the understanding of the role of acid volatiles (HCl, CO_2, etc.) as weathering agents.

Substances in the solid earth are organized as minerals, that is, as stable amorphous or crystalline compounds, and as rocks which are usually made up of two or more minerals. Stability is a concept relative to T, P, and other substances present.

Most minerals are carbonates, silica, silicates, halides, aluminum silicates, oxides, sulfides, and sulfates. Aluminum silicates appear in rocks and the important clays. Clay–clay interactions play a significant role in the distribution of cations in soils and in pore waters of sediments.

Crystal systems and the roles of their equilibria as well as nonequilibria in nature are examined and the phase rule is introduced.

The broad geochemical cycle is examined. It consists of weathering, sedimentation, subduction, and uplift. The rough invariance of reservoir sizes is verified and stability mechanisms are discussed. Perturbations due to anthropogenic activities as well as the natural activity for calcium are considered.

Equations for the kinetics of cycles, coupled to an extended form of the phase rule, are discussed. These equations are vital for an eventual quantitative understanding of the geochemical history of water bodies, the transfer of chemicals among reservoirs, and their responses to perturbations.

REFERENCES

Barth, T. F. W. (1952). *Theoretical Petrology*, Wiley, New York.

Behrens, E. W., and L. S. Land (1972). *J. Sedim. Petrology* **42**, 155.

Berner, R. A. (1971). *Principles of Chemical Sedimentology*, McGraw-Hill, New York.

Berner, R. A. (1978). *Am. J. Science* **278**, 1475.

Bischoff, J. L., and T.-L. Ku (1971). *J. Sedim. Petrology* **41**, 1008.

Broecker, W. S. (1971). *Quatern. Res.* **1**, 188.

Broecker, W. S., Y.-H. Li, and T.-H. Peng (1971). In *Impingement of Man on the Oceans*, D. W. Hood, Ed., Interscience, New York, p. 287.

Brown, H. (1953). In *The Atmosphere of the Earth*, G. P. Kuiper, Ed., University of Chicago Press, Chicago, p. 258.

Chave, K. E. (1965). Science **148**, 1723.

Clarke, F. W. (1959). "The Data of Geochemistry," *Geol. Surv. Bull.* **770**, Washington.

Corliss, J. B., J. Dymond, L. I. Gordon, J. M. Edmond, R. P. von Herzen, R. D. Ballard, K. Green, D. Williams, A. Bainbridge, K. Crane, and T. H. van Andel (1979). *Science* **203**, 1073.

Drever, J. (1974). In *The Sea*, E. D. Goldberg, Ed., Wiley, New York, p. 337.

Drever, J. I. (1977). *Sea Water: Cycles of the Major Elements*, Dowden, Hutchinson and Ross, Stroudsburg, Pa.

Edmond, J., C. Measures, R. E. McDuff, L. H. Chan, B. Grant, L. I. Gordon, and J. B. Corliss, Manuscript.

Garrels, R. M. (1965). *Science* **148**, 69.

Garrels, R. M., and C. L. Christ (1965). *Solutions, Minerals, and Equilibria*, Harper & Row, New York.

Garrels, R. M., and F. T. Mackenzie (1971a). *Evolution of Sedimentary Rocks*, Norton, New York.

Garrels, R. M., and F. T. Mackenzie (1971b). *Nature* **231**, 382.

Garrels, R. M., and F. T. Mackenzie (1972). *Mar. Chem.* **1**, 27.

Garrels, R. M., F. T. Mackenzie, and C. Hunt (1973). *Chemical Cycles and the Global Environment*, Kaufmann, Los Altos.

Gass, I. G., P. J. Smith, and R. C. L. Wilson (1971). *Understanding the Earth*, M.I.T. Press, Cambridge.

Gilluly, J., A. C. Waters, and A. O. Woodford (1959). *Principles of Geology*, Freeman, New York.

Goldberg, E. D. (1965). In *Chemical Oceanography*, J. P. Riley and G. Skirrow, Eds., Vol. 1, Academic Press, New York.

Goldberg, E. D. (1971). In *Man's Impact on Terrestrial and Oceanic Ecosystems*, M.I.T. Press, Cambridge, p. 261.

Goldberg, E. D., and G. O. S. Arrhenius (1958). *Geochim. Cosmochim. Acta* **13**, 153.

Goldschmidt, V. M. (1933). *Fortschr. Mineral* **17**, 112.

Goldschmidt, V. M. (1954). *Geochemistry*, Clarendon Press, Oxford.

Gregor, C. B. (1980). *Proc. Konink. Nederland Akad. Wetenschop. Ser. B.* **82**, 173.

Halley, E. (1715). *Trans. R. Soc.* **29**, 296.

Harbaugh, J., and G. Bonham-Carter (1970). *Computer Simulation in Geology*, Wiley, New York.

Heath, G. R. (1974). In *Geologic History of the Ocean*, W. W. Hay, Ed., *Soc. Econ. Paleont. Mineral., Spec. Publ.*, Tulsa.

Helgelson, H. C., and F. T. Mackenzie (1970). *Deep-Sea Res.* **17**, 877.

Hemley, J. J. (1959). *Am. J. Sci.* **257**, 241.

Hemley, J. J., C. Meyer, and D. H. Richter (1961). *U.S. Geol. Survey Profess. Papers* **424-D**, 338.

Hemley, J. J., and W. R. Jones (1964). *Econ. Geol.* **59**, 538.

Holdren, Jr., G. R., and R. A. Berner (1979). *Geochim. Cosmochim. Acta* **43**, 1161.

Holland, H. D. (1965). *Proc. Natl. Acad. Sci.* **53**, 1173.

Holland, H. D. (1972). *Geochim. Cosmochim. Acta* **36**, 637.

Holland, H. D. (1978). *The Chemistry of the Atmosphere and Oceans*, Wiley, New York.

Hutchinson, G. E. (1954). In *The Earth as a Planet*, G. P. Kuiper, Ed., University of Chicago Press, Chicago.

Jones, M. M., and R. M. Pytkowicz (1973). *Bull. Soc. R. Sci. Liege*, **42**, 125.

Kittel, C. (1959). *Introduction to Solid State Physics*, Wiley, New York.

Krauskopf, K. B. (1956). *Geochim. Cosmochim. Acta* **10**, 1.

Krauskopf, K. B. (1967). *Introduction to Geochemistry*, McGraw-Hill, New York.

Lasaga, C. (1980). *Geochim. Cosmochim. Acta* **44**, 815.

Lerman, A., F. T. Mackenzie, and R. M. Garrels (1975). *Geol. Soc. Am. Memoir* **142**, 205.

Li, Y. (1972). *Am. J. Sci.* **272**, 119.

Loewengart, S. (1975). *Israel J. Earth Sci.* **24**, 15.

Mackenzie, F. T., and R. M. Garrels (1965). *Science* **150**, 57.

Mackenzie, F. T., and R. M. Garrels (1966). *Am. J. Sci.* **264**, 507.

Mackenzie, F. T., and R. Wollast (1977). In *The Sea*, Vol. 6, E. D. Goldberg, Ed., Interscience, New York, p. 739.

Madelung, E. (1918). *Physik Z.* **19**, 524.

Manheim, F. T. (1976). In *Chemical Oceanography*, Vol. 6, 2nd Ed., J. P. Riley and G. Skirrow, Eds., Academic Press, New York, p. 115.

Mason, B. (1966). *Principles of Geochemistry*, Wiley, New York.

M.I.T. (1970). *Man's Impact on the Global Environment*, M.I.T. Press, Cambridge.

Morse, J. W., and R. A. Berner (1972). *Am. J. Sci.* **272**, 548.

Murray, C. N., and J. P. Riley (1971). *Deep-Sea Res.* **18**, 533.

Odum, P. (1961). *Fundamentals of Ecology*, Saunders, Philadelphia.

Olausson, E. (1971). In *The Micropaleontology of the Oceans*, B. M. Funnel, Ed., Cambridge University Press, Cambridge, p. 375.

Onsager, L. (1931). *Phys. Rev.* **37**, 405.

Peterson, M. N. A. (1966). *Science* **154**, 1542.

Prigogine, I. (1955). *Thermodynamics of Irreversible Processes*, Interscience, New York.

Prigogine, I. (1978). *Science* **201**, 777.

Pytkowicz, R. M. (1963). *Deep-Sea Res.* **10**, 633.

Pytkowicz, R. M. (1965a). *J. Geol.* **73**, 196.

Pytkowicz, R. M. (1965b). *Limnol. Oceanogr.* **10**, 220.

Pytkowicz, R. M. (1967). *Geochim. Cosmochim. Acta* **31**, 63.

Pytkowicz, R. M. (1968). In *Oceanography Marine Biology Annual Review*, Vol. 6, H. Barnes, Ed., Allen and Unwin, London, p. 83.

Pytkowicz, R. M. (1970). *Geochim. Cosmochim. Acta* **34**, 836.

Pytkowicz, R. M. (1971). *The Chemical Stability of the Oceans,* with an appendix by M. S. Longuet-Higgins, Oregon State University Technical Report, No. 214.

Pytkowicz, R. M. (1972). In *The Changing Chemistry of the Oceans,* D. Dyrssen and D. Jagner, Eds., Almqvist and Wiksell, Stockholm, p. 147.

Pytkowicz, R. M. (1973a). *Am. J. Sci.* **273**, 515.

Pytkowicz, R. M. (1973b). *Swiss J. Hydrol.* **35**, 7.

Pytkowicz, R. M. (1975a). *Earth Sci. Revs.* **11**, 1.

Pytkowicz, R. M. (1975b). *Limnol. Oceanogr.* **20**, 971.

Pytkowicz, R. M. (1978). *Thalassia Yugo.,* 255.

Pytkowicz, R. M. (1979). *Geochim. J.* **13**, 15.

Revelle, R., and R. Fairbridge (1957). In *Treatise on Marine Ecology and Paleoecology 1,* J. W. Hedgpeth, Ed., *Geol. Soc. Am. Memoir* **67**, Washington, p. 239.

Ronov, A. B. (1968). ''Probable Changes in the Composition of Seawater During the Course of Geologic Time,'' Lecture, VII International Sedimentological Congress, Edinburgh, Scotland.

Ronov, A. B., and A. Yaroshevsky (1968). In *The Earth's Crust and Upper Mantle,* P. J. Hart, Ed., American Geophysical Union, Washington, p. 37.

Rubey, W. W. (1951). *Geol. Soc. Am. Bull.* **62**, 1111.

Ryther, J. H. (1963). In *The Sea,* Vol. 2, M. N. Hill, Ed., Wiley, New York, p. 347.

Sayles, F. L., and W. S. Fyfe (1973). *Geochim. Cosmochim. Acta* **37**, 87.

Sayles, F. L., and F. T. Manheim (1973). *Geochim. Cosmochim. Acta* **39**, 103.

Sayles, F. L., T. R. S. Wilson, D. N. Hume, and P. C. Mangelsdorf, Jr. (1973). Science **181**, 154.

Siever, R. (1968a). *Sedimentology* **11**, 5.

Siever, R. (1968b). *Earth Planet. Sci. Lett.* **5**, 106.

Siever, R., and N. Woodford (1973). *Geochim. Cosmochim. Acta* **37**, 1851.

Siever, R., and N. Woodford (1979). *Geochim. Cosmochim. Acta* **43**, 717.

Sillen, L. G. (1961). In *Oceanography,* M. Sears, Ed., *Am. Assoc. Adv. Sci.,* No. 67, Washington, p. 549.

Sillen, L. G. (1967). *Science* **156**, 1189.

Stumm, W., and J. J. Morgan (1970). *Aquatic Chemistry: An Introduction Emphasizing Chemical Equilibria in Natural Waters,* Interscience, New York.

Sverdrup, H. V., M. W. Johnson, and R. H. Fleming (1942). *The Oceans: Their Physics, Chemistry and General Biology,* Prentice-Hall, Englewood Cliffs, N.J.

Urey, H. (1952). *The Planets: Their Origin and Development,* Yale University Press, New Haven.

van Andel, T. H., and T. C. Moore, Jr. (1974). *J. Geol.* **2**, 871.

Weyl, P. K. (1966). *Geochim. Cosmochim. Acta* **30**, 663.

Whitfield, M. (1979). In *Activity Coefficients in Electrolyte Solutions,* R. M. Pytkowicz, Ed., CRC Press, Boca Raton, p. 153.

Wollast, R. (1974). In *The Sea,* Vol. 5, E. D. Goldberg, Ed., Interscience, New York, p. 359.

SUGGESTED READINGS

Ahrens, L. H., Ed., *Origin and Distribution of the Elements,* Pergamon Press, New York, 1968.

Bolin, B., *Sci. Am.* **223**, 124 (1970). Carbon cycle.

Ben-Yaakov, S., *Geochim. Cosmochim. Acta* **36**, 1395 (1974). Diffusion of seawater ions.

Brancazio, P. J., and A. G. W. Cameron, Eds., *The Origin and Evolution of Atmospheres and Oceans,* Wiley, New York, 1964.

Broecker, W. S., T.-H. Peng, and R. Engh, Radiocarbon **22**, 256 (1980). Radiocarbon in the oceans and modeling of the carbon system.

Burton, J. D., and P. S. Liss, *Geochim. Cosmochim. Acta* **37**, 1761 (1973). Additional removal of silica from the oceans.

Calvert, S. E., Nature **219**, 919 (1968). Silica balance in the oceans and diagenesis.

Chave, K. E., *J. Geol.* **62**, 266 (1954). Biogeochemistry of magnesium.

Chestworth, W., J. Dejou, and P. Larroque, *Geochim. Cosmochim. Acta* **45**, 1235 (1981). Weathering of basalt.

Degens, E. T., *Geochemistry of Sediments,* Prentice-Hall, Engelwood Cliffs, N.J., 1965.

Drever, J. I., *J. Sedim. Petrol.* **41**, 951 (1971). Weathering in igneous terrain.

Fowler, W. A., *Proc. Natl. Acad. Sci.* **52**, 524 (1964). The origin of the elements.

Goldberg, E. D., and J. J. Griffin, *J. Geophys. Res.* **69**, 4293 (1964). Sedimentation rates and mineralogy.

Heezen, B. C., and I. G. MacGregor, *Initial Reports of the Deep Sea Drilling Project,* Vol. 20, U.S. Government Printing Office, Washington, 1974.

Holland, H. D., *Proc. Natl. Acad. Sci.* **53**, 1173 (1965). History of seawater and effect on the atmosphere.

Holland, H. D., *Geochim. Cosmochim. Acta* **36**, 637 (1972). Geological history of seawater.

Junge, C. E., M. Schidlowski, R. Eichman, and H. Pietrite, *J. Geophys. Res.* **80**, 4542 (1975). Carbon cycle and isotope geochemistry. O_2 evolution.

Lloyd, R. M., *Science* **156**, 1228 (1967). ^{18}O composition of marine sulfate.

Mackenzie, F. T., and R. M. Garrels, *Science* **150**, 57 (1965). Reactivity of silicates with seawater.

Mackenzie, F. T., and R. M. Garrels, *Am. J. Sci.* **264**, 507 (1966). Chemical balance between rivers and oceans.

Mackenzie, F. T., R. M. Garrels, O. P. Bricker, and F. Buckley, *Science* **155**, 1404 (1967). Silica control in the oceans.

Mackenzie, F. T., and R. Wollast, in *The Sea,* Vol. 5, E. D. Goldberg, Ed., Interscience, New York, 1974, p. 739. Global cycles.

Mackenzie F. T., and R. Wollast, in *Dahlem Conference,* W. Stumm, Ed., Dahlem Conferences, Berlin, 1977, p. 45. Global cycles and man.

McCrea, J. M., *J. Chem. Phys.* **18**, 849 (1950). Isotopic composition of carbonates and paleotemperatures.

Ronov, A. B., VII International Sedimentological Congress, Edinburgh, 1968. Changes in seawater composition.

Schidlowski, M., and C. E. Junge, *Geochim. Cosmochim. Acta* **45,** 589 (1981). Coupling of C, S, and CO_2 cycles.

Schink, D., *J. Geophys. Res.* **79,** 2243 (1974). SiO_2 in pore waters.

Siever, R., *Earth Planet. Sci. Lett.* **5,** 106 (1968). Clay–seawater equilibrium.

Tzur, Y., *J. Geophys. Res.* **76,** 4208 (1971). Pore water diffusion.

Urey, H., *The Planets: Their Origin and Development,* Yale University Press, New Haven, 1952.

Van Denburgh, A. S., and J. H. Feth, *Water Resources Research* **1,** 537 (1965). Chloride balance in rivers.

CHAPTER THREE

AN OCEANIC OVERVIEW

This chapter is dedicated to colleagues in the fundamental physical, earth, and biological sciences, as well as to engineers, who lack the time to work their way through oceanographic texts and the background to select the subjects that are relevant to this book. Material on lakes is presented at the end of this chapter. References are presented for more specialized study.

This chapter covers introductory to state-of-the-art material. A challenging and rewarding feature of the oceans is that they do not respect the disciplinary compartments so dear to many scientists. Everything happens at once and the study of any one aspect, such as marine chemistry, can only be elucidated in terms of other processes. In turn, chemistry helps shed light on marine biology and geology, and hydrographic transport. Of course, there are specialized topics within each discipline, such as metal complexation, which are relatively self-contained and are discussed later. I, therefore, selected the following topics for this chapter.

1. A geographic summary is used to provide a framework within which to discuss oceanic processes. As an example of its importance, features that alter the bottom topography affect the pressures at which chemical reactions occur and change their extent and rates. Furthermore, they alter the motions of waters within oceans, seas, and lakes.

2. Currents and water masses are examined from a descriptive point of view, because transport is one of the key quantities that control the distributions of chemicals in the oceans. The hydrobiochemical equation and its applications are studied because they reveal the interactions between transport and the uptake or release of substances such as phosphates, nitrates, carbon dioxide, and oxygen. Metals are covered in Chapter 8, Volume II.

3. The energy budget of the earth is examined because it provides an important application of the first law of thermodynamics, the sources of energy for motion in the oceans, the radiation as well as the temperatures required by marine life. It also yields the background for the study of the fossil fuel burning problem.

4. The hydrological cycle is examined in some detail to complete the survey of water transport in nature, an important geochemical process triggered by the energy uptake of the earth.

5. Marine biology and geology are reviewed briefly because of their role in the distribution of chemicals in the sea, which in turn can shed light on life and sedimentation.

3.1 A GEOGRAPHICAL SUMMARY

3.1.1 Oceans: Land and Water Distribution

One can recognize five major oceans: Atlantic, Pacific, Indian, Arctic, and Southern. The Southern Ocean is not accepted by many as a separate ocean because there are no land barriers limiting it to the north. It has some features, however, which make it convenient to consider it separately. As one heads south in the Indian, Atlantic, and Pacific Oceans, in the southern hemisphere, the surface temperature of the waters decreases gradually at first. However, roughly at 60°S there is a sharp gradient of temperature with latitude. This gradient results from the meeting of the Antarctic Surface with the Subantarctic Water. Furthermore, approximately at this latitude, there is the formation of a relatively low-salinity cold-water mass, the Antarctic Intermediate Water Mass, which travels northward while sinking to about 800–1000 m. This water mass consists of a mixture of Antarctic Surface with the Subantarctic Water. Therefore, two characteristic features occur at the northern boundary of the Southern Ocean. In addition to the oceans, one finds a number of seas such as the Mediterranean, Bering, the American Mediterranean (Carribean), the Baltic, the Black, and the Red Seas.

Note that the map in Figure 3.1 is a central cylindrical perspective projection which exaggerates the area and shape at high latitudes. I use it because, at lower latitudes, it provides an image of the earth surface to which we are accustomed. However, projections that better reflect the actual area of the surface of the earth are available. Only the cores (central parts) of the currents are sketched to gain in clarity what is lost in detail.

It is of interest to observe that the oceans are interconnected by the West Wind Drift (the Antarctic Circumpolar Current). This current, turbulent mixing, and the water masses that move in the deep oceans provide, on a time scale of about 1000 yr, the means to maintain oceanic waters as one well-stirred system. This shows, for example, that the reactions between seawater and minerals in the deep oceans make themselves felt at the surface roughly 1000 yr after the event. Thus, the waters that bathe our beaches bear the imprint of processes that occurred centuries ago.

In Table 3.1, the distributions of land and water in the two hemispheres are presented. One sees that roughly two-thirds of the earth's surface is covered by water and that the water to land ratio is higher in the Southern than in the Northern Hemisphere.

3.1.2 Effect on Climate

The specific heat, that is, the amount of heat necessary to raise the temperature of 1 g of a substance by 1°C, is 1 cal/g · °C for water versus 0.2 cal/g · °C for most silicate minerals. This means that more heat is necessary to change the temperature of 1 g of seawater than 1 g of land. If we consider in addition the larger area of the oceans and the convective overturn of the oceanic waters which brings new seawater constantly to the surface, we realize that the oceans warm or cool more slowly than land. They act, therefore, as temperature stabilizers on the earth. This effect is greater in the Southern than in the Northern Hemisphere due to the larger proportion of water in the former.

The larger land area in the Northern Hemisphere causes a greater chemical denudation (weathering

followed by transport of solutes in river water to the oceans) than in the Southern Hemisphere (Table 3.2). In addition, with the higher concentration of population in the Northern Hemisphere, most of the human perturbation of the natural fluxes of elements into the atmosphere and the oceans occurs north of the equator.

3.1.3 Area, Depth, and Volume

The area, mean depth, and volume of the oceans are presented in Table 3.3 and the percentage areas of depth zones are shown in Table 3.4. Menard and Smith (1966) considered the Atlantic and the Arctic Oceans together but I subtracted from their area for the Atlantic that area given by Kossina (1921) for the Arctic. I then used Kossina's values for the average depths of these two oceans.

The depth distribution is important from a chemical point of view because the pressures encountered in deep waters and submarine sediments affect the extent and the rates of chemical reactions. Furthermore, photosynthesis can only occur in the near-surface zone where light can penetrate. In deeper waters, therefore, oxidative reactions such as respiration and decay predominate. This leads to the release of chemicals that are eventually returned to and fertilize near-surface waters. Of course oxidation occurs near the surface but it is slower than photosynthesis during fertile periods. The photic zone, in which light is sufficient for plant growth, may vary in depth from a few meters to about 100 m depending on the transparency of the water.

3.1.4 Topographic Features

The main topographical features of the oceans are illustrated in Figure 3.2. The slopes are greatly exaggerated for clarity. A trough is long with gently sloping sides, whereas a trench is narrow and steep. A seamount is an individual mountain, a rise is a long broad elevation, and a trench is long with steep sides. The continental shelves usually extend to a depth of about 120 m and tend to be wider along eastern coastlines. The slopes are at times cut by submarine canyons. Sills occur above ridges or rises and separate basins or troughs from oceans and seas. These features are important because they change the motions of waters, the transport of chemicals, bottom life, and the nature of sediments. As examples, ridges can be sites of upwelling of new seafloor materials and sills can cause stagna-

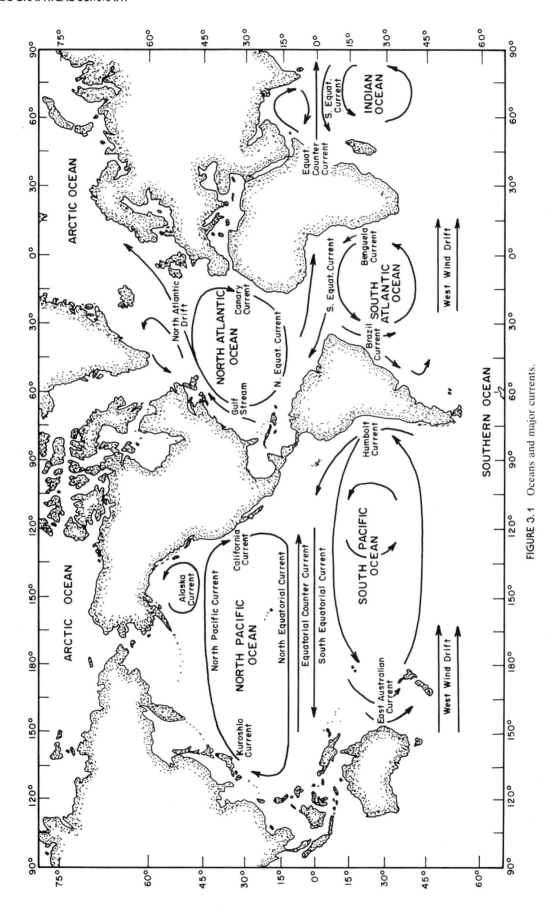

FIGURE 3.1   Oceans and major currents.

TABLE 3.1 Distribution of land and water in the two hemispheres.

	Northern Hemisphere	Southern Hemisphere	All Oceans	All Land
Water (10^6 km²)	154.7	206.4	361.1	
Land (10^6 km²)	100.3	48.6		148.9
Water (%)	60.7	80.9	70.8	
Land (%)	39.3	19.1		29.2

tion, with the utilization of all the oxygen, and the formation of sulfides. Life, except for some forms of bacteria, cannot exist in sulfide-bearing waters.

3.2 WATER MOTIONS

The scope of physical oceanography transcends the study of current and water masses. It includes transport of waters, the study of tides, waves, inertial motions, transfer of momentum from the atmosphere, the resolution of motions between advective trends and fluctuations, and, in general, all hydrodynamic aspects of the sea. We shall be concerned only with the qualitative and, at times, the semiquantitative transport of chemicals by advection, the net water motion resulting from wind stress and density differences, and with eddy diffusion, the turbulent mixing.

TABLE 3.3 Area, mean depth, and volume of the oceans including adjacent seas [based on Kossina (1921); Menard and Smith, (1966)].

Ocean	Area (10^8 km²)	Average Depth (km)	Volume (10^8 km³)
Pacific	1.80	3.94	7.09
Atlantic	0.93	3.93	3.65
Indian	0.74	3.84	2.84
Arctic	0.14	1.20	0.17
World Ocean	3.61	3.81	13.75

3.2.1 Main Currents

Surface currents and deep water masses transport chemicals to regions where changing conditions can affect chemical reactions, biological utilization, and geochemical processes. The reviews of currents and water masses of motions in the oceans in this section and in the next ones are descriptive introductions. Quantitative treatments can be found in a number of specialized texts (e.g., Sverdrup et al., 1942; Pickard, 1963; Von Arx, 1962).

The main surface currents are shown in Figure 3.1. They are primarily generated by the wind drag on the sea surface. This wind stress is transmitted to deeper layers within the water and decreases gradually with depth. The energy is obtained ultimately from solar radiation, which causes pressure and temperature gradients to be formed in the atmosphere. The effects of wind stress in general are not

TABLE 3.2 Annual solute load delivered to the oceans by rivers [based upon Garrels and Mackenzie (1971), except for the last column which was obtained from Livingstone (1963)].

Continents	Area (10^8 km²)	Flux/Area (tons/km²)	Total Flux (10^8 tons)	River Water Flux (10^{15} L/yr)
Northern				
Asia	0.47	32	14.9	11.05
North America	0.21	33	7.0	4.55
Europe	0.11	42	4.6	2.50
			26.5	18.10
Southern[a]				
Africa	0.30	24	7.1	5.90
South America	0.20	28	5.5	8.01
Australia	0.10	2	0.2	0.32
			12.8	14.23

[a] I counted Africa and South America as being in the Southern Hemisphere although substantial parts of their drainage basins are north of the equator, to demonstrate that, even with this distortion, most weathering occurs in the Northern Hemisphere.

TABLE 3.4 Percentage area of depth zones in the oceans [from Kossina (1921)].

Depth Interval	%	Depth Interval	%
0–200	7.6	4000–5000	33.0
200–1000	4.3	5000–6000	23.1
1000–2000	4.2	6000–7000	1.1
2000–3000	6.8	>7000	0.1
3000–4000	19.6		

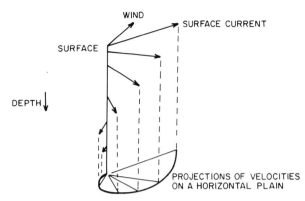

FIGURE 3.3   Ekman spiral.

often beneath the first few hundred meters and the deeper circulation arises from other causes. An exception to this rule is the Antarctica Circumpolar Current, that is, the West Wind Drift.

3.2.2   Coriolis Force and Wind Stress

Aridtjof Nansen, the famous Arctic explorer, first observed that water does not move in the same direction as the wind and G. G. Coriolis, a French mathematician, explained this observation. Newton's first law states that a particle, once set in motion, will follow a straight line in space with regard to coordinates fixed by stars, unless some external force is applied. Consider a particle of seawater that follows a straight path in space and at the same time, a horizontal path on the earth's surface. It will, because of the rotation of the earth, be deflected to the right in the Northern Hemisphere and to the left in the Southern Hemisphere in terms of coordinates fixed to the earth. The Coriolis "force" acts as if it were a real force acting at right angles to the direction of motion. It is not, however, a physical force and it does no work.

The horizontal angle between the wind direction and the direction of motion at the sea surface is roughly 45° and increases with depth because each layer is deflected relative to the one above it. Thus, if a homogeneous ocean existed, one would obtain an Ekman spiral (Figure 3.3).

The long-range surface currents caused by winds are known as drifts. The large circular currents are known as gyres. Thus, at midlatitudes in Figure 3.1, one finds the subtropical gyres. The clockwise gyres in the Northern Hemisphere and the anti-clockwise gyres in the Southern Hemisphere result from the Coriolis force. Transient currents occur as the result of tides, shifting winds, and so on, and are imposed as perturbations on the drifts.

3.2.3   Upwelling

An important transient process is known as upwelling. Prevailing winds are from the north during the summer in the Northern Hemisphere and from the south in the Southern Hemisphere. They cause surface waters off the west coasts of continents to drift away from the shore. This leads to a replacement by colder waters from below (Figure 3.4) which are usually nutrient rich because of the oxidation of settling organic debris. This causes coastal waters to become very fertile and promotes fisheries. Upwelling usually occurs from a depth of about 300–400 m.

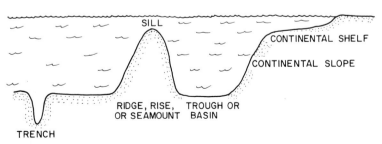

FIGURE 3.2   Topographical features in the oceans.

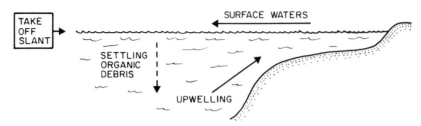

FIGURE 3.4 Upwelling.

3.3 DISTRIBUTION OF TEMPERATURE, SALINITY, AND DENSITY: WATER TRANSPORT

3.3.1 Distributions of Temperature and Salinity

The distributions of the temperature and the salinity in the Western Atlantic Ocean are roughly sketched in Figure 3.5 to illustrate their main features. Details can be found, for example, in the reports of the Meteor expedition (Wüst, 1935).

3.3.2 Density of Seawater

The temperature and the salinity control the density and the specific gravity of seawater which in turn

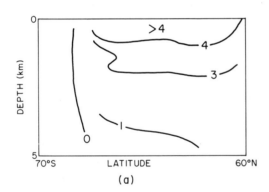

(a)

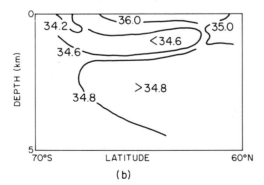

(b)

FIGURE 3.5 Rough sketches of (a) temperature and (b) salinity distributions in the Western Atlantic Ocean.

drive the deep oceanic circulation. I should remind the readers that the density, expressed in g/cm^3, is the mass per unit volume, whereas the specific gravity is the ratio of the density of seawater at a given temperature and pressure to that of pure water at 4°C and 1 atm. Oceanographers use the specific gravity, which will be denoted by s_T. It is a dimensionless number which is numerically almost equal to the density ρ_T. It is not exactly equal to ρ_T because the isotopic composition of the pure water used to obtain s_T may vary so that its density is not exactly unity.

The specific gravity of seawater of most oceanic waters lies in the range 1.0255–1.0285. In practice one can decrease the number of figures used for large numbers of oceanic computations by using the quantity

$$\sigma_T = (\rho_T - 1) \times 10^3 \qquad (3.1)$$

and one commonly says that the density of seawater (actually the sigma-t value) ranges from 25.5 to 28.5. To be exact one should use the symbol $\rho_{T,S,P}$ since the in situ specific gravity depends on temperature, salinity, and pressure. The subscript S is omitted to simplify the notation. The effect of pressure is neglected because in synoptic oceanography (the determination, presentation, and interpretation of the distribution of properties in the oceans) one usually compares properties such as the temperature, the salinity, and the values of σ_T at the same depth and, therefore, pressure to infer the motions of waters. The effect of pressure on the specific gravity is not negligible when vertical profiles are studied. As an example, a seawater sample at 0°C and 35‰ salinity would have a specific gravity of 1.02813 ($\sigma_T = 28.13$) but at 4000-m compression would increase it to 1.04849 ($\sigma_{T,S,P} = 48.49$) (Pickard, 1963).

The statement of the previous paragraph regarding the use of surfaces of constant depth for the

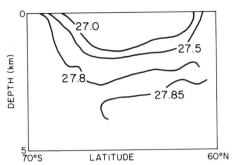

FIGURE 3.6 Sketch of the sigma-*t* distribution in the Western Atlantic Ocean.

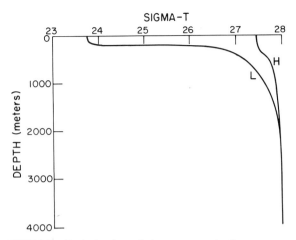

FIGURE 3.7 Typical values of signa-*t* versus depth.

comparison of properties is not strictly true. A theorem in physical oceanography states that water masses tend to flow along isentropic surfaces (adiabatic flow which maintains the entropy of the water mass constant) and that these surfaces are approximately the same as the isopleths (surfaces of constant density). One finds in the oceans that depth surfaces can be used instead of isopleths as a fair approximation for depth below circa 2000 m (see Figures 3.6 and 3.7). This is not the case, however, for upper waters because of large changes in specific gravity with the large gradients of T and S. Still, constant depth surfaces are often used for rough oceanographic inferences in descriptive synoptic work done with the temperature, salinity, and chemical parameters of near-surface waters.

Values of σ_T can be obtained from the temperature and the salinity from the Hydrographic Tables (Knudsen, 1901) or from improved recent equations (Chen and Millero, 1977; Millero et al., 1980).

In this section and in the next one I treat temperature, salinity, and the specific gravity as descriptive tools. Another application, which transcends the scope of this book, is the calculation of currents and volume transport (dynamic oceanography). In it one often uses specific volume anomalies, defined by

$$\delta_T = \alpha_{S,T,P} - \alpha_{35,0,P} \qquad (3.2)$$

$\alpha_{S,T,P}$ is the in situ specific volume (the inverse of the specific gravity) and $\alpha_{35,0,P}$ is the specific volume of a defined water of 35‰ salinity at 0°C and P atm. I use the symbol α in this chapter to conform to oceanographic usage. Examples of the quantitative use of σ_T, often denoted by sigma-T, can be found in Fomin (1964) and Reid (1973).

In essence, lighter waters overlaying a surface of constant density immersed in the oceans tend to pile up higher than denser waters in order to exert the same pressure on this surface. The resulting small variations in sea level heights lead less dense waters to flow toward depressions.

In Figure 3.7 one observes that the specific gravity presents sharp gradients at depths between approximately 200 and 1000 m. This feature is known as the permanent pycnocline (halocline when due to a salinity gradient and thermocline when resulting from a temperature gradient). In addition, a seasonal pycnocline (the summer thermocline) is formed at intermediate latitudes at depths around 50 m due to the warming of near-surface waters.

The density stratification, with a positive gradient $d\rho_T/dz$ that is found in the oceans, indicates a stability to vertical mixing which increases with the magnitude of the gradient. Thus, the depth of penetration of the wind drag decreases with increasing $d\rho_T/dz$. Pycnoclines, therefore, act as effective barriers to the transfer of momentum between the wind-mixed layer above the pycnocline and the waters beneath it.

Before leaving this section, I will differentiate between the in situ and the potential temperature. The potential temperature is the one that is measured if a seawater sample is brought adiabatically (without exchanging heat with its surroundings) to the surface. It will cool while being raised because it will expand as the hydrostatic pressure is released. The corresponding density is known as the potential density, which should be used when the relative stability of deep waters is examined.

3.3.3 Water Masses

While the near-surface circulation is primarily wind-driven, that of deeper waters results from the higher densities found at high latitudes (Figure 3.6). It is known as the thermohaline circulation and results from waters seeking their density level. Differences in surface density which lead to sinking are low temperatures at high latitudes and salinity gradients resulting primarily from evaporation. Waters do not need to have low temperatures and high salinities simultaneously to sink but only a higher density than that of underlying waters. Thus, intermediate waters, which have a natural depth in the density stratification at about 1000 m, are cold but have a salinity minimum relative to those of shallower and of deeper waters. The low salinity results from high precipitation (rain and snow) and the melting of sea ice.

Conservative Properties. Conservative properties play an important role in the study of water masses. Let us consider the conservative nature of T and $S‰$. The temperature and the salinity are conservative properties which can be used to track a water mass. The term conservative refers to the fact that T and $S‰$, once the waters left the sea surface, are affected only by mixing. These properties are not subject to other processes such as biological activity and geochemical reactions except to a very minor extent. Thus, the salinity and the temperature of a mixture of water types i [waters that come from a given source with characteristic values of T_i and $(S‰)_i$] are given by

$$T = \sum_i f_i T_i \qquad (3.3)$$

and

$$S(‰) = \sum_i f_i (S‰)_i \qquad (3.4)$$

f_i represents the fractions of water types i in the water mass with properties T and $S‰$. The salinity is introduced later in this chapter and, at this point, it is enough to say that it represents to a good approximation the salt content of a given seawater.

Since T and $S‰$ depend only on the properties at the sources of the waters i and on changes in T and $S‰$ due to mixing, they can serve at times to help us elucidate the pathways of water types. Oxygen,

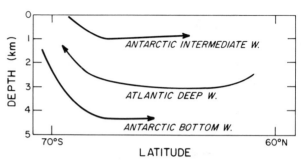

FIGURE 3.8 Cores of the water masses in the Atlantic Ocean.

on the other hand, is subject to changes in the water column due to biological oxidation. It is nonconservative but can also be used in water mass analysis from the standpoint that it decreases in the direction of older waters due to its consumption. These topics will be examined further as we proceed with this chapter.

Two properties help us analyze water masses, core properties and T–$S‰$ diagrams. First, one seeks extremum values in the conservative properties. A maximum or minimum in vertical profiles of salinity or temperature must originate at or near the surface of the sea because they cannot be caused by mixing of different waters at depths (such mixing yields values intermediate between those of the waters that mix) because T and $S‰$ are conservative. Thus, such extrema can be used to characterize the cores of water masses. Extrema in conservative properties are weakened in the direction of motion of the water masses due to vertical and horizontal mixing with adjacent waters.

Atlantic Ocean. The water masses in the Atlantic Ocean are shown in Figure 3.8. Their presence has been confirmed by studies in dynamic oceanography.

The Antarctic Intermediate Water, corresponding to a salinity minimum, sinks near the Antarctic Convergence. At the convergence there is a sharp surface temperature gradient in a north–south direction, roughly at 60°S. It propagates northward with its core roughly at 1000 m. This gradient and the Intermediate Water result from the mixing of Antarctic Surface Water with Sub-Antarctic Water.

The Atlantic Deep Water, which corresponds to salinity maximum, is formed at high northern latitudes in the Atlantic Ocean and carries on its upper reaches a relatively warm but highly saline contribution from the Mediterranean Sea. The high salinity results from evaporation. The Antarctic

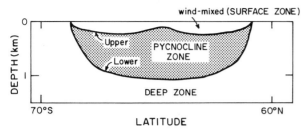

FIGURE 3.9 Sketch of the boundaries of the pycnocline versus latitude. The strong pycnocline only extends down to circa 1000 m. *Source:* Based on Gross, M. G., *Oceanography: A View of the Earth,* 2nd Ed., copyright © 1977, p. 159. Reprinted by permission of Prentice-Hall, Inc., Englewood Cliffs, N.J.

Bottom Water is formed primarily during winter in the Weddel Sea (in the Atlantic sector of the Southern Ocean). These waters feed each other by mixing mostly at high latitudes, thus closing cycles.

The permanent pycnocline is weakened and finally destroyed when the latitude increases indicating low stability as the result of the vertical propagation of these waters (Figure 3.9).

Pacific Ocean. In the Pacific Ocean, one does not find a deep salinity maximum similar to that of the Atlantic Deep Water except in the far south. Instead, one finds a relatively shallow salinity maximum formed in the northern reaches of the Pacific which indicates the formation of the Arctic Intermediate Water. The picture that emerges is shown in Figure 3.10.

The Pacific Deep Water appears to be Atlantic Deep Water caught in the West Wind Drift, which circles around Antarctica at all depths and enters the Pacific Ocean between Antarctica and Australia. The corresponding salinity maximum is not found at latitudes north of 40°S, indicating that this deep water blends with Antarctic Bottom Water, also supplied primarily by the West Wind Drift, and

loses its characteristic extremum. Some of the Antarctica Bottom Water may be supplied by the Ross Sea, in the Pacific sector of the Southern Ocean, where intense freezing occurs in winter.

Southern Ocean. A more detailed view of the cores of the water masses in the Pacific sector of the Southern Ocean is shown in Figure 3.11 (Pytkowicz, 1968). I should point out that all the water masses are moving eastward (into the plane of the page) and that the arrow drawn toward the weakening of the salinity maximum for the Pacific Deep Water may be due to mixing with adjacent waters rather than actual net motion (advection) toward the south.

Much more comprehensive north–south sections around Antarctica may be found in the reports of the Discovery expedition (Deacon, 1937) although they do not reveal the Returning Antarctica Intermediate Water (Figure 3.11), which will be discussed later.

The circulation of the oceans plays a vital role in the distributions of chemicals in the sea. Let us consider an example.

EXAMPLE 3.1. THE DISTRIBUTION OF CARBONATE IONS. First let us consider as a hypothetical limiting case a stagnant ocean. I shall assume that this ocean does not become anoxic so as to treat one process at a time.

The solubility of calcium carbonate increases

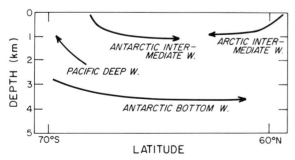

FIGURE 3.10 Water masses of the Pacific Ocean.

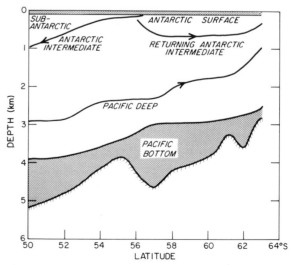

FIGURE 3.11 Cores of the water masses at high southern latitudes in the Pacific and the Southern Oceans, along 106°W (Pytkowicz, 1968).

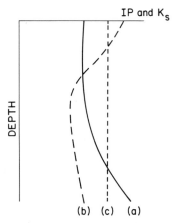

FIGURE 3.12 IP and K_s in (a) a stagnant ocean, (b) the actual ocean, and (c) a very well-stirred ocean.

with higher pressures and lower temperatures. Therefore, the ion product $IP = [Ca^{2+}] [CO_3^{2-}]$ would equal K_s, the stoichiometric solubility product, and would increase with depth. The relation $IP = K_s$ would result from the long exposure of the seawater to calcareous tests, since hundreds of thousands of years are required for equilibration by molecular diffusion. This result is shown in Figure 3.12.

Next, let us contrast the actual ocean to the one above. In the actual ocean supersaturated upper waters do not have the time to precipitate inorganically and undersaturated deep waters are not exposed to sediments long enough to dissolve them until equilibrium is reached.

EXAMPLE 3.2. WHAT MAY HAPPEN TO THE OCEANIC CIRCULATION DUE TO AN INCREASE IN THE BURNING OF FOSSIL FUELS? The answer will be a first approximation since the network of changes is exceedingly complex. The CO_2 as released to the atmosphere would increase and it would retain more infrared radiation from the earth. Thus, the temperature of surface waters, as well as that of the continents, would increase. This would stratify further the water column and would decrease the vertical mixing. Also, the temperature of high latitude waters would rise and the thermohaline circulation would slow down.

Other Approaches. The older theory of the circulation of deep waters, below 2000 m and not including bottom waters, is contrasted to the more recent view obtained theoretically by Stommel (1968) in Figure 3.13. He predicted intense streams in the

western margins of the oceans which feed waters toward the east and the poles instead of a uniform propagation at all longitudes. The intense western streams have been confirmed by observations but, on the basis of the propagation and direction of weakening of core properties, I prefer the older view for the less intense spreading of the deep waters in midoceans as well as in their eastern margins.

The analysis of water masses is facilitated by the use of temperature–salinity (T–$S‰$) diagrams. An example is shown in Figure 3.14, plotted by me (Pytkowicz, 1968) during cruise 14 of the *USNS Eltanin,* in July and August of 1964. Each point represents the temperature and the salinity at a given depth. Nodes indicate what is known as water types and the segments between the nodes reflect the vertical mixing of the water types that bracket them.

Figure 3.14 represents a T–$S‰$ diagram for a given station. If we follow water types along their direction of motion, the nodal points will tend to converge due to mixing and the resulting weakening of extrema (Figure 3.15). Thus, water types at depth do not represent the actual source characteristics (at or near the sea surface where the waters originated). One can use the term water mass instead of water type for the entity characterized by an attenuated extremum and consider it a mixture of water types. Then, the term type can be applied to the water at its source. In practice both labels are used interchangeably.

3.3.4 Advection–Diffusion Equation

It can be shown (e.g., Sverdrup et al., 1942) that the local time variation of a conservative property X, in an infinitesimal cube at a fixed position within the oceans, is given by

$$\frac{\partial X}{\partial t} = \frac{\partial}{\partial x}\left(\frac{A_x}{\rho}\frac{\partial X}{\partial x}\right) + \frac{\partial}{\partial y}\left(\frac{A_y}{\rho}\frac{\partial X}{\partial y}\right)$$
$$+ \frac{\partial}{\partial z}\left(\frac{A_z}{\rho}\frac{\partial X}{\partial z}\right) - \frac{\partial(v_x X)}{\partial x}$$
$$+ \frac{\partial(v_y X)}{\partial y} + \frac{\partial(v_z X)}{\partial z} \tag{3.5}$$

x and y are horizontal axes while z is the vertical axis positive when pointing downward. v_x, v_y, and v_z are the components of the advective velocity. Advection is the net motion such as that of currents. A_x, A_y, and A_z are the eddy diffusion or mixing coefficients.

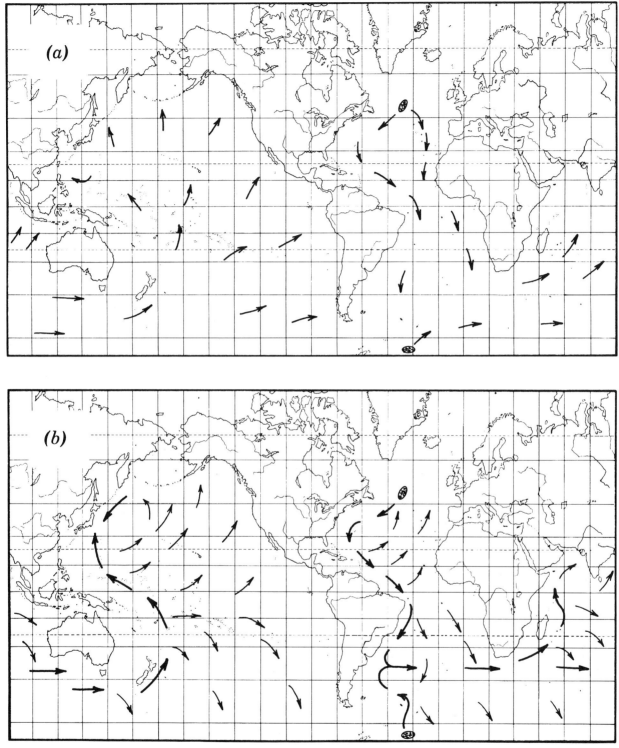

FIGURE 3.13 The propagation of deep waters, below circa 2000 m and above the bottom waters, according to Stommel (1958).

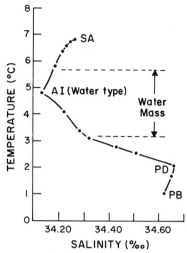

FIGURE 3.14 A *T–S* diagram north of the Antarctica Convergence (Pytkowicz, 1968). SA: Sub-Antarctic Water; AI: Antarctic Intermediate Water; PD: Pacific Deep Water; PB: Pacific Bottom Water.

dX/dt in Equation (3.5) is the local time rate change in the value of *X* per unit volume. The density drops out when unit volumes are not used. The next three terms indicate that part of the change due to mixing with adjacent waters, and the last three correspond to changes caused by advection through the cube. $\partial X/\partial i$ is the gradient of *X* in the direction *i* (*i* = *x*, *y*, *z*).

The expression of eddy diffusion by terms such as those shown in Equation (3.5) is obtained by analogy to the equation for molecular diffusion. Eddy diffusion coefficients, however, are several orders of magnitude larger than coefficients of molecular diffusion. The molecular diffusion is the motion of solutes in water at rest caused by concentration or chemical potential gradients. Eddy diffusion is thought of by hydrodynamicists in terms of fluctuation theory.

When the advective velocity and the eddy diffusion vary little along the three axes, Equation (3.5) is reduced to

$$\frac{\partial X}{\partial t} = \frac{A_x}{\rho}\frac{\partial^2 X}{\partial x^2} + \frac{A_y}{\rho}\frac{\partial^2 X}{\partial y^2} + \frac{A_z}{\rho}\frac{\partial^2 X}{\partial x^2}$$
$$- v_x\frac{\partial X}{\partial x} + v_y\frac{\partial X}{\partial y} + v_z\frac{\partial X}{\partial z} \qquad (3.6)$$

I will return to this equation when the modeling of chemical distributions in the oceans is examined.

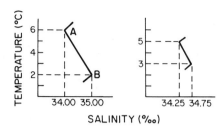

FIGURE 3.15 Hypothetical *T–S‰* diagrams at two positions, 1 and 2, along the direction of motion of two water types *A* and *B*.

3.3.5 Water Fluxes and Residence Times Within the Oceans

Broecker and Li (1970) arrived at the box model shown in Figure 3.16, based upon the dissolved stable carbon and the decay of carbon 14 in oceanic waters. These authors represented the oceans by three boxes only because they did not have enough measured parameters to solve the relevant equations for the fluxes that would occur if a larger number of boxes were considered. They represent all the Surface Waters by the box SW, considered Intermediate, Deep, and Bottom Waters as one in the Atlantic (ADW) as well as in the Pacific plus Indian Oceans (PIDW).

Broecker and Li (1970) did not attempt to distinguish the exchange of SW with ADW as well as with PIDW which occurs in the poorly stratified regions at high latitudes. Exchange across the pycnocline is sluggish at low latitudes, per unit area of the oceans, due to the vertical stability within the pycnocline. It may, however, be significant when extended over the whole oceans.

The residence times found by Broecker and Li (1970) and those found by Bolin and Stommel (1961), who also used box models, are presented in Table 3.5. The residence time of seawater is the average time a water molecule spends in a given reservoir or the renewal time for the waters in the reservoir.

For rough computations, one may use a mean residence time of 1600 yr for the Pacific Ocean according to the more recent work of Broecker (1974) or 1000 yr for all the oceans. Broecker derived and used the equation

$$J_{mix} = \frac{\lambda\,\Delta z\,A_{ocean}}{\dfrac{C^*_{surf}/C_{surf}}{C^*_{deep}/C_{deep}}} \qquad (3.7)$$

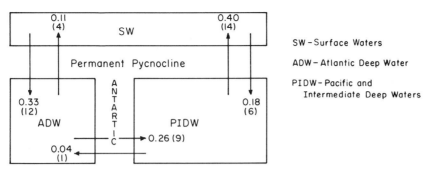

FIGURE 3.16 Box model for the exchange of waters between the wind-mixed layer and the deep oceans [after Broecker and Li (1970)]. The numbers in parentheses are the fluxes in sverdrups (10^6 m^3/sec) and the other ones represent ocean volumes/1000 yr.

to calculate the residence time. $\lambda = (1/8200)$ yr is the fraction of carbon 14 atoms decaying each year; C^*/C is the ratio of the number of carbon 14 atoms to the number of stable carbon atoms which is 19% smaller at depth than near the surface in the Pacific Ocean; J_{mix} is the upward as well as the downward mixing flux between deep waters and the surface layer; Δz is the average thickness of the deep layer (Deep and Bottom Waters); and A_{ocean} is the area of the ocean.

The mean depth of the oceans is 3800 m and Broecker selected 600 m for the depth of the surface (wind-mixed layer), leaving $\Delta z = 3200$ m. He then calculated $J_{mix} = (225$ cm/yr$) \times A_{ocean}$. By dividing $J_{mix}/(\Delta z \, A_{ocean})$ Broecker obtained $\tau = 1600$ yr for the residence time of deep Pacific waters.

A problem arises in the calculations of Broecker because the ratio of C^*/C in the surface layer to that measured in deep waters applies only to great depths. On the other hand, the depth range for which the 19% increase in C^*/C was applied included the faster moving intermediate waters. The increase should in actuality be much lower for intermediate waters (the cores of intermediate waters range from 800 to 1000 m). Therefore, $\tau = 1600$ m probably is an upper limit for the residence time of all waters below the pycnocline.

TABLE 3.5 Residence times (years) for oceanic waters.

Box	Broecker and Li (1970)	Bolin and Stommel (1961)
Antarctic Intermediate Waters		250
Pacific and Indian Deep Waters	1360	1200
Atlantic Deep Water	400	

I prefer about 200 m, the mean upper limit of the pycnocline, as the transition between the wind-mixed (surface) layer because the pycnocline at its onset already presents an effective barrier to vertical mixing (e.g., Pytkowicz, 1964). For rough geochemical computations, I set $\tau = 1000$ yr as the average value for all the oceans. All deep waters must eventually pass through the surface layer because the oceans are well-stirred systems, as is shown later in this chapter. Therefore, the residence time of waters in the WML (wind-mixed layer) must be

$$\tau_{WML} = \frac{200}{3600} \times 1000 \text{ yr} \cong 56 \text{ yr} \qquad (3.8)$$

For the Pacific Ocean it is about 90 yr.

The vertical mixing motion of the waters corresponds to $3600/1000 = 3.6$ m/yr for all the oceans. In other words, 3.6 m of water from below 200 m move up and 3.6 m of surface water move down each year. Note that the radiocarbon data provides the net result of advection (water masses sinking at high latitudes) and mixing so that the 3.6 m/yr is actually equivalent to results of both types of processes. The number of 3600 yr is simply the mean depth, 3800 m, minus the depth of the WML, that is, 200 m.

3.4 ENERGY BUDGET OF THE EARTH

3.4.1 Conservation of Energy and Entropy Increase

The energy budget is important to the understanding of the forces that drive the oceanic circulation and the distribution of chemicals, as well as the relationships between light and biochemical processes. It also affects the hydrological cycle and, consequently, weathering. Furthermore, the earth is a steady-state system rather than an equilibrium one, from the standpoint of its internal (total) energy, because entropy is released to the universe as the result of irreversible work and because there are energy and material fluxes (see Chapter 4). Thus, the earth is an interesting case of the first law of thermodynamics, as well as of irreversible thermodynamics, since its average annual internal energy content changes very little. I shall, therefore, examine the energy budget in some detail.

Our planet can be characterized thermodynamically by V, the volume, and T (see Chapters 1 and 4). Since V and T are invariant on the average, functions of state such as the entropy do not change in time. What happens is that solar energy supplies the driving force for a terrestrial energetic steady state which, as all steady states, causes the degradation of energy and increases the entropy of the universe. The term degradation means that the same amount of energy still exists but, due to less contrast, a smaller fraction is available to do work. At the same time, the energy of the sun decreases.

The entropy generation and the stability condition in the case of the earth do not follow the Prigogine theorem for steady states near equilibrium (Prigogine, 1955) but rather the Prigogine theorem for systems with an inherent stability far from equilibrium (Prigogine, 1978).

The argument above is a simplification in that a net amount of heat resulting from radioactive processes in the interior of the earth is lost to space so that there is a slight gain in the entropy of the earth. This energy loss, however, is very small compared to the solar input.

3.4.2 Global Budget

The average solar energy that reaches the surface of the earth (including the air envelope) is about 8.3×10^{-3} cal/sec · cm^2. The return fluxes are shown in Figure 3.17 with details presented in Table 3.6. These fluxes are based upon data presented by Von Arx (1962), Weyl (1970), and Gross (1972).

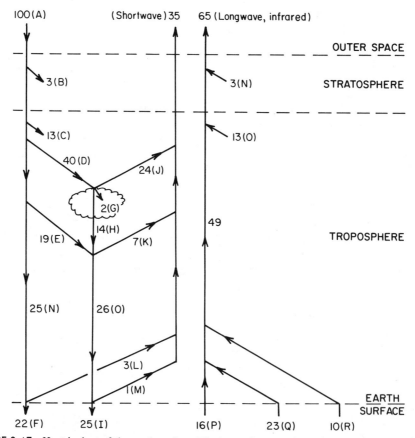

FIGURE 3.17   Heat budget of the earth surface. The letters in parentheses are explained in Table 3.7.

TABLE 3.6 Heat budget of the earth surface (in percentages) corresponding to Figure 3.17.

A. Incoming solar radiation		100	
	Components of Downward Fluxes	Downward Fluxes	Upward Fluxes
B. Absorbed by ozone		3	
C. Absorbed by water vapor and dust		13	
D. Interacts with clouds[a]		40	
E. Back scattered by clouds			−4
F. Adsorbed by clouds	2		
G. Down scattered by clouds	14		
H. Scattered by air, dust and water vapor		19	
I. Upward scatter (In O) Down scattered	12 (part of 0)		−7
J. Direct solar radiation reaching land and sea		25	
K. Direct solar radiation adsorbed by land and sea	22		
L. Direct solar radiation reflected by land and sea			−3
M. Diffuse radiation that reaches land and sea		26[b]	
N. Diffuse radiation absorbed by land and sea	25		
O. Diffuse radiation reflected by land and sea			−1
P. Returned by the stratosphere			−3
Q. Radiated by water vapor and dust			−13
R. Net long-wave radiation (large amount of infrared radiation is bounced back and forth between the atmosphere and the earth surface)			−16
S. Latent heat			−23
T. Sensible heat			−10
		+100	−100

[a] Note that upward fluxes have an algebraic minus sign but that in each box they are a positive part of the net flux. Thus, D = E + F + G or 40 = 24 + 2 + 14.

[b] Not to be counted when adding downward fluxes for the overall balance on the earth because it is a secondary flux resulting from primary fluxes already listed above (E + H − I).

The solar energy, a total of about 700 cal/cm^2 · day (Dietrich, 1963) drives the physical, geological, biological, and chemical processes on land and in the sea. It controls the temperature of the earth surface and the contrast in temperature between high and low latitudes which drives the winds and the oceans; provides the energy requirements for life and many chemical and geochemical processes;

TABLE 3.7 Heat budget of the oceans (Dietrich, 1963), expressed as percentages of the radiation absorbed (295 cal/cm^2 · day).

Direct plus diffuse light	+100
Back radiation	−42
Direct conduction to atmosphere	−7
Lost through evaporation	−51
Heat gained by chemical and biological processes	0.1[a]
Heat gained by friction	0.05
Heat gained from radioactive decay	0.000017
Heat gained from the earth's interior	0.03
Heat lost by advection and mixing	0 regional ≠ 0

[a] And lost in reverse processes such as oxidation.

and drives the hydrological cycle and, consequently, part of the geochemical cycles, and so on.

About 295 cal/cm^2 · deg, or 42% of the total incoming solar radiation reaches the oceans (Dietrich, 1963). Twenty-seven percent of the incoming radiation is absorbed in the first centimeter of seawater, 62% in the top meter, and only 0.45% reaches the 100-m depth, the bottom of the photic zone in clear waters. This energy is reradiated with the following balance shown in Table 3.7.

Of course, the amounts of energy gained are eventually lost in the forms presented in the second through fourth lines of Table 3.7. In addition, an amount of radiation equal to 8% of that absorbed is reflected directly or in a diffuse manner at the sea surface. This is not shown in the table. It is surprising to observe what a small fraction of the energy absorbed is utilized by life. This is due in part to nutrient limitation, the depletion of nitrate and/or phosphate in the photic zone.

The electromagnetic spectrum is shown in Figure 3.18. Solar radiation spans the light spectrum while the energy released by the earth contains a large fraction of infrared radiation as the result of the earth's temperature being lower than the sun's. Fortunately for us, most of the ultraviolet component of the solar radiation is absorbed in the ozone (O_3) layer of the stratosphere, which is present 12–21 miles above sea level. Intense ultraviolet radiation would end life. The ozone is at equilibrium through the reactions

$$O_2 + UV \text{ radiation} \rightarrow O + O \qquad (3.9)$$

$$O_2 + O \rightarrow O_3 \qquad (3.10)$$

$$2O_3 \rightarrow 3O_2 + \text{heat} \qquad (3.11)$$

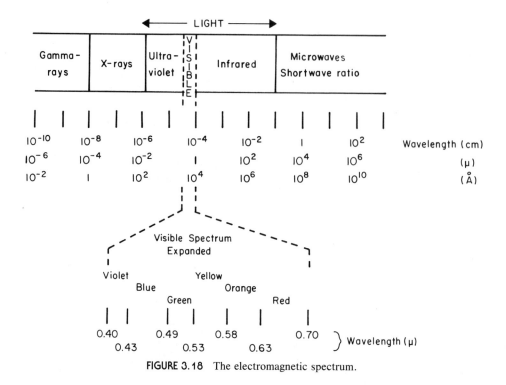

FIGURE 3.18 The electromagnetic spectrum.

The oxygen atoms formed by (3.9) are very reactive and combine with molecular oxygen to form ozone (3.10). The ozone is unstable and breaks down to oxygen (3.11).

Carbon dioxide and water vapor in the atmosphere play an important role in the radiation balance because of the absorption of ultraviolet radiation from the sun and of the infrared radiation radiated from the earth. This absorption of infrared radiation is known as the greenhouse effect and helps maintain the temperature of our environment within acceptable limits. Humankind is now adding considerable amounts of carbon dioxide to the atmosphere through the burning of fossil fuels (oil, coal, natural gas) and may induce a warming of the earth surface through an enhancement of the greenhouse effect. Later on I shall examine the differential absorption of solar radiation in seawater. The term differential refers to the fact that the extent of the absorption depends on the wavelength of the radiation although I can mention at this time that blue light penetrates the deepest in clear waters and reaches about 100 m, the depth of the photic, or surface productive, zone.

3.4.3 Regional Budget

So far I have dealt with the overall budget. Now I will examine regional energy budgets before pro-

ceeding to the water transport they cause. In regional terms one finds that the energy absorbed per cm^2 of the earth (atmosphere, land, and sea) passes through a maximum at about 10°N (Table 3.8). This happens in part because the equatorial clouds and the ice and snow covers at high latitudes are strong reflectors. The reflecting power (albedo or reflectivity) decreases the gross radiation absorbed in these regions (second column in Table 3.8). The decreasing trend toward high latitudes, however, results primarily from the relatively low solar angle toward the poles.

TABLE 3.8 Annual mean heat radiation (in $cal/cm^2 \cdot min$) (after Houghton, 1954).

North	Radiation Absorbed by Atmosphere, Land, and Sea	Long-Wave Radiation to Space	Net Absorption
0	573	488	85
10	578	502	76
20	574	503	71
30	532	492	40
40	444	469	−25
50	352	442	−90
60	261	419	−158
70	192	400	−208
80	147	385	−238
90	117	380	−263

TABLE 3.9 Poleward atmospheric and oceanic transport of heat across latitude circles, in 10^{19} cal/day (Houghton, 1954).

Degrees North Latitude	Heat Flux	Degrees North Latitude	Heat Flux
0	0.0	50	9.61
10	4.05	60	6.68
20	7.68	70	3.41
30	10.46	80	0.94
40	11.12	90	0.0

TABLE 3.10 Poleward transport of heat by the atmosphere and the oceans across latitude circles, in 10^{19} cal/day.

Degrees North Latitude	North Atlantic	North Pacific	Atmosphere
0	0.6	0.0	0.0
10	1.2	1.0	1.9
20	1.6	0.9	5.2
30	1.9	0.6	8.0
40	1.1	0.0	10.0
50	0.9	0.0	8.7
60	0.4	0.0	6.3

Source: Von Arx, W. C. *An Introduction to Physical Oceanography,* © 1962. Addison-Wesley, Reading, MA. Table 6-1. Reprinted with permission.

Note that the quantities in Table 3.8 are fluxes (cal/cm^2 · min). Thus, they do not reflect the higher surface areas of the earth at low latitudes. Still, the net gain of energy at low latitudes and loss at high ones means that there must be a net transport of heat by the atmosphere and the oceans toward high latitudes (see Table 3.9). The breakdown of this transport into oceanic and atmospheric fluxes is presented in Table 3.10.

The nonuniform heat absorption (and radiation) by the atmosphere, the sea, and the land provides the forces that generate winds and wind-driven oceanic surface currents, as well as the thermohaline gradients that cause the motions of deep water masses. In terms of regions within the oceans at average annual steady states the following equation must be satisfied,

$$Q_1 + Q_2 + W = Q_3 + Q_4 + Q_5 \quad (3.12)$$

where Q_1 = absorbed solar radiation
Q_2 = net heat brought by surface currents and deep water masses
W = work done by wind drag on the ocean
Q_3 = heat radiated
Q_4 = heat lost by evaporation
Q_5 = sensible heat transferred to the atmosphere

There is no term for the frictional heat generated during the turbulent motions of the waters because it appears in the terms on the right of Equation (3.12). I should add that I neglected geothermal heating in this section.

As the result of the processes described above, the annual average temperature of the sea surface ranges from slightly below zero at 70°S and 60°N to about 28°C at the equator.

Equation (3.12) is an expression of the conserva-

tion of energy, that is, of the first law of thermodynamics (see Chapter 4). From the standpoint of the second law, one observes that the annual average thermodynamic state of the system (atmosphere + land surface + oceans) remains constant, if one neglects slow geological and climatic trends. The state can be defined by P, T or V, T plus the mole fractions or, even better, the chemical potentials of the components. This means that the entropy of the system also remains constant. The system is at a steady state rather than at equilibrium and, therefore, it must generate entropy. I conclude, therefore, that

$$\frac{dS}{dt} \text{ (system)} + \frac{dS}{dt} \text{ (surroundings)}$$

$$= \frac{dS}{dt} \text{ (surroundings)} > 0 \quad (3.13)$$

The surroundings include the sun and space. Note that in this argument I am not including heat exchange between the crust of the earth and the deeper mantle.

3.4.4 Evaporation, Precipitation, and the Latitudinal Transport of Oceanic Waters

The key to the thermohaline circulation is that the solar energy absorbed by the atmosphere and the ocean affects the change in temperature and salinity of surface waters with latitude. The salinity gradients respond to the solar input through evaporation and precipitation.

The mean annual precipitation exceeds evapora-

TABLE 3.11 Average annual evaporation, precipitation, net evaporation, surface salinity, temperature, and σ_T (Wüst, 1954) for all oceans.

Zone (deg)	Precipitation (cm/yr)	Evaporation (cm/yr)	Net Evaporation (cm/yr)	Salinity (‰)	Temperature (°C)	σ_T
70–65N	34	12	−22	33.4	2.1	26.71
65–60	65	20	−45	32.35	3.7	25.73
60–55	77	34	−43	32.66	5.2	25.84
55–50	105	55	−50	33.41	7.0	26.19
50–45	112	66	−46	33.69	9.2	26.08
45–40	102	84	−18	34.14	13.2	25.70
40–35	86	108	22	35.11	17.6	25.48
35–30	74	125	51	35.50	20.5	25.03
30–25	63	132	69	35.76	22.7	24.62
25–20	57	137	80	35.66	24.6	23.98
20–15	70	135	65	35.14	26.0	23.16
15–10	103	132	29	34.76	26.9	22.59
10–5	187	126	−61	34.43	27.4	22.18
5–0	146	113	−33	34.73	27.2	22.47
0–5	105	125	20	35.07	26.9	22.85
5–10	109	137	28	35.25	26.5	23.09
10–15	94	139	45	35.42	25.8	23.42
15–20	76	137	61	35.62	24.6	23.96
20–25	68	133	65	35.74	23.0	24.51
25–30	65	123	58	35.68	21.1	24.99
30–35	70	110	40	35.46	18.5	25.52
35–40	90	96	6	35.04	15.6	25.89
40–45	110	78	−32	34.54	11.8	26.29
45–50	117	56	−61	34.14	7.7	26.58
50–55	109	39	−70	33.96	4.4	26.94
55–60S	84	12	−72	33.94	1.7	27.18
70–0N	101.0	110.6	9.6	34.71	21.06	24.01
0–60S	91.45	102.1	10.7	35.03	17.99	24.99

tion for latitudes above 40° north and south and evaporation is greater than precipitation between 40° and the equator (except for a belt between 0° and 10° north). This causes the surface salinity to be high in the tropical regions and decrease in the equatorial region and toward high latitudes (Table 3.11).

The salinity effect causes a decrease in density with increasing latitude but is compensated by the lowering temperature so that sigma-t increases. There is, therefore, a sinking of intermediate and deep waters at high latitudes. An exception is that of Mediterranean waters. They become quite saline due to high evaporation and low runoff and, after crossing the Straits of Gibraltar, sink toward the upper zone of the Atlantic Deep Water.

Next, I will estimate part of the hydrological cycle within the oceans based upon Table 3.11, since the excess net precipitation at high latitudes and in

the Equatorial Zone must flow into the Tropical Zones. The excess precipitation in cm/year in each latitudinal segment for all the oceans minus the net evaporation (Table 3.11) times the area of the segment yields the net precipitation in g/yr (Table 3.12) if we assume roughly 1 g-H_2O/cm³-SW. Table 3.12 yields a net evaporation of -3.15×10^{19} g/yr from the oceans between 70°N and 60°S. This is slightly smaller than -3.6×10^{19} g/yr obtained from Figure 3.19 in part due to the uncertainty in the numbers, although the waters present above 70°N may contribute to the difference.

In Figure 3.19 the water fluxes between latitudinal zones, calculated from Table 3.12, are shown. The numbers represent very rough order of magnitude estimates of that part of the flow due to evaporation and precipitation, because we have no reliable coverage above 70°N and 60°S. An improved estimate can be made by actually calculating the

TABLE 3.12 Area of latitudinal zones and net precipitation.

Zone (deg)	Water Area (10^{16} cm^2)	Net Precipitation (10^{19} g/yr)			Zone (deg)	Water Area (10^{16} cm^2)	Net Precipitation (10^{19} g/yr)
90–85N	0.979						
85–80	2.545						
80–75	3.742				70–75S	0.522	
75–70	4.414				75–70	2.604	
70–65	2.456	0.0540			70–65	6.816	
65–60	3.123	0.140			65–60	10.301	
60–55	5.399	0.232			60–55	12.006	0.864
55–50	5.529	0.276			55–50	13.388	0.937
50–45	5.512	0.304			50–45	14.693	0.896
45–40	8.411	0.151	1.16		45–40	15.833	0.507
40–35	10.029	−0.220			40–35	16.483	0.0989
35–30	10.806	−0.551			35–30	15.782	−0.631
30–25	11.747	−0.811			30–25	15.438	−0.895
25–20	13.354	−1.068			25–20	15.450	−1.004
20–15	14.981	−0.974			20–15	16.147	−0.985
15–10	16.553	−0.480	−4.0		15–10	17.211	−0.774
10–5	16.628	1.014			10–5	16.898	−0.473
5–0	17.387	0.574	1.59		5–0	16.792	−0.336
Total	154.695	−1.359				206.364	−1.795

Source: Sverdrup, H. V., M. W. Johnson, and R. H. Fleming, *The Oceans: Their Physics, Chemistry, and General Biology,* © 1942, renewed 1970, p. 13. Reprinted by permission of Prentice-Hall, Inc., Englewood Cliffs, NJ.

river fluxes into the latitudinal oceanic zones. Note that NR is almost equal to the global stream flux.

Complete water fluxes for all the oceans at all depths have not been established although the Atlantic Ocean is best known. Table 3.13 presents values in the Atlantic Ocean (according to Sverdrup et al., 1942).

These types of estimates are useful for calculations of chemical transports and would be of even greater value if made for the different oceans.

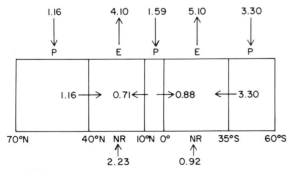

FIGURE 3.19 Net water fluxes due to evaporation and precipitation in 10^{19} g/yr (*P:* precipitation; *E:* evaporation; *NR:* net river over river flow into regions of net precipitation). Fluxes due to cooling and freezing and to the wind stress are not included.

Let us examine two cases of transport of chemicals by waters.

EXAMPLE 3.3. MERIDIONAL TRANSPORT OF TPO$_4$ IN THE ATLANTIC OCEAN

$$T\text{PO}_4 = [\text{H}_3\text{PO}_4] + [\text{H}_2\text{PO}_4^-]$$
$$+ [\text{HPO}_4^{2-}] + [\text{PO}_4^{3-}] \tag{i}$$

It is well known that TPO$_4$ has low values near the sea surface because of utilization by organisms, passes through a maximum at intermediate levels, and decreases a little toward the bottom. The surface values vary with season and location. As a rough illustration, the phosphate content (in μg-atom/L) of Atlantic waters is:

Surface	Intermediate	Deep	Bottom Zone
0.2	3.2	2.8	2.6

One obtains for the fluxes of TPO$_4$, in 10^9 μg-atom/ sec, from the water transport at the equator (see Table 3.13),

TABLE 3.13 Meridional transport of water in the Atlantic Ocean (10^6 m³/sec).

	At 30°S	
	Moving North	Moving South
Surface currents	23	17
Intermediate	9	
Deep		18
Bottom	3	

	At the Equator	
	Moving North	Moving South
Surface currents	6	
Intermediate	2	
Deep		9
Bottom	1	

Surface	Intermediate	Deep	Bottom Zone
1.2(N)	6.4(N)	25.2(S)	2.6(N)

where the symbols N and S stand for north and south. The net transport is 15×10^9 μg/sec toward the south.

When an equivalent calculation is made for 30°N, one obtains:

$TPO_4 = 0.2$ μg-atom/L		1.0 μg-atom/L
Equator	-15×10^{19} μg-atom/sec	-10.2
30°S	-1.26	$+3.54$

where the minus sign represents southward transport.

It is obvious that the values at the equator and at 30°S are not compatible because they indicate a major sink for TPO_4 between these two latitudes, while the sedimentation of phosphates in biogenic apatites and in other minerals is slight. The TPO_4 inputs for the example are reasonably close to reality and, therefore, it appears that the water transport data are off. Still, Example 3.3 serves to illustrate a method and reveal inconsistencies in our knowledge.

EXAMPLE 3.4. THE TRANSPORT OF CALCIUM. The North Atlantic Ocean receives the largest supply of calcium from rivers because of the large continental area that acts as a drainage basin. This calcium sediments in part as biogenic $CaCO_3$ but most of it dis-solves below the carbonate compensation depth. The calcium that has not been taken up by organisms and the dissolved calcium enter the Indian and the Pacific Oceans in which the river flux is smaller. Because of this gradient between oceans, a net amount of calcium is constantly added to the latter two oceans. From a net standpoint a steady-state calcium budget obeys the relations:

$$\text{Atlantic,} \quad R - S - T = 0 \quad \text{(i)}$$

$$\text{Pacific and Indian,} \quad R - S + T = 0 \quad \text{(ii)}$$

R is the river input, S is the sedimentation, and $-T$ is the transport to other oceans.

According to Olausson (1971) the input of calcium into the oceans is

Atlantic		Pacific	Indian	
North	South			
32.2	3.7	5.5	4.6	(10^{13} g/yr)
70	10	12	10	(% of total)

so that 70% of all calcium flux enters the North Atlantic. The accumulation in the sediments of calcium, in percentages of the total deposition, is

Atlantic		Pacific	Indian
North	South		
35	20	25	20

Thus, we see that the positive input into the Pacific and Indian Oceans is reflected in a surplus sedimentation in these two oceans.

Note that in any segment of the earth the internal energy E is constant on a yearly average so that $\Delta E = Q - W = 0$. The symbol Q represents the heat absorbed and W is the work done by the system. Thus, in such a segment, $Q = W$. For the earth surface as a whole, $Q = 0$ because the energy absorbed from the sun equals the heat released by the earth. Therefore, $W = 0$. This simply means that, for each water parcel that sinks z m there is one that upwells z m, and that for each liter evaporated one precipitates. Furthermore, the photosynthetic energy is released during oxidation and the chemical pathways are cyclic.

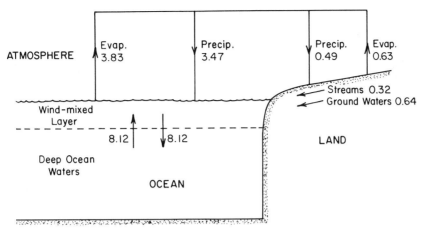

FIGURE 3.20   The hydrological cycle [based in part on Garrels and Mackenzie (1971)] in 10^{20} g-H$_2$O/yr.

3.4.5   A Further Note on the Hydrological Cycle

The components of the water cycle in nature are to a large measure the result of the unequal distribution of heat on the surface of the earth.

The general hydrological cycle in nature is presented in Figure 3.20. The flux of waters across the wind-mixed layer boundary with deep waters was calculated from $J_{mix} \times A_{ocean} = 8.12 \times 10^{20}$ cm^3/yr, which corresponds roughly to 8.12×10^{20} g-H$_2$O/yr.

The water content of the various reservoirs is shown in Table 3.14. The residence times in Table 3.14 were obtained from $\tau = R/J$ where R is the reservoir content and J is the total flux into or out of it.

3.4.6   Absorption of Light

I will examine the absorption of visible radiation in seawater before proceeding to a brief review of life in the oceans because light is necessary for marine photosynthesis.

The percentage of light of different wavelengths absorbed by pure water is shown in Figure 3.21. Carefully filtered seawater has an absorption similar to that of pure water, with little transmission of light in the red and in the infrared and considerable transmission in the blue. The percentage of transmission of light versus depth is shown for filtered seawater in Figure 3.22. The light that enters the sea surface is absorbed or scattered at wavelengths that depend on the suspended load. Ultraviolet and infrared as well as red radiation are absorbed readily, whereas blue-green light of wavelengths in the range 0.4–0.6 μm penetrates to greater depths (ca. 100 m) as is shown in Figure 3.22. Only about 1% of the incident radiation reaches this depth. In waters that are turbid due to high productivity or a high particulate load, however, green-blue light is scattered and red light is the one that penetrates the seawater the furthest (0.63–0.7 μm). This shift alters photosynthetic rates since photosynthetic pigments are dominant at >0.6 μm, whereas accessory pigments absorb at less than 0.6 μm (Parsons and Takahashi,

TABLE 3.14   Water content of natural reservoirs in 10^{20} g.

Reservoir	Water Content	Residence Time (yr)
Oceans	13,750	3.590
Wind-mixed layer	722	1.89
Deep waters	13,028	1,600
Pore waters of sediments	3,300	
Ice	200	
Rivers and lakes	0.3	0.83
Atmosphere	0.13	0.081

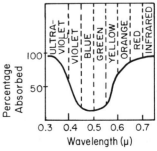

FIGURE 3.21   Sketch of the fraction of light transmitted below the upper meter of clear water.

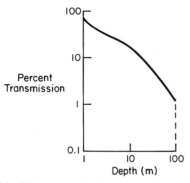

FIGURE 3.22 Light transmission versus depth [after Weyl (1970)] in filtered seawater.

1973). The light compensation depth may be as shallow as 1 m in very turbid waters.

The amount of light which impinges on the surface depends, of course, on latitude, season, and time of day as well as on the cloud cover. For a given intensity and spectral distribution of light reaching the sea surface and for a given turbidity of the waters, there will be a depth (the light compensation depth) above which photosynthesis exceeds oxidation. Photosynthesis exceeds the breakdown of biological matter above this depth and the reverse is true below it. The light intensity at this depth is known as the compensation intensity. The light compensation depth ranges from about 1 m in very turbid coastal waters to 100 m in clear oceanic waters, as we saw above.

Waters above the light compensation depth constitute the photic zone.

EXAMPLE 3.5. WHAT HAPPENS TO NUTRIENTS SUCH AS TPO_4 AND TO OXYGEN IF THE TURBIDITY IS SUCH THAT THE LIGHT COMPENSATION DEPTH IS 10 m WHILE THE WML EXTENDS TO 200 m (THE DEPTHS OF THE PERMANENT PYCNOCLINE)? The depth of the photic zone is 10 m and photosynthesis predominates over oxidation in it. Thus, $[O_2]$ is high and TPO_4 is low. I am assuming that the photic zone is not warm enough to unload a significant fraction of its oxygen to the atmosphere. This unloading happens in some instances because the solubility of oxygen decreases with increasing temperatures.

Below 10 m, oxidation is dominant since there is essentially no photosynthesis. Thus, oxygen is utilized and TPO_4 is released in the breakdown of settling dead organic matter. The oxygen deficit and the nutrients released at depth above 200 m are within the WML and are, therefore, recycled to the photic zone. The recycled TPO_4 and TNO_3 act as fertilizers.

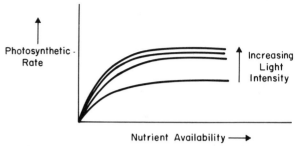

FIGURE 3.23   Light, nutrients, and photosynthetic rate.

EXAMPLE 3.6.   WHAT HAPPENS WHEN THE PHOTIC ZONE EXTENDS TO 100 m BUT THE SEASONAL SUMMER THERMOCLINE STARTS AT 50 m? Those fractions of the dead organisms and fecal pellets that are not oxidized above 50 m yield nutrients that become essentially unavailable to the upper 50 m where most of the photosynthesis takes place. The reason for this is that the pycnocline is an effective barrier to the mixing of waters.

Although light is the controlling factor for photosynthesis, when there is a plentiful supply of nutrients, the chemicals necessary for the maintenance of life and low temperatures can at times be limiting. This is shown in Figure 3.23. Eventually one reaches a light saturation state beyond which a further increase in intensity not only does not increase the photosynthetic rate further but may actually harm life and slow down photosynthesis (Parsons and Takahashi, 1973).

3.5   LIFE AT SEA

3.5.1   Biomass and Ecosystems

The biomass is the standing stock of living matter of interest in a given reservoir. The part of the crust of the earth which contains living organisms and a substantial amount of their products is known as the biosphere. It plays an important role in the control of chemical fluxes in nature. As an example, the atmospheric oxygen is primarily generated by the very slight excess photosynthesis over oxidation in the marine biota, as was shown earlier.

An ecosystem is a community of organisms that interact with each other and their chemical and energetic environments. Its material content is relatively self-contained and stable. Examples are a lake or a forest plus solar radiation and other energy fluxes. A river, if it contributes a large fraction of the organisms present in a lake, is part of the lacustrian ecosystem.

Ecosystems vary with the time of day, the seasons, and trends. They do have a measure of stability which increases with the number of species present. Thus, a single crop can be wiped out by a virus disease that is characteristic of that crop but mixed crops will survive partially. Nonperturbed ecosystems tend toward steady states rather than equilibria on a yearly average because energy and matter flow through them. Single organisms, however, are transient. Ecosystems are mostly cyclic through the regeneration of nutrients, for example, the Amazon Basin, but not entirely so as can be seen by the case of a river flowing through a lake and bringing organisms and nutrients to it while removing them in the outflow.

The oceans can be considered as ecosystems or one may make distinctions between plankton living near the sea surface and the benthos on the seafloor. This distinction is not clear-cut because benthic organisms feed in part on settling plankton residues. One may also consider the surface fauna and waters of the Southern Ocean as partly separated from those of the Pacific Ocean by the Antarctica Convergence.

Note that the biomass does not keep on growing continuously. Net growth shows a diurnal effect because photosynthesis ceases at night when the opposite effects, death with decay and respiration, take over, and a seasonal effect because most growth occurs during the spring bloom and summer growth. During winter, at least in high latitudes, oxidation prevails and the TPO_4 of the waters increases due to the mineralization of organisms. The oxygen–phosphate seasonal patterns can be seen in Redfield (1948) and in Pytkowicz (1964).

The total mass of living matter in the oceans, the marine biomass, is of the order of 2×10^9 g-C, expressed as grams of carbon. The complex of organisms plus the environmental factors with which they interact is known as the ecosystem (organisms, light, temperature, nutrients, etc.). The distributions of nutrients and their geochemical cycles will be discussed later on.

3.5.2 Light and Photosynthetic Efficiency [based in part on Curl and Small (manuscript)]

One measure of the photosynthetic efficiency is the ratio of P, the gross photosynthesis, to L, the light absorbed by the chloroplasts where photosynthesis occurs. This efficiency depends on the color of the light and increases toward the red.

EXAMPLE 3.7 THE PHOTOSYNTHETIC EFFICIENCY OF RED LIGHT (650 μm). One mole-quantum, that is, $N_A \times$ (energy of the quantum) has an energy equivalent of 44 kcal/mole-quantum. The energy required to reduce 1 mole of CO_2 in the reaction $CO_2 + H_2O = CH_2O + O_2$ is roughly 110 kcal/mole.

If 10 quanta are required to reduce 1 mole of CO_2 then

$$\frac{P}{L} \times 100\% = \frac{110 \text{ kcal/mol}}{(44 \text{ kcal/mole-quantum}) \times 10 \text{ quanta}} \cong 25\% \quad \text{(i)}$$

The average P/L of the oceans is about 19% and it sets an upper limit to productivity if nutrients are not growth limiting.

The quantum of light is simply $h\nu$ where h is the Planck constant ($h = 6.6252 \times 10^{-27}$ erg · sec in the cgs system) and ν is the light frequence in sec^{-1}. Quantum and photon are synonyms.

3.5.3 Classification of Organisms

Organisms may be classified, with regard to their role in the ecosystem and the biochemical cycle, into the following:

1. *Producers.* They synthesize new organic matter from inorganic chemicals such as carbon dioxide, water, and salts (autotrophic nutrition). Examples include phytoplankton and seaweeds.

2. *Consumers.* These organisms absorb the organic matter synthesized by the producers for their body building and energetic requirements (heterotrophic nutrition). An example includes nekton (the fish).

3. *Reducers.* They use organic matter, and in this sense are consumers, and break it down into water, carbon dioxide, and inorganic salts (mineralization). Examples include bacteria and molds.

The marine and lake organisms are also classified according to their role in the ecosystem and to their habitat into the following:

1. *Plankton.* Small (microscopic in large measure) organisms that float or drift with the waters, although some have a slight motion of their own. They are present in communities. Phytoplankton are mainly plant cells although some species are heterotrophic and difficult to distinguish from animals. Zooplankton are animals that graze on phytoplankton.

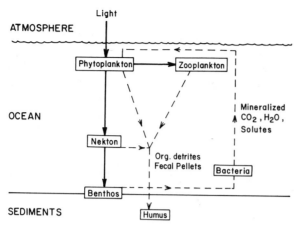

FIGURE 3.24 Organic pathways. Solid lines represent transfer of organic matter between trophic levels.

2. *Seaweeds.* They are large plants, floating or attached. There are green, brown, and red algae. Some green algae are coralline because they secrete $CaCO_3$ and contribute to coral reefs.

3. *Nekton.* These are animals larger than the zooplankton. They control their own motion (fishes of the sea).

4. *Benthos.* These are organisms that live on or in the sea bottom.

The term pelagic is used for organisms present in the open oceans.

The phytoplankton constitute the first trophic level because they build up the organic matter that is used to form the next levels, that is, the zooplankton and the nekton plus benthos. They are the basis for all marine life, the pasturage of the sea. The organic pathways are shown schematically in Figure 3.24.

The main planktonic forms are described next. This presentation is not required for an understanding of this book but provides useful background knowledge and a feeling for the variety of plankton.

Phytoplankton

Diatoms. These are unicellular algae in capsules (frustules) impregnated with silica. They are found everywhere in open surface waters, especially in the colder ones. They range in size from 10 μm to 1 mm. These organisms are important in the silica budget of the oceans.

Dinoflagellates. These are unicellular algae enveloped by cellulose membrane. They are second in

importance to diatoms and are present in all oceans below the euphotic zone. They are, therefore, heterotrophs (consumers).

Coccolithophores. These are very small organisms (down to 0.005 mm in diameter) with calcareous ($CaCO_3$) disks (coccoliths) on their surface. They exist mostly in tropical and semitropical waters although there are some cold water species. They are mainly photosynthetic, although some heterotrophic forms are found below the euphotic zone. Their tests form part of the globigerina ooze, an important calcareous sediment.

Silicoflagellates. These are very small with a siliceous skeleton and one long flagellum (whiplike extension).

Blue-green Algae. These are simple cellular structures with a chitin wall. May be rounded single cells, filaments, plates, etc. Are of local importance in the oceans.

Zooplankton. These graze on phytoplankton and spend feeding periods (nights) in the euphotic zone. They migrate downward several hundred meters during the day to protect themselves from predators, and to lower their metabolic rates in colder waters during their rest period. The smaller ones are protozoa (microscopic animals, often one celled).

Foraminifera. These are surrounded by calcareous shells (tests) with holes through which they push temporary extensions when the organisms are active. Foraminifera are, in general, less than 1 mm in diameter or length although some species reach a diameter of several millimeters. Foraminifera are most common in warm waters and are a major source of calcareous sediments.

Radiolaria. These are spherical pelagic organisms, 50 μm to several millimeters in diameter, with siliceous skeletons.

Tintinnidae. These live within chitineous envelopes (lorica).

Crustacea. These are the most important zooplankton and are metazoa, that is, they are larger, more complex organisms than protozoa. Crustacea include copepods which constitute the bulk of the zooplankton. Several thousand species have been identified. Euphausiids include the important shrimplike krill.

3.5.4 Photosynthesis and Productivity

The photosynthetic reactions are very complex and involve a large number of steps (Calvin and Baasham, 1962). In their simplest forms they can be represented by the equations

$$CO_2 + H_2O \xrightarrow{\text{light}} \underset{\text{carbohydrate}}{CH_2O} + O_2 \qquad (3.14)$$

or

$$6CO_2 + 6H_2O \xrightarrow{\text{light}} \underset{\substack{\text{sugar} \\ \text{(glucose)}}}{C_6H_{12}O_6} + 6O_2 \qquad (3.15)$$

in terms of the production of carbohydrates. Sugars are examples of carbohydrates. A more detailed formulation, of the formation of phytoplankton matter, which includes the uptake of phosphorus and nitrogen, is

$$106CO_2 + 122H_2O + 16HNO_3 + H_3PO_4$$
$$= (CH_2O)_{106}(NH_3)_{16}H_3PO_4 + 138O_2 \qquad (3.16)$$

This equation yields $\Delta O_2/\Delta CO_2 = -1.23$ instead of -1.00, the ratio from Equation (3.14). The N/P ratio in reaction (3.16) is roughly 16, an important number in the modeling of biochemical fluxes in the oceans.

Let us consider the constituents of reaction (3.16). Nitrogen and phosphorus in oceans and lakes can become limiting nutrients since excessive utilization/supply may exhaust one of them. This can lead in extreme conditions to a great reduction or to the cessation of life. Carbon is usually present in excess and is not limiting. Oxygen may become depleted in semi-isolated environments such as some lakes, fjords with shallow sills, and basins at the bottom of the sea. When oxygen is completely depleted, bacteria turn to sulfate as a source of energy and produce sulfide. Plankton cannot exist under such anoxic conditions.

Anoxic regimes can be accelerated by the oversupply through streams of phosphates and nitrates from sewage, agricultural fertilizers, and household activities (e.g., the use of detergents). This oversup-ply accelerates life as well as the excessive utilization of oxygen for the oxidation of the resulting organic matter. The resulting overproduction and the replacement of oxygen by sulfide is known as eutrophication.

The replacement of oxygen by sulfide changes the chemical nature of sediments as we shall see in Chapters 7 and 8, Volume II.

Next, let us consider the energy associated to Equation (3.16). We saw that one measure of the photosynthetic efficiency is the productivity ratio of the energy that reaches chloroplasts. Thus, it is a measure of the physiological efficiency.

The quantum efficiency, on the other hand, is the fraction of the radiation absorbed by the sea which is used for photosynthetic processes. It is a broad measure of the marine biological capacity to utilize energy. The quantum efficiency is about 1 to 2%, that is, 1.1 to 2.2×10^4 kcal of solar radiation must be absorbed by the sea to reduce 1 mole of CO_2, because this reduction requires 110 kcal/mole. The quantum efficiency is greatest for blue light, in contrast to the photosynthetic efficiency because of the greater absorption of blue light by seawater.

EXAMPLE 3.8. A ROUGH ESTIMATE OF THE QUANTUM EFFICIENCY OF PRIMARY PRODUCTION IN THE SEA

I.   Total radiation that reaches the surface of the earth (Weyl, 1970): 8.3×10^{-3} cal/cm^2 · sec.

II. Radiation that penetrates the sea and the land: 3.9×10^{-3} cal/cm^2 · sec.

III. Ratio of ocean surface to global surface: $361.1/510 = 0.708$.

IV. Radiation absorbed by the sea: $0.708 \times$ (II) $= 2.8 \times 10^{-3}$ cal/cm^2 · sec.

Land and sea are not uniformly distributed with latitude but the energy absorbed per unit area changes with distance from the equator. Furthermore, the absorption per unit area of land and water are different. These effects were neglected for simplicity.

V.   Gross photosynthesis in the sea (Pytkowicz, 1967): 1260×10^{14} g-C/yr.

VI. Energy required for photosynthesis per mole: 110 kcal/mole of CO_2 reduced.

VII. Total photosynthetic energy: (VI) $\times$ (VII) $= 1.40 \times 10^{18}$ kcal/yr.

VIII. Radiation taken up by the oceans: (IV) × area in cm^2 × sec/yr = 2.36×10^{20} kcal/yr.

IX. Quantum efficiency: (VII)/(VIII) = 0.6%.

The quantum efficiency of the sea is approximately equal to that of good crops on land. It appears that the most efficient photosynthetic production is a long-term low-level one rather than an intense rate of formation of organic matter. Thus, because of the low photosynthetic efficiency, the potential food production from the oceans may not be as large as one may think. This is especially true when one realizes that the mass of each trophic level is, as a rough rule of thumb, about 10% of that of the preceding level. Nekton, which include edible fish, as a result of this reduction, have a mass that is 1% of that of the phytoplankton and correspond indirectly to a 0.01–0.02% quantum efficiency.

The energy that drives the global biota is very small when compared to the solar input. It is also small relative to the energies of the hydrological and geochemical cycles, which are driven more efficiently by the energy of the sun and by the thermal energy from the interior of the earth.

Humans receive the smallest fraction of the food energy in nature because they are at the apex of the biotic chain. The physiological external (muscular) work that an animal can do is only part of its intake because of respiration and elimination.

Yet, humans can generate work, that is, utilize energy many times larger than their food intake, in contrast to other animals. The reason for this is the utilization of available external energy (nuclear, hydraulic, fossil fuel, hydrogen). As an example, a driver does much less work than his car.

The difference between the utilization of nonfood energy source and the physiological external source is a foundation of our civilization and also explains why humans, through mining as an example, have become significant geological agents.

EXAMPLE 3.9. THE HUMAN PER CAPITA CONSUMPTION OF ENERGY IN THE UNITED STATES IN 1970.

I. Units used:
    1 barrel of oil equivalent
    (B/DOE) = 5.8×10^6 Btu
    1 Btu   = 1/3.412 kW (kilowatt-hour)

1 Btu	= 1055 J = 252 cal
	= 32.5 (B/DOE)/day
fuel consumption	= 32.5
	= 4.75×10^7 kcal/day (natural gas, oil, coal, geothermal,* nuclear)

Note that only part of the energy consumed becomes useful work due to losses. Furthermore, hydroelectric energy was not included. Thus, the results are approximate.

II. Population $\cong 2.03 \times 10^8$ persons.

III. Per capita annual energy consumption (fuels plus hydroelectric) = 365 × 4.75 × $10^7/2.03 \times 10^8$ = 81.5 kcal/person-yr.

IV. Dietary supply = 3.33 kcal/person-day = 1.22 kcal/person-yr.

Thus, we see that the consumption of fuels in the United States in 1970, excluding hydroelectric power, is 81.5/1.22 = 66.8 times the food energy.

It is interesting to observe that the photosynthetic process belongs to the class of redox reactions to be studied in Chapter 7, Volume II. The CO_2 is reduced and O_2 is oxidized. The oxygen evolved during its course comes from the water, as can be seen in Equations (3.15) and (3.16).

One must distinguish between total productivity, the total grams of carbon taken up by organisms per unit time, and net productivity (which is the difference between the total productivity and the part of it used for respiration). For primary productivity (the production of phytoplankton) the net is, on the average, roughly 90% of the total when the same units are used for both quantities. The primary net productivities of different types of marine regions, according to Ryther (1960), are presented in Table 3.15.

3.5.5 Fertility and Chemistry, A First View

From a chemical point of view it is interesting to observe that the mean net productivity is exceedingly high in the upwelling zone. This happens because upwelled waters are rich in nutrients, such as phosphates, which permit higher yields of phytoplankton than in other regions. The nutrients themselves form a cycle as settling organic debris are

*Does not strictly belong to the fuel category

TABLE 3.15 Primary productivity [from Ryther (1960)].

Province	Percentage of Oceanic Area	Area (10^6 km^2)	Mean Net Productivity (g-C/m$^2 \cdot$ yr)	Total Net Productivity (10^9 tons-C/yr)
Open ocean	90	326	50	16.3
Coastal zone	9.9	36	100	3.6
Upwelling zone	0.1	3.6	300	0.1
Total				20.0

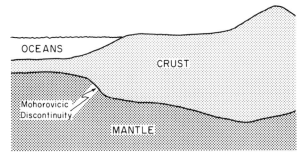

FIGURE 3.25 The crust of the earth.

mineralized and upwelling waters return the nutrients to the surface.

Note that the net primary productivity does not accumulate organic matter (increase the standing stock of plankton) year in and year out since most of the organisms formed in spring and in summer decay during fall and winter. The carbon fixed by photosynthesis and its subsequent passways were examined in Chapter 2. A further study of the role of nutrients will be undertaken later in this chapter. I may mention at this point that a net photosynthetic rate of 200×10^{14} g-C/yr corresponds to 16.7×10^{14} moles-C/yr and, according to the molar proportions in Equation (3.16), to an annual utilization of $(0.067/5.7) \times 10^{14} = 0.0117 \times 10^{14}$ moles-P/yr and 0.175×10^{14} moles-N/yr.

Actually, the amounts of C, P, and N need not be as great as it may appear from the above numbers because the phytoplankton population is recycled about five times in a year. Thus, the amounts of nutrients required are of the order of C = 40, P = 0.0023, and N = 0.035 in 10^{14} moles/yr. Still, this shows the need for large quantities of inorganic phosphate and nitrate in the photic zone if photosynthesis is to proceed.

The term oxidation will be used interchangeably with mineralization when dealing with the biota. Most of the organic matter incorporated into phytoplankton is oxidized in the photic or euphotic zone (roughly the upper 100 m of the oceans). In other words, photosynthesis exceeds oxidation during daytime in the growing season in the photic zone. The excess organic matter reaches deeper waters as organic detritus and fecal pellets or becomes part of the next trophic levels. The organic content of the deeper waters is either oxidized and eventually returned to the surface, chiefly by waters upwelling at high latitudes, or is settled as refractory organic matter (resistant to bacterial degradation) in the sediments. The flux of this refractory component into the sediments, although a very slight fraction of the primary productivity, is very important because

a corresponding flux of oxygen which is not used to oxidize this matter is released to the atmosphere (see Chapter 2).

3.6 THE OCEAN BOTTOM

3.6.1 The Crust

We saw in Chapter 1 that the earth is constituted of a core at its center, surrounded by the mantle which underlays the crust on which we live. An expanded view of the crust is shown in Figure 3.25. It is thinner under the oceans and becomes deeper under the continents, with the Mohorovicic discontinuity separating the crust from the mantle. A layer of submarine sediments, not shown in the figure, covers the oceanic crust. The features of the crust are inferred in part from the velocity of seismic waves, which depend on the density of materials present in their path, and in part on direct sampling.

Most of the crust, whether oceanic or continental, contains basalt, a quenched igneous rock with silicates of calcium, magnesium, and iron. In the continental crust there are intrusions of granite, a slowly crystallized rock containing aluminum and potassium silicates. Of course, many other rock types are present, as we saw earlier.

It is generally accepted that the crust consists of enormous blocks which move relative to each other and which are enriched with new crustal material that upwells at oceanic ridges. This leads to seafloor spreading and, therefore, subsidence beneath the continents as is shown in Figure 3.26.

3.6.2 Submarine Sediments

Submarine sediments are quite complex since they depend on terrigenous material such as clays brought from the continents, remnants of calcareous and siliceous tests, products of diagenetic (post-

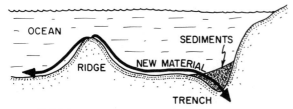

FIGURE 3.26 Seafloor spreading and subsiding.

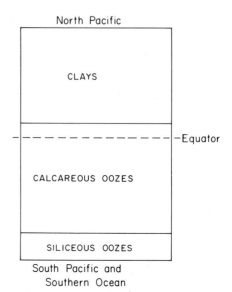

FIGURE 3.27 A sketch of the distribution of sediments in the Pacific Ocean.

depositional) alteration, oxides, and so on. Geological charts of the sea bottom, showing predominant sediments, can be found in standard texts such as those by Gross (1972) and Anikovchin and Sternberg (1973).

The main features, in the Pacific Ocean as an example, are sketched in Figure 3.27. Note that this figure is a predominance diagram and not an exclusive one in that the calcareous zone, for instance, is high in fragments of calcareous tests but also contains other types of sediments. Manganese nodules, of considerable interest and potential value, are widely distributed, especially in the northern part of the Pacific Ocean. The decrease in carbonates relative to clays in the North Pacific is due to a low productivity of calcareous tests relative to the rates of carbonate dissolution and terrigenous input. This rate of dissolution is enhanced by the high CO_2 content of the waters (see Chapter 5, Volume II). The siliceous belt which is also present to some extent in the North Pacific reflects the high production of siliceous tests.

3.7 COMPOSITION OF SEAWATER, CONSERVATIVE PROPERTIES, CHLORINITY, AND SALINITY

3.7.1 Historical Overview and Some Important Concepts

Humans became aware of the presence of salt in the oceans quite early in history through the two approaches that still guide our quest for knowledge: the search for practical uses and the endeavor to understand nature.

From a practical point of view humans linked salt deposits needed for the preservation and seasoning of food and their well-being to the solar evaporation of seawater. Furthermore, they knew that there was more than one component in sea salt. In effect, the ancient Chinese, although unaware of the chemistry involved, allowed rainwater to percolate through

salt mounds to remove magnesium sulfate, an effective laxative.

It is interesting to observe that salt was not readily available in medieval Europe in spite of the large amount present in the oceans (about 35 g/kg-SW). Rich families in Britain placed salt halfway down the table and only important people, sitting "above the salt," were allowed to use it. We have not come much further than that in our practical use of seawater as a reservoir of chemicals.

First I shall mention some practical aspects of marine technology. At present, in addition to table salt, we only extract freshwater, magnesium, and bromine in significant amounts from seawater. The reasons for this are the technical problems and the expense to separate individual constituents from the complex mixture of salts present in the oceans and the high cost of energy to transport seawater to the processing site. I am confident, however, that the ongoing research on the interactions between solutes in seawater and the increasing need for fresh water will lead eventually to an expanded and viable extractive technology.

We still depend to a great extent on organisms and sediments to concentrate and synthesize inorganic and organic chemicals that we can use as food or as sources of minerals. Examples of materials of sedimentary origin that have actual or potential value are phosphorites, manganese nodules, and limestone.

There are other areas of practical interest such

as: the effects of salts released by bursting bubbles and present as aerosols on the condensation of water in the atmosphere; the effects of $MgSO_4$ on the attenuation of sound in seawater; the effects of the pathways and the removal of pollutants from the sea; the effects of the relationships between distributions of nutrients and fishery grounds; and so on.

I should state that a large fraction of these and other applications have required an increase in our basic knowledge of the oceans.

The quest for an understanding of the fundamental chemistry of the oceans started a long time ago. The ancient Greeks and the Arabs of the first millenium A.D. were already intrigued by the cause of the saltiness of the oceans.

Robert Boyle (ca. 1670) observed that silver nitrate, when added to seawater, caused a heavy precipitate to appear, whereas only a slight turbidity was produced in fresh water. We now know that silver reacts with the chloride present in natural waters and that insoluble silver chloride is formed. Boyle concluded correctly that rivers and groundwaters leach chemicals from soils and transport them to the oceans where they accumulate gradually and become concentrated (Boulton, 1966).

Our knowledge of the complex mixture of salts present in seawater expanded greatly during the nineteenth century, the golden age of analytical chemistry, when Marcet (1819), Forchhammer (1865), and Dittmar (1884) measured the major oceanic salts. We now know that the major constituents of seawater are present as ions, with the exception of boric acid which is primarily in the molecular state. These major constituents are:

Chloride	Cl^-	Bicarbonate	HCO_3^-
Sodium	Na^+	Bromide	Br^-
Sulfate	SO_3^{2-}	Boric acid	$B(OH)_3$
Magnesium	Mg^{2+}	Strontium	Sr^{2+}
Calcium	Ca^{2+}	Fluoride	F^-

The ionic nature of seawater salts is very important because, by affecting the distribution of electric charges in seawater, it affects most physicochemical properties of the oceans. As one example, ions transport electric current and provide an important tool for measuring the salt content of seawater, its electrical conductance.

Marcet (1819) determined several major constituents in 14 samples collected from the Atlantic and Arctic Oceans and from several seas and found that the same ones were present everywhere. Fur-

thermore, he concluded that they were present nearly in the same proportions all over the oceans, even though the total amount of salt did vary. This conclusion, which may be called the law of constant proportions, was subsequently confirmed by the extensive determinations of Forchhammer (1865) and the careful chemical analyses by Dittmar (1884).

Dittmar, an outstanding analytical chemist at the Andersonian University, Glasgow, analyzed 77 samples of seawater collected during the 1873–1876 expedition of H.M.S. *Challenger* for dissolved salts and gases. The voyage of the *Challenger,* a small ship of 2300 tons which covered nearly 70,000 miles, is a milestone in the scientific study of the oceans.

The law of constant proportions implies, for example, that if the magnesium to calcium ratio at point A is five it will be essentially the same at point B even though the total calcium plus magnesium might be 1500 mg/kg-SW at A and 1600 mg/kg-SW at B. This law as we shall see later on, has been of great practical value. It permits us to measure routinely only a small group of major constituents and then to calculate the other ones from the known proportions to the measured ones.

The variation in the total salt content is easy to understand. It results from dilution by rainfall, snow, river, or melt water, and from concentration due to evaporation or freezing of seawater, which can be quite extensive at high latitudes and leaves behind it concentrated brines which sink and mix with deeper waters.

The constant relative composition of the major constituents is more intriguing and led to an important insight. The early view of the oceans recognized the existence of surface currents, which were utilized by ships. As an example, Benjamin Franklin (1786), when Postmaster General, found that British mail packets required two weeks longer to cross the Atlantic from England to the United States than American ships. This happened because American captains, due to their whaling experience, were familiar with the Gulf Stream and did not fight the current. Franklin, therefore, asked a Captain Folger of Nantucket to draw the Gulf Stream and had the resulting chart printed and distributed to mariners. Part of a letter of his on the subject is reproduced in Figure 3.28.

It was thought, however, that the deep oceans, in spite of currents present on the surface, were stagnant and devoid of oxygen and life (the azoic zone).

Let us see how the law of constant proportions

314   MARITIME OBSERVATIONS.

ration, and more easily manageable than the first, and perhaps may be as effectual.*

Vessels are sometimes retarded, and sometimes forwarded in their voyages, by currents at sea, which are often not perceived. About the year 1769 or 70, there was an application made by the board of customs at Boston, to the lords of the treasury in London, complaining that the packets between Falmouth and New-York, were generally a fortnight longer in their passages, than merchant ships from London to Rhode-Island, and proposing that for the future they should be ordered to Rhode-Island instead of New-York. Being then concerned in the management of the American post-office, I happened to be consulted on the occasion; and it appearing strange to me that there should be such a difference between two places, scarce a day's run asunder, especially when the merchant ships are generally deeper laden, and more weakly manned than the packets, and had from London the whole length of the river and channel to run before they left the land of England, while the packets had only to go from Falmouth, I could not but think the fact misunderstood or misrepresented. There happened then to be in London, a Nantucket sea-captain of my acquaintance, to whom I communicated the affair. He told me he believed the fact might be true; but the difference was owing to this, that the Rhode-Island captains were acquainted with the gulf stream, which those of the English packets were not. We are well acquainted with that stream, says he, because in our pursuit of whales, which keep near the sides of it, but are not to be met with in it, we run down along the sides, and frequently cross it to change our side: and in crossing it have sometimes met and spoke with those packets, who were in the middle of it, and stemming it. We have informed them that they were

* Captain Truxten, on board whose ship this was written, has executed this proposed machine; he has given six arms to the umbrella, they are joined to the stem by iron hinges, and the canvas is double. He has taken it with him to China. February 1786.

MARITIME OBSERVATIONS.   315

were stemming a current, that was against them to the value of three miles an hour; and advised them to cross it and get out of it; but they were too wise to be counselled by simple American fishermen. When the winds are but light, he added, they are carried back by the current more than they are forwarded by the wind: and if the wind be good, the subtraction of 70 miles a day from their course is of some importance. I then observed that it was a pity no notice was taken of this current upon the charts, and requested him to mark it out for me, which he readily complied with, adding directions for avoiding it in sailing from Europe to North-America. I procured it to be engraved by order from the general post-office, on the old chart of the Atlantic, at Mount and Page's, Tower-hill; and copies were sent down to Falmouth for the captains of the packets, who slighted it however; but it is since printed in France, of which edition I hereto annex a copy.

This stream is probably generated by the great accumulation of water on the eastern coast of America between the tropics, by the trade winds which constantly blow there. It is known that a large piece of water ten miles broad and generally only three feet deep, has by a strong wind had its waters driven to one side and sustained so as to become six feet deep, while the windward side was laid dry. This may give some idea of the quantity heaped up on the American coast, and the reason of its running down in a strong current through the islands into the bay of Mexico, and from thence issuing through the gulph of Florida, and proceeding along the coast to the banks of Newfoundland, where it turns off towards and runs down through the Western islands. Having since crossed this stream several times in passing between America and Europe, I have been attentive to sundry circumstances relating to it, by which to know when one is in it; and besides the gulph weed with which it is interspersed, I find that

FIGURE 3.28   Excerpt from a letter written by Benjamin Franklin (1786).

was one factor that helped dispel the idea of stagnant deep waters. Imagine a glass jar filled with water, into which we drip a blue and a yellow dye (Figure 3.29). If the dyes are added fast, there will be large zones in which the water is either blue or yellow instead of the green that results from mixing. In other words, the two dyes will not be present in the same proportions throughout the water. If, on the other hand, the dyes are added slowly and with stirring, most of the water will have the same tone of green and will contain the same proportions of yellow and blue.

The example above is somewhat analogous to what happens in the oceans with regard to the major constituents as various rivers bring different pro-

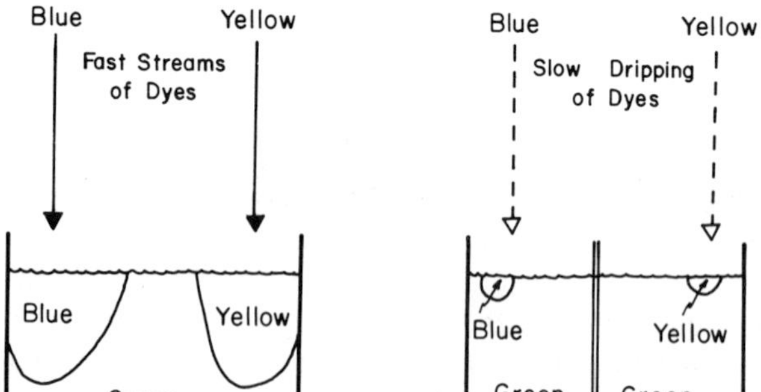

FIGURE 3.29   The mixing of waters and the law of constant proportions illustrated.

portions of the major constituents. The law of constant proportions shows, therefore, that the oceans are well stirred at all depths rather than stagnant in the abyssal zone and that this stirring is fast relative to the addition of major salts by rivers. We now know that the mixing is such that waters remain in the deep oceans about 1000 yr before being returned to the surface. This mixing is fast relative to the addition of chemicals by rivers. The concept of the azoic zone was further dispelled when life and oxygen, which is necessary to most organisms, were found in the deep oceans.

A stirring (mixing) which is fast relative to the rate of addition of major solutes is necessary but is not sufficient to ensure constant proportions. One could, for example, have a fast local reaction in the container shown in Figure 3.29 which would consume one of the dyes. Such reactions do not occur for the major constituents although they can happen for nutrients such as TPO_4. Active photosynthesis can change the $TPO_4/[Na^+]$ ratio. The law of constant proportions leads, therefore, to two conclusions:

1. The rates of addition of major constituents by rivers are slow relative to the mixing time of the oceans.
2. The concentrations of the major constituents are relatively insensitive to local biological and geochemical processes.

The second conclusion means that the major constituents change very little from place to place in the oceans in spite of biological, geological, and chemical processes. This is why their concentrations are known as conservative properties. These considerations do not imply that these constituents do not undergo such reactions but, rather, that their concentrations are large enough to mask the effects of marine processes.

The law of constant proportions does not apply to minor constituents such as the principal nutrients (phosphate and nitrate) and trace metals (mercury, zinc, lead, cadmium, etc.) or to dissolved gases that are involved in marine life (oxygen and carbon dioxide).

It should be realized that a constant proportion at any one time does not imply the same proportion in the past or in the future. There can be a slow evolution of the composition of seawater which is not significant in 1000 yr, the average oceanic mixing time. The degree of evolution is a function of how closely seawater is in a chemical steady state (see Chapter 2). It appears likely, for example, that the total inorganic dissolved carbon dioxide, TCO_2, ratio to the chlorinity Cl‰ will increase significantly over the next 100 yr due to the burning of fossil fuels. During this burning, CO_2 is released to the atmosphere and part of it is absorbed by the oceans (Broecker et al., 1971; Pytkowicz and Small, 1977).

The scope of marine chemistry expanded greatly since the 1930s as the subdisciplines of chemistry came of age and as marine-oriented geologists as well as biologists learned that they could obtain insights into their fields through chemical processes. Among other pioneers, A. C. Redfield in the United States and H. W. Harvey in Great Britain related biology to the chemistry of the oceans, K. Buch in Finland and H. Wattenberg in Germany introduced inorganic equilibria into the study of the oceans and mineral–seawater equilibria, and the American chemist, T. G. Thompson, opened up an amazing variety of subjects to marine chemistry.

Since World War II the range of research related to marine chemistry has expanded so greatly that it is difficult to select fairly names of leading chemical oceanographers. Research now covers physical, inorganic, analytical, organic, and radiochemistry, plus geochemistry and biochemistry, and a fair part of it is done in nonoceanographic institutions and departments. Topics covered are not limited to basic research, the foundation of our understanding and utilization of the oceans, but include such practical subjects as: the extraction of minerals from seawater; the effects and fates of pollutants; the chemical aspects of marine fisheries; the impact of marine aerosols on meteorology and climatology; corrosion; sound attenuation; and so on. Unfortunately, the pendulum of federal support for research is swinging excessively toward the applied side. This may limit the national foundation for the technology of the future, an effect that can be observed by comparing the appropriations for basic work with the prosperity of nations.

3.7.2 Sampling at Sea

Nonoceanographers are often intrigued by the way in which samples for chemical analyses are collected at sea. I shall, therefore, examine briefly the principle of the method, which is illustrated in Figure 3.30.

A string of bottles, which are held open at both

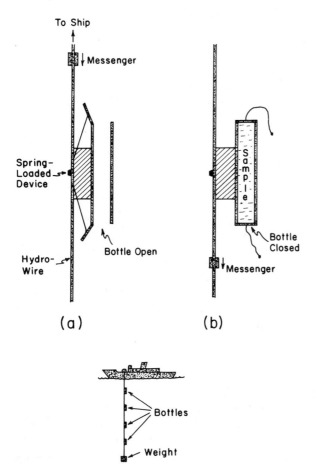

FIGURE 3.30 Principle of operation of bottles to collect seawater.

the next bottle shut, and so on. The array is then brought back to the shipboard.

The actual depth of sampling is calculated by comparing temperature readings on two thermometers that are attached to each bottle, one protected and the other one unprotected from the water pressure at depth. Mercury is exchanged from the bulb of the unprotected thermometer because of the combined effects of cooling and pressure, whereas mercury returns to the bulb of the protected one because of cooling done. The difference in the temperature readings yields the pressure and, therefore, the sampling depth. Note that the temperature of the unprotected thermometer is an apparent one.

3.7.3 Classification of Matter in Seawater and Its Composition

The forms of matter present in seawater are shown in Table 3.16. These forms reflect the fact that rivers, winds, and submarine volcanism bring to the oceans all the naturally occurring elements and many of their compounds, as the result of physical and chemical weathering, biological processes, and magmatic reactions. Furthermore, gases are taken up from the atmosphere or generated in situ and marine life produces a host of organic chemicals.

Major Constituents. A few dissolved constituents, known as the major constituents, account for over

TABLE 3.16 A classification of matter present in the oceans.

	Examples
Water (the solvent)	
Particulate	
Living	Phytoplankton
Dead organic	Fecal pellets
Inorganic	Quartz fragments
Colloidal	
Organic	Proteins
Inorganic	Clays
Solutes	
Major constituents	Na^+ (sodium ions)
Minor constituents	
Nutrients	Phosphate ions
Trace metals	Zinc ions
Others	Iodide ions
Dissolved gases	Oxygen
Dissolved organics	Methane
Stable isotopes	Carbon 13
Radioactive constituents	Carbon 14

ends, is attached to the weighted hydrowire and lowered from the deck of the oceanographic vessel. Each bottle is held open by a mounted spring-loaded device to which wires coming from the bottle caps are attached. The bottles are flushed constantly during the descent until they reach predetermined depths. These depths are read on a meter wheel, but are only known approximately at this stage, unless some form of an acoustic pinger is used, because the wire often does not hang straight but has a curvature (the wire angle). The hydrowire is stopped when the bottles are spaced at nearly the desired depths.

After a wait during which the thermometers attached to the bottles reach thermal equilibrium with their surroundings, a metal cylinder known as a messenger is released and slides along the wire. It trips the spring-loaded device on the topmost bottle which is then shut. At the same time a messenger originally attached to this bottle is released and trips

TABLE 3.17 The major constituents of seawater and their concentrations in ppm (parts per million by weight), w_f (weight fractions), and g-atom/kg-SW (gram atoms per kilogram of seawater) at 19.000‰ chlorinity. Based upon the compilation by Pytkowicz and Kester (1971).

Constituent	ppm	w_f	g-atom/kg-SW
Cl^-	18,971	0.55025	0.5351
Na^+	10,565	0.30657	0.4596
SO_4^{2-}	2,660	0.07712	0.02769 (S)[a]
Mg^{2+}	1,269	0.03650	0.05220
Ca^{2+}	404.2	0.01172	0.01008
K^+	391.4	0.01135	0.01001
HCO_3^{-} [b]	142.3	0.003569	0.002332 (C)
Br^-	65.99	0.001911	0.0008258
$B(OH)_3$ [c]	26.1	0.000187	0.000422
Sr^{2+}	7.790	0.000326	0.00008240
F^-	1.3	0.000037	0.000068

[a] (C), (S) indicate that the g-atom/kg-SW refers to carbon and sulfur.

[b] Actual values for the titration alkalinity expressed as if it were all bicarbonate.

[c] Boric acid is present primarily in the dissolved molecular form.

99% of the dissolved load of seawater. Furthermore, these major constituents are known as conservative because they are present in seawater in almost constant proportions even when the total content of salt changes (see Table 3.17).

EXAMPLE 3.10. REACTIVITIES, RESIDENCE TIMES, AND DISTRIBUTIONS OF ELEMENTS IN THE OCEANS. Let us consider three constituents 1, 2, and 3 with the following properties:

Element	A (g)	$J_{AB} = J_{BC}$ (g/yr)	B (g)	τ_{AB} (yr^{-1})	τ_{BC} (yr^{-1})
1	10^{22}	10^{14}	10^{22}	10^8	10^8
2	10^{20}	10^{16}	10^{23}	10^4	10^7
3	10^{10}	10^8	10^9	10^2	10

A_i, B_i, and C_i are the contents of element in the weathering, in the oceanic, and in the sedimentary reservoirs, J is the flux, and τ is the residence times.

First, we see that the weathering reactivities ($1/\tau_{AB}$) increase in the order 1, 2, 3 although the order of increasing fluxes from weathering is $(J_{AB})_3$, $(J_{AB})_1$, $(J_{AB})_2$. This apparent discrepancy is due to the different sizes of the reservoirs A_1, A_2, A_3. I remind you that the reactivities are just the rates of weathering per unit mass in the reservoir so that

$(J_{AB})_1$, for example, is $(k_{AB})_1 A_1$ with $(k_{AB})_1 = 1/(\tau_{AB})_1$.

It may appear at first that 1 is more reactive to sedimentation from the oceans than 2 since $B_1 < B_2$. The reactivities actually increase in the order 1, 2, 3 because $(\tau_{AB})_1 < (\tau_{AB})_2 < (\tau_{AB})_3$ so that $(k_{AB})_1 > (k_{AB})_2 > (k_{AB})_3$.

Elements 1 and 2 are definitely conservative since their τ is much larger than 10^3 yr, the mixing time of the oceans, whereas 3 is nonconservative.

Minor Constituents. Minor constituents comprehend essentially all the remaining naturally occurring elements, with the exception of gases, which are often treated separately. Some minor constituents and their concentrations are shown in Table 3.18. One should keep in mind that the concentrations are tentative because it is difficult to measure such small amounts without introducing contaminants. The numbers presented in Table 3.18 are subject to revision, represent average values for

TABLE 3.18 Selected minor constituents of seawater and their average concentrations, based on the compilation by Pytkowicz and Kester (1971).

Constituent	μg/kg-SW	μg-atom/kg-SW
Aluminum	6	0.2
Arsenic	4	0.05
Barium	21	0.15
Cadmium	0.116	0.00104
Cesium	0.4	0.003
Chromium	0.17	0.0032
Cobalt	0.15	0.0025
Copper	1.3	0.020
Gold	0.040	0.2×10^{-3}
Iodine	54	0.43
Iron	4.8	0.086
Lithium	190	27.4
Lead	0.05	0.2×10^{-3}
Manganese	1.2	0.022
Mercury	0.03	0.1×10^{-3}
Nickel	3.6	0.061
Nitrogen	288	20.6
Phosphorus	31	1.0
Platinum	?	?
Radium	8×10^{-8}	4×10^{-10}
Silicon	2050	73.0
Silver	0.16	0.0015
Tin	0.8	0.007
Titanium	1	0.5×10^{-3}
Uranium	3	0.01
Zinc	10.7	0.164

quantities that vary over wide ranges, and are shown simply to illustrate the orders of magnitude of the minor constituents for a comparison with the major ones.

Silicon is present in large amounts on the average but is not included with the major constituents because it is not conservative. In effect, it has been found to vary from essentially zero in fertile surface waters, from which it is removed by siliceous organisms, to over 5000 $\mu g/kg$-SW in deep waters.

The nitrogen shown in Table 3.18 corresponds to that present in NH_4^+, NO_2^-, and NO_3^-. In addition there are gases such as N_2, O_2, CO_2, H_2, CH_4 (methane), CO, and the noble gases.

3.7.4 Chlorinity Explained

The total salt content of seawater and its changes must be known in order to understand the transport of chemicals by hydrographic processes, and to characterize the electrolyte content of the oceans for physicochemical, biochemical, and geochemical studies. As an example, deep waters with different salt contents have different ionic strengths which alter chemical reactions, may have different surface sources, and probably have undergone different extents of mixing with adjacent waters.

The determination of the concentrations of all the major constituents is a difficult and tedious process. The constant proportions of these constituents led to the realization that, once the ratios of their concentrations to that of any one of them were established, only this reference constituent would have to be measured. The concentrations of the other constituents could then be calculated from the known ratios.

Chloride could have been a fine reference ion because its concentration can be measured rapidly in a chloride solution by adding $AgNO_3$ and precipitating Cl^- as AgCl. Silver, unfortunately, also precipitates two other halogens from seawater, bromide and iodide (silver fluoride is quite soluble). Chlorinity was introduced to get around this problem, while taking advantage of the titration by silver nitrate. The reactions on addition of $AgNO_3$ to seawater are

$$Ag^+ + Cl^- \rightarrow AgCl \downarrow \qquad Ag^+ + Br^- \rightarrow AgBr \downarrow$$

$$Ag^+ + I^- \rightarrow AgI \downarrow \qquad (3.17)$$

At this point I shall digress briefly to refresh the memories of those readers from disciplines other than chemistry for whom this subject is only a vague recollection from freshman days.

The g-atom of an element is, as I mentioned earlier, simply that amount of the element which contains its atomic weight expressed in grams (one gram-atomic weight or one gram-atom). One g-atom of any element always contains 6.0232×10^{23} atoms (Avogadro's number) and elements always react together in multiples of g-atom. Thus, in reaction (3.17) 1 g-atom of Ag reacts with 1 g-atom of Cl. In the case of the dissolution of Na_2SO_4

$$Na_2SO_4 \rightarrow 2Na^+ + SO_4^{2-} \qquad (3.18)$$

2 g-atom of Na^+ are formed for each g-atom of SO_4^{2-}.

One rule that is universally valid is that the same number of g-equiv react with each other in a chemical process. The g-equiv for ions is simply the g-atom or the mole divided by the charge. Thus, 1 g/atom of Na^+ contains only 1 g-equiv, whereas 1 mole of SO_4^{2-} has 2 g-equiv. The term mole is used loosely for SO_4^{2-} for complexes that form an ion. The alternative would be to introduce an additional term g-complex. Equation (3.18) shows that the above rule is obeyed since 2 g-atom (1 g-equiv Na^+/g-atom = 2 g-equiv Na^+) react with 1 mole (2 g-equiv SO_4^{2-}/mole = 2 g-equiv SO_4^{2-}). Note that g-atom = g-equiv for univalent ions.

I return now to reactions (3.18). One can write the equation

$$\frac{w_{Ag}/kg\text{-SW}}{AW_{Ag}} = \frac{[Cl^-]}{AW_{Cl}} + \frac{[Br^-]}{AW_{Br}} \frac{[I^-]}{AW_I}$$

$$(3.19)$$

w_{Ag} is the weight of silver necessary to precipitate the halogens in 1 kg-SW (one kilogram of seawater), AW represents the gram-atomic weight of the subscribed element, and the concentrations in brackets are expressed in ‰ (parts per thousand by weight or g-element/kg-SW). I do not distinguish between mass and weight in this work.

Equation (3.19) simply expresses the fact that the number of equivalents of Ag^+ which must be added to remove the Cl^-, Br^- and I^- from 1 kg-SW is equal to the number of equivalents of these halogens which are present in that kg-SW.

Equation (3.19) may be rewritten as

$$Cl_e(\text{‰}) = w_{Ag} \frac{AW_{Cl}}{AW_{Ag}} = [Cl^-] + [Br^-] \frac{AW_{Cl}}{AW_{Br}}$$

$$+ [I^-] \frac{AW_{Cl}}{AW_I} \qquad (3.20)$$

Equation (3.20) represents the original definition of chlorinity (Forch et al., 1902), known as the chloride equivalent and denoted by the symbol $Cl_e(\text{‰})$. This definition states that the chlorinity is the total weight of chloride, bromide, and iodide, in g/kg-SW, when bromide and iodide are converted into their equivalent weights of chloride. This definition is clarified with an example.

EXAMPLE 3.11. CHLORINITY ILLUSTRATED

I. Suppose that we know a priori that a given sample of seawater contains

18.971‰ Cl^- 0.06599‰ Br^- 0.000054‰ I

II. From Equation (3.22) we would calculate

$$Cl_e(\text{‰}) = (18.971 \to 0.02928 + 0.000015)\text{‰}$$

$$= 19.000\text{‰} \qquad (i)$$

III. The weight of Ag^+ needed to precipitate these halogens from 1 kg-SW would be

$$w_{Ag} = 19.000 \frac{AW_{Ag}}{AW_{Cl}} = 57.810 \text{ g-}Ag^+ \qquad (ii)$$

IV. Note that 57.810 g-Ag^+ precipitate 18.971 g-Cl^- plus 0.06599 g-Br^- plus 0.000054 g-I^-, but that they would also precipitate 19.000 g-Cl^- (the chloride equivalent of the halogens present in 1 kg-SW). This occurs because the above amounts of bromide and iodide contain the same number of ions as 0.02928 plus 0.000015 g-Cl^- [see Equation (i)]. I^- can be neglected when $Cl_e(\text{‰})$ is given to three decimal places.

V. In actual practice we would not determine Br^- but would find for the particular seawater sample under study that the endpoint of the titration would occur upon addition of

57.810 g-Ag^+ to 1 kg-SW. We would then calculate from

$$Cl_e(\text{‰}) = w_{Ag} \frac{AW_{Cl}}{AW_{Ag}} \qquad (iii)$$

that $Cl_e(\text{‰}) = 19.000\text{‰}$.

$Cl_e(\text{‰})$ has a drawback because it depends on atomic weights [see Equation (3.22)] that undergo periodic revisions as the isotopic composition of the elements becomes better known. This could lead to a periodic recalculation of large quantities of oceanic data. To avoid such a problem, Jacobsen and Knudsen (1940) redefined the chlorinity, hopefully once and for all, as follows: Chlorinity is the weight of silver necessary to precipitate the chloride, bromide, and iodide in 0.3286707 kg-SW. I shall use the symbol $Cl_J(\text{‰})$ for the chlorinity defined in this manner.

It turns out that

$$Cl_J(\text{‰}) = 1.00043 \ Cl_e(\text{‰})_{1902}$$

$$Cl_J(\text{‰}) = Cl_e(\text{‰})_{1939}$$

$$Cl_J(\text{‰}) = 1.0000201 \ Cl_e(\text{‰})_{1963} \qquad (3.21)$$

where the subscripts correspond to the 1902, 1939, and 1963 weights. Thus, $Cl_J(\text{‰})$ varies while $Cl\text{‰}$ remains the same by definition. Do keep in mind, however, that in practice $Cl_J(\text{‰})$ is only measured to five significant figures so that the difference between $Cl_e(\text{‰})$ for 1939 and 1963 is negligible. The relationship based on the 1939 atomic weights arises from the definitions of $Cl_J(\text{‰})$ and $Cl_e(\text{‰})_{1939}$ (Jacobsen and Knudsen, 1940). In other words, $Cl_J(\text{‰})$ is and will remain the mass of silver required to precipitate the chlorine equivalent in 1 kg-SW, in terms of the 1939 atomic weights.

3.7.5 Determination of the Chlorinity

The titration procedure used for the primary determination of the chlorinity is outlined briefly. Details can be found, for example, in Strickland and Parsons (1965).

During the titrations one adds $AgNO_3$ to the seawater after dissolving K_2CrO_4 in it. The silver chromate is more soluble than the silver halides and, therefore, it only starts to precipitate after the silver halides have been removed. The Ag_2CrO_4 imparts a full yellow color to the seawater when it starts to precipitate at the halide endpoint. One then calculates the chlorinity from the weights of seawater and the silver added at the endpoint.

For routine determinations of the chlorinity at sea, one uses the electrical conductance of seawater determined in a conductance bridge and a known chlorinity–conductance relation (Cox et al., 1967).

3.7.6 The Defined Salinity

A salinity related to but not exactly equal to the total salt content of seawater was first defined, based on the analytical procedure that had been used in sea-going work, by Forch et al. (1902). According to that definition, salinity is the weight of dissolved matter, in g/kg-SW, when bromine and iodine are expressed as their equivalents of chlorine, all carbonate is converted to oxide, and the organic matter is destroyed. I shall use the symbol $S_m(\text{‰})$ to represent the salinity thus defined.

Forch et al. (1902) then correlated values of $S_m(\text{‰})$ and $Cl_e(\text{‰})$ obtained for a small number of near-surface samples of seawater which were biased by the effect of freshwater influx into the Baltic Sea. From such a correlation they obtained the expression

$$S_F(\text{‰}) = 1.8050 \, Cl_e(\text{‰}) + 0.030 \quad (3.22)$$

$S_F(\text{‰})$ is essentially equal to the directly measured salinity $S_m(\text{‰})$ but differs slightly from it because $S_F(\text{‰})$ was determined from a linear fitting of $S_m(\text{‰})$ versus $Cl_e(\text{‰})$. Thus, $S_F(\text{‰})$ could be obtained from the measured $Cl_e(\text{‰})$ and Equation (3.22). Equation (3.22) is known as the Knudsen relation.

A newer definition of the salinity, the practical salinity $S\text{‰}$ (Wooster et al., 1969), was obtained from the expression

$$S\text{‰} + 1.80655 \, Cl_J(\text{‰}) \quad (3.23)$$

This relationship was derived in such a manner as to preserve the numbers calculated by Equation (3.22) as closely as possible for open oceanic waters while rejecting any coefficients such as the 0.030 in the former relation. The term 0.030 in Equation (3.22) was attributed to the influence in earlier data of rivers that drain into the Baltic.

The practical salinity $S\text{‰}$ is determined as follows. Clean natural seawater is diluted or evaporated so that it has the same electrical conductivity at 15°C as a solution containing 32.4357 g-KCl/kg-soln. This Standard Seawater is found by titration to have a $Cl_J(\text{‰}) = 19.374$. Its defined $S\text{‰}$ is nearly 35.000.

A salinometer, actually a conductivity bridge, takes over at this point. The practical salinity scale is defined as a function of the conductivity ratios r_{15} of waters of unknown salinities to the primary Standard Seawater of the previous paragraph by means of

$$S\text{‰} = a_0 + a_1 r_{15}^{0.5} + a_2 r_{15} + a_3 r_{15}^{1.5}$$
$$+ a_4 r_{15}^2 + a_5 r_{15}^{2.5} \quad (3.24)$$

with $a_1 = 0.0080$, $a_1 = -0.1692$, $a_2 = 25.3851$, $a_3 = 14.0941$, $a_4 = -7.0261$, $a_5 = 2.7081$, and $\Sigma a_i = 35.0000$. The last significant figure is uncertain. Thus, for $r_{15} = 1$, the salinity is 35.000. This equation was obtained from a measured r_{15} versus $Cl_J(\text{‰})$ relationship at 15°C and Equation (3.23).

At other temperatures one has relationships between $S\text{‰}$ and r_T, the conductivity ratio at T°C.

In practice one determines r_T between an unknown seawater and a primary or secondary standard seawater with a salinometer. There, the International Oceanographic Tables (UNESCO, 1966) yield $S\text{‰}$. If desired, the value of $Cl\text{‰}$, the practical chlorinity, is calculated from

$$S\text{‰} = 1.80655 \, Cl\text{‰} \quad (3.25)$$

Note that Equation (3.25) is not identical to (3.23). The reason for this is that the procedure used above was almost, but not quite, circular. $Cl_J(\text{‰})$ was used in conjunction with Equation (3.23) to determine $S\text{‰}$ versus r_{15} (or r_T). This led to curve fittings that were not perfect to obtain the salinity–conductivity ratios so that we do not return exactly to $Cl_J(\text{‰})$ when backtracking to obtain $Cl\text{‰}$. Therefore, Equation (3.25) yields values of $Cl\text{‰}$ which are not exactly $Cl_J(\text{‰})$. This subject is quite a morass of unnecessary confusion.

EXAMPLE 3.12. $S\text{‰}$, $Cl_J(\text{‰})$, AND $Cl(\text{‰})$ (K. KILHO PARK, CLASS NOTES)

$S\text{‰}$ (from r_{15})	$Cl_J(\text{‰})$ (Measured)	$Cl\text{‰}$ (Calculated)
21.809	12.326	12.072
29.386	16.721	16.266
34.312	19.390	19.019

TABLE 3.19 Chlorinity ratios of the major constituents of seawater, expressed in (g/kg-SW)/‰ and, therefore, nondimensional [based on the compilation by Pytkowicz and Kester (1971)]. Cl_J(‰) is used.

Constituent	Ratio	Constituent	Ratio
Cl^{-a}	0.99847	HCO_3^-	0.00749
Na^+	0.5561	Br^-	0.003473
SO_4^{2-}	0.1400	$B(OH)_3$	0.00137
Mg^{2+}	0.06680	Sr^{2+}	0.00041
Ca^{2+}	0.02127	F^-	0.000067
K^+	0.0206		

[a] Calculated from the definition of chlorinity and the bromide/chlorinity ratio.

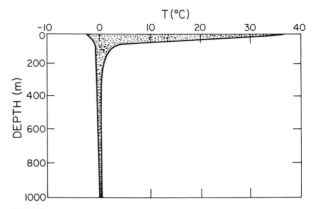

FIGURE 3.31 Vertical variability of temperature.

Properties that depend on the salinity are: density, thermal expansion, specific heat, compressibility, adiabatic temperature change with pressure, thermal conductivity, viscosity, electrical conductivity, surface tension, dissociation of acids, solubilities of gases and minerals, and so on. The study of such properties constitutes a sizable part of physical, chemical, and geological oceanography, but, except for some topics, transcends the scope of this book.

3.7.7 Chlorinity Ratios and the Total Salt Content

The chlorinity ratios of the major constituents are presented in Table 3.19. These data apply to the major oceanic basins but may not be valid in estuarine waters, pore waters of the sediments, and special regions where intense local processes may change significantly the proportions of the major constituents. Examples of the latter are brines in which evaporation has caused partial precipitation of salts, including some tidal pools and areas of heavy $CaCO_3$ deposition such as the Bahama Banks. Most rivers are so dilute that they may be assumed to act like distilled water, from the standpoint of the relative composition of estuaries, when they mix with seawater at least for mixtures containing up to about 20% seawater–80% river water (Pytkowicz et al., 1975). I should emphasize that, in such cases, the defined salinity may lose its meaning and we may not be able to interpret the chlorinity, measured by titration, and the conductance of the water in terms of the amounts of dissolved salts.

It is interesting to observe that S‰ ranges from about 32 to 37‰ in the near-surface layers of the oceans and from about 34.8 to 35‰ in the intermediate waters. The temperature varies from

-1.8–36°C in the former case to roughly 0°C at 1000 m (see Figures 3.31 and 3.32).

Thus, the density of seawater decreases rapidly from the surface to about 400 m but increases very slowly at greater depths. This indicates that the upper water stratification is relatively large and can serve as a barrier to the vertical transport of solutes by mixing. High latitudes are exceptions since there is considerable thermal uniformity in the water column. Small inaccuracies in the meaning of S‰ do not affect descriptive aspects of oceanography.

Note that P varies from 1–100 atm in these diagrams and, therefore, varies by a larger factor than T and S‰. This does not necessarily imply a greater effect on chemical reactions in the upper 1000 m because this effect is exponential and usually picks up with depth only at over 2000–3000 m.

For oceanic waters the defined salinity, S‰, can be used to estimate Σ_M(‰), the major constituent content of seawater. This can be done as follows:

1. Table 3.17 corresponds to Cl‰ = 19.000‰.

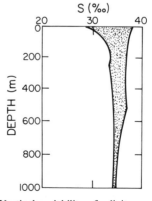

FIGURE 3.32 Vertical variability of salinity.

---

2. $S‰$ can be obtained from Equation (3.25) and is 34.324‰.

3. $\Sigma_M(‰)$ can be obtained by adding the concentrations (in ppm) of the major constituents in Table 3.17 and dividing by 10^3. One finds 34.504‰.

4. Hence

$$\Sigma_M(‰) = \frac{34.504}{34.324}\, S(‰) = 1.0052\, S(‰)$$

(3.26)

The contribution of the minor constituents to the total dissolved inorganic load of seawater is difficult to estimate because their concentrations are quite variable and because we are not yet certain of the chemical forms in which they are present in the oceans. In terms of their average concentrations in elemental form, I estimate that they may contribute about 0.003‰ to $\Sigma_M(‰)$ and would only change Equation (3.26) by one unit in the last significant figure. It is likely, therefore, that we can calculate $\Sigma_T(‰)$, the total dissolved inorganic salts of seawater, from

$$\Sigma_T(‰) = 1.005\, S‰ \qquad (3.27)$$

We thus see that the salinity is within about 0.5% of the total dissolved inorganic salt content of seawater if we loosely call $B(OH)_3$ a salt. Furthermore, insofar as the proportions of the major ions and boric acid remain constant, the salinity defines both the composition and the ionic strength. Thus, the salinity can be used in physicochemical studies to fix the composition variable of open ocean seawater considered as an ionic medium due to its high and constant relative concentrations of the major constituents.

I should add that slight geographical and depth variations in the major ion proportions do occur [e.g., review by Pytkowicz and Kester (1971)] although their examination is beyond the scope of this work.

3.7.8 Water Type, Water Mass, and the T–S‰ Diagram

A water type is by definition a water of a given T and $S‰$. It exists only in a given uniform region of the oceanic surface at a given time. Thus, the mixture of Antarctica Surface and Sub-Antarctica Wa-

ters which will yield the Antarctica Intermediate Water constitute roughly a water type, let us say A, during part of the winter. Water types can only exist at the surface because they mix at depth.

Water type A mixes with other waters when it sinks and forms a water mass, the Antarctica Intermediate Water. The Antarctica Intermediate Water in the Atlantic results from A mixing, after sinking, with shallower waters as well as with the Atlantic Deep Water. The Atlantic Deep Water Mass results in turn from water types formed in the North Atlantic Ocean and the Mediterranean which mix with shallower waters and Antarctica Bottom Water.

Let us consider only two water types A and B which mix during advection in order to understand the T–$S‰$ diagram (see Figure 3.33).

Water types A and B sink and mix while moving either in the same or in different directions. The result of the mixing is a water mass AB (see Figure 3.33) of which X is a representative point. The values T_A, $S_A(‰)$, T_B, $S_B(‰)$ are not found below the sea surface because these properties of A and B gradually converge due to mixing. Still, these quantities can be used to analyze water masses in terms of water types through equations such as $T = f_A T_A + f_B T_B$ and $S‰ = f_A S_A(‰) + f_B S_B(‰)$. The terms f_A and f_B represent the fractions of A and B present at a point in the ocean at which the temperature is T and the salinity is $S‰$.

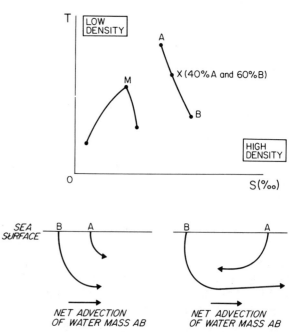

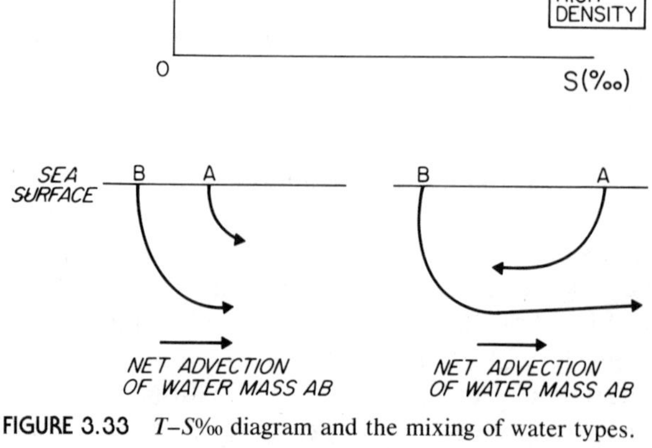

FIGURE 3.33 T–$S‰$ diagram and the mixing of water types.

The oceanic situation is more complex than that indicated in this T–S‰ diagram because a number of water types combine to form each water mass. Water masses may have core properties (maxima or minima in T or S‰) relative to the water masses above and beneath them. If there are core properties, then they can be used to characterize water masses in T–S‰ diagrams. A case in point is the Antarctica Intermediate Water which has a salinity minimum. Water mass M in the T–S‰ diagram has a temperature maximum.

Before proceeding to the modeling of the oceans, let us examine the relationship between different sets of concentration units in seawater.

EXAMPLE 3.13. CONCENTRATION UNITS IN SEA-WATER. Consider a seawater of $S = 35$‰ which contains 1.294 g-Mg^{2+}/kg-SW. The atomic weight of Mg is 24.31. The specific gravity at 1 atm and 25°C is 1.0233 g/cm^3, that is, sigma-t = σ_t = 23.3.

The following units can be used, among others:

$$\text{moles/kg-SW} = \frac{1.294}{24.31} = 0.0532 \qquad \text{(i)}$$

Actually one should also correct for the increase in specific gravity to 1.0274 g/cm^2 due to the lower temperature, about 0°C, at depth

$$\text{molality} = \frac{\text{moles-mg}^{2+}}{\text{kg-H}_2\text{O}}$$

$$\frac{0.0532}{1 \text{ kg-SW} = 0.035 \text{ kg-salts}} = 0.0552 \; m \qquad \text{(ii)}$$

$$\text{molarity} = \frac{\text{moles-Mg}^{2+}}{\text{L-SW}} = \frac{0.0532}{1/1.0233}$$

$$= 0.0545 \; M \qquad \text{(iii)}$$

$$\text{normality} = \frac{\text{equiv-Mg}^{2+}}{\text{L-SW}} = 2 \times \text{molarity}$$

$$= 0.1089 \; N \qquad \text{(iv)}$$

If the water sample is lowered to 10,000 m then the volume decreases by about 4%. The concentrations become

0.0532 mole/kg-SW	0.0552 m
0.0499 M	1.040 N

3.8 MODELING OF CHEMICAL DISTRIBUTIONS IN THE OCEANS

Processes that control chemical distributions in the oceans are, more often than not, nonequilibrium ones. As such, they can cast light on important processes that prevent equilibration. It is, therefore, an important topic in a book which emphasizes not only the roles of equilibria, but also those of rate processes in natural waters.

3.8.1 Typical Oceanic Distributions

The fact that the wind-mixed layer (WML) is well stirred does not mean that there are no horizontal gradients of salinity, temperatures, and solutes (dissolved chemicals), as we saw earlier. The residence time of waters in the WML is on the order of 56 yr. This time is quite long relative to hydrological, biological, and thermal processes which can produce patchiness. Examples of processes that can cause variations are:

1. *Diurnal.* Tidal effects, peaking of photosynthesis during the day, changes in pH due to photosynthesis and CO_2 exchange across the sea surface because of life and thermal variations.
2. *Seasonal.* Supply of nutrients by upwelling, effects due to temperature and climatic changes, river discharges.
3. *Trends.* Gradual changes in population and agricultural and industrial practices.
4. *Irregular.* Irregular shifts in winds and cloud cover.

Horizontal and Isentropic Variations. A typical case of horizontal surface patchiness is shown in Figure 3.34. The actual oxygen concentration also varies from place to place but I prefer to show the percentage of saturation because it indicates whether the oxygen tends to enter or leave the sea. The 105% regions correspond to the warming of waters, which decreases the solubility of O_2, and to the production of O_2 during photosynthesis. The 91% saturation results from upwelling of waters in which the oxygen has been depleted by the oxidation of organic matter followed by surface water transport toward the southwest.

An example of the surface variation of CO_2, a

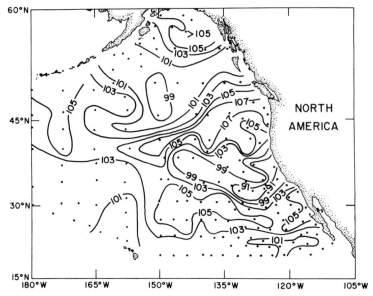

FIGURE 3.34 Surface patchiness of oxygen percent saturation in the Pacific Ocean in summer (Kester and Pytkowicz, 1968).

subject that will be discussed in Chapter 2, Volume II, can be found in Skirrow (1965). The near-surface distribution of phosphate is shown in Figure 3.35. Low values occur in subtropical gyres where the waters downswell. The sinking of the waters occurs because in the northern hemisphere, for example, the gyres are clockwise and, due to the Coriolis

force, have a component of motion to the right of the direction of motion. This causes a piling up toward the center with a consequent sinking which drags phosphate from the surface. Mixing with waters richer in TPO_4 during sinking plus some oxidation of organic matter increases the phosphate downward.

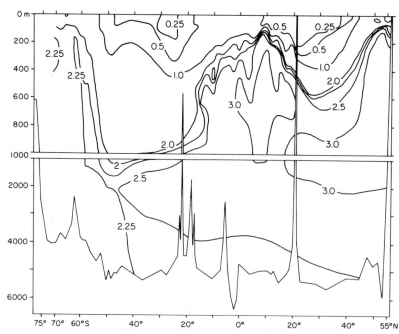

FIGURE 3.35 Values of the dissolved inorganic phosphate in the upper waters of the Pacific Ocean at 160°W (Reid, 1973). The units are µg-atom/L.

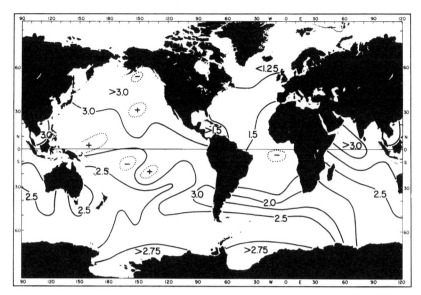

FIGURE 3.36 Dissolved phosphate gradients at 2000-m depth(Redfield, 1958).

Redfield (1958) found in the world ocean gradients of phosphate compatible with the Stommel (1958) circulation patterns proposed by Stommel (1958) and described earlier in this chapter.

The inorganic dissolved phosphate TPO_4 increases in the direction of motion because it is released by the oxidation of organic matter carried by the waters or settling from the photic zone. The results are in agreement with carbon 14 dating in that the oldest Deep–Intermediate Waters are found in the Northwest Pacific Ocean and contain the most phosphate. The primary source of these waters is the surface of the North Atlantic surface of the ocean in which the lowest TPO_4's are found (see Figure 3.36).

Oxygen, below the photic zone, decreases roughly toward older waters because of the oxidation of organic matter. The age of a water depends not only on advection but also on mixing. Thus, a negative gradient of oxygen does not characterize the advective direction alone. The decrease in oxygen concentration in the oceans should follow approximately the trends of TPO_4 increase determined by Redfield (1958).

Not everyone agrees with significant oxidation in deep waters. Menzel (1968) concluded that particulate organic carbon is independent of depth and latitude and that oxygen correlates linearly with $S‰$. Hence, he concluded that oxygen is conservative at depth and that its gradients are due to physical processes alone. His conclusion was biased by the low gradients and precision of measurements of particulate organic carbon as well as by the use of (O_2)–$S‰$ correlations over relatively small distances.

However, when the whole oceans are looked at, there are sizable deviations of (O_2)–‰ correlations, as is shown below.

EXAMPLE 3.14. OXYGEN–SALINITY RELATIONSHIPS BELOW ABOUT 3 km.

Ocean	Latitude	O_2 (mL/L)	$S‰$	Depth (km)	$(O_2)/S‰$
Atlantic	60°N	7	34.9	Surface	0.2006
	20°N	6	34.9	3	0.1719
	40°S	5	34.9	3	0.1433
Pacific	70°S	4	34.7	3	0.1153
	0°	2.5	34.6	3	0.0723
	40°N	2.5	34.6	3	0.0723

We see, therefore, that the deep oxygen is nonconservative and oxidation does occur. The core of the deep waters is at about 3 km and these waters originate in large measure at 60°N at the surface of the Atlantic Ocean; hence, the choices of depths and latitudes. The progression of the waters in Example 3.14 is from the North Atlantic to the South Atlantic and from there, via the Antarctic Circumpolar Waters, into the Indian and Pacific Oceans. Craig (1971) also concluded that oxygen utilization does occur in deep waters. The horizontal decrease in O_2 is usually slight over narrow ranges of latitude and this may cause the deep oxygen to appear quasiconservative. An exception to this, at somewhat shallower depths (1–2 km) is that of waters below a very fertile surface. This can induce a rain of organic matter or a large amount of it brought from the surface source. Thus, the $[O_2]$ of Antarctic Intermediate Water drops by several mL/L over a few degrees of latitude from its surface source (Redfield, 1942; Pytkowicz, 1968).

Vertical Variations. Vertical variations of concentrations are quite interesting because they provide insights into hydrographic processes, photosynthesis and respiration, scavenging by adsorption of solutes on suspended solids, redox conditions, and so on.

Conservative solutes such as Na^+ simply increase with depth like the salinity. There is a very slight vertical gradient which increases somewhat in the depth ranges of the summer and the permanent pycnoclines and may be larger in regions of heavy rainfall and near river outlets.

Nutrients such as TPO_4, TNO_3, and TCO_2, and trace metals (Chapter 8, Volume II), as well as gases such as O_2 and CO_2, can show marked changes with depth due to photosynthesis, oxidation, and, for gases, also exchange across the sea surface.

Typical vertical profiles of O_2 and TPO_4 are shown in Figure 3.37, where TPO_4 is the total inorganic phosphate (H_3PO_4, $H_2PO_4^-$, HPO_4^{2-}, PO_4^{3-}).

The oxygen is constant to a depth of 200–300 m in winter due to wind mixing in unstratified waters. In summer production tends to increase the dissolved oxygen. This can lead to O_2 supersaturation which is further enhanced by the warming of the waters. As a result of the supersaturation, more oxygen leaves the waters than the amount produced by photosynthesis. The reverse occurs in winter (Redfield, 1948; Pytkowicz, 1964). These results are expanded upon next.

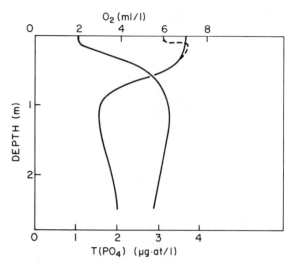

FIGURE 3.37 Typical vertical profiles of O_2 and TPO_4. The broken line shows summer data.

The method of Redfield (1948), adapted by Pytkowicz (1964) to the open ocean, is based on the fact that there is a fairly well-defined ratio between oxygen consumed and TPO_4 released during oxidation, as we shall soon see.

Before examining my results let us consider some interesting features of life and nonconservative gases such as O_2 and CO_2. First, $CO_2 + H_2O = CH_2O + O_2$ does not actually represent an equilibrium assemblage of chemicals in oceans or lakes. $CO_2 + H_2O \rightarrow CH_2O + O_2$ predominates in the photic zone during the growing season, whereas the reverse goes to completion at depth and, during winter, in the photic layer. Thus, we have a transient system in terms of days, weeks, or even months and a steady-state system from the standpoint of annual averages.

Second, nonconservative gas would reach equilibration with the atmosphere, except for thermal lags, if life did not exist. Radiation can warm up seawater before it has a chance to unload the oxygen from the now supersaturated water. In reality, the thin surface layer of water bodies equilibrates rapidly with atmospheric gases but organisms prevent equilibration beneath the microlayer by utilizing or releasing O_2 and CO_2.

The near-surface oxygen patterns off Oregon (Pytkowicz, 1964) are shown in Figure 3.38. A quantitative plot, without the biological oxygen, can be found in the original paper.

The grey area shows the amount of oxygen lost between the onset of the phytoplankton bloom and September. The summer pycnocline started at a

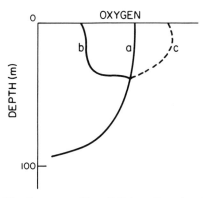

FIGURE 3.38 Oxygen profiles off Oregon [based on Pytkowicz (1964)]. (*a*) Winter, (*b*) actual summer, and (*c*) summer profile if there was no loss to the atmosphere.

TABLE 3.20 Seasonal oxygen changes off Oregon (Pytkowicz, 1964). Q_B is the oxygen production since the January–April average and Q_E is the corresponding oxygen exchange. Q_E is negative for loss to the atmosphere. The units are mL-O_2 for a column 1 cm^2 × 45 m deep. Q_O is the total oxygen observed in the water column.

Up to	Q_B	Q_E	Q_O
January–April average	0	0	29.8
June	3.0	−3.2	29.6
September	4.9	−7.3	27.4

depth of about 45 m. Some results are shown in Table 3.20.

Eddy diffusion coefficients could be calculated because the waters must reach the surface through mixing to unload oxygen. The values obtained were:

Depth (m)	10	20	30
A_z (g/cm · sec)	2.3	1.3	0.1

One observes the decrease in A_z and, therefore, the decrease of vertical mixing when the stratification induced by the thermocline is approached.

The entry coefficient E is given by

$$J = E[(pO_2)_{atm} - (pO_2)_{SW}] \qquad (3.28)$$

was found to be 3×10^6 mL-O_2/month · m^2 · atm.

The most important conclusion from this work is the variety of information that can be obtained from the study of nonequilibrium systems in nature.

The oxygen minimum at about 1000 m probably results from two processes. First, there is a high utilization for the oxidation of the large amount of organic matter in Intermediate Waters (Redfield, 1942; Pytkowicz, 1968). One observes a substantial decrease in oxygen even before waters leave Antarctica. This indicates extensive oxidation of matter that sank with the Intermediate Waters from the very fertile waters near the Antarctica Convergence. The second mechanism is the coupling of slow advection and eddy diffusion which lead to a limited rate of resupply of oxygen. We shall examine later the coupling of advection, mixing, and oxidation (Wirtky, 1962). The increase in oxygen at depths below 1000 m is mostly due to the replenishment of this gas by the sinking Deep and Bottom

Waters and less oxidation in them because the bulk of it occurred in intermediate and shallow levels.

The TPO_4 is nearly a mirror image of the oxygen. This is not entirely so for two reasons: the TPO_4 at the surface is not volatile, in contrast to O_2, and at depth it involves two components

$$TPO_4 = (TPO_4)_p + (TPO_4)_{ox} \qquad (3.29)$$

The subscript p refers to the preformed quantity present when the water left its surface source. The subscript ox indicates the oxidative amount, that is, the quantity released when oxygen is consumed during oxidation since the water left the surface. $(TPO_4)_{ox}$ should be a mirror image of the oxygen with increasing depth. $(TPO_4)_p$, however, has its maximum at a greater depth than $(TPO_4)_{ox}$. This is the reason why TPO_4 reaches its largest value at a depth greater than that at which oxygen passes through a minimum. Later I shall show that these concepts apply as well to a water type as to the more usual case of a water mass resulting from a mixture of water types.

These considerations regarding the relationships between oxygen and phosphate can be extended to nitrate and to TCO_2, the total carbon dioxide. Only TPO_4 is being used to illustrate principles because it has the simplest relation to oxygen. In the case of nitrogen, we must consider NH_4^+, NO_2^-, and NO_3^-, whereas for carbon we have TCO_2 and TA (see Chapter 3, Volume II).

In this section little was said about another type of variation, the seasonal one. In general, one finds that nutrients are depleted from the near-surface waters, due to photosynthesis, during the productive season from the spring bloom through summer. They are returned to the waters during the remainder of the year. Examples can be found in Harvey (1957).

3.8.2 The Hydrobiochemical Equation

We have considered the relationships among biological, chemical, and hydrographic processes qualitatively. Let us examine next the key equation that brings out the quantitative aspects.

The terms of the equation are

$$\frac{\partial[N]}{\partial t} = \text{biological release} + \text{mixing}$$

$$+ \text{ advection} + \text{ gas uptake} \qquad (3.30)$$

$[N]$ represents the concentration of a nutrient or a nonconservative gas. The term $\partial[N]/\partial t$ is the local rate of change with time, that is, the change at a given point in the ocean and not at a point that moves with the water mass.

Redfield et al. (1963) added a biological source term R to the advective–diffusive equation and neglected gases. They then obtained the equation

$$\frac{\partial[N]}{\partial t} = R + \sum_i \frac{A_i}{\rho} \frac{\partial^2[N]}{\partial i^2} - \sum_i v_i \frac{\partial[N]}{\partial i}$$

$$(3.31)$$

with $i = x, y, z$. The more rigorous form of this equation would contain A_i/ρ within the second derivative $\partial/\partial i^2$, and v_i within $\partial/\partial i \cdot A_i$, ρ, and v_i change with position. A gas term source can be easily added to (3.31).

Even the simplified form (3.31), with A_i, ρ, and v_i held constant, is difficult to handle as we shall see later in this chapter. Most workers who applied Equation (3.31) either used a finite increment approach (e.g., Riley, 1951; McGill, 1964) or made assumptions about the invariance of some of the terms (e.g., Wirtky, 1962). The finite increment form means that the ocean is subdivided into finite boxes of edges Δx, Δy, and Δz and that $\partial[N]$ is replaced by $\Delta[N]$.

3.8.3 The Redfield Model

This model (e.g., Redfield et al., 1963) is often thought by biologists to be an oversimplification and yet time after time it has been shown to work and work well. Essentially, it consists in the use of ratios of nutrients present in marine organisms to elucidate changes in dissolved nutrients and gas concentrations during photosynthesis and oxidation, the exchange of gases, and hydrographic pro-

cesses. Since no attention is paid as to whether the nutrients (P, N, Si, etc.) are present in fats, carbohydrates, or proteins, the reason why the model works probably results from the essentially complete degradation of organic matter in the regions studied. Differential breakdown of various substances versus depth is incompatible with the model.

The argument presented by Redfield et al. (1963) is the following one. The proportions of carbon, nitrogen, and phosphorus are, when normalized to $TPO_4 = 1$,

	TCO_2	TNO_3	TPO_4
Zooplankton	103	16.5	1
Phytoplankton	108	15.5	1
Average	106	16	1

The units in the ratios TCO_2/TPO_4 and TNO_3/TPO_4 are in g-atom. TNO_3 represents $[NH_4^+] + [NO_2^-] + [NO_3^-]$.

When TPO_4, mostly as HPO_4^{2-}, is released by oxidation, 16 nitrogen atoms, essentially as NO_3^-, and 106 carbon atoms, as TCO_2, are also released. The following reactions occur:

$$CH_2O + O_2 = CO_2 + H_2O \qquad (3.32)$$

$$NH_3 + 2O_2 = HNO_3 + H_2O \qquad (3.33)$$

Thus, two atoms of oxygen oxidize one atom of carbon, whereas four atoms of oxygen oxidize one atom of nitrogen. The net result is that, per atom of phosphorus released during oxidation, 276 atoms of oxygen are utilized as is shown below:

$$2 \times 106 \text{ atoms } C + 4 \times 16 \text{ atoms of } N$$

$$= 276 \text{ atoms of } O_2 \qquad (3.34)$$

Equation (3.34) is obtained from Equations (3.32) and (3.33) plus the stoichiometry tabulated above. As an example, each atom of nitrogen utilizes four atoms of oxygen. As there are 16 atoms of N per atom of P, nitrogen utilizes 4×16 atoms of oxygen. The reverse occurs during photosynthesis.

On a g-atom/L-SW scale

$$\frac{\Delta[O_2]}{\Delta(TPO_4)_{ox}} = -276 \qquad (3.35)$$

In the usual units of μg-atom TPO_4/L and mL-O_2/L-SW, the above ratio becomes

$$\frac{\Delta[O_2]}{\Delta(TPO_4)_{ox}} = -3 \qquad (3.36)$$

$\Delta[TPO_4]_{ox}$ does not include changes in preformed phosphate. The mL-O_2 refer to STP (standard temperature = 25°C and pressure = 1 atm conditions). Expected changes in seawater follow the proportions

$$\Delta(TPO_4)_{ox} : \Delta(TNO_3)_{ox} : \Delta(TCO_2)_{ox} : \Delta(O_2)$$
$$= 1 : 16 : 106 : -276 \qquad (3.37)$$

It is important to note that the ratio -3 seldom occurs for the variations with depth of $[O_2]$ versus the actual measured TPO_4 unless the TPO_4 is corrected for the varying preformed phosphate. Let us examine this point starting with Equation (3.38)

$$TPO_4 = (TPO_4)_p + (TPO_4)_{ox} = (TPO_4)_p$$
$$+ \tfrac{1}{3}\{[O_2]_e - [O_2]\} \qquad (3.38)$$

applied to points at a vertical station. I remind you that $TPO_4 = [H_3PO_4] + [H_2PO_4^-] + [HPO_4^{2-}] + [PO_4^{3-}]$. $[O_2]_e$ is the equilibrium (saturation value of the oxygen at a given temperature and salinity). The term in brackets is known as the apparent oxygen utilization (AOU) since the waters left their surface sources where equilibrium with the atmosphere existed. It will be shown later that the AOU is the actual oxygen utilization.

Rearranging Equation (3.38),

$$[O_2] = -3(TPO_4) + 3(TPO_4)_p + [O_2]_e$$
$$= -3(TPO_4)_{ox} + [O_2]_e \qquad (3.39)$$

provides the relationship between $[O_2]$ and the other quantities at some point within the oceans. I remind you that $[O_2]$ and TPO_4 are the measured values of the in situ oxygen and phosphate and $(TPO_4)_p$ is the calculated preformed phosphate. $[O_2]_e$ is calculated from tables that yield the oxygen solubility versus T and S.

EXAMPLE 3.15. CALCULATION OF THE AOU, $(TPO_4)_p$, AND $(TPO_4)_{ox}$. Assume that measured values were $[O_2] = 5$ mL/L and $(TPO_4) = 3$ μg-atom/L for a deep sample brought to shipboard. Furthermore, for the in situ T (°C) and $S‰$, tables

yield $[O_2]_e = 6$ mL/L (see Chapter 5, Volume II). Then, as

$$TPO_4 = (TPO_4)_p + \tfrac{1}{3} \text{ AOU} \qquad (i)$$

$$\text{AOU} = 6 - 5 = 1 \text{ mL/L utilized} \qquad (ii)$$

$(TPO_4)_{ox} = \tfrac{1}{3}$ AOU $= 0.333$ μg-atom/L released
$(TPO_4)_p = TPO_4 - (TPO_4)_{ox} = 2.666$ μg-atom/L brought from the surface.

The preformed phosphate is a valuable quantity because it is a conservative one and can be used in conjunction with T and $S‰$ to track water masses (Pytkowicz, 1968). The AOU indicates the net use of oxygen regardless of the sources and fractions of the water types present.

EXAMPLE 3.16. THE MEANING OF THE AOU AND THE PREFORMED PHOSPHATE IN WATER MASSES (RMP). Consider two water types A and B with the source properties

$$[O_2]_{eA} = 6 \text{ mL/L} \qquad [O_2]_{eB} = 7 \text{ mL/L}$$

At a point X in the oceans they are present in the fractional amounts $f_A = 0.4$ and $f_B = 0.6$. The solubility of oxygen is almost linear for temperatures and salinities that are not too far apart. This is usually the case on mixing surfaces. Thus, the solubility in water mass at X is

$$[O_2]_{eX} = 0.4 \times 6 + 0.6 \times 7 = 6.6 \text{ mL/L} \quad (i)$$

We do not know f_A and f_B a priori but do obtain 6.6 mL/L from tables of $[O_2]_e$ versus the in situ T and S. Richards (1965) and Kester and Pytkowicz (1971) showed that surface values of $[O_2]$ in winter, when deep waters are formed, are essentially $[O_2]_e$, that is, solubilities.

Let us assume that 3 mL/L and 4 mL/L have been utilized in A and B by the time they reach X.

Then, the value $[O_2]_X$ actually measured at X also is

$$[O_2]_X = 0.4 \times 3 + 0.6 \times 3 = 3.0 \text{ mL/L} \quad (ii)$$

Thus, as

$$\text{AOU}_{AX} = 6 - 3 = 3 \text{ mL/L}$$

$$\text{AOU}_{BX} = 7 - 3 = 4 \text{ mL/L} \qquad (iii)$$

The AOU_X of the water mass is

$$AOU_X = 0.4 \times 3 + 0.6 \times 4 = 3.6 \text{ mL/L}$$

$$\text{(iv)}$$

The same result can be obtained from $[O_2]_{eX} - [O_2]_X$ where $[O_2]_{eX}$ is obtained from tables and $[O_2]_X$ from measurements.

In general terms,

$$AOU_X = [O_2]_{Xe} - [O_2]_X$$

$$= \sum_i f_i [O_2]_{ei} - \sum_i f_i [O_2]_i$$

$$= \sum_i f_i \{[O_2]_{ei} - [O_2]_i\} = \sum_i f_i \, AOU_i \quad \text{(v)}$$

The AOU has a nice feature when utilization in different waters is considered because it eliminates the effect of the sources. Thus, one may think at first that the same extent of utilization occurred in A and B as $[O_2]_{AX} = [O_2]_{BX} = 3$ mL/L. However, $AOU_{AX} = 3$ mL/L while $AOU_{BX} = 4$ mL/L.

Similar reasoning leads to

$$(TPO_4)_{pX} = \sum_i f_i \, (TPO_4)_{piX} \quad \text{(vi)}$$

Equation (3.39) shows us that $\Delta[O_2]/\Delta TPO_4 = -3$ only occurs when $(TPO_4)_p + [O_2]_e$ does not vary or varies little with depth. There is no reason for such a constancy although it, as well as constant ratios of TN/TPO_4, do occur. I determined a ratio of -3.00 in September off Oregon (Pytkowicz, 1964), while Postma (1964) obtained a ratio of 3.7. Redfield (1942) in the Atlantic and Richards (1965) in the Cariaco Trench found $\Delta TN/\Delta TPO_4 = 15$. These results lend indirect credence to the Redfield hypothesis and provide tools for studying oceanic processes such as O_2 exchange (Pytkowicz, 1964) and others that we shall examine soon.

In actuality the ratio 3 is expected for $\Delta[O_2]/\Delta(TPO_4)_{ox}$ with depth rather than for $\Delta[O_2]/\Delta(TPO_4)$. This cannot be verified directly because then Equation (3.41) leads to a circular argument. Still, the verification can be made for other couples such as TCO_2/AOU (Culberson and Pytkowicz, 1970).

Some of the considerations discussed above are illustrated in Figures 3.39 and 3.40. Thus, we observe that a mixture of water types does not yield a slope of -3 unless the two types have the same value of $(TPO_4)_p + (O_{2e})$. On the other hand, the ratio $[O_2]_e - [O_2]/(TPO_4)_x$ is expected to be -3 regardless of the history of the waters.

Earlier I cited several examples in which

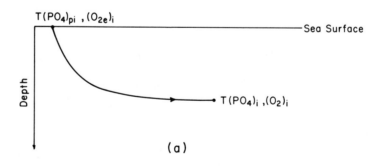

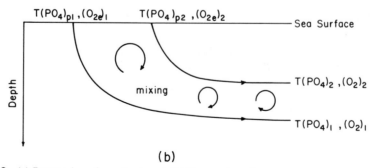

FIGURE 3.39 (a) Progression of a water type i. (b) Progression of two water types A and B which mix.

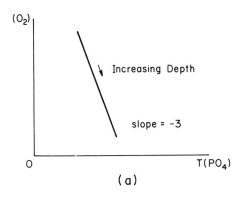

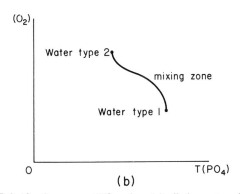

FIGURE 3.40 O_2 versus TPO_4 when (a) all the waters in the depth range of interest have a same constant $(TPO_4)_p = (O_{2e})$ or are of the same type i and (b) when two water types with different constants are mixing.

Redfield ratios were found for nutrient/nutrient and oxygen/nutrient ratios without correction for preformed quantities but advised the reader that these were exceptional cases. Culberson and Pytkowicz (1970), however, made actual corrections for preformed quantities and normalizations due to variations in salinity for $\Delta TCO_2/AOU$ with depth. The normalizations are required because (O_2), as an example, can change in part due to dilution resulting from mixing as is shown below.

EXAMPLE 3.17. NORMALIZATION PROCEDURE. Consider two waters M and N at different depths. We observe that $[O_2]_M = 6$ mL/L with $S_M(‰) = 34‰$ and that $[O_2]_N = 4$ mL/L with $S_N(‰) = 32$ at a shallower depth. Note that I reserve the symbols i, A, B for water types.

In the normalization procedure we attribute a small part of the low value of $[O_2]_N$ to dilution resulting from rainwater, as is indicated by the low value of $S_N(‰)$.

The normalization procedure consists in correct-

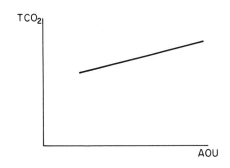

Observed corrected $\Delta TCO_2/AOU = 106/285$

Redfield $\Delta TCO_2/AOU = 106/276$

FIGURE 3.41 The TCO_2–AOU correlation below 1000 m in the Eastern Tropical Pacific [based on Culberson and Pytkowicz (1970)].

ing the values of (O_2) and of other nonconservative parameters to a normalizing salinity, usually 35‰. We obtain the normalized values by the expressions (35/34)6 mL/L (35/32)4 mL/L, with the results:

Water type	Original $[O_2]$ (mL/L)	Original $S‰$	Normalized $[O_2]$ (mL/L)
M	6.00	34	6.18
N	4.00	32	4.38

35‰ is used because it is approximately the highest value and, therefore, the least subject to dilution by rain or rivers, in normal seawaters.

Culberson and Pytkowicz (1970) found in the Eastern Tropical Pacific, after the appropriate normalizations, the relationship shown in Figure 3.41.

Note that the Redfield ratios cannot apply everywhere. As an example, consider what happens at the oxygen minimum layer shown earlier by examining Figure 3.42. From A at about 100-m depth to B at roughly 800 m there can be a slope of -3. However, at the oxygen minimum near 1000 m the slope must be zero and at D it is infinite. Below the oxygen minimum layer the Redfield ratio is, for unknown reasons, a little larger than 3.

3.8.4 Thoughts on Nutrients and Special Cases Related to the Redfield Model

First, I shall mention an interesting observation by Ketchum et al. (1958). It is known that algae grown in nitrogen-deficient cultures have low TN/TPO_4

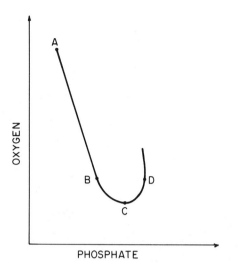

FIGURE 3.42 Oxygen and phosphate correlation including the oxygen minimum layer.

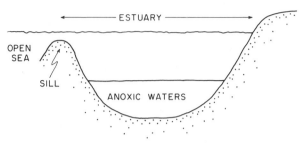

FIGURE 3.43 An anoxic estuary.

ratios. Yet, Ketchum et al. (1958) found that algae in extremely low TN waters in Long Island Sound had the normal 15–16 ratio. This appears to contradict the Redfield hypothesis but what happens is that the nitrogen deficiency is only apparent. In actuality there is a sufficient supply of remineralized nitrate but it is used up as soon as it arrives.

A second line of thought, based upon the work of Redfield et al. (1963), concerns the availability of nutrients versus the composition of plankton. The key data are shown below:

	TPO_4	TN	TCO_2	$[O_2]$
Ratio in plankton	1	16	106	
Ratio of change in seawater due to life	1	16	106	−276
Ratio of availability in seawater	1	15	1000	200–300

The middle line is different from that in Redfield et al. which is $1:15:105:-235$ because the -235 value is biased by uptake of O_2 from the atmosphere near the sea surface. This uptake by hydrodynamic transport from the actual surface masks in part the utilization and, therefore, the AOU.

We see that there is a surplus of carbon but that nitrate and phosphate can be depleted and become limiting constituents for life. Oxygen is not usually limiting due to uptake from the atmosphere into sinking waters. However, it can become limiting in poorly ventilated basins and fjords. A typical situa-

tion of this kind is shown in Figure 3.43. Dissolved oxygen is depleted if the estuary is not flushed with dense oxygenated waters which sink to the bottom and replenish it. Eventually oxygen is not available for oxidation and sulfur-reducing bacteria use the reduction of SO_4^{2-} instead. This leads to the generation of H_2S and life practically ceases to exist. These anoxic environments favor the precipitation of heavy metal sulfides. A review of anoxic waters was presented by Richards (1965).

Note above that the ratio of the availability of nitrate to phosphate about equals the ratio in plankton. This striking parallelism led Redfield (1958) to consider two alternatives as explanations; that organisms adapted to what was available in the water or that organisms control the proportions of nitrate and phosphate in seawater. He favored the latter one.

I consider the first mechanism more likely. My reasoning is as follows. When plankton in the photic layer approach the point at which they deplete nutrients, they must learn to live with what is available. The observation of Gordon Riley which was mentioned earlier, indicates the availability ratio can be 16 to 1 due to transport even if the TN/TPO_4 ratio in the water has changed. Thus, there really is no simple nutrient ratio–plankton ratio relationship needed by the Redfield explanation.

Another important consideration is that of the inter-relationships between nutrients, oxygen, and carbon dioxide. Culberson and Pytkowicz (1970) showed that TCO_2 changes are inversely correlated to those of the AOU and obey the Redfield ratios. This means that, when oxygen is utilized in the oxidation of organic matter, CO_2 is liberated in the amounts expected from the photosynthetic–oxidative reaction. This CO_2 is then distributed among the species CO_2, H_2CO_3, HCO_3^-, and CO_3^{2-} which are the components of TCO_2, and this distribution obeys the equilibrium relations of the carbonic acid system. These equilibria are fast relative to the supply or removal of TCO_2. The Redfield $(TCO_2)_{ox}$–AOU correlation only occurs if there is no extensive dissolution or removal of $CaCO_3$ or if a correction is made for this process. This is necessary because $\Delta CaCO_3$ changes the TCO_2 but not the AOU. I remind you that the changes in TCO_2 and AOU are not dictated by equilibria; only the speciation is.

The TCO_2 is straightforward and is the sum of the concentrations of the inorganic CO_2 species. It changes due to oxidation and $\Delta CaCO_3$. It is usually expressed in mmole/kg-SW. TA, the titration alkalinity, is more complex but for the time being I shall assume that its changes are due only to the dissolution or precipitation of calcium carbonate. In reality, there are also changes due to titration by acids released during oxidation. TA is expressed in meq/kg-SW. Thus, if 1 mmole of $CaCO_3$ dissolves, TCO_2 changes by 1 mmole but TA changes by 2 meq. The simplest relationship between $\Delta(TCO_2)$, $\Delta(TCO_2)_{ox}$, ΔAOU, and ΔTA is

$$\Delta TCO_2 = r_{C/AOU} \, \Delta AOU + 2 \, \Delta TA \quad (3.40)$$

$r_{C/AOU}$ is the Redfield ratio for carbon and AOU. I assume that these quantities have been normalized by the procedures of Culberson and Pytkowicz (1970).

EXAMPLE 3.18. SIMPLIFIED EXAMPLE OF THE ΔTCO_2–ΔAOU CORRELATION INCLUDING THE EFFECT OF $CaCO_3$ DISSOLUTION. We find at two depths in a given station:

106/276 is the Redfield ratio $r_{C/AOU}$ when O_2 is expressed in μg-atom/L instead of in the more usual mL/L.

3.8.5 Applications of the Redfield Model

Some Comments on the Value of the AOU. Gradients in conservative properties $[T, S, (TPO_4)_p]$ only yield the direction of motion of waters along extremum values. These originate at the surface and are weakened by mixing along the direction of motion. Changes elsewhere cannot be interpreted unambiguously for salinity. Let us consider, as an example, that $S\permil$ increases toward the north along an isentropic surface. We may interpret this gradient as being due to a low-salinity water moving to the north and mixing with higher salinity waters or vice versa. What makes extremum values useful, in the few cases in which they appear, is that they do start at the surface (e.g., the salinity minimum of the Antarctic Intermediate Water) relative to shallower and deeper waters so that there is no ambiguity in the advective direction.

Nonconservative properties, on the other hand, yield the direction of "motion," at least in the sense of the transition from younger to older waters and do so at all depths. This transition is indicated by the decrease in $[O_2]$ and the increase in AOU, TPO_4, TN, and TCO_2.

The AOU $= [O_2]_e - [O_2]$ is a better property than $[O_2]$ for water mass and biological analyses because it eliminates the effects of the sources on interpretations of utilization as we saw earlier.

Depth	TCO_2 (mmole/kg-SW)	AOU (μg-atom/kg-SW)	TA (meq/kg-SW)
500	2.410	550	2.320
1000	2.390	☐	2.318

Let us calculate, as an exercise, AOU_{1000} at 1000 m if the Redfield ratios are obeyed. As a simplifying assumption, let us neglect the effects of nitric and phosphoric acids released by oxidation on TA. Thus, changes in TA reflect only $CaCO_3$ dissolution or precipitation.

From $r_{C/AOU} = 106/276$ and Equation (3.40) we have

$$AOU_{1000} = AOU_{500} - \frac{276}{106} \times 0.016$$

$$= 544 \; \mu\text{g-atom/L}$$

Care should be exercised in the interpretation of AOU data for patterns of motion. Waters underlying regions of very high fertility may have large AOU's and yet be younger than adjacent waters that are below unproductive regions. Then, the gradient of AOU reflects the oxidation of settling debris rather than different ages. This type of behavior can happen when one moves away from fertile continental shelves. Thus, nonconservative properties also have ambiguities and should be used in conjunction with conservative properties.

Redfield in the Atlantic. The paper by Redfield (1942) on the processes determining oxygen, phosphate, and other organic derivatives is rather long and hard, although valuable. A summary follows.

A large fraction of the TPO_4 and of the AOU in deep waters of the Atlantic Ocean may be attributed to the decomposition of organic matter already brought from the high latitude sources. A significant fraction of this oxidation occurs within a few degrees from the source.

North of 40°S, in the Tropical Atlantic, there is an appreciable amount of local organic settling and oxidation. The mineralization occurs primarily in the upper 700 m.

The O_2 minimum and the TPO_4 maximum are due to the slow isentropic motion of waters in conjunction with the extensive oxidation of large organic contents brought from the surface sources.

Wirtky (1962) introduced the effects of vertical advection and diffusion but did not consider the oxidation of source organic matter.

The Southern Ocean. I presented the cores of the water masses in the Southern Ocean earlier, based on Pytkowicz (1968). T–S‰ diagrams north of the Antarctica Convergence did not show clearly the cores of the Antarctic Intermediate Water (salinity minimum) and the Pacific Deep Water (salinity maximum) as can be seen in Figure 3.44a.

The Antarctic Convergence was at 57°S during the 1964 Antarctic winter when I collaborated with Lamont Geological Observatory scientists aboard the USNS *Eltanin*. The convergence occurs where Sub-Antarctic and Antarctic Surface Waters meet. They produce the Antarctic Intermediate Water which sinks and moves toward the north.

The waters south of the Antarctic Convergence are of special interest because of their high fertility, among other reasons. Unfortunately, T–S‰ diagrams lose resolution there and I used T–$(TPO_4)_p$ diagrams instead (see Figure 3.45b. Nodes resulting from a T maximum and a TPO_4 discontinuity led me to locate what I termed the Returning Antarctic Intermediate Water.

This water mass can act as a fertilizing agent by carrying organic debris down to where they were remineralized, and returning the nutrients to the surface. Of course, the Pacific Deep Water also contributes nitrate, phosphate, and so on.

The Northeast Pacific. The AOU and the $(TPO_4)_p$ were used by Pytkowicz and Kester (1966) to study

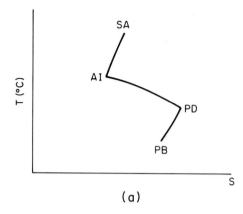

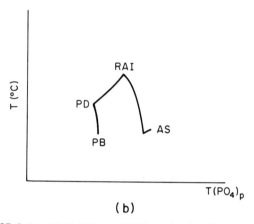

FIGURE 3.44 (*a*) T–S‰ at 160°W north of 57°S and (*b*) T–$p(PO_4)$ at 160°W, south of 57°S (Pytkowicz, 1968a). AS represents the Antarctica Surface Water, SA the Sub-Antarctic, AI the Antarctic Intermediate, PD the Pacific Deep, and PB the Pacific Bottom Water.

the circulation in the Deep–Intermediate Waters of the Northeast Pacific.

The 0–1000 m region was well known. For waters below 2500 m the classical view was that of a sluggish southward transport as part of a clockwise gyre due to the Coriolis force (Sverdrup et al., 1942) while Stommel (1958) advocated a northeast motion. McAllister (1962) concluded that there was a general northward motion below 1200 m but did not look at individual depth ranges.

We examined the AOU gradients along the surfaces of $\sigma_t = 27.42$, 27.60, and 27.70 which correspond closely to 1000-, 1500-, and 2000-m depths. The trends on all the surfaces were roughly the same so I only show the results for 1000 m in Figure 3.45. This is a two-dimensional representation of three-dimensional processes as, otherwise, there would be a surplus of water at B, which may be a

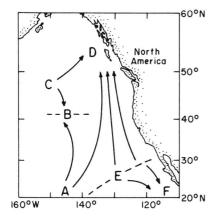

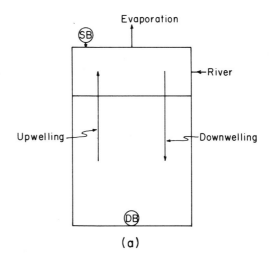

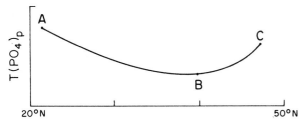

FIGURE 3.45 (a) The AOU at 1000 m and (b) the $T(PO_4)_p$ in the Northeast Pacific (Pytkowicz and Kester, 1966). B is a convergence and E a divergence.

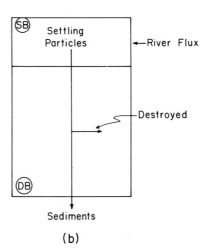

FIGURE 3.46 (a) Solute fluxes in the two-box model. (b) Particulate matter fluxes. SB is the surface and DB the deep box (or ocean).

region of convergence or a new water mass at B, and a deficit in region E where waters diverge. A and C are water masses.

The plot so far does not tell us whether the waters in the region B have a maximum AOU only as the result of mixing between A and C or if there is an additional water mass present. This question is solved by means of the preformed phosphate as shown in Figure 3.46b. If B were simply a mixture of A and C then it would be on the straight line $\overline{AC}$. The occurrence of a minimum value shows that B is a third water mass, presumably coming from the west. Thus, the AOU and $(TPO_4)_p$ permitted us to analyze the waters present to a greater extent than what can be done by physical oceanographic methods alone.

There are further applications of the Redfield model, which at times use empirical extensions. Examples are: Postma (1964), Sugiura (1965), Park (1967), Alvarez–Borrego (1973), and Johnson and Pytkowicz (1979). An analysis of changes in TCO_2 and AOU, as a precursor of later papers on the entry of fossil fuel CO_2 into the oceans, can be found in Chen (1978).

3.8.6 Broecker Model

Broecker (1974) considered the chemical modeling of the oceans in terms of the simplest possible box model; the surface box (SB) and the deep box (DB). The corresponding solute fluxes are shown in Figure 3.46a while in Figure 3.46b the particulate fluxes of a given element are presented. His purpose was to estimate fluxes and budgets within and through the oceans.

From the conservation of mass, J_{partic} is

$$J_{partic} = J_{river}c_{river} + J_{mix}c_{deep} - J_{mix}c_{surf}$$

$$(3.41)$$

J_{river} is the annual river flux into the oceans and J_{mix} is the annual mixing flux of water across the pycnocline. This flux is the same upward as downward because there is no net flow of water in mixing. c_i is the average composition of reservoir i (river, SB, DB). The symbol J_{partic} is the total particulate flux which may include soft and hard parts of organisms.

From the above relation Broecker derived

$$f = \frac{1}{1 + (J_{mix}/J_{river})(c_{deep}/c_{river} - c_{surface}/c_{river})}$$

(3.42)

and

$$fg = \frac{1}{1 + J_{mix}c_{deep}/J_{river}c_{river}}$$

(3.43)

g is the fraction of an element reaching SB which is removed by settling as particulate matter, after being dissolved in DB and upwelled. The symbol f represents the fraction of a given element in particular form which survives destruction and is incorporated into the sediments.

Through the use of estimated quantities, Broecker found fg, the fraction of an element removed per oceanic mixing cycle, to be 0.01 for phosphorus. Thus, 1% of the phosphorus in the oceans is removed into the sediments per residence time of the oceanic waters, for which Broecker uses 1600 yr (I prefer 1000). The value of g is 0.95, that is, 95% of the phosphorus reaching SB is taken down in particulate matter. $J_{mix}/J_{river} = 20$ according to Broecker with $J_{mix} \cong 2$ m/yr.

The relatively large removal of TPO_4 suggests a high reactivity and, hence, a relatively short residence time. Broecker calculated $\tau = \tau_{SW}/fg = 2 \times 10^5$ yr when $\tau_{SW} = 1600$ yr is used. This is indeed short when compared to about 2×10^8 yr for sodium. Note that τ corresponds to a steady state in which rivers contribute and offset the 1% of the phosphorus lost to the sediments during τ.

An interesting point is made to the effect that nutrients and shell materials (biolimiting and biointermediate) have shorter residence times (3×10^4 to 6×10^5 yr for P, Si, Ca) than biounlimited constituents (2×10^7 to 2×10^8 yr for S and Na). This suggests that biological uptake is one of the important removal mechanisms of elements from the oceans. The term biounlimiting means that the element in question is never used up to such an extent that its scarcity limits life.

3.8.7 Riley in the Atlantic Ocean: A Box Model

Riley (1951) examined the factors that control the distributions of O_2, TPO_4, and TN in the Atlantic in terms of advection, diffusion, and oxidation. He made no corrections for preformed quantities so that his O_2/TPO_4 and O_2/TN ratios are not expected to be the correct oxidative ones.

Riley divided the Atlantic into large cubes and applied the hydrobiochemical equation in finite increment form to them. In such an equation, derivatives such as $\partial N/\partial x$, where N is a general property, are replaced by $\Delta N/\Delta x$. A steady state was assumed, that is, $\partial N/\partial t$ was set as zero.

Advective velocities were obtained by the usual geostrophic calculations and were used in conjunction with variations in salinity to obtain eddy diffusion coefficients. The water flux equation, that is, the hydrobiochemical equation without the biological source term, was applied for this. Then, the source term R could be calculated as being numerically equal to the physical removal because the system was assumed to be at a steady state.

Riley discussed the changes in oxygen and nutrients with depth as functions of types of biological processes, for example, plankton versus bacteria metabolisms. One of his most useful results, elaborated upon later by Wirtky (1962) was that the oxygen consumption rate decreases roughly exponentially with depth, below around 1000 m.

It is implicit in the work of Riley (1951) and Wirtky (1962) that physical and biological processes must be called on concurrently to explain chemical distributions. Thus, although Riley made the mistake of not using Redfield's concepts, he extended the scope of Redfield's work in which the biological effects were overemphasized. McGill (1964) also analysed the Atlantic, along lines roughly parallel to those of Riley.

Some average values obtained by Riley in surface to intermediate waters are

σ_t	26.5	26.7	26.9	27.1	27.3	27.5	27.7
R	−67.3	−25.2	−16.4	−17.1	−11.4	−4.2	−1.3

where R is the rate of oxygen utilization in 10^{20} mL-O_2/L-SW · sec. The σ_t values correspond roughly to the depth range 100–1500 m.

3.8.8 The Continuum Model of Wirtky

Continuum models are preferable to box models because most oceanic properties vary continuously. Still, the solutions of such models are difficult and may require assumptions that are not required by the simpler box models. The choice between approaches depends on the problem on hand, the data available, and the mathematical difficulties in the solution of the relevant equations.

Wirtky (1962) was primarily interested in explaining the shape of the dissolved oxygen versus depth curves and the oxygen minimum depth. He considered hydrographic and biological processes concurrently but omitted the effect of the oxidation of material brought from the surface sources.

Earlier there had been two extreme views of the oxygen minimum. First, there was the biological approach of Redfield (1942) to the effect that the minimum was formed primarily near the surface source of intermediate waters by intense oxidation of rich suspended organic matter. This was followed by the oxidation of settling organic debris while the waters moved to lower latitudes. The view of Redfield (1942) has been well documented as the biological part of the answer, but the physical aspects were underemphasized. The proponents of the second approach, the physical (e.g., Wust, 1954), claimed that the oxygen minimum resulted from a sluggish circulation and slow oxygen replacement at the O_2 layer. Wirtky did not require such a sluggish motion to make his model work but felt that the biological explanation, without a contribution from advection and eddy diffusion, was not sufficient to explain the shape of the oxygen versus depth curve. I agree with this conclusion but consider that he should have taken the effects of the past history of the waters into account when he solved the $[O_2]$ versus z problem as a one-dimensional model. Still, some interesting features emerged from the calculations.

Wirtky set the hydrobiochemical equation in the reduced form

$$A_z \frac{\partial^2 [O_2]}{\partial z^2} - v_z \frac{\partial [O_2]}{\partial z} = R_z^{(s)} \qquad (3.44)$$

He assumed a steady state, that is, $\partial A/\partial t = 0$. Furthermore, he set A_z and v_z as constant with depth, and considered the horizontal terms to be negligible. This last assumption is often a good one if A_x and A_y are negligible and if the gradients of $[O_2]$ in the x and the y directions are small. This is so provided that what is a purely local process is being considered in the z direction, if horizontal terms are nearly constant for a distance around the site of interest. It fails, however, when an integrated effect over very large values of the horizontal coordinates

must be taken into account. $R_0^{(s)}$ in Equation (3.44) is the utilization, that is, it is a sink term.

The distribution of temperature yielded $v_z/A_z = (1 - 2 \times 10^{-5})$/cm. The term $R_z^{(s)}$ was obtained by fitting the exponential curve

$$R_z^{(s)} = R_0^{(s)} e - \alpha z \qquad (3.45)$$

with $R_0^{(s)} = 10^{-8}$ mL/L $\cdot$ sec to the data of Riley (1951). The term R_0 was obtained by curve fitting below the wind-mixed layer to avoid seasonal effects.

The solution of Equation (3.44) was

$$[O_2] = C_1 + C_2 \exp(- \frac{v_z}{A_z} z) + C_3 \exp(-\alpha z) \qquad (3.46)$$

with

$$C_3 = \frac{R_0^{(s)}}{A_z \alpha (\alpha - v_z/A_z)} \qquad (3.47)$$

C_1 and C_2 were the observed values of $[O_2]$ at 600 and 3000 m. A problem that was not made explicit is that, although v_z/A_z was known, A_z was not. Thus, there are three adjustable parameters, $R_0^{(s)}$, α, and A_z. They are enough to allow the fitting of observed and theoretical curves (see Figure 3.47) without proving that Equation (3.44) is the correct one for the $[O_2]$ distribution. In any event, we know that this is the case because a three-dimensional problem was reduced to a one-dimensional case.

Wirtky concluded that he could explain the observed curve in terms of A_z, v_z, and R because the neglect of A_z or v_z would yield a curve without the concavity present at about 2 to 3 km. A partial ex-

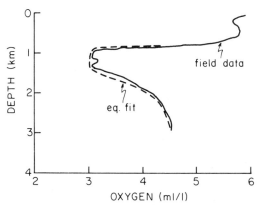

FIGURE 3.47 The fit of Wirtky's equation to field data (Wirtky, 1962).

tension to two dimensions was made at the end of the paper.

Wirtky provides insights into vertical distributions of oxygen but the fitting procedure used means that his explanation is possible but not necessarily the right one. We do know that his explanation may be the right one for the shape of the curve. It cannot, however, be the quantitative explanation because the development of the oxygen minimum starts already in the Southern Ocean and is a three-dimensional problem.

The model of Wirtky encouraged further research not only with oxygen but also with isotopes (e.g., Craig, 1969, 1970; Kroopnick, 1970; Culberson, 1972).

3.8.9 The Work of Kuo and Veronis

Kuo and Veronis (1970) studied the three-dimensional circulation problem with oxygen and ^{14}C. Their work will only be mentioned briefly because their primary purpose was physical.

These authors attempted to determine the relative roles of advection and eddy diffusion from the measured O_2 and ^{14}C distributions, used in conjunction with the hydrobiochemical equation and the circulation model of Stommel (1958) for abyssal velocities. The relative values of advection and mixing were obtained by fitting the equation to the data.

Later, Kuo and Veronis (1973) repeated the work with an improved numerical procedure and more recent abyssal oxygen data. Some key results in terms of rough averages, under abyssal conditions, were $A_z = 6 \times 10^6 \, cm^2/sec$, $v_z = 1.5 \times 10^{-5} \, cm/sec$ (upwelling), $R = 0.002 \, mL/L \cdot yr$ and a water transport in the Antarctic Circumpolar Drift of $35 \times 10^6 \, m^3/sec$. The problem of strong oxidation near the source pertains to intermediate waters and was not encountered in abyssal waters.

3.8.10 Anoxic Waters

A different set of chemical processes arises in regions where the oxygen is depleted and bacteria produce H_2S by using SO_4^{2-} as a source of energy sulfide (Richards, 1965).

Anoxic environments are important geochemically because they lead to the sedimentation of the sulfides of many heavy metals. Such environments can occur in fjords, estuaries with shallow sills, trenches, and basins at the ocean bottom. Another significant class of sulfide-bearing waters is the pore

waters of some organic-rich submarine and lake sediments, below the depth to which oxygen can penetrate to a sufficient extent to maintain oxidizing conditions.

Richards (1965), after presenting the general photosynthetic–oxidative reaction shown earlier, states that the next oxidative reaction, once all the oxygen is consumed, is the denitrification

$$(CH_2O)_{106}(NH_3)_{16}H_3PO_4 + 84.8HNO_3 = 100CO_2$$
$$+ 42.4N_2 + 148.4H_2O + H_3PO_4 \quad (3.48)$$

Possibly, the oxidation of ammonia

$$5NH_3 + 3HNO_3 = 4N_2 + 9H_2O \quad (3.49)$$

also occurs.

Sulfate reduction occurs last and becomes the source of energy for bacteria. The reaction is

$$(CH_2O)_{106}(NH_3)_{16}H_3PO_4 + 53SO_4^{2-}$$
$$= 106CO_2 + 53S^{2-} + 16NH_3$$
$$+ 106H_2O + H_3PO_4 \quad (3.50)$$

The usual Redfield ratios relative to O_2 obviously do not apply under these circumstances although TN/TPO_4 is still 16. A new relation appears as $[NH_3]$ versus $[S^{2-}]$ yields a straight line of slope $16/53 = 0.302$.

Some fjords become anoxic due to eutrification. This term means overutilization of oxygen in the oxidation of extra large amounts of organic matter. This excess matter results from the input of nutrients produced during human activities (fertilizers, detergents, sewage). As an example, basins in the Baltic Sea lost their oxygen for a period. Other anoxic fjords such as Saanich Inlet in Vancouver Island are due to infrequent flushing by oxygen-rich seawater.

3.9 LAKES

3.9.1 General

In the preceding chapters I developed the general features of the oceans. This could be done in a relatively easy way because of the relative invariance of the main constituent aspects of seawater.

The broad physical, chemical, biological, and geological features of lakes and rivers are exceedingly variable and a fairly extensive survey would require too much space, while not being in the mainstream of this book. For these reasons, I shall only touch on selected descriptive aspects of these systems next and shall treat the equilibrium and kinetic aspects in specialized chapters.

The large variability in the composition of lakes can be visualized when one contrasts moraine lakes, fed by glaciers on rocky terrains, and lakes in fertile agricultural regions.

3.9.2 Origins of Lake Basins

1. Lakes formed by the melting of present and past glaciers. They are supplied primarily by glacial flour derived from the physical and chemical weathering of rocks exposed to the glacier. The flour may consist mainly of aluminum silicates in igneous regions or of calcareous materials in limestones and dolomites. Cyclic salts are brought by snow and rain.
2. Tectonic lakes when basins resulting from tectonism (warping, faulting, folding) are filled with water.
3. Lakes formed by volcanic activity when craters and lava depression (physical or chemical) result in basins.
4. Solution lakes formed by basins that result from dissolution by rain and rivers.
5. River lakes resulting from the partial damming of streams.

3.9.3 Dimensions

This section is based mainly on Hutchinson (1957). Let us define the following terms:

$$\text{mean depth } \bar{z} = \frac{V}{A}$$
$$\text{maximum depth} = z_M$$
$$\text{length} = l$$
$$\text{volume} = V$$
$$\text{area} = A$$

The dimensions of some large lakes are shown in Table 3.21.

Lakes plus rivers contain only 0.002% of the volume of water in the hydrosphere but play a major role in life, including human societies in their

TABLE 3.21 Dimensions of some lakes [selected from Hutchinson (1957)].

Lake	$\bar{z}$ (m)	z_M (m)	A (km^2)	V (km^3)
Tanganyika	572	1,470	34,000	1,900
Nyasa	273	706	30,800	8,400
Superior	145	307	83,300	12,000
Victoria	40	79	68,800	2,700

physiological and geographical aspects, and weathering.

3.9.4 Physical Aspects

One important distinction between freshwater lakes, rivers, and oceans is that the former have a temperature of maximum density at 4°C. Seawater does not have a temperature of this kind at the usual salinities.

Consider a lake with a deep summer temperature of 4°C and a near-surface one of 12°C. Furthermore, let the surface temperature reach 2°C during winter. The surface water is stable because it is less dense than the deep water and a thermal inversion, that is, $dT/dz > 0$, occurs. Eventually the surface may freeze. Such an inverse thermal stratification can only occur in the oceans if $dS‰/dz$ is large enough to cause a positive density gradient in spite of the colder surface water.

Freshwater lakes have pycnoclines but, in contrast to the oceans, they are always thermoclines. The thermal structures of lakes depend on the climate, latitude and radiation input, mean depth, volume, and so on. Shallow and small volume lakes reflect more closely changes in air temperature than deeper and larger ones. The part of a lake that is above the thermocline is known as the epilimnion, whereas the deeper part is the hypolimnion.

Note that in most lakes the density is essentially only a function of temperature since the salt content is minimal. The main hydrological features are the winter overturn, thermal convection, eddy diffusion, and wind-driven transport.

3.9.5 The Main Constituents of Lake Waters

In summer the seasonal thermocline can block the effective exchange of solutes across it, unless a lake is shallow and the thermocline is dissipated by winds.

The salinity in limnology is the total dissolved inorganic matter in the water obtained by evaporation, which yields the total weight of solids per unit mass or volume of the lake water, and then ignites the residue so that it contains only nonvolatile components. This last step eliminates organic matter. CO_2 from carbonates is lost and the CaO can be used as in the seawater salinity. For $Ca(HCO_3)_2$ ignition leads to $CaO + CO_2 + H_2O$. Some workers, for example, Clarke (1959), include SiO_2, Al_2O_3, and Fe_2O_3 in the definition of salinity. The salinity can vary over wide limits, from about 100 mg/L for the average lake to 226,000 mg/L for the Dead Sea.

The main ions present in lakes are Na^+, K^+, Mg^{2+}, Ca^{2+}, CO_3^{2-} (actually mostly HCO_3^-), SO_4^{2-} and the halides. They are brought to the lakes by rivers and rain. In contrast to the oceans, however, the concentrations in most lakes follow the order Ca > Mg > Na > K, although this order can be reversed for soft waters.

Lakes can be classed chemically as bicarbonate, in which $[HCO_3^-] > [SO_4^{2-}] > [Cl^-]$, sulfate, or chloride. The first type is the most common.

only on processes within them but also on the chemical compositions of inflows and outflows.

In general, one should distinguish between rivers, lakes, and groundwaters. I use the term groundwater conventionally to indicate flowing subterranean waters, semistationary water lenses, and the soil waters that occupy the interstitial spaces in soils. The latter can have bulk values or can be waters of hydration. Soil waters often have inputs balanced by evaporation. They are rich in decaying organic matter when under fertile regions.

Some typical lake compositions are shown in Table 3.22. Hard lakes are relatively rich in calcium because of limestone and/or dolomite drainage. The mean composition of lakes as a function of the weathering sources is shown in Table 3.23.

Carbonate lakes may have salinities ranging from 10 to 250 mg/L. For salinities of 10.5 and 246.2 in mg/L the composition is

Salinity	Na^+	K^+	Mg^{2+}	Ca^{2+}	CO_3^{2-}	SO_4^{2-}	Cl^-	Specific Conduct $\times$ 10^6 at 25°C
10.5	0.7	0.3	0.4	2.5	4.4	1.5	0.7	20
246.2	16.6	6.0	9.9	59.0	103.2	34.9	16.6	400

Two types of lakes are found from the standpoint of fluxes of matter, closed and open.

1. Closed lakes have sluggish flow of rivers through them and the composition is dominated by the precipitation of least soluble salts during evaporation.

2. Open lakes have river fluxes through them and have compositions which depend not

We see that the relative composition of the major constituents of lakes changes with the salinity in contrast to that of the oceans. This probably results from secondary changes in the nature of drainage basins, evaporation, and a more intimate contact with the sediments.

Silica in lake waters can range from 0 to over 70 mg/L. Its total amount and vertical profile depend on sources such as weathering of aluminum silicates

TABLE 3.22 Composition of some lakes in terms of mean equivalent proportions (Hutchinson, 1957).

	Mean of Four Wisconsin Lakes (Soft)	Mean of Two Swedish Lakes (Hard)	Large Alpine Lake (Hard)	Mean of North German Lakes (Soft)
Na^+	10.9	13.6	4.5	43.0
K^+	4.8	2.2	1.9	6.7
Mg^{2+}	37.7	16.9	25.4	14.3
Ca^{2+}	46.9	67.3	68.2	36.0
CO_3^{2-}	69.6	74.3	85.4	42.4
SO_4^{2-}	20.5	16.2	10.8	14.1
Cl^-	9.9	9.5	3.9	43.5

TABLE 3.23 Mean composition of lakes as a function of weathered sources (Hutchinson, 1957) in equivalent proportions of each element.

	Igneous Rocks	Sedimentary Rocks	Waters from Igneous Rocks	Mean River
Na	20.1	4.8	30.6	15.7
K	9.8	8.0	6.9	3.4
Mg	33.1	34.0	14.7	17.4
Ca	37.1	53.2	48.3	63.5
CO_3		93.8	73.3	73.9
SO_4		6.2	14.1	16.0
Cl			12.6	10.1

and sinks of which the main one is the utilization by diatoms and other siliceous organisms.

The trace metal chemistry is subject to the intimate contact between sediments and supernatant waters, especially in shallow lakes.

One chemical problem of freshwaters is the seasonal and regional variability of the relative composition as well as the dilution of the solutions. This means that such waters cannot be used as ionic media for equilibrium calculations (see Chapter 5, and Chapters 1 and 5 in Volume II).

3.10 RIVER INPUTS, RELAXATION TIMES, AND OCEANIC CONTENTS

3.10.1 General and Composition

Rivers are a crucial link between the continents and the oceans although wind (aeolian) transport is also significant. The water fluxes and compositions of rivers depend on the drainage basin, the season, and human activities. The variability can be en-

visioned by examining the compositions of rivers presented by Livingstone (1963) (see Table 3.24). The inorganic solute composition of a world-average river, which can be used for geochemical modeling, is presented in Table 3.25.

Garrels and Mackenzie (1971) examined the SiO_2 content of the continental-average rivers and distinguished between the HCO_3^- caused by the weathering of igneous rocks and the HCO_3^- caused by the weathering of carbonate rocks. The results reflected the geological nature of the continents. The continents follow the order (high SiO_2) Africa–South America–Asia–North America–Europe (high $CaCO_3$) with carbonate weathering faster than silicate.

The world-average river has the composition shown in Table 3.25 which totals about 3.09 meq-ions/L the flux of suspended matter (inorganic and organic) brought to the oceans by rivers each year; this is within the range 85 to 217 × 10^{14} g.

TABLE 3.25 Composition of the world-average river based on the data of Livingstone (1963) and presented in ppm as well as meq/L.

	meq/L	ppm
Na^+	0.270	6.3
K^+	0.059	2.3
Mg^{2+}	0.342	4.1
Ca^{2+}	0.750	15
Cl^-	0.220	7.8
SO_4^{2-}	0.234	11.2
NO_3^-	0.016	1
HCO_3^-	0.958	58.4
Fe^{2+}	0.024	0.67
SiO_2	0.218	13.1
Total	3.092	

TABLE 3.24 Average composition of river waters for the different continents according to the data compiled by Livingstone (1963) in ppm.

	North America	South America	Europe	Asia	Africa	Australia
Na^+	9	4	5.4	9.3	11	2.9
K^+	1.4	2	1.7	—	—	1.4
Ca^{2+}	21	7.2	31.1	18.4	12.5	3.9
Mg^{2+}	5	1.5	5.6	5.6	3.8	2.7
Fe	0.16	1.4	0.8	0.01	1.3	0.3
SiO_2	9	11.9	7.5	11.7	23.2	3.9
Cl^-	8	4.9	6.9	8.7	12.1	10
SO_4^{2-}	20	4.8	24	8.4	13.5	2.6
NO_3^-	1	0.7	3.7	0.7	0.8	0.05
HCO_3^-	68	31	95	79	43	31.6

The world-average river composition can be used (a) to describe the result of weathering, (b) illustrate the effect of oceanic reactions in the conversion of river solutes into submarine sediments, (c) show in a very general manner the effect of mixing in a world-average estuary on the properties of the resulting waters, and (d) illustrate the effect of rivers on the S‰–Cl‰ relation.

Groundwaters also reflect the composition of their drainage basins but can have additional properties. As an example, they often have a higher CO_2 load than river waters as the result of the oxidation of buried organic matter and are, therefore, the stronger weathering agent.

3.10.2 Cyclic Salts

Oceanic salts are cyclic in that they are ejected from the sea surface by bursting bubbles and eventually contribute significantly to the river solutes. The salts may fall out on the oceans and the continents as dry particles or while dissolved in rain and snow. Chesselet et al. (1972) observed the fractionation of cyclic salts in aerosols, rain, and in the East Antarctica Plateau. They found that the $[Cl^-]/[Na^+]$ was about that of seawater but that $[Na^+]/[K^+]$, $[Na^+]/[Ca^{2+}]$, and $[Ca^{2+}]/[Mg^{2+}]$ were smaller than the oceanic ratios. The mechanism for the difference in the enrichment of various ions in marine aerosols has not been entirely ascertained. It may be related in part to equilibria and kinetics in surface-active organic films on the bursting bubbles.

It is important for weathering and residence time calculations to take into account the cyclic content of rivers. Garrels and Mackenzie (1971) presented the following values:

Constituent	Na^+	K^+	Mg^{2+}	Ca^{2+}	Cl^-	SO_4^{2-}	HCO_3^-
Percentage of cyclic constituent by weight in rivers	35	15	7	0.7	55	6	0.2

3.10.3 Inputs, Relaxation Times, Residence Times, and Oceanic Contents

Oceanic steady-state contents B_0 have usually been interpreted in terms of residence times τ_B. I shall show next that this is not sufficient since two other factors, rates of weathering and relaxation times, must also be taken into consideration.

The steady-state solution of Equation (2.74) is

$B_0 = k_{AB}A_0/k_{BC}$. Thus, B_0 depends not only on $k_{BC} = 1/\tau_B$ but also on A_0, the amount in the weathering reservoir, and $k_{AB} = 1/\tau_A$. The term τ_A is the residence time of the weathering reservoir. $k_{AB}A_0$, of course, is the flux J_{AB}. For a given A_0, the magnitudes of B_0 are affected by k_{AB} and k_{BC}, the reactivities, as is shown below.

k_{AB}/k_{BC}	k_{BC}	$\tau_B = 1/k_{BC}$	B_0
Large	Large	Short	Medium
Large	Small	Long	Large
Small	Large	Short	Small
Small	Small	Long	Medium

Thus, we see that the reactivity on land indeed affects B_0. The oceanic content of course is also affected by A_0.

$\tau_B = 1/k_{BC}$ is the relaxation time during the build-up time of the oceanic content to $1 - 1/e$ of its final value B_0 and also is the residence time at the steady state. The numerical value of τ_{BC} is the same during these two stages. We see from Equation (2.74) that a long τ_{BC} implies a long time for the attainment of the steady state and, from the solution of this equation, that a long τ_{BC} implies a large B_0. This is shown in Figure 3.48 and simply means that the amounts of a given substance in the ocean have time to build up to larger values of B_0 when the removal reactivity k_{BC} is small.

I should mention at this point that the rigorous value of τ_B is

$$\tau_B = \frac{B_0}{J_{AB} - J_{\text{cyclic}}} \qquad (3.51)$$

where J_{cyclic} is the flux of cyclic salts present in rivers after leaving the oceans to the atmosphere and being returned to land by rain and snow.

Characteristic values of τ_B, sorted by me from Goldberg (1965) for classes of constituents, are presented in Table 3.26. Na has a long τ_B because it is taken up very slowly by clays, whereas in the case of Cl probably τ_B is also large because it is con-

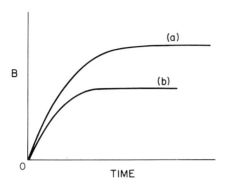

FIGURE 3.48 (a) large τ_{AB} leads to a large B_0 and a slow attainment of the steady state. For (b), the reverse is true.

TABLE 3.27 Inversions in the oceanic content B_i^0 versus τ_i correlation found in Goldberg (1965). River fluxes $(J_i)_{AB}$ from B_i^0/τ_i.

Element	τ_i (10^6 yr)	B_i^0 (10^{20} g)	$(J_i)_{AB}$ (10^{14} g/yr)
Na	210	147.8	0.704
Mg	22	17.8	0.809
Ca	1	5.6	5.6
K	10	5.3	0.53
Sr	10	0.11	0.011
Si	0.035	0.052	1.48
Li	12	0.0023	0.00019

Source: Reprinted with permission from E. D. Goldberg, *Chemical Oceanography*, Vol. 1, pp. 163–196, © 1965, Academic Press, Inc. (London) Ltd.

trolled by the slow formation of evaporites. Nutrients such as P are probably removed in buried organic debris, including remnants of hard parts of organisms, a process with intermediate τ_B's. Trace metals can be taken out relatively fast as oxides and hydroxides in silica and clays, sulfides formed in anoxic environments, and organic matter.

From the standpoint of inputs, the long τ_B of Na$^+$ means a small k_{AB} as $\tau_B = 1/k_{AB} = 1/k_{BC}$ at the steady state. Thus, the long τ_B indicates a low average weathering reactivity of igneous rocks plus evaporites. The reactivity is the rate of weathering per unit mass of the reservoir under study, that is, J_{AB}/A_0. Thus, we see that B_0 does not depend only on $\tau = 1/k_{BC}$. The good correlation found often between τ and B_0 appears to indicate that k_{BC} varies more than k_{AB} in going from element to element. Of course, this interpretation is in terms of the ratios of the reactivities (weathering and sedimentation per unit mass) alone but A_0 also enters into the values of B_0.

TABLE 3.26 Oceanic residence times τ_B and oceanic contents B_0 for classes of constituents.

Constituent	Class	τ_B (10^6 yr)	B_0 (10^{20} g)
Na	Major	210	147.8
K	Major	10	
Mg	Major	22	5.3
Ca	Major	1	17.8
Mg	Major	17.8	5.6
P	Nutrient	0.2[a]	
Si	Nutrient	0.6[a]	
Al	Trace metal	0.0031	0.00014
Cu	Trace metal	0.043	4.1×10^{-5}
Pb	Trace metal	0.00056	4.1×10^{-7}

[a] Broecker (1974).

Typical examples of inversions in the B_0 versus τ_B correlations, found in the data of Goldberg (1965), are shown in Table 3.27. They are due to the neglect of J_{AB}.

The data in Table 3.27 are arranged in order of decreasing oceanic contents and the two inversions, in which τ_B is too small (for Ca and Si) correspond to large river inputs.

In the case of solute–solid equilibrium, the differential equation (Pytkowicz 1971b, 1972) is

$$\frac{dB}{dt} = k_{AB}A^0 - k_{BC}[B_i - (B_i)_e] \quad (3.52)$$

where $(B_i)_e$ is the saturation value. Note that an additional sink is needed if $B_e > B$ and a stationary ocean is to be reached. $B_i - (B_i)_e$ expresses a tendency toward equilibration which may result from the interaction of the freshwater or seawater solutes with settling detrital solids (e.g., clays), the formation of an authigenic precipitate (e.g., aragonitic oolites in the Bahama Banks), or sediments in which case k_{BC} is akin to a diffusion coefficient.

The solution of Equation (3.52) is

$$B_i = (B_i)_e (1 - e^{-k_Bct}) + A^0 \frac{k_{AB}}{k_{BC}} (1 - e^{-k_Bct})$$

$$(3.53)$$

At the steady state ($t \to \infty$) this equation is reduced to

$$\frac{B_0 - (B_i)_e}{J_{AB}} = \frac{B^0 - (B_i)_e}{(J^0)_{BC}} = \frac{1}{k_{BC}} \quad (3.54)$$

and to

$$B_0 = B_e + \frac{k_{AB}}{k_{BC}} A_0 \qquad (3.55)$$

Equation (3.54) shows that the residence time has an additional term B_e relative to the earlier model and that $\tau = B_0/(J_i)_{AB}$ can only be used when there is no tendency toward a solid–solution equilibrium. From Equation (3.57) we see that the equilibrium $B_0 = (B_i)_e$ can only occur for the stationary value B_0 if the last term is negligible when compared to B_e.

3.10.4   A Note on Estuaries

Estuaries vary enormously in their physical, chemical, biological, and geological features. In addition they are subject to tidal and seasonal effects. I shall not attempt to elaborate on these features but will refer the reader to Church (1975) and Burton and Liss (1976). These two references cover chemical features that shall be discussed in relevant chapters of the present work.

I shall only mention at this time that estuaries are sites of mixing of river and seawater, in which the dense seawater often wedges under the river water. There are, however, estuaries in which entering waters are less dense than deep waters, as in the somewhat extreme case of the Mediterranean Sea. The dynamics of mixing do not yield simple gradients of salinity because of structural features, winds, tides and so on.

The relative composition of the world-average estuary remains that of seawater up to very high dilutions (about 80% freshwater) because rivers act almost as distilled water dilutants. This is due to their very slow salinity (Pytkowicz et al., 1975). This simplifies the study of chemical equilibria because estuarine waters remain as ionic media. Of course, estuaries fed by rivers with large compositional differences from those of the oceans, and relatively concentrated solutes will not act in this simple way. Furthermore, the complexity of inorganic and organic dissolved and particulate constituents as well as the changes in pH, ionic strength and adsorption, desorption, coagulation, and so on, which occur during mixing make the study of estuaries very difficult. Also, interactions with sediments that may have low Eh's add further problems when rivers enter estuaries.

EXAMPLE 3.19.   ESTUARIES AND THE RELATIVE COMPOSITION OF WATERS.   I shall present here an example which differs from that of Pytkowicz et al. (1975) of the effect of estuarine mixing on the composition of waters. I shall again utilize the world-average river.

Consider the data:

river $[Cl^-] = 7.8 \times 10^{-3}$‰

$\qquad\qquad\qquad\qquad [Ca^{2+}] = 15 \times 10^{-3}$‰

ocean $[Cl^-] = 19.000$‰ $[Ca^{2+}] = 0.4$‰

I am treating the chlorinity as if it were the chloride ion concentration for simplicity. Consider now a mixture of river and seawater with $[Cl^-] = 10$‰. Then,

fraction of river water, $f_{RW} = 9/19 = 0.4737$

fraction of seawater, $f_{SW} = 10/19 = 0.5263$

Thus,

$$[Ca^{2+}]_{EW} = 0.4737 \times 7.8 \times 10^{-3}$$
$$+ 0.5263 \times 0.4 = 0.2142‰$$

where the subscript EW indicates estuarine waters.

Next, we find that a pure seawater of $[Cl^-] = 10$‰ contains

$$[Ca^{2+}]_{SW} = (10/19) \times 0.4 = 0.2105‰$$

Thus, the difference in $[Ca^{2+}]$ is only $[(0.2142 - 0.2105)/0.2142] \times 100\% = 1.73\%$. The same would of course be true of $[Ca^{2+}]/[Cl^-]$ at $[Cl^-] = 10$‰. Thus, the relative composition varies little for the calcium/chlorine ratio.

SUMMARY

An important feature of the oceans is the concurrent hydrographic transport, and biological, geological, as well as chemical, processes.

The geography of the oceans and the oceanic surface currents are examined. These currents are primarily wind driven and occur above the pycnocline.

Features such as the effect of the oceanic specific heat on the climate and the effect of large land masses on erosion and discharge into the oceans are considered.

Seawater below the pycnocline moves because of density variations that form a number of moving water masses present at different depths. The circulation of the oceans is linked through the Antarctic Circumpolar Water.

Temperatures and salinity are conservative quantities in that they are not affected by biological and geochemical processes below the sea surface. Thus, they serve to trace the motion and the mixing of waters. Nonconservative properties, such as oxygen and phosphate, also are tracers of water masses but in addition provide insight into the hydrochemistry of life.

The hydrobiochemical equation describes the local time change in a property as a function of advection and eddy diffusion, and as a source function, thus permitting oceanic processes to be quantified. An alternative approach is the more rudimentary of box models. It yields an average oceanic overturn of about 1000 yr.

The first and second laws of thermodynamics are discussed and the heat budget is presented for the earth surface and the sea. Their effect on water motions and the transport of chemicals as well as the effect of radiation on life are discussed. The water balance of the atmosphere and the moving oceans is established.

The concepts of biosphere ecosystems, trophic levels, and biomass are introduced. The effects of light and nutrients as well as those of diurnal and seasonal changes in growth conditions are examined.

The classifications of marine organisms are presented in terms of producers, consumers, plankton, nekton, and benthos. The key equation for photosynthesis and oxidation is

$$106CO_2 + 122H_2O + 16HNO_3 + H_3PO_4$$
$$= (CH_2O)_{106} (NH_3)_{16}H_3PO_4 + 138O_2$$

This equation regulates biologically induced changes in the oceans as the net trophic levels decrease by a factor of 10.

The crust of the earth is examined especially from the standpoint of submarine sediments.

Chlorinity and salinity are defined and related to the law of constant proportions. This law implies that the oceans are well stirred. The natures of the oceanic solutes and concepts related to reactivities, relaxation times, and so on, are considered. The present $Cl\text{‰}-S\text{‰}$ relation is $S\text{‰} = 1.80655\ Cl\text{‰}$.

Water types and water masses are defined and illustrated and typical oceanic distributions of solutes are presented. Then, modeling procedures for these distributions are introduced. These models yield information on biological activity, the exchanges of gases across the sea surface, and the motion of waters. The valuable quantities AOU and preformed phosphate are evaluated and applied. The Redfield ratios are developed for anoxic waters.

The properties of lakes are treated next. The properties of lakes show much greater variability than those of oceans.

Finally, the relationships between river inputs, relaxation times, and oceanic contents are examined. Residence times are the steady-state cases of relaxation times.

APPENDIX: A NOTE ON LARGE PROJECTS

At this point, I wish to call your attention to the existence of large sea-going projects, since they are an important part of contemporary oceanography. As an example, IDOE, the International Decade for Oceanic Research of the National Science Foundation, sponsored GEOSECS. This project was multi-university, in which ships occupied a large number of stations in the oceans and measured a variety of chemical parameters such as the concentrations of radioactive and stable isotopes, gases, nutrients, the CO_2 system components, and organic matter. Results and interpretations can be found, for example, in volume **16**, 1972 and volume **23**, 1974 of *Earth and Planetary Science Letters.*

An interesting program was carried over several years by Belgian scientists, on the entry of waters from the Schelde (in Flemish) or Escaut (in French) estuary into the North Sea, as a baseline for potential effects of pollution (Nihoul and Wollast, undated). The work was multidisciplinary and the propagation of river water, its progressive mixing, the fate of riverborne chemicals, the biological uptake, and sedimentation were followed and interpreted. The coordination of the project by C. J. Nihoul centered on the hydrobiochemical equation and progressed simultaneously through theoretical models, for example, the mathematical description of dispersion. It also centered on observations.

These large endeavors have been useful for projects that required logistical support beyond the means of individuals or small teams.

REFERENCES

Alvarez-Borrego (1973). Ph.D. Thesis, Oregon State University, Corvallis.

Anikovchin, W. A., and R. W. Stenberg (1973). *The World Ocean,* Prentice-Hall, Englewood Cliffs, N.J.

Bolin, B., and H. Stommel (1961). *Deep-Sea Res.* **8**, 95.

Boulton, R. (1966). *The Works of the Honourable Robert Boyle, Esq.,* Vol. 1, Phillips and Taylor, London, p. 282.

Broecker, W. S. (1974). Chemical Oceanography, Harcourt, Brace, Jovanovich, New York.

Broecker, W. S., and Y.-H. Li (1970). *J. Geophys. Res.* **75**, 3545.

Broecker, W. S., and Y.-H. Li (1971). In *Impingement of Man on the Oceans,* D. W. Hood, Ed., Interscience, New York, p. 287.

Bubic, S., and M. Branica (1973). *Thalassia Yugo.* **9**, 47.

Burton, J. D., and P. S. Liss (1976). *Estuarine Chemistry,* Academic Press, New York.

Calvin, M., and J. A. Baasham (1962). *The Photosynthesis of Organic Compounds,* Benjamin, New York.

Chen, C.-T. (1978). *Science* **201**, 735.

Chen, C.-T., and F. J. Millero (1977). *Deep-Sea Res.* **24**, 365.

Chesselet, R., J. Morrell, and P. B. Menard (1972). In *The Changing Chemistry of the Oceans,* D. Dyrssen and D. Jagner, Eds., Almqvist and Wiksell, Stockholm, p. 93.

Church, T. M., Ed. (1975). *Marine Chemistry in the Coastal Environment,* American Chemical Society, Washington.

Clarke, F. W. (1959). "The Data of Geochemistry," *Geol. Surv. Bull.* **770.**

Cox, R. A., F. Culkin, and J. P. Riley (1967). *Deep-Sea Res.* **14**, 203.

Craig, H. (1969). *J. Geophys. Res.* **74**, 5491.

Culberson, C. (1972). Ph.D. Thesis, Oregon State University, Corvallis.

Culberson, C., and R. M. Pytkowicz (1970). *J. Oceanogr. Soc. Japan* **26**, 95.

Curl, H., and L. Small, Undated Manuscript.

Deacon, G. E. R. (1937). *Discovery Reports* **15**, 1.

Dietrich, G. (1963). *General Oceanography,* Wiley, New York.

Dittmar, W. (1884). *Report on the Scientific Results of the Exploring Voyage of H.M.S. Challenger,* Vol. 1, *Physics and Chemistry,* Part 1.

Forch, C., M. Knudsen, and S. P. L. Sorensen (1902). *D. Klg. Danske Vidensk. Selsk, Skrifter* **6** *Raekke, Naturvidensk. og. Mathem.,* Afd xii, I.

Forchhammer, G. (1865). *Phil. Trans.* **155**, 203.

Franklin, B. (1786). *Trans. Am. Phil. Soc.* **2**, 314.

Garrels, R. M., and F. T. Mackenzie (1971). *Evolution of Sedimentary Rocks,* Norton, New York.

Goldberg, E. D. (1965). In *Chemical Oceanography,* Vol. 1, Academic Press, New York, p. 163.

Gross, M. G. (1977). *Oceanography: A View of the Earth,* 2nd Ed., Prentice-Hall, Englewood Cliffs, N.J., p. 159.

Harvey, H. W. (1957). *The Chemistry and Fertility of Sea Waters,* Cambridge University Press, Cambridge.

Houghton, H. G. (1954). *J. Meteor.* **11**, 1.

Hutchinson, G. E. (1957). *A Treatise on Limnology,* Vol. 1, Wiley, New York.

Jacobsen, J. P., and M. Knudsen (1940). *Publ. Sci. Assoc. Oceanogr. Phys.* No. 7, Liverpool.

Johnson, K. S., and R. M. Pytkowicz (1979). In *Activity Coefficients in Electrolyte Solutions,* Vol. 2, R. M. Pytkowicz, Ed., CRC Press, Boca Raton. p. 1.

Kester, D. R., and R. M. Pytkowicz (1968). *J. Geophys. Res.* **73**, 5421.

Ketchum, B. H., R. F. Vaccaro, and N. Corwin (1958). *J. Mar. Res.* **17**, 282.

Knudsen, M. (1901). *2me Conference International pour l'Exploration de la Mer,* Report Supplement **9.**

Kossina, E. (1921). *Berlin Univ. Institut Meereskunde Heft* **9.**

Kroopnick, P. M., W. G. Denser, and H. Craig (1970). *J. Geophys. Res.* **75**, 7668.

Kuo, H.-H., and G. Veronis (1970). *Deep-Sea Res.* **17**, 79.

Kuo, H.-H., and G. Veronis (1973). *Deep-Sea Res.* **20**, 871.

Livingstone, D. (1963). *Geol. Surv. Professional Paper,* No. 44-G.

Marcet, A. M. (1819). *Phil. Trans.* **109**, 161.

McGill, D. A. (1964). In *Progress in Oceanography,* Vol. 2, M. Sears, Ed., Macmillan, New York, p. 127.

Menard, H. W., and S. M. Smith (1966). *J. Geophys. Res.* **71**, 4305.

Menzel, D. (1968). *Deep-Sea Res.* **15**, 237.

Miller, F. J., C.-T. Chen, A. Bradshaw, and K. Schleicher (1980). *Deep-Sea Res.* **27**, 255.

Olausson, E. (1971). In *The Micropaleontology of the Oceans,* B. M. Funnel, Ed., Cambridge University Press, Cambridge, p. 375.

Park, K. (1967). *Limnol. Oceanogr.* **12**, 353.

Parsons, T., and M. Takahashi (1973). *Biological Oceanographic Processes,* Pergamon Press, New York.

Pickard, G. L. (1963). *Descriptive Physical Oceanography,* Pergamon Press, New York.

Postma, H. (1964). *Neth. J. Sea Res.* **2**, 258.

Prigogine, I. (1955). *Thermodynamics of Irreversible Processes,* Interscience, New York.

Prigogine, I. (1978). *Science* **201**, 777.

Pytkowicz, R. M. (1964). *Deep-Sea Res.* **11**, 381.

Pytkowicz, R. M. (1968). *J. Oceanogr. Soc. Japan* **24**, 21.

Pytkowicz, R. M. (1971a). *Limnol. Oceanogr.* **16**, 39.

Pytkowicz, R. M. (1971b). *Oregon State University Technical Report* No. 214, 1971b.

Pytkowicz, R. M. (1972). In *The Changing Chemistry of the Oceans,* D. Dyrssen and D. Jagner, Eds., Almqvist and Wiksell, Stockholm, p. 147.

Pytkowicz, R. M., and D. R. Kester (1966). *Deep-Sea Res.* **13**, 373.

Pytkowicz, R. M., and R. Gates (1968). *Science* **156**, 690.

Pytkowicz, R. M., and D. R. Kester (1971). In *Oceanography Marine Biology Annual Review*, Vol. 9, H. Barnes, Ed., Allen and Unwin, London, p. 11.

Pytkowicz, R. M., E. Atlas, and C. H. Culberson (1975). In *Marine Chemistry in the Coastal Environment*. T. M. Church, Ed., *A.C.S. Symposium, Series 18,* American Chemical Society, Washington, p. 1.

Redfield, A. C. (1942). *Paper. Phys. Oceanogr. Meteor* **9**, 1.

Redfield, A. C. (1948). *J. Mar. Res.* **7**, 347.

Redfield, A. C., B. H. Ketchum, and F. A. Richards (1963). In *The Sea,* Vol. 2, M. N. Hill, Ed., Interscience, New York, p. 26.

Reid, J. L. (1973). *Northwest Pacific Ocean Waters in Winter,* John Hopkins University Press, Baltimore.

Richards, F. A. (1965). In *Chemical Oceanography,* Vol. 1, J. P. Riley and G. Skirrow, Eds., Academic Press, New York, p. 611.

Riley, G. A. (1951). *Bull. Bingham. Oceanogr. Coll.* **12**, 1.

Stommel, H. (1958). *Deep-Sea Res.* **5**, 80.

Strickland, J. D. H., and T. R. Parsons (1965). *A Manual of Seawater Analysis,* 2nd Ed., *Res. Bd. Canada Bull.* **125.**

Sugiura, Y. (1965). *Bull. Soc. Franco-Japonaise d'Oceanogr.* **2**, 7.

Sverdrup, H. V., Johnson, M. W., and R. H. Fleming (1942). *The Oceans: Their Physics, Chemistry and General Biology,* Prentice-Hall, Englewood Cliffs, N.J.

UNESCO (1966). *UNESCO Technical Papers in Marine Science* No. 4.

Von Arx, W. C. (1970). *An Introduction to the Marine Environment,* Wiley, New York.

Wirtky, K. (1962). *Deep-Sea Res.* **9**, 11.

Wooster, W. C., A. J. Lee, and G. Dietrich (1969). *Deep-Sea Res.* **16**, 321.

Wüst, G. (1935). *Deutsche Atlantische Exped. Meteor, 1925–1927, Wiss. Erg.,* Vol. 2.

Wüst, G. (1954). *Arch. Meteor. Geophys. Biokim. (A)* **7**, 305.

SUGGESTED READINGS

Alvarez-Borrego, S., L. I. Gordon, L. B. Jones, P. K. Park, and R. M. Pytkowicz, *J. Oceanogr. Soc. Japan* **28**, 71 (1972). O_2–CO_2 nutrient relationships.

Anderson, G., *Limnol. Oceanogr.* **9**, 284 (1964). Seasonal and geographic distribution of productivity.

Armstrong, F. A. J., in *Chemical Oceanography,* Vol. 1, J. P. Riley and G. Skirrow, Eds., Academic Press, New York, 1965, p. 323. Phosphorus in the oceans.

Bien, G. S., N. W. Rakestraw, and H. E. Suess, *Limnol. Oceanogr.,* A. C. Redfield 75th Anniversary Vol., **R25** (1965). ^{14}C and deep water motions.

Bolin, B., and H. Stommel, *Deep-Sea Res.* **8**, 95 (1961). Abyssal circulation.

Boyle, E., *Geochim. Cosmochim. Acta* **38**, 1729 (1974). Chemical mass balance in estuaries.

Broecker, W. S., *J. Geophys. Res.* **71**, 5827 (1966). Oceanic rate of mixing.

Broenkow, W. W., *Limnol. Oceanogr.* **10**, 40 (1965). Mixing model for nutrients.

Carritt, D. E., and J. H. Carpenter, in *Physical and Chemical Properties of Sea Water, NAS-NRC Publ.* **600,** 67, Washington, 1959.

Connors, D. N., and D. R. Kester, *Mar. Chem.* **2**, 301 (1974). Specific gravity–chlorinity–saline conductivity of seawater.

Cox, R. A., in *Chemical Oceanography,* Vol. 1, J. P. Riley and G. Skirrow, Eds., Academic Press, New York, p. 73. Physical chemistry of seawater.

Craig, H., in *Earth Science and Meteorites,* compiled by J. Geiss and E. D. Goldberg, Amsterdam-North Holland, 1963, p. 103. Radiocarbon, mixing rates, and residence times.

Craig, H., *J. Geophys. Res.* **76**, 5078 (1971). O_2 consumption in abyssal waters.

Culkin, F., and R. A. Cox, *Deep-Sea Res.* **13**, 789 (1966). Na, K, Mg, Ca, and Sr in seawater.

Dugdale, V. A., and R. C. Dugdale, *Limnol. Oceanogr.* **7**, 170 (1962). Nitrogen metabolism in lakes.

Fofonoff, N. P., in *The Sea,* Vol. 1, M. N. Hill, Ed., Interscience, New York, 1963, p. 3. Physical properties of seawater.

Fonselius, S. H., *J. Geophys. Res.* **68**, 4009 (1963). Anoxic basins.

Goldberg, E. D., in *Impingement of Man on the Oceans,* D. W. Hood, Ed., Interscience, New York, 1971, p. 75. Atmospheric transport.

Goldberg, E. D., and G. O. S. Arrhenius, *Geochim. Cosmochim. Acta* **13**, 152 (1958). Chemistry of pelagic sediments.

Grill, E. V., and F. A. Richards, *J. Mar. Res.* **22**, 51 (1964). Nutrient regeneration.

Hoffman, G. L., and R. A. Duce, *J. Geophys. Res.* **77**, 5161 (1972). Fractionation of alkali and alkaline earths in the atmosphere.

Hutchinson, G. E., *Am. Sci.* **61**, 269 (1973). Eutrophication.

Knauss, J. A., *J. Geophys. Res.* **67**, 3943 (1962). Deep circulation.

Knudsen, M., *Hydrographical Tables,* Tutein og Koch, Copenhagen, 1901.

Kroopnick, P. M., Ph.D. Thesis, University of California, San Diego, 1971. O_2, C, and their isotopes in the oceans.

Lerman, A., and B. F. Jones, *Limnol. Oceanogr.* **18**, 72 (1973). Transport between lake sediments and brims.

Lisitzin, A. P., *Soc. Econ. Paleontol. Mineral., Tulsa, Spec. Publ.* **17**, 1972. Sedimentation.

Mann, C. R., et al. *Deep-Sea Res.* **20**, 791 (1973). Marine silicate distribution.

McGill, D. A., in *Progress in Oceanography,* Vol. 2, M. Sears, Ed., Macmillan, New York, 1964, p. 127. Distribution of O_2 and TPO_4 in the Atlantic Ocean.

Menard, H. W., *Marine Geology of the Pacific,* McGraw-Hill, New York, 1964.

Millero, F. J., in *Marine Chemistry in the Coastal Environment,* T. M. Church, Ed., *ACS Symposium Series* **18,** Washington, 1975, p. 25. Physical chemistry of estuaries.

Millero, F. J., A. Gonzalez, P. G. Brewer, and A. Bradshaw, *Earth Planet. Sci. Lett.* **32,** 468 (1976). Density of deep waters.

Millero, F. J., A. Gonzalez, and G. K. Ward, *J. Mar. Res.* **34,** 61 (1976). Density of seawater.

Millero, F. J., and A. Poisson, *Deep-Sea Res.* **28,** 625 (1981). Equation of state of seawater.

Morris, A. W., and J. P. Riley, *Deep-Sea Res.* **13,** 699 (1966). Chlorinity ratios in seawater.

Park, K., *Science* **146,** 56 (1964). Effect of $CaCO_3$ on the electrical conductivity of seawater.

Park, K., *Limnol. Oceanogr.* **12,** 353 (1967). Nutrient regeneration.

Pattenden, N. J., R. S. Cambray, and K. Playford, *Geochim. Cosmochim. Acta* **45,** 93 (1981). Composition of surface microlayer.

Richards, F. A., J. J. Anderson, and J. D. Cline, *Limnol. Oceanogr.* **16,** 43 (1971). Chemistry in anoxic waters.

Riley, J. P., and Tongudai, M., *Chem. Geol.* **2,** 263 (1967). Chlorinity ratios.

Riley, J. P., and R. Chester, *Introduction to Marine Chemistry,* Academic Press, New York, 1971.

Shulenberger, E., and J. L. Reid, *Deep-Sea Res.* **28,** 901 (1981). Pacific shallow oxygen maximum.

Stefansson, U., and F. A. Richards, *Limnol. Oceanogr.* **8,** 394 (1964). Nutrients off the Columbia River.

Suguira, Y., and H. Yoshimura, *J. Oceanogr. Soc. Japan* **20,** 14 (1964). Marine O_2 and TPO_4.

Sverdrup, H. V., M. W. Johnson, and R. H. Fleming, *The Oceans: Their Physics, Chemistry and General Biology,* Prentice-Hall, Englewood Cliffs, N.J., 1942.

Tully, J. P., and F. G. Barber, in *Progress in Oceanography,* M. Sears, Ed., Macmillan, New York, 1961. Model of Sub-Arctic Pacific.

Turekian, K. K., *Oceans,* Prentice-Hall, Englewood Cliffs, N.J., 1968.

UNESCO, *International Oceanographic Tables,* 2 vols. UNESCO Office of Oceanography, National Institute Oceanography, Wormley, England, 1966.

Wooster, W. S., A. J. Lee, and G. Dietrich, *Deep-Sea Res.* **16,** 321 (1969). Redefinition of salinity.

Wooster, W. S., and G. H. Volkmann, *J. Geophys. Res.* **65,** 1239 (1960). Deep Pacific circulation.

ZoBell, C. E., and C. H. Oppenheimer, *J. Bacteriol.* **60,** 771 (1950). Pressure and marine bacteria.

CHAPTER FOUR

THERMODYNAMICS FOR EARTH SCIENTISTS

4.1 CONCEPTS

Thermodynamics is a difficult subject which can be bypassed in some aspects of aquatic chemistry.

I should mention, however, that if time permits, a study of this chapter can provide understanding and, therefore, guide the judgment of the reader regarding which concepts and equations apply to a given problem.

I considered in selecting material for this chapter, not only what is directly relevant to the rest of this book but also other material the reader may find of use in other aspects of the aquatic sciences. Other advantages of such an approach are completeness and a logical development of the subjects. Subjects directly relevant to electrolytes are developed in the next chapter. Emphasis is placed on simple systems in this chapter because the extension to multicomponent systems is conceptually straightforward. Still, some material on multicomponent solutions is presented here when needed and, to a considerable extent, in Chapter 5.

The relatively short presentation of selected thermodynamic material in this chapter will help the reader to bypass textbooks on this subject. Such texts contain considerable unnecessary material for the study of solutions.

Thermodynamics deals with energy–matter interactions from a macroscopic point of view. Thus, it deals with macroscopic properties of the system of interest such as pressure, temperature, volume, potential energy, internal energy, enthalpy, entropy, and so on. In chemistry, thermodynamics provides guidelines for predicting the direction and the extent of reactions, based on energy considerations. A mechanical analog is the prediction of the behavior of a body in the gravitational field. If it is free, it will fall because of its potential energy. Similarly, in chemical thermodynamics a reaction will proceed if such an event will decrease the energy of the chemical system.

Let us look at some definitions. A system is that part of the universe which is under scrutiny. The remainder is its surroundings. Intensive properties of a system are those properties that are independent of its mass. Examples are the temperature and the density. Extensive properties, such as the volume and the potential energy, depend on the amount of matter held in the system.

Homogeneous systems contain no discontinuities of intensive properties. As examples, solutions or crystals of a substance are homogeneous systems because the density and the temperature either do not vary or vary continuously within them. Homogeneous systems are also known as phases. One should recognize that pure crystals of halite (NaCl) at the bottom of a beaker constitute a phase because the intrinsic properties are the same among all the crystallites. In this case, therefore, discontinuities can exist.

Heterogeneous systems contain two or more phases. Liquid water in contact with water vapor is a two-phase heterogeneous system because there is a discontinuity between the intrinsic properties of the two phases.

The state of a system is defined by its macroscopic properties. In a gas, for example, one does not attempt to define the state of each molecule by its position and momentum but rather one uses observable properties such as pressure, temperature, and volume.

An open system can exchange matter and heat with and do work on its surroundings. A closed system can only exchange heat and do work, an adiabatic system can do work, and an isolated system does not interact with its surroundings. Examples include

Open system	Lake, humans
Closed system	Metal cylinder and piston with a gas that expands
Adiabatic system	Insulated metal cylinder and piston with a gas that expands
Isolated system	Hot coffee in a perfect thermos

Note that absolutely adiabatic and isolated systems are idealizations since energy leaks even through the insulated wall.

4.2 FIRST LAW, INTERNAL ENERGY, AND ENTHALPY

The first law of thermodynamics states that the total energy of the universe is constant.

4.2.1 Internal Energy

The internal energy E of a system is the total energy, in all forms, contained within it. If the system loses an amount of internal energy ΔE by doing work, then according to the first law, its surroundings gain ΔE. An example of such a system is a car loaded with fuel (the fuel plus car is the system) which loses the internal energy contained in the fuel as it moves. Another example is the reaction of copper with oxygen

$$Cu + \tfrac{1}{2} O_2 \rightarrow CuO \qquad (4.1)$$

$\Delta E = -37,000$ cal/mole. By this we mean that 37,000 calories are released to the environment when 1 mole of copper oxide (CuO) is formed from 1 g-atom of copper and $\frac{1}{2}$ mole of oxygen.

4.2.2 Work, Heat, and Reversibility

Let us see how E is related to W, the work done by the system, and Q, the heat absorbed by the system during a process.

Suppose that a gas is allowed to expand by moving a piston in a cylinder. In a reversible (infinitely slow) isothermal expansion with the cylinder immersed in a large thermostatic bath, the absorption of heat would compensate at all times for the heat transformed into and used as work. The gas performs the work

$$W = P_{ext}\Delta V \qquad (4.2)$$

on its surroundings which are at pressure $P_{ext} = P_{int}$. At any instant in this conceptual reversible process, the expansion can become a compression if P_{int} is decreased by an infinitesimal to $P_{int} - dP$.

From the first law,

$$\Delta E = E_B - E_A = Q - W = Q - P_{ext}\Delta V \qquad (4.3)$$

$-W$ appears because ΔE decreases when work is done by the system ($W > 0$).

EXAMPLE 4.1. FIRST LAW AND THE EARTH. The internal energy of the earth is essentially constant, in terms of thousands of years. Thus,

$$\Delta E = Q - P_{ext}\Delta V = 0 \qquad (i)$$

The earth does some external $P_{ext}\,\Delta V$ such as when mountains are built and tides rise. This work, however, is compensated for by erosion and low tides when geological short periods are considered.

We can, therefore, set $W = 0$ and (i) yields $Q = 0$ as was discussed earlier. This relation, of course, is not exact if the interior of the earth is slowly cooling.

EXAMPLE 4.2. If a battery delivers 209.2 J of energy and a motor connected to it does 104.6 J of work, what are the extent and direction of the heat exchange between the battery and the surroundings?

$$\Delta E = -\frac{209.2}{4.1840} = -50 \text{ cal} \qquad (i)$$

where 4.1840 is the conversion factor from J to cal.

$$W = -\frac{104.6}{4.1840} = -25 \text{ cal} \qquad (ii)$$

$$Q = \Delta E - W = -25 \text{ cal} \qquad (iii)$$

Thus, 25 cal are released by the battery.

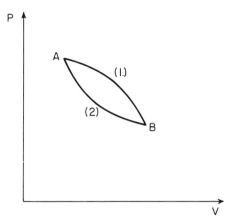

FIGURE 4.1 Two different paths to go from state A to B. P is the pressure and V the volume.

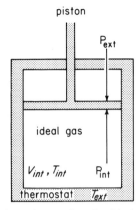

FIGURE 4.2 Isothermal expansion of an ideal gas. P_{int}, V_{int}, and $(T_K)_{int}$ are the values of these parameters for the ideal gas while P_{ext} and T_K correspond to the environment.

In differential form the first law is

$$dE = \delta Q - \delta W \qquad (4.4)$$

The symbol δ indicates that δQ and δW depend not only on the initial and the final states, that is, on V_A and V_B as well as T, but also on the path followed (Figure 4.1).

Note that the gas may go from state A to state B by many paths, two of which are presented in the figure. A and B are states because, anticipating later considerations, the phase rule tells us that a system with one component and one phase has only two degrees of freedom. Thus, the specification of two variables such as P and V fixes the state of the system.

The work done by the system in going from A to B, an expansion against the external pressure, is

$$W = \int_A^B P_{ext} dV \qquad (4.5)$$

and is given by the area under curves such as (1) or (2) of Figure 4.1, which describes the path followed. W is positive because work is done by the gas on the surroundings. $P_{ext} dV$ represents the work because, if the expansion is made by pushing a piston of cross section A through a distance dr, the work is the scalar product $\mathbf{F} \, d\mathbf{r} = PA \, dr = P \, dV$ where $\mathbf{F}$ is the vector force.

Let us show next that Q and W are functions of the path. Consider the internal energy change in a cyclic path ABA (see Figure 4.1). The change $\Delta E_{AB} + \Delta E_{BA}$ must be zero because E_A is a function of state defined by P and T. Therefore, $Q_{ABA} = W_{ABA}$.

The work W_{ABA} depends on the curves $\widehat{AB}$ and $\widehat{BA}$ because it is the area between them. Thus, W_{ABA} is a function of the path and so is Q_{ABA} as $W_{ABA} = Q_{ABA}$. This conclusion about the virtual nature of W and Q can be generalized to open paths although then $Q \neq W$.

Examples of closed paths are steam engines and the hydrological cycle.

A reversible process is one that proceeds infinitely slowly, always at equilibrium with its surroundings, in such a manner that it can be reversed at any stage without changing the properties of the environment by more than an infinitesimal amount. It is an abstraction because all natural processes occur at finite speeds but it is a critical concept because it pertains to the efficiency of human-made and natural engines. I have not yet explained the equilibrium process sufficiently so I will illustrate the reversible process by using as an example the simple expansion and contraction of an ideal gas.

Assume that an ideal gas is in a container in contact with a thermostat and that it can expand by moving a frictionless piston (Figure 4.2). Let us keep the ideal gas in thermal and pressure equilibrium with its surroundings at all times, that is, let the internal temperature and pressure be within dT and dP of the external values. The gas is made to expand reversibly and isothermally by always keeping $P_{ext} = P_{int} - dP$ and $T_K = $ constant.

The work done on the gas is given by

$$-W = \int_A^B P_{ext} dV_{int} = nRT_K \int_A^B \left(\frac{1}{V_{int}}\right) dV \qquad (4.6)$$

because the ideal gas obeys the law

$$P_{int}V_{int} = nR(T)_{int} \qquad (4.7)$$

and $P_{int} = P_{ext}$ is true within an infinitesimal dP. R is the gas constant and T is the absolute temperature. The term n represents the number of moles of gas. Thus, the work is

$$-W = nRT \ln \frac{V_B}{V_A} \qquad (4.8)$$

EXAMPLE 4.3.   What is the maximum work done by 3 moles of an ideal gas expanding at 298°K from 1 to 2 L? What is the work done against an external pressure of 1 atm? What is Q for these processes? What are the energy and the heat wasted in the second process?

The maximum (reversible) work done by the gas is, when it expands,

$$W_{rev} = 3RT \ln \frac{V_B}{V_A} = 5.1522 \times 10^{10} \text{ ergs} \quad (i)$$

Now, if the external pressure is kept at 1 atm instead of at $P_{int} - dP$ then

$$W^* = P(V_B - V_A) = 1 \text{ L} \cdot \text{atm}$$
$$= 9.7484 \times 10^8 \text{ ergs} \qquad (ii)$$

We see that the irreversible work is smaller than the reversible.

As $E = $ constant for an isothermal process with an ideal gas, $\Delta E = 0$, and

$$Q_{rev} = W_{rev} = 5.1522 \times 10^{10} \text{ ergs}$$
$$= 1.2310 \times 10^3 \text{ cal absorbed} \qquad (iii)$$

$$Q^* = W^* = 9.7484 \times 10^8 \text{ ergs}$$
$$= 23.299 \text{ cal absorbed} \qquad (iv)$$

As $\Delta E_{rev} = \Delta E^*$ in an isothermal process,

$$Q^* - Q_r = W_r - W^* \qquad (v)$$

$$Q^* - Q_{rev} = 5.0547 \times 10^{10} \text{ ergs} = 1207.7 \text{ cal}$$
$$(vi)$$

This is the excess heat evolved and wasted by frictional dissipation in the irreversible expansion.

$$W_{rev} - W^* = -5.0547 \times 10^{10} \text{ ergs} \quad (vii)$$

is the amount of work that was not done by the system in the irreversible process, that is, the part of the energy available that was wasted.

Irreversible work will also follow Equation (4.5) but now P_{int} is larger than and differs from P_{ext} by more than an infinitesimal. Then, if we reverse the expansion to a compression through a change $-\Delta P_{int}$, the work done by the system is $P_{int} dV$ with $dV < 0$ instead of $P_{ext} dV$. Thus, the work done by the system is negative and that done by the environment is positive. The work done from the two directions differs in absolute value.

The reversible work is the only one that has a definite path in a diagram such as the one shown in Figure 4.1 because P and V, being related through Equation (4.7), allow (4.6) to be used. Any other path will not yield a definite work given by Equation (4.6) because, although the work is still given by the integral of $P_{ext} dV$, the external pressure $P_{ext} \neq P_{int}$. Therefore, Equation (4.7) cannot be used in the integration.

An irreversible expansion is obtained if we set, for example, $P_{ext} = P_{int} - \Delta P$. This shows that the reversible path yields the maximum useful work available in an expansion of the gas from V_A to V_B since P_{ext} is larger at every stage for the reversible expansion than for an irreversible expansion.

It is interesting to note that the heat Q absorbed by the gas for its temperature to remain constant during an expansion must equal W, since the internal energy E of an ideal gas does not depend on P and V but only on T. This occurs because an ideal gas contains no potential energy due to the absence of attraction between the particles, but contains only kinetic energy which is fixed by n and T.

EXAMPLE 4.4.   ADIABATIC COOLING AND HEATING. Adiabatic changes in temperature play a role in the study of the oceans. First, assume that the container in which a seawater sample is brought to shipboard is perfectly insulated thermally and is of the type closed by rubber caps.

As $Q = 0$,

$$\Delta E = -W \quad \text{or} \quad \Delta E = -W^* \qquad (i)$$

depending on how slowly the sample is brought up. Usually $\Delta E = -W$ is a good approximation. W is the work of expansion done by the sample during decompression. The decrease in E leads to an adiabatic cooling which is, per 1000-m decrease in depth (Sverdrup et al., 1942) as follows.

Values of $-\Delta T/1000$ m as a function of T (°C) and S‰.

S‰	T (°C)	0	2	4	6
34		0.043	0.060	0.077	0.094
36		0.047	0.064	0.081	0.097

If the in situ temperature was 2°C at $z = 6000$ m and $S = 35$‰, the adiabatic cooling would be $6 \times 0.062 = 0.372$. The temperature of the water, measured in the laboratory, assuming no heat exchange, would be $2 - 0.372 = 1.628$°C.

This temperature, 1.628°C, is known as the potential temperature while thermometers attached to the outside of the bottle register 2°C, the actual in situ temperature. It also is the temperature of a water at depth if no heating occurred when it sank. The potential temperature is used in physical oceanography in studies of heat transfer from the earth to the sea in deep trenches.

Note that ΔE is not fixed by z because the value of E at the surface depends on the adiabatic cooling, that is, on ΔT, which occurs when the sample is raised. Thermodynamically, the adiabatic change in T is represented by Equation (4.29), for a reversible process, although a simplified expression was used in the pioneering work of Kelvin and in the data tabulated above.

At this point I shall again mention reversibility versus irreversibility because these are difficult concepts that bear repeating in various ways. Frictional losses in a mechanical process increase with the rate of the process. Processes that occur infinitely slowly are called reversible because, as only an infinitesimal amount of energy is dissipated, the energy stored in one direction can be used to reverse the process and bring the system back to its initial state. Thus, the heat δQ_{rev} taken from the environment during a reversible expansion dV_{exp} of a gas can be returned to the environment during a compression $dV_{comp} = -dV_{exp}$.

Spontaneous processes occur at finite rates and are irreversible due to energy dissipation. In the example of the previous paragraph an irreversible expansion and compression would lead to $dV_{comp} <$ $-dV_{exp}$ because part of the heat taken from the environment was lost in the friction of the piston, and became unavailable for $P\,dV$ work.

Organisms, as an example, undergo spontaneous processes and, therefore, are unable to do a maximum useful work. If a man lifts a weight from A to B, in practice he must use more energy than if he could do it reversibly.

4.2.3 Enthalpy and the Silica Thermometer

A useful function is the enthalpy H, defined by

$$H = E + PV \qquad (4.9)$$

H is a state function because E, P, and V are, also. Here, as in the case of E, we are interested in changes rather than in absolute values of H.

In terms of finite increments,

$$\Delta H = \Delta E + P\,\Delta V + V\,\Delta P \qquad (4.10)$$

At constant pressure and by the use of the first law,

$$\Delta H = \Delta E + P\,\Delta V = \Delta E - W = Q \qquad (4.11)$$

Thus, the change in ΔH in a process that occurs at constant pressure is the reversible heat absorbed by the system. This is why H is often referred to as the heat content. I remind the reader that, in the absence of asterisks, Q and W follow reversible paths. Incidentally, bodies contain thermal energy but the term heat is best reserved for the amount of thermal energy transferred in a process.

Let us examine the usefulness of H. LeChatelier's principle states that systems respond to stresses in such a way as to minimize changes imposed on them. Therefore, a rise in T at a constant P favors reactions with $\Delta H > 0$. Such heat-absorbing reactions are known as endothermic, whereas exothermic processes release heat and have $\Delta H > 0$. The symbol ΔH^0 is used when the reactants and products are in their standard states.

ΔH_f^0 is the heat of formation, that is, the ΔH for the formation of a compound at its standard state from its elements also in their standard states.

$\Delta H_f^0 = -19.754$ kcal/mole at 298.16°K and -20.981 kcal/mole at 1000°K for $H_2 + S = H_2S$ (Kelley, 1960). This reaction, therefore, is exothermic and the reverse process is favored by an increase in temperature at 298.16°K and even more so at 1000°K.

A cylinder filled with a real gas with a piston open to the atmosphere on one side, to hold $P = 1$ atm, has $\Delta H > 0$ as it cools upon expansion. It is an endothermic system and the expansion is favored by an increase in temperature.

As I mentioned earlier, standard states will be discussed later but readers can ignore it for the time being and still gain insight into the concepts that are being presented.

The dissolution of $CaCO_3$ is exothermic as are those of CO_2, O_2, and H_2SO_4. That of sugar in water is endothermic and is, therefore, favored by an increase in temperature, an observation well known by cooks.

EXAMPLE 4.5. THE SOLUBILITIES OF CO_2 VERSUS TEMPERATURE AND ITS DISTRIBUTIONS IN NEAR-SURFACE OCEANIC WATERS.   Selected values of the solubility of CO_2 in seawater of Cl‰ = 19, extracted from Murray and Riley (1971), are, in (moles/atm) $\times 10^4$,

T (°C)	0	6	12	18	24
Solubility	638	510	416	347	292

Therefore, the dissolution of CO_2 is exothermic because the solubility is inhibited by an increase in temperature.

The consequence of this is that the solubility of CO_2 is increased with latitude. In terms of solubility, and disregarding the effects of life on $[CO_2]$, this gas should enter the oceans from the atmosphere at high latitudes and be released at low ones. Such a behavior was indeed observed by Takahashi (1961). Roughly, the pCO_2 by STP volume in surface waters was 254 ppm at 55°N and 336 ppm at 10°S, whereas that of the atmosphere was 310–320 ppm.

Enthalpies are useful in calculating the effect of temperature on equilibria at low as well as at high temperatures which are of interest to the geochemists, if the K's of these reactions are not known a priori at the desired temperatures. This can be done because $\partial \ln K/\partial T = \Delta H^0/RT^2$. Krauskopf (1967), for example, calculated for the gas–water reaction $CO_2 + H_2O = CO_2 + H_2$ that, in going from 25° to 417°C, the equilibrium constant decreases from 182,000 to 10.

EXAMPLE 4.6. K_{sO} VERSUS T AND THE SILICA GEOTHERMOMETER. Krauskopf (1967) examined the reaction (Siever, 1962)

$$SiO_2 \text{ (quartz)} + 2H_2O = H_4SiO_4 \text{ (silicic acid)} \quad \text{(i)}$$

which is an important reaction since it shows the relative stabilities of inert quartz and reactivity silica. These minerals play a role in Al–silicate equilibria in natural waters. The term log K versus T, obtained from the equation

$$\frac{\partial \ln K}{\partial T} = \frac{\Delta H^0}{RT^2} \quad \text{(ii)}$$

was found to be

$$\log K = -\frac{\Delta H^0}{2.303R}\left(\frac{1}{T}\right) + C \quad \text{(iii)}$$

with $\Delta H^0 = $ constant. The temperature integration is more complex when ΔH^0 varies with the temperature, since the heat capacity, which in itself may vary, must be taken into consideration. This case will be examined later.

Siever found, from the solubility of quartz between 25° and 200°C,

$$\log K_{sO} = -\frac{\Delta H}{2.303R}\left(\frac{1}{T}\right) + C \quad \text{(iv)}$$

that is,

$$\log K_{sO} = -A\frac{1}{T} + C \quad \text{(v)}$$

where C is a constant of integration, valid for the range 25–200°C. Equation (v) is that of a straight line for log quartz solubility versus $1/T$. The temperature is in degrees Kelvin, the slope is 1.132, and $\Delta H^0 = 5.18$ kcal/mole as $2.303R = 0.00458$ kcal/°K $\cdot$ mole. A similar procedure can be used for SiO_2 (amorphous silica) although the upper limit for the known solubility temperature is about 140°C. A solubility versus $1/T$ plot yields the silica thermometer for geochemists for use in inaccessible hot zones (e.g., hydrothermal sources) since the dissolved silica, if at saturation, yields the temperature. For example, 500 ppm of amorphous silica indicate a temperature of about 400°K (Siever, 1962).

4.2.4 Specific and Molar Heats

ΔH^0 for reactions can then be obtained if they are tabulated, or can be calculated from known specific heat data as functions of temperature.

The specific heat capacity or specific heat of a system is the amount of heat that must be transferred to 1 g of the component (for a single-component system) to raise its temperature by 1° (Celcius or Kelvin). The molar heat capacity or molar heat is the amount of heat that must be transferred to 1 mole to raise the temperature by 1°C or 1°K.

The heat capacity depends on the conditions under which the heat is transferred from the environment. Let us assume that only mechanical $P\,dV$ work is done by the system. Then the change in E at constant pressure is

$$dE = dQ_P - P\,dV \qquad (4.12)$$

The symbol dQ_P was used instead of δQ_p because dV is only a function of the initial and final states as is dE while P is fixed. Therefore, dQ_P is determined by these states and is unique. We have then

$$dH = dE = P\,dV = dQ_P \qquad (4.13)$$

$$C_P = \frac{dQ_P}{dT} + \left(\frac{\partial H}{\partial T}\right)_P \qquad (4.14)$$

C_P is the heat capacity at constant P.

At constant volume there is no $P\,dV$ work done and

$$dE = dQ_V \qquad (4.15)$$

which yields

$$C_V = \frac{dQ_V}{dT}\left(\frac{\partial E}{\partial T}\right)_V \qquad (4.16)$$

The definitions of specific heat are quite general and may be applied either to homogeneous or heterogeneous systems (for example, ice and blood which contain cells and colloids such as proteins) but the molar heats refer to a given component in a given phase. Molar heats pertain to 1 mole or g-atom of a substance. We will reserve the symbols c_P and c_V for specific heats and use C_P and C_V for molar heats.

An important relation between C_P and C_V will be derived next. From the definition of H,

$$C_P = \left(\frac{\partial E}{\partial T}\right)_P + P\left(\frac{\partial V}{\partial T}\right)_P \qquad (4.17)$$

However,

$$\left(\frac{\partial E}{\partial T}\right)_P = \left(\frac{\partial E}{\partial T}\right)_V + \left(\frac{\partial E}{\partial V}\right)_T\left(\frac{\partial V}{\partial T}\right)_P \qquad (4.18)$$

Equation (4.18) shows that heating the system at constant pressure leads to the same dE as heating it at constant volume and then bringing it at constant temperature to the original pressure. Introducing Equations (4.14) and (4.16) into (4.18), we obtain

$$C_P = C_V + \left[P + \left(\frac{\partial E}{\partial V}\right)_T\right]\left(\frac{\partial V}{\partial T}\right)_P \qquad (4.19)$$

For an ideal gas $(\partial E/\partial V)_T = 0$ and

$$C_P = C_V + P\left(\frac{\partial V}{\partial T}\right)_P \qquad (4.20)$$

The heat capacity has several uses. We shall see later in Equation (4.201) that it enters the relationship between the variation of the chemical equilibrium K with T when ΔH varies with temperature. It is also used in the study of the expansion of gases as we shall see after a brief digression.

First, however, I shall mention the well-known use of differences in heat capacities to contrast changes in temperature of the lithosphere and the hydrosphere when they absorb the same amount of solar heat. Thus, the earth surface, with a smaller C_P, warms up more than the sea surface during the day but cools faster at night. This causes the sea breeze during the day which is replaced by a breeze from the land at night when the colder and denser land air intrudes under the warmer marine air.

From a seasonal standpoint, the temperature changes little in the Equatorial Zone because the sun remains nearly in a vertical plane but a progressively larger change occurs as the latitude increases. These increases are not uniform. In Mediterranean (marine West Coast and the Mediterranean Sea) climates, westerlies predominate and, due to the circulation of the waters, there is a relatively small seasonal change in oceanic temperatures. Therefore, the climate is reasonably equable. Thus, the mean annual temperature on the Oregon coast ranges from 40 to 60°F while the change at the same latitude on the East Coast of the United States varies between 26 and 70°F.

4.2.5 Expansion of Gases

The expansion of gases is of interest in a variety of applications ranging from meteorology to engines. Furthermore, it illustrates thermodynamics and serves as a foundation for the study of aqueous solutions. The isothermal expansion of an ideal gas was treated earlier and we found that it obeys the relation

$$PV = nRT \qquad (4.21)$$

Next we shall examine the adiabatic expansion, which is needed for the study of the Carnot cycle and the second law of thermodynamics. In an adiabatic process $Q = 0$ and, if only mechanical $P\,dV$ work is done, one obtains from the first law

$$dE + P\,dV = 0 \qquad (4.22)$$

However, at constant pressure,

$$dE = \left(\frac{\partial E}{\partial T}\right)_V dT + \left(\frac{\partial E}{\partial V}\right)_T dV \qquad (4.23)$$

As $(\partial E/\partial V)_T = 0$ for an ideal gas, Equation (4.22) becomes, when the Equation (4.19) is introduced into Equation (4.23),

$$C_V\,dT + P\,dV = 0 \qquad (4.24)$$

because $dE = (\partial E/\partial T)_V = C_V\,dT$.

For 1 mole of gas, $P = RT/V$ which is introduced in Equation (4.24). Then, a division of the terms by T yields

$$C_V\,\frac{dT}{T} + R\,\frac{dV}{V} = 0 \qquad (4.25)$$

Integrating Equation (4.25) for an adiabatic change from A to B

$$C_V \ln\frac{T_B}{T_A} + R \ln\frac{V_B}{V_A} = 0 \qquad (4.26)$$

For 1 mole of ideal gas $(\partial V/\partial T)_P = R/P$ and Equation (4.20) yields

$$C_P - C_V = R \qquad (4.27)$$

Introducing Equation (4.27) into (4.26) and rearranging

$$\ln\frac{T_A}{T_B} = (\alpha - 1)\ln\frac{V_B}{V_A} \qquad (4.28)$$

with $\alpha = C_P/C_V$. Thus,

$$\frac{T_A}{T_B} = \left(\frac{V_B}{V_A}\right)^{\alpha-1} \qquad (4.29)$$

Since $T_A/T_B = P_A V_A/P_B V_B$ from the ideal gas law,

$$P_1 V_1^{\alpha} = P_2 V_2^{\alpha} \quad \text{or} \quad PV^{\alpha} = \text{constant} \qquad (4.30)$$

in contrast to the condition $PV = $ constant for isothermal processes.

From Equation (4.24) we see that, in an adiabatic process,

$$\delta W = P\,dV = -C_V\,dT \qquad (4.31)$$

or, integrating for the expansion from state A to state B,

$$W = C_V(T_A - T_B)(T_A - T_B) \qquad (4.32)$$

Thus, in an adiabatic expansion $T_B < T_A$ due to the cooling of the gas and $W > 0$, showing that work was done by the system. Remember that we are using the convention that positive work is that which is done by the system.

4.3 SECOND LAW, ENTROPY, AND FREE ENERGY

4.3.1 Carnot Cycle

The Carnot cycle is an important idealization which yields the maximum efficiency of engines and leads to the key concept of entropy. Note that the term "engine" can be extended to organisms and waters in motion as work done in them at the expense of heat. Earlier we saw that the system undergoing idealized reversible processes could perform maximum work relative to irreversible paths. We shall see now that this maximum work cannot equal the energy put into the system. As an example, the reversible work performed by a steam engine is always smaller than the heat supplied to the boiler.

Imagine, if you will, a frictionless engine

operated by the expansion and contraction of an ideal gas. This gas can be placed in turn in contact with an infinite reservoir of thermal energy at temperature T_H or with a colder reservoir at temperature T_C.

Engines, to remain in action, operate in cycles. The Carnot cycle operates as follows (see Figure 4.3). In the path AB the ideal gas expands reversibly from V_A to V_B while in thermal equilibrium with the reservoir at temperature T_H (Figure 4.3a). Next, an insulating jacket is placed over the gas cylinder and the gas expands adiabatically from V_B to V_C (Figure 4.3b). In the third step, the gas is compressed isothermally while in contact with the reservoir at temperature T_L, and, finally, it is compressed adiabatically back to the original state (P_A, V_A).

The work done by the gas and the heat absorbed by the gas are:

Step AB

$$W_{AB} = RT_H \ln \frac{V_B}{V_A} > 0 \qquad (4.33)$$

$$Q_{AB} = W_{AB} > 0 \qquad (4.34)$$

Step BC

$$W_{BC} = C_V(T_L - T_H) < 0 \qquad (4.35)$$

$$Q_{BC} = 0 \qquad (4.36)$$

Step CD

$$W_{CD} = RT_L \ln \frac{V_D}{V_C} < 0 \qquad (4.37)$$

$$Q_{CD} = W_{CD} < 0 \qquad (4.38)$$

Step DA

$$W_{DA} = C_V(T_H - T_L) > 0 \qquad (4.39)$$

$$Q_{DA} = 0 \qquad (4.40)$$

According to our conventions, $W > 0$ when heat is absorbed by the system.

The total work done by the gas in the cycle is

$$W = RT_H \ln \frac{V_B}{V_A} + RT_L \ln \frac{V_D}{V_C}$$

$$= Q_{AB} + Q_{CD} \qquad (4.41)$$

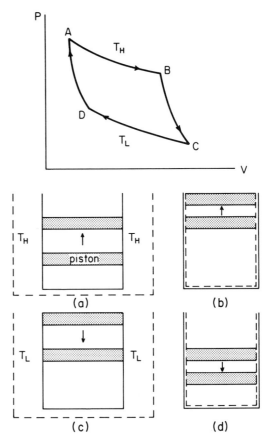

FIGURE 4.3 The Carnot cycle. The broken lines indicate irreversible pathways.

as $W_{BC} = -W_{DA}$. Thus, the work done by the gas during the cycle equals the heat absorbed at T_H minus the heat rejected at T_L, as $Q_{AB} > 0$ and $Q_{CD} < 0$. It is obvious, therefore, that even in the ideal Carnot engine one cannot have 100% conversion of heat into work. This result applies to any form of energy converted into work.

Let us calculate next the efficiency of the Carnot cycle. From Equation (4.28),

$$\left(\frac{V_D}{V_A}\right)^{\alpha-1} = \frac{T_H}{T_L} = \left(\frac{V_C}{V_B}\right)^{\alpha-1} \qquad (4.42)$$

Therefore,

$$\frac{V_D}{V_A} = \frac{V_C}{V_B} \quad \text{and} \quad \frac{V_B}{V_A} = \frac{V_C}{V_D} \qquad (4.43)$$

Introducing the last equality into Equation (4.41)

$$W = R(T_H - T_L) \ln \frac{V_B}{V_A} \qquad (4.44)$$

Dividing Equation (4.44) by $Q_{AB} = W_{AB} = RT_H \ln (V_B/V_A)$,

$$\frac{W}{Q_{AB}} = \frac{T_H - T_L}{T_H} = \frac{Q_{AB} + Q_{CD}}{Q_{AB}} > 0 \quad (4.45)$$

It is seen, therefore, that the maximum conversion of heat into work available when an amount of heat Q_{AB} is transferred to the gas is $(T_H - T_L)/T_H$. A 100% efficiency would occur only if the heat sink was at absolute zero (0°K), a condition not realizable in practice. Furthermore, this efficiency is only valid for reversible processes which can only be approximated but not achieved in a finite time.

From Equation (4.45) it follows that

$$\frac{Q_{AB}}{T_H} + \frac{Q_{CD}}{T_L} = 0 \quad (4.46)$$

for the system, with $Q_{AB} > 0$ and $Q_{BC} < 0$. The same relationship is valid for the surroundings in a reversible process although now we would have $Q_{AB} < 0$ and $Q_{BC} > 0$.

4.3.2 Entropy

Equation (4.46) defines a function of state, the entropy S, which is a measure of energy dissipated in a host of processes. It applies not only to the conversion of heat into work but can be generalized to encompass the concept of order. Systems tend toward randomness and it can be shown that the entropy of a system decreases with an increase of order. We must do work to build and maintain structures such as houses and living organisms against this tendency toward the breakdown of order.

The entropy undergoes changes according to the relations

$$dS = \frac{dQ_{\text{rev}}}{T} \qquad \Delta S = \frac{Q_{\text{rev}}}{T} \quad (4.47)$$

The meaning of S will be discussed further in the next section. For the time being I will simply point out that $dS = dQ_{\text{rev}}/T$, is not a function of the path since dQ_{rev} is unique. It is only δQ_{irrev} which can have a number of pathways. Thus, Q_{rev} is a function of state and so is S.

Let us assume that the work done by the gas in the Carnot cycle was stored reversibly in, for example, an electric cell. The cell then could be used

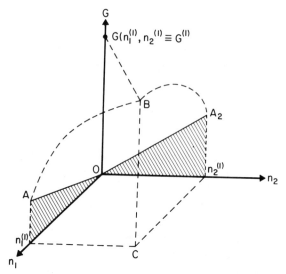

FIGURE 4.4 Partial molal quantities illustrated.

to do reversible work on the gas and behave as an ideal refrigerating cycle $ADCBA$, the reverse Carnot cycle.

I have shown above that maximum work is done by systems when reversible processes lead them from state A to state B, such as in the expansion of an ideal gas. It can also be shown by a similar reasoning that a minimum work has to be done on systems to return them from B to A if the path used is still reversible. Thus, a reversible isothermal compression of a gas from B to A requires less work and releases less heat than an irreversible compression would.

Let us now imagine that step CD in the Carnot cycle (Figure 4.4) was done irreversibly (at finite speed). Then $W^*_{CD} > W_{CD}$ and $Q^*_{CD} > Q_{CD}$ where asterisks indicate irreversible processes. For the overall cycle $ABCDA$, the work W^* is smaller than Q_{rev} and

$$\frac{W^*}{Q_{AB}} = \frac{Q_{QB} + Q^*_{CD}}{Q_{AB}} < \frac{T_H - T_L}{T_H} \quad (4.48)$$

as $Q^*_{CD} < 0$.

Thus, the system still absorbs Q_{AB} at T_H but releases a larger absolute amount of heat $|Q^*_{CD}|$ at T_L. Bars indicate the magnitude independently of the sign. Consequently, less heat is available to generate work. Therefore, the area within the cycle is smaller since CD irreversible is above CD reversible. Note that this argument was developed for CD irreversible but could as well have been extended to an irreversible path AB, with reference to Figure 4.4. The difference

$$W^{*} - W_{\text{rev}} = Q_{CD} - Q^{*}_{CD} \qquad (4.49)$$

represents a permanent loss of energy, that is, degradation. This occurs because if work W^* was stored in an electrochemical cell then the cell, by driving a reversible refrigerating cycle, would not have the energy W required to return the system to state A.

EXAMPLE 4.7. A MAN AS AN IRREVERSIBLE ENGINE. A man burns up calories when he climbs a mountain and also when he comes down. A reversible engine, on the other hand, would retrieve the work done to lift it, that is, the increase in potential energy Mgh, as kinetic energy when it returned from the mountain. This comparison illustrates the irreversibility of our motions as we do work in both directions. M in the expression above is the mass, g is the acceleration of gravity, and h is the gain in elevation. Mg is the weight.

A man weighing 80 kg (176 lb) who walks up 1000 m (about 3300 ft) gains a potential energy of 8×10^9 ergs or $(8 \times 10^9)(3.6 \times 10^7) = 222$ cal. This is the reversible work he must do. However, hikers burn roughly 600 cal/hr and require 90 min to gain 1000 m in elevation. Thus, they use up 900 instead of 222 cal. The excess is due to friction and to our metabolic inefficiency. A faster hiker is even less efficient because he acts in a more irreversible way.

On the downhill the hiker would only burn about 200 cal/hr in spite of the conversion of potential into kinetic energy. If he were moving reversibly no metabolic energy would have to be expended by him.

Note that a water parcel that sinks in a lake or in the ocean does not convert all of its potential into kinetic energy because of frictional losses.

Equation (4.49) leads toward but is not a complete measure of the degradation of energy. Let us consider a falling weight which is suspended by pulleys and which, through friction, converts its potential energy into heat which is transferred to a reservoir at temperature T_H. Next, let the heat be transferred onto a reservoir at temperature T_L. This procedure is equivalent to a direct transfer from the weight to the reservoir T_L. Therefore, since the sum of the irreversibilities of two steps (transfer of heat to T_H followed by transfer to T_L) must be greater

than that for each step, it follows that the irreversibility of the transfer to T_H must be smaller than that to T_L (Lewis and Randall, 1961).

The argument presented above was used simply to illustrate that the amount of heat degraded is not per se a measure of irreversibility. Rather, for the same amount of heat transferred, the degree of irreversibility increases with a decrease in the temperature of the thermal sink. Thus, we expect the degree of irreversibility to be of the form Q/θ where θ is a function of the temperature that increases with increasing temperature. Let us call Q/θ the entropy. Next, we ask whether Q^*, the irreversible heat, or Q_{rev} should be used in this expression.

Consider the free expansion of an ideal gas, caused by opening a stopcock between the gas container and an evacuated container. Let the gas be in contact with an infinitely large thermal energy reservoir of temperature T. The expansion of the system, that is, of the gas, from V_A to V_B requires no $P\,dV$ works since $P_{\text{ext}} = 0$. The temperature T of the gas is kept constant due to the contact with the infinite thermal reservoir. Thus, a spontaneous process occurred fast and the entropy of the system (the gas) must have increased.

To calculate the corresponding change in entropy let us now compress the gas reversibly and isothermally (still at temperature T) back from V_B to V_A by means of a weight in contact with the thermal reservoir which exerts a pressure $P_{\text{int}} + dP$, where P_{int} is the internal gas pressure. The work done on the gas will be

$$W = nRT \ln \frac{V_A}{V_B} = -Q_{\text{rev}} \quad \text{with} \quad W < 0 \quad (4.50)$$

$-Q_{\text{rev}}$ is the heat released by the gas. The change in entropy of the weight–thermal reservoir system is Q/θ and the change in the entropy of the gas during the compression is $-Q/\theta$ because a reversible process causes no change in the entropy of the system (gas) plus surroundings (reservoir plus weight). Thus, the change in the entropy of the gas during the free expansion was Q/θ, that is, the heat that would have been absorbed by the gas if the expansion had been infinitely slow, divided by θ. The reversible heat transfer Q is used, rather than Q^*, because the entropy is a function of state.

I shall not dwell on temperature scales in this volume but will simply state that the Kelvin scale was so defined as to make $\theta = T^{\circ}\text{K}$. Thus, we conclude that the change in entropy of a system, when it proceeds reversibly or irreversibly from state A to state B, is given by

$$\Delta S = S_B - S_A = \frac{Q}{T} \qquad (4.51)$$

where Q_{rev} is the heat absorbed by the system when it proceeds reversibly from A to B.

EXAMPLE 4.8.   ENTROPY CHANGE IN A BATTERY.   If a battery in operation evolves 1000 cal of heat at 298°K and another releases 1000 cal at 273°K, which one is more reversible? What are the respective entropy changes?

The second battery is more irreversible. This is so because part of the 1000 cal released at 298°K could be used in a reversible heat engine (an idealization) operating between $T_H = 298°K$ and $T_L = 273°K$ so that a maximum of

$$\frac{209 - 273}{298} \times 1000 = 83.9 \text{ cal}$$

could be recovered.

Since the entropy change is the heat wasted isothermally divided by the temperature,

For the first process $\Delta S = 1000/298$
                                             $= 3.36 \text{ cal/°K}$
For the second process $\Delta S = 1000/273$
                                             $= 3.66 \text{ cal/°K}$

4.4   SECOND LAW OF THERMODYNAMICS AND FREE ENERGY

4.4.1   The Second Law

The second law of thermodynamics states that the entropy of the universe is increasing. This is a consequence of our observations that in spontaneous processes the entropy of systems plus their surroundings increases and only in idealized reversible processes does the entropy remain constant.

One of the simplest examples of a spontaneous process is the flow of heat from a hot to a cold object until both reach the same intermediate temperature. Since thermal contrast is required to do work (in the Carnot engine no work can be done when $T_H = T_L$), this process leads to a decrease in available energy.

We observe in nature that all systems when left

to themselves approach a final state of rest (equilibrium), and only leave this state when external forces are used. A gas tends to expand into a vacuum, warm bodies transfer heat to and reach thermal equilibrium with cold ones, and my files tend to become random instead of well organized when I am too lazy to work on them. Thus, the structure of matter and the contrast in energy are constantly decreasing in the universe. The total energy of the universe is constant but its availability to do work becomes smaller with time.

Entropy is a quantity so defined that its change is a measure of the degradation of energy or a measure of a loss of structure. A gas confined to one side of a reservoir has structure which it loses when it spreads throughout the reservoir. The second law of thermodynamics simply states that spontaneous processes increase the entropy of the universe. In other words, the entropy of the universe is actually increasing.

It can be shown, in the case of thermal energy, that the gain of entropy of a system is represented by (e.g., Lewis and Randall, 1961)

$$\Delta S = S_B - S_A = \frac{Q}{T} \qquad (4.52)$$

where Q is the heat absorbed by the system at temperature T in going reversibly from state A to state B. S_A and S_B are functions of the states A and B of the system because S is a function of state, independent of the path used to go from A to B. The measure of ΔS in terms of Q, that is, Q/T, is an idealization, however, because a reversible Q cannot be achieved in nature.

No energy is dissipated in a reversible process, that is, in one that proceeds infinitely slowly, and the entropy of the universe (system plus surroundings) does not change. This means that the ΔS of the system, given by Equation (4.52), must cancel the ΔS of its surroundings, that is,

$$\Delta S_{system} + \Delta S_{surroundings} = 0 \qquad (4.53)$$

Thus, in a reversible process entropy is transferred but is not produced. Conceptually, a weight on a pulley can be lowered reversibly in an idealized frictionless system and its potential energy stored as heat in a reservoir. The heat can then be used to raise the weight back to its original position and there has been no decrease in available energy. Of course, reversible processes are abstractions which do not occur in the real world but serve to guide our thoughts.

When an irreversible (spontaneous or actual) process occurs, energy is dissipated and the entropy of the universe increases. Therefore,

$$\Delta S_{\text{system}} + \Delta S_{\text{surroundings}} > 0 \qquad (4.54)$$

It is interesting to examine the thermal efficiency of a heat engine to clarify the mechanical relevance of entropy. Suppose that an amount of heat $(Q_{\text{rev}})_H$ is taken from a hot source and $(Q_{\text{rev}})_L$ is transferred to a cold sink while an engine operates reversibly. From the second law

$$\frac{(Q_{\text{rev}})_H}{T_H} - \frac{(Q_{\text{rev}})_L}{T_L} = 0 \qquad (4.55)$$

The minus sign in front of $(Q_{\text{rev}})_L$ is needed because the heat leaves the system. From the first law the work done is

$$W_{\text{rev}} = (Q_{\text{rev}})_H - (Q_{\text{rev}})_L \qquad (4.56)$$

since the internal energy of the whole system (source, engine, sink) is constant. From (4.55) and (4.56) one obtains

$$\frac{W_{\text{rev}}}{(Q_{\text{rev}})_H} = \frac{T_H - T_L}{T_H} \qquad (4.57)$$

Thus, even the maximum (ideal) thermal efficiency of an engine is smaller than unity because only part of the heat $(Q_{\text{rev}})_H$ can be used for work and the remainder goes to the cold sink.

For an irreversible engine it can be shown that

$$\frac{W_{\text{irrev}}}{(Q_{\text{irrev}})_H} < \frac{T_H - T_L}{T_L} \qquad (4.58)$$

4.4.2 Entropy and Probability

The entropy is not only usable in energy transfer but can be related to the probability Ω of a system through the relation

$$S = k \ln \Omega \qquad (4.59)$$

EXAMPLE 4.9.   Two books of equal size are on the left side of a bookshelf which has room for four volumes. There are only two ways in which this can

be done, by having one or the other book on the left of the other. Then,

$$S = k_B \ln 2 \qquad (i)$$

k_B is the Boltzmann constant.

Next, we can place the two books in any of four positions on the shelf which is divided in four equal spaces. Then,

$$S = k_B \ln 10 \qquad (ii)$$

In effect, if the books are labeled A and B, we may have $A\,B\,\_\,\_,\ A\,\_\,B\,\_,\ A\,\_\,\_\,B,\ \_\,A\_\,B,\ \_\,\_\,A\,B$ and the permutations of A and B.

If A and B can be slid freely in the shelf or placed anywhere in the room then Ω becomes enormous.

The natural tendency of A and B upon handling is to become scattered sooner or later since very few books stay in the same shelf for decades or centuries. They tend to spread away from the simplest structure represented by the low probability in (i). Energy, such as that of a librarian, is required to maintain the order.

The same tendency toward randomness is true of the human body, buildings, and minerals (crystals) that are weathered.

EXAMPLE 4.10.   ENTROPY CHANGES IN NATURE. The average annual entropy of the crust of the earth, of an ocean, or of a large lake changes very little over a time period that is short relative to their lifetime, since their state defined by P, T and their chemical composition are relatively invariant. Still, irreversible work is occurring and entropy is being produced. This entropy gain is transferred to space so that the entropy of the universe is increasing, while solar and internal heat is used to maintain the structure and composition of the three systems.

Next, I will define in a preliminary way some quantities that are needed to convert the laws of thermodynamics into working tools for chemistry, and use them to illustrate the concepts presented so far.

4.4.3 Some Concepts and Illustrations

EXAMPLE 4.11.   A FORMAL ILLUSTRATION OF PART OF EXAMPLE 4.10.   The earth receives from the sun a heat flux of J_Q cal/cm$^2 \cdot$ day at an average surface temperature of $T°$C. What are the resulting ΔE, ΔH, and ΔS of the earth?

This is a steady-state situation because as much heat is released from the earth to space as is received from the sun. Otherwise, the temperature of our environment would have grown and would have reached unlivable values. Thus, since the state of the earth does not change in annual average, $\Delta E = \Delta H = \Delta S = 0$.

The entropy of the universe increases, however, because heat taken from the sun at high temperatures is released to space at exceedingly low ones.

EXAMPLE 4.12.   A RETURN TO THE PROBLEM OF A MOUNTAIN CLIMBER.   A hiker burns a net 5000 cal in a climb and only eats at the end of the day. What is the change in his thermodynamic state just before

Substance	H$_2$O	CO$_2$	NaCl	CaCO$_3$	Ca^{2+}	Cl$^-$
H_f^0	−68.3	−94.0	−151.9	−288.45	−129.9	−161.63

dinner relative to the state right after breakfast? Assume that the change in atmospheric pressure is negligible. First,

$$\Delta E = Q^* - W^* = -5000 \text{ cal} \qquad (i)$$

If the reversible work to bring the hiker to the top of the mountain is $W_{rev} = 3500$ cal then Q_{rev} is −1500 cal, the heat lost to the colder air (20°C air versus a 37°C body temperature).

Thus, his change of state, in addition to $\Delta E = -5000$ cal, is $\Delta H = (Q_{rev})_P = -1500$ cal and

$$\Delta S = -\frac{1500}{310} + \frac{1500}{293} = 0.281 \text{ cal/°C} \qquad (ii)$$

Of course, his actual work was higher than 3500 cal since he did not move at an infinitesimal speed. His task at night camp is not only to rebuild his heat balance but also to reverse the gain in entropy. A large gain signifies randomness and death. Thus, he needs calories and a negentropy flow from food and a fire.

I mentioned briefly the standard heat of formation ΔH_f^0 earlier. It is the increase in ΔH when 1 mole of the substance is formed from its elements. Calculations are made relative to a $\Delta H_f^0 = 0$ for elements in their standard states (e.g., H$_2$ at 25°C and 1 atm pressure). It is important to distinguish ΔH_f^0 from ΔH^0.

EXAMPLE 4.13.   HEATS OF FORMATION AND CHANGES IN ENTHALPIES.   Note that ΔH_f^0 differs in general from ΔH^0 for a reaction since in the latter we may already start with compounds or ions as reactants, whereas in the former the reactants are the elements in their standard states. Thus, we can have ΔH^0 for CO$_2$ + H$_2$O = H$_2$CO$_3$ but ΔH_f^0 is obtained from C(solid) + 3O(gas) + 2H(gas) where O represents oxygen. Actually, the calculation is based on tables which contain the values of ΔH_f^0 for C + 2O = CO$_2$ and 2H + O = H$_2$O so that ΔH_f^0(H$_2$CO$_3$) = ΔH_f^0(CO$_2$) + ΔH_f^0(H$_2$O). On the other hand ΔH^0 = ΔH_f^0(H$_2$CO$_3$) − ΔH_f^0(H$_2$O) − ΔH_f^0(CO$_2$). The quantity ΔH^0 can also be measured directly in a constant-pressure calorimeter.

Some examples of heats ΔH_f^0 in kcal/mole, at 25°C, are (Berner, 1971; Glasstone, 1969; National Bureau of Standards, 1952):

Therefore, $\Delta H_f^0 = -288.45$ for the reaction

$$Ca_{(s)} + C_{(s)} + 1.5O_{2(g,STP)} = CaCO_{3(s)} \qquad (i)$$

and that of

$$Na_{(s)} + 0.5Cl_{2(g,STP)} = NaCl_{(s)} \qquad (ii)$$

is −151.9 kcal/mole. The values of ΔH^0 for dissolution, that is, for CaCO$_{3(s)}$ = Ca^{2+} + CO$_3^{2-}$ and for NaCl$_{(s)}$ = Na$^+$ + Cl$^-$, differ from the ΔH_f^0 values because now we must take into account the heats of formation of the hydrated ions and because the directions of the reactions were reversed.

The ΔH_f^0 of Na$^+$ and Cl$^-$ in aqueous solutions are −57.28 and −40.02 kcal/mole, respectively. Therefore,

$$NaCl_{(s)} \rightarrow Na^+_{(aq)} - Cl^-_{(aq)} \qquad (iii)$$

has

$$\Delta H^0 = 151.9 - 57.28 - 40.02$$
$$= 54.6 \text{ kcal/mole} \qquad (iv)$$

For the reaction

$$CaO_{(s)} + CO_{2(g)} = CaCO_3 \qquad (v)$$

$$\Delta H^0 = \Delta H_f^0(CaCO_{3(s)}) - \Delta H_f^0(CaO)_{(s)}$$
$$- \Delta H_f^0(CO_{2(g)}) \qquad (vi)$$

Thus, we see that enthalpy changes for reactions can be obtained from heat of formation data.

Heats of formation and changes in enthalpy in general can be used in a number of ways, some of which are relevant to the earlier discussion of the enthalpy and others which are associated with the entropy. We shall soon examine a quantity called the free energy F for which $\Delta F = \Delta H - T \Delta S$ at constant temperature.

Heats of formation are also useful for predicting the heat of combustion of substances when direct values are not available, and free energies of formation.

EXAMPLE 4.14. The heat of combustion of methane is 212.7 kcal/mole since

$$CO_{2(g)} + 2H_2O_{(1)} = CH_{4(g)} + 2O_{2(g)} \qquad (i)$$

with $\Delta H_f^0(CO_2) = -17.9$, $\Delta H_f^0(O_2) = 0$, $\Delta H_f^0(H_2O) = -68.3$, and $\Delta H_f^0(CH_4) = -17.9$ kcal/mole. Thus, the heat of combustion is -212.7 kcal/mole.

ΔF represents the change in free (available) energy F during a process. It is important because ΔF^0 at the standard state equals $RT \ln K$ so that it is related to equilibria. Spontaneous reactions have $\Delta F^0 < 0$ for the system since energy is expended by it. Another useful relation is $\Delta F = \Delta H - T \Delta S$.

EXAMPLE 4.15. If, for a given reaction at 300°K, $\Delta H = -30$ kcal/mole, determined calorimetrically, and $\Delta S = 50$ cal/mole · °K, then

$$\Delta F = 30 - 15 = 15 \text{ kcal/mole} \qquad (i)$$

The process is not spontaneous, that is, it would not occur unless driven since $\Delta F > 0$.

In conjunction with ΔS_f^0, the molar entropy of formation, $\Delta F_f^0 = \Delta H_f^0 + T \Delta S_f^0$, can be calculated and used to yield the relative stabilities of substances. Thus, one finds at atmospheric pressure that, of the two forms of $CaCO_3$, calcite is more stable than aragonite since it has the larger negative ΔF_f^0 (-269.98 versus -269.75 kcal/mole).

Applications of thermodynamics to seawater as a binary water–sea salt system, in terms of changes in free energy, entropy, enthalpy, partial molar volume, expansibility, and compressibility during hydrographic processes can be found in Fofonoff (1962). Leyendekkers (1976) examined rigorously the contributions of the components of sea salt to the above properties. I shall not pursue these topics which refer primarily to strong or "completely" dissociated electrolytes because my emphasis is placed on chemical equilibria and kinetics. The one exception to this rule, besides the overviews of relevant broad topics in oceanography and geochemistry, will be the study of activity coefficients of strong electrolytes in Chapter 5. This study is necessary for an understanding of activities, activity coefficients, and equilibria.

Geochemical and aquatic applications of equilibrium thermodynamics and rates of reactions can be found in Garrels and Christ (1965), Krauskopf (1967), Stumm and Morgan (1970), Berner (1971), Broecker and Oversby (1971), Holland (1978), and in this book.

4.5 FREE ENERGY EQUILIBRIUM AND THE OCEANS

4.5.1 Free Energy

We have already examined three thermodynamic functions E, S, and H and will define two more which depend only on the state of the system.

$$A = E - TS \qquad (4.60)$$

is known as the work function or the Helmholtz free energy. In a reversible isothermal process (4.51) and (4.60) yield

$$\Delta A = \Delta E - Q_{rev} = W_{rev} \qquad (4.61)$$

Thus, A is the maximum work (of expansion, electrical, etc.) that can be done by a system at constant temperature.

EXAMPLE 4.16. FREE ENERGY AND THE ORIGIN OF THE UNIVERSE. Let us consider a cosmological application of classical thermodynamics as a mental exercise (RMP). You are probably all familiar with the various hypotheses for the origin of the Universe: the Big Bang, the pulsating, and the steady-state universe. Let us consider the Big Bang in which the primordial cosmic particles coalesced because of their potential energy. This caused an enormous increase in temperature and an explosion which led to the formation of stars and planets.

The universe in this case is the sum total of all matter and energy that exist. It is, therefore, an isolated system in that no matter or energy can leave or enter it because there is nothing but the universe.

Let us consider next the pre-Bang shrinking. For it to be a spontaneous process it is necessary that $dA < 0$ at any instant. Now,

$$A = dE = T \, dS = S \, dT \qquad \text{(i)}$$

Because the system is isolated, $dE = 0$ if the conservation of energy holds. The temperature increases so that $-S \, dT < 0$. The entropy increases because a random system coalesces, that is, acquires contrast with the rest of space.

$$F = H - TS \qquad (4.62)$$

is the Gibbs free energy or simply the free energy. At constant temperature and pressure it can be shown that $\Delta F = \Delta A + P \, \Delta V$. Thus, F is the maximum non-$P \, \Delta V$ work that can be done by a system at constant temperature and pressure.

We see, therefore, that the change in the internal energy E of a system does not represent the work that the system can do because part of ΔE corresponds to an exchange of heat. The functions A and F are the ones that yield the work potential of the system.

F is a valuable function because it helps us ascertain the direction and the extent to which chemical reactions may proceed, as we shall see later on in this chapter.

4.5.2 Free Energy and General Equilibrium Conditions

In this section, mechanical and general equilibrium conditions are considered to set the stage for hydrodynamic and chemical equilibria. The latter will be mentioned only briefly at this point.

Let us examine in detail how the free energy is related to mechanical equilibrium. In mechanical systems, such as a free-falling object, spontaneous processes occur when the systems contain available potential energy. When the object comes to rest on the surface of the earth, its remaining potential energy is no longer available for a further change of state (change in height). The condition for a spontaneous process is

$$\Delta E_P = Mgh < 0 \qquad (4.63)$$

At mechanical equilibrium (a ball at the bottom of a well or a loaded spring at rest) the net force acting on the system is zero and, therefore, the work and the change in energy involved in a slight displacement from equilibrium are zero. In other words, $dE_P = 0$ and E_P is a minimum. By the way, these mechanical systems could be treated as thermodynamic ones if we also took into consideration heat released or absorbed. We will, however, concentrate on the thermodynamic properties of chemical systems containing large numbers of very small particles rather than the thermodynamic properties of single solid objects.

Let us seek the thermodynamic conditions for spontaneous processes, and for equilibria in isolated (no heat, work, or matter exchange) and closed systems (no exchange of matter) of fixed composition for which only $P \, dV$ work can be done. For an irreversible process $\delta Q^*/T < dS$, where δQ^* is the heat absorbed irreversibly. If we assume that only $P \, dV$ work is done then $\delta Q^* = dE + P \, dV$ and, for a spontaneous process,

$$\frac{dE + P \, dV}{T} < dS \qquad (4.64)$$

A reversible process, which consists of a series of equilibrium states, yields the equilibrium condition,

$$\frac{dE + P \, dV}{T} = dS \qquad (4.65)$$

For an isolated system (no work, heat, or mass exchange) E and V are constant and, therefore,

$$(dS)_{E,V} \geq 0 \quad \text{(isolated system)} \qquad (4.66)$$

The sign $>$ applies to a spontaneous process while $=$ refers to the equilibrium condition (maximum entropy) in an isolated system. One concludes that, at equilibrium, $S_{E,V}$ has a maximum value because $(dS)_{E,V}$ during the approach to equilibrium is larger than zero, that is, $S_{E,V}$ increases.

It is interesting to observe that a process that occurs at an infinitesimal rate provides a necessary condition for reversibility but that infinite reservoirs at temperatures T_H and T_L are required to provide also a sufficient condition (RMP). For example, an infinitely slow isothermal expansion of an ideal gas absorbs an amount of heat Q/T_H and the changes in entropy are $-Q/T_H$ for the surroundings and Q/T_H for the system. In reality, however, there has been a decrease dT_H in the temperature of the surroundings since its heat capacity is finite in spite of its infinite size. If we proceed with a Carnot cycle, there is also an increase dT_L in the temperature of the infinitely large low-temperature reservoir. We neglected these infinitesimal changes in the earlier sections. If, however, the reservoirs are finite then changes $\Delta T_H < 0$ and $\Delta T_L > 0$ must be taken into consideration.

Now, if we attempt a Carnot cycle in an isolated system that contains finite size reservoirs at T_H and T_L, there will be changes $\Delta T_H < 0$ and $\Delta T_L > 0$. We can even conceive of a partial Carnot cycle in which T_H drops to T_h during the first expansion, because of the heat absorbed by the engine. Then, T_h tends to T_L during the adiabatic expansion, and T_L rises to T_1 during the first compression step. Let us assume that T_1 reaches the value T_h during this step. Then, the system will have reached thermal equilibrium with a maximum S compatible with the constraints on the system and no more work can be done. Thus, we found an infinitely slow process that occurs at thermal and pressure equilibrium in a finite isolated system. It is not reversible, however, since eventually there is no thermal contrast and S reached a maximum (RMP).

EXAMPLE 4.17. FREE ENERGY OF A HETEROGENEOUS PROCESS (e.g., GLASSTONE, 1969). Consider the reaction

$$CuSO_4 \cdot 3H_2O_{(s)} + 2H_2O_{(g)} = CuSO_4 \cdot 5H_2O_{(s)}$$

$$(i)$$

The free energy of this reaction is

$$-\Delta F = RT \ln \left[\frac{1}{(P_w)_e} - \frac{1}{P_w} \right] \qquad (ii)$$

$-\Delta F = 657$ cal/mole H_2O and work is done because water vapor goes from its equilibrium value with liquid water to equilibrium with the salt hydrate system.

Let us next consider the equilibrium conditions obtainable from A and F. Introducing

$$T\,dS \geq dE + P\,dV \qquad (4.67)$$

into

$$\Delta A = dE - T\,dS - S\,dT \qquad (4.68)$$

and into

$$dF = dE + P\,dV + V\,dP$$
$$- T\,dS - S\,dT \qquad (4.69)$$

one obtains for the free energies

$$(dA)_{V,T} \leq 0 \qquad (dF)_{P,T} \leq 0 \qquad (4.70)$$

at constant pressure and at constant volume. The equal sign applies to equilibrium and $<$ applies to spontaneous processes.

EXAMPLE 4.18. FREE ENERGY CHANGE IN A REACTION. As an example of the free energy we have, for the reaction $H_{2(g)} + \frac{1}{2}O_{2(g)} = H_2O_{(l)}$, $\Delta F^0 = -56.7$ kcal. Hydrogen and oxygen are in their standard states ($P = 1$ atm, $T = 25°C$) and water is pure and also at 25°C and 1 atm. Thus, we see that the formation of water, which is exceedingly slow in the absence of a catalyst (a substance that accelerates a reaction without actually being consumed or formed, e.g., a platinum powder), is spontaneous as $\Delta F^0 < 0$.

The dissolution of acetic acid, $HAc = H^+ + Ac^-$, has $\Delta F^0 = -RT \ln K < 0$ at the standard state ($\Delta F^0 < 0$, $K > 1$). Therefore, it will proceed away from it until equilibrium is reached. The ΔF for a reaction, once the reactants and products leave the standard state, differs from ΔF^0 and becomes zero once equilibrium is reached. I shall examine this topic, as well as the conventional standard states used, shortly. Briefly, in anticipation of the more extended explanation, what happens is that the actual ΔF away from the standard state is $\Delta F = RT \ln(K/Q_r)$ where Q_r is the reaction quotient $[H^+][Ac^-]/[HAc]$. At the standard state $Q_r = 1$ by definition because substances have unit activities, whereas at equilibrium $Q_r = K$ and $\Delta F = 0$. If $Q < K$ then $\Delta F < 0$ and the reaction proceeds as written (ΔF is written above for $HAc \to H^+ + Ac^-$) until $Q_r = K$ and it proceeds in the reverse direction if $Q_r > K$. The standard state for solutes is subtle and will be discussed at length later on.

EXAMPLE 4.19. STANDARD CHANGES IN FREE ENERGY, ENTHALPY, AND ENTROPY AS WELL AS THE THERMODYNAMIC EQUILIBRIUM CONSTANT FOR A CHEMICAL REACTION ILLUSTRATED. Let us assume that ΔH^0 obtained in a calorimeter at constant P for the reaction

$$A + B = C + D \tag{i}$$

is 10 kcal/mole at 298.16°K (25°C).

Next, ΔF^0 is measured in a galvanic cell from E^0, the standard electromotive force, and is found to be -20kcal/mole at 298.16°K. Then, as

$$\Delta F^0 = -RT \ln K \tag{ii}$$

the equilibrium constant is

$$K = -\frac{\log^2}{2.3026RT} = \frac{0.30103}{2.3026 \times 1.98726 \times 298.16}$$
$$= 2.206 \times 10^{-4} \tag{iii}$$

Do not confuse K with °K since the latter is the absolute temperature. R is 1.98726 cal/deg · mole.
Then

$$\Delta S^0 = \frac{\Delta H^0 - \Delta F^0}{T} = 0.1006 \text{ kcal/mole} \cdot \text{deg} \tag{iv}$$

The negative ΔF^0 and the positive ΔS^0 show that the reaction is spontaneous from left to right when the reactants and products are in their standard state.

In chemical reaction

$$\Delta F = -RT \ln K + RT \ln Q_r \tag{v}$$

where Q is the reaction quotient

$$Q_r = \frac{\{C\}\{D\}}{\{A\}\{B\}} \tag{vi}$$

in which the activities $\{\ \}$ apply to the actual concentrations rather than to the standard state concentrations.

The system tends to the equilibrium K, that is, toward $Q_r = K$ as $\Delta F = 0$. This last relation is the equilibrium condition for a closed system at constant P and T. If $Q_r > K$ then the reaction (i) proceeds to the left as $F > 0$ for the reaction to the right and vice versa. In the standard state it always would proceed to the right as we saw earlier.

Thus, we see that thermodynamics yields the direction of reactions and also their extent Q_r to K.

With additional mass balance and activity coefficient information we can obtain the actual concentrations [ ] in the two states.

Free energy changes can be determined by a variety of techniques such as the electromotive force, reaction isotherms, solubilities, spectroscopy, and vapor pressures. Examples can be found in standard texts.

Note that Equations (4.67) and (4.71) concern closed systems, that is, systems that do not exchange matter with their surroundings. Equation (4.66) is even more limiting since it applies to isolated systems that do not interact in any manner with their surroundings. Equation (4.70), if we only consider $P\,dV$ work, corresponds to restricted systems, that is, to systems that exchange energy but not work.

Types of systems are as follows:

1. *Restricted.* Exchange energy (e.g., thermal) but no work is done on the surroundings.
2. *Adiabatic.* No energy exchange but work can be done.
3. *Closed.* Energy is exchanged but not matter.
4. *Isolated.* No exchange of any type and no work on the surroundings.

The general equilibrium and spontaneous reaction criteria are tabulated in Table 4.1.

4.5.3 Hydrothermodynamics

Equations of motion and conservation of momentum (Von Arx, 1962) meet the conditions set by the laws of thermodynamics (Reid, 1964).

TABLE 4.1 Criteria for equilibria and spontaneous reactions under a variety of conditions for closed systems of fixed composition and for which only $P\,dV$ work can be done.

$(dS)_{E,V} \geqslant 0$
$(dA_H)_{T,V} \leqslant 0$
$(dF)_{P,T} \leqslant 0$
$(dS)_{H,V} \geqslant 0$
$(dE)_{S,V} \leqslant 0$
$(dH)_{S,P} \leqslant 0$

First Law. This law will be approached through some definitions. First, there is a function E of position and time in the fluid which varies when a small parcel of water meanders through the oceans. The existence of such a parcel is conceptual since it cannot be followed by an observer. E, the internal energy, as well as other extensive properties, is defined per unit mass of seawater in this type of work because a mole has no meaning for a sea salt composed of many electrolytes.

The value of E along the moving parcel of seawater changes due to:

1. The change in position in the gravitational field (change in potential energy).
2. The kinetic energy which can be converted into potential energy but which is also dissipated as heat, due to frictional forces.
3. The flux of chemical energy through diffusion of sea salt across the parcel surface when the parcel reaches water of a different salinity.
4. Work done on or by the water parcel, such as $P \, \Delta V$ work.
5. There is furthermore a quantity Q^* which represents the flux of heat across the parcel surface due to conduction and radiation.

Again, the first law, $\Delta E = Q^* - W^*$ is valid. It is usually expressed in terms of the specific processes listed above. W^* is obtained from the first four processes.

It is important to realize that ΔE, ΔH, ΔA, and ΔF can be calculated exactly, at least in principle, when seawater moves, let us say, from a state $[T_A, P_A, (S\text{‰})_A]$ to $[T_B, P_B, (S\text{‰})_B]$ over some time interval Δt. This applies to a parcel or to an integration over a large body of water. The state can be defined by three properties because, according to the phase rule, there are two components (water and sea salt) and one phase (seawater) so that the number of degrees of freedom is $f = 2 - 1 + 2 = 3$.

On the other hand, the heat taken up, Q^*, and the work done, W^*, being spontaneous (irreversible) processes, cannot be ascertained by equilibrium thermodynamics. They can only be calculated from transport equations, such as diffusive ones, and fall in the domain of irreversible thermodynamics.

The end product of this exercise is to solve a differential equation compatible with the equations of motion, the conservation of momentum, and the

first law of thermodynamics to obtain the time and position variation of the energies associated to the five processes stated above. Some readers with a background in mathematics may wish to follow further the solution of this system of equations. I shall outline this topic based on Reid (1964). The extensive properties in what follows are understood to be specific ones (per g or mL of seawater) unless otherwise indicated.

The principles of hydrothermodynamics are:

1. The conservation of mass (equation of continuity).
2. Conservation of momentum.
3. Conservation of energy.
4. Dissipation of energy.

This last principle corresponds to the second law and is an inequality that places restrictions on diffusion, heat conduction, dissipation or energy by viscosity and chemical reactions, and so on. These restrictions will not be treated here but pertain to a positive change of entropy and, in the case of a steady state, to a minimum rate of entropy generation (Prigogine, 1955).

Continuity Condition. Let $\mathbf{v}\,(x, y, z, t)$ be the velocity of the fluid (seawater in this case) across an arbitrary surface A and m_t be the net mass of fluid that crosses A per unit time. Then

$$m_t = \int_A \boldsymbol{\eta}\rho\mathbf{v} \, dA \qquad (4.71)$$

where $\boldsymbol{\eta}$ is the unitary vector perpendicular to the surface. The relation $(m_t)_{\text{out}} = (m_t)_{\text{in}}$ applies because of the conservation of mass, if no sink is present within the volume V under consideration.

The velocity vector $\mathbf{v}$ is defined by Equation (4.71) in which the density ρ is in turn defined by

$$m = \int_A \rho(x, y, z, t) \, dV \qquad (4.72)$$

m is the mass present in V at a given time.

The equation of continuity is

$$\frac{dm}{dt} = -\int_A \boldsymbol{\eta}\rho\mathbf{v} \, dA \qquad (4.73)$$

This equation and the others are derived with coordinates and properties that follow a parcel of water.

The extensive properties are specific, that is, they are expressed per unit volume or per unit mass. As an example, the internal energy E is in cal/g rather than cal/mole and is expressed in terms of a generalized sea salt.

Equation (4.73) can be placed in the form (Reid, 1964)

$$\frac{\partial \rho}{\partial t} + \nabla \cdot \rho \mathbf{v} = 0 \qquad (4.74)$$

Note, that, for a general vector property $\mathbf{G}$, $-\nabla G$ is the gradient of G. $\nabla \cdot \mathbf{G}$ is its divergence, and $\nabla \times \mathbf{G}$ is its curl. The operator ∇ is

$$\nabla = \mathbf{i}\frac{\partial}{\partial x} + \mathbf{j}\frac{\partial}{\partial y} + \mathbf{k}\frac{\partial}{\partial z} \qquad (4.75)$$

$\mathbf{i}$, $\mathbf{j}$, and $\mathbf{k}$ are the unit vectors along the three axes. Another form of the equation of continuity is

$$\nabla \cdot \mathbf{V} = -\frac{1}{\rho}\frac{D\rho}{Dt} = \frac{1}{\alpha}\frac{D\alpha}{Dt} \qquad (4.76)$$

where α is $1/\rho$. The operator D/Dt applied to $\mathbf{G}$ is

$$\frac{DG}{Dt} = \frac{\partial G}{\partial t} + u\frac{\partial G}{\partial x} + v\frac{\partial G}{\partial y} + w\frac{\partial G}{\partial z} \qquad (4.77)$$

with $u = dx/dt$, $v = dy/dt$, $w = \partial z/dt$. Thus,

$$\frac{DG}{Dt} = \frac{\partial G}{\partial t} + \mathbf{v} \cdot \nabla G \qquad (4.78)$$

$\partial G/\partial t$ is the local variation in time, that is, the variation for fixed coordinates.

The equation of continuity simply states that no seawater is created in a volume V surrounded by a surface A within the body of the oceans.

Continuity of Salt (Conservative Constituents). The salinity is given by

$$S(‰) = \sum_i s_i \qquad (4.79)$$

where the summation is extended over the major constituents with concentrations s_i (for example, in mg/g-SW). The net diffusion of salt across the surface of a moving parcel is

$$\sum(t) = \int_{V(t)} \rho S‰ \, dv \qquad (4.80)$$

Equation (4.73) leads to a diffusive flux of salt $\mathbf{J}_S$ at a point x, y, z at time t, due to the salt gradient. Thus, the flux across A is

$$\int_{A(t)} \boldsymbol{\eta} \mathbf{J}_S \, dA \qquad (4.81)$$

If there is no source (dissolution) or sink (precipitation) within V, the divergence theorem leads to

$$\rho\frac{DS}{Dt} + \nabla \cdot \mathbf{J}_S = 0 \qquad (4.82)$$

Next, let us consider the advective and the nonadvective transport. The former is related directly to the velocity $\mathbf{v}$ and is $\rho\, S‰\mathbf{v}$ (advective flux of salt). The nonadvective flux is $\mathbf{J}_S$, which is observed in a macroscopic scale but is associated to phenomena in a macroscopic value. In the equations of hydrothermodynamics the condition of continuity (of total mass) is the only one that does not contain the nonadvective flux, because of the manner in which $\mathbf{v}$ is defined.

Nonconservative Constituents. We are still following a parcel of water along its track. For a nonconservative property

$$\frac{\partial \rho N}{\partial t} + \nabla \cdot [\rho N)\mathbf{v} + \mathbf{c}_N)] = R_s \qquad (4.83)$$

where R_s is the sink function, ρN is the concentration of the nonconservative property (g/cm^3 since N is in g/g-SW) and $\mathbf{c}_N$ is defined by

$$\mathbf{J}_N = \rho N \mathbf{c}_N \qquad (4.84)$$

Thus, $\mathbf{c}_N$ is the diffusion of the constituent current and $\mathbf{v} + \mathbf{c}_N$ is the total velocity of the particles of N.

Equation (4.83) is equivalent, for the conservative case, to

$$\frac{\partial \rho S‰}{\partial t} + \nabla \cdot [\rho S‰ (\mathbf{v} + \mathbf{c}_S)] \qquad (4.85)$$

where the subscript S indicates the salinity.

Conservation of Momentum. The equation of absolute motion, with ordinates fixed in space, is

$$\mathbf{G}(t) = \int_{V(t)} \rho \mathbf{v} \, dV = \int_m \mathbf{v} \, dm \qquad (4.86)$$

$G(t)$ is the total momentum of a fluid of absolute velocity $\mathbf{v}$ in a volume V of constant mass M. Taking into account Newton's law of motion, $d\mathbf{G}/dt = \mathbf{J}$, one eventually obtains

$$\frac{\partial \rho v_i}{\partial t} + \frac{\partial}{\partial x_j}(\rho v_i v_j - p_{ij}) = \rho \mathbf{g} \quad (4.87)$$

where ρv_i is the density of momentum, $\rho v_i v_j$ is the advective momentum flux, p_{ij} is the nonadvective flux of momentum due to the hydrostatic pressure and viscous forces, and $\mathbf{g}$ is the acceleration of gravity.

p_{ij} is the scalar of $\mathbf{p}$ on the face of a tetrahedron normal to the x axis (x = constant). The tetrahedron replaces the actual parcel surface to simplify the mathematical treatment. The total force on this face is $0.5 l_2 l_3 p_{ix}$ where l_2 and l_3 are the lengths of the two perpendicular sides of the triangle (face of the tetrahedron) for which x = constant.

Equation of Motion on a Rotating Earth.

The equation is

$$\rho \frac{D'\mathbf{v}}{Dt} = \rho(\mathbf{g} + \mathbf{f}_m + \mathbf{f}_s) - 2\rho \mathbf{\Omega} \times \mathbf{v}'$$

$$- \Delta P + \frac{\partial \tau_{ij}}{\partial x_j} \quad (4.88)$$

where $\mathbf{v}'$ is the magnitude of the relative velocity, $2\rho\mathbf{\Omega}v'$ is the apparent force due to the rotation of the earth $\mathbf{\Omega}$, $\mathbf{f}$ is the force per unit volume produced by tides, P is the pressure, $\mathbf{g}$ is the acceleration of gravity, and τ is the viscous force.

Introduction of the First Law.

I discussed the first hydrothermodynamic law in conceptual terms earlier. Formally, from the equations that were derived above and from modified forms of those equations, the first law can be expressed as

$$\rho \frac{D'}{Dt}\left(\frac{1}{2}v'^2 + E\right) = -\boldsymbol{\nabla} \cdot \mathbf{Q} - \boldsymbol{\nabla} \cdot \mu \mathbf{J}_s$$

$$+ \rho(\mathbf{g} + \mathbf{f}) \cdot \mathbf{v}' + \frac{\partial p_{ij}\mathbf{v}'_i}{\partial x_j} \quad (4.89)$$

or

$$\frac{1}{2}\rho \frac{D'v'^2}{Dt} + \rho\mathbf{v}' \cdot (\mathbf{g} + \mathbf{f}) + \mathbf{v}'_i \frac{\partial p_{ij}}{\partial x_j} \quad (4.90)$$

These equations permit the study of heat transfer and energy of hydrodynamic processes in a manner compatible with the first law of thermodynamics. They refer to the hydrodynamics of the simplified chemical system water–sea salt.

Other Thermodynamic Forms.

Generalized forms of H, S, A, and F which take into account mixing and, for the last three quantities, the degradation of energy, can also be defined and applied.

The specific thermodynamic properties of water–sea salt are needed for many physical processes. Examples are advection and eddy diffusion as related to the internal energy, salt diffusion or mixing and the entropy and free energy, heat transfer and the enthalpy, and evaporation and the internal energy as well as the free energy.

It is well to remember that these processes are irreversible so that the changes in the thermodynamic quantities are functions of the initial and final states. Work done by a parcel of water and heat absorbed are, however, functions of the paths followed by the processes. The properties are usually defined in terms of a moving water parcel since they are invariant at a given site except when near the sea surface.

The degradation of energy and the second law were studied by Reid (1964), Eckart (1940), and Fofonoff (1962). They also examined the second law which they placed in nonequilibrium hydrothermodynamic format. H, S, A, and F were then applied to oceanic problems of which I shall only present one example because the results are used mainly, although not only, for hydrodynamic studies.

EXAMPLE 4.20. AN APPLICATION OF THE ENTROPY. One application of the above concepts concerns the specific entropy s' (Fofonoff, 1962), which varies according to

$$ds' = \frac{\partial s'}{\partial T}dT + \frac{\partial s'}{\partial P}dP + \frac{\partial s'}{\partial S\text{‰}}dS\text{‰}$$

$$= \frac{c_P}{T}dT - \frac{\partial v'}{\partial T}dP - \frac{\partial \mu}{\partial T}dS\text{‰} \quad (i)$$

c_P is the specific heat at constant pressure and v' is the specific volume at constant P.

If there is complete mixing within a layer of water then s' and $S\text{‰}$ must be the same within it.

Then

$$\partial s' = \frac{c_P}{T} \, dT - \frac{\partial v'}{\partial T} \, dP = 0 \qquad \text{(ii)}$$

or

$$\left(\frac{\partial T}{\partial P}\right)_{s'} = \frac{T \, \partial v'/\partial T}{c_P} = \Gamma \qquad \text{(iii)}$$

Γ is known as the adiabatic lapse rate of T as a function of T, P, and $S\permil$. It is used in the study of thermal inversions in deep trenches, in the computation of the speed of sound, and in the calculation of the static stability of waters.

The potential temperature discussed earlier is related to Γ by

$$T_f = T_i \int_{P_i}^{0} \Gamma \, dP \qquad \text{(iv)}$$

i represents the initial value at depth and f the value at the surface.

Fixed Coordinates.   Let us next consider the position in the ocean rather than coordinates tied to a specific water parcel (RMP). In deep waters there is essentially a steady-state situation which, at a given position, can always be defined by T, P, and $S\permil$ even though advection and eddy diffusion are still occurring. If we look at the oceans as consisting of a simultaneous large number of small cubes at all positions available, then, if we neglect seasonal and other changes in the surface layer, the thermodynamic properties of the ocean as a whole can be considered to be at a steady state. This steady state actually is defined by all the P, T, $S\permil$ in the cubes $dx\,dy\,dz$ or $\Delta x\,\Delta y\,\Delta z$ which exist in the oceans. These ensembles of cubes (infinitesimal or finite increment ones) define in turn the thermodynamic properties E, H, A, and F of the oceans.

I presented two views of the oceans and will conciliate them next. Any energy loss from a water parcel is transmitted to another so that total energy is conserved. An example is that of water which loses potential energy by sinking but is replaced at its original height. Thus, from an energy standpoint the two views concur.

The second (positional) view leads directly to a constant entropy since the state of the overall ocean is not changed. In the water parcel method, entropy is generated when there is heat exchange and greater thermal uniformity because of mixing. This entropy gain cannot be undone by the oceans alone. What happens then is that the entropy generated is transmitted to space and the thermal contrast is reconstituted by further solar radiation.

A First Note on Chemical Reactions.   Chemical reactions of solutes in the sea are linked to hydrodynamic transport if they are slow and cannot reach equilibrium during the motion of waters. These motions change P and T and, therefore, the thermodynamic equilibrium constant. Furthermore, mixing changes $S\permil$ so that stoichiometric constants are also altered. In general, the lack of equilibration is not only due to physical processes, but also to biological and geochemical processes. An example of a reaction that does not reach equilibrium is $Ca^{2+} + CO_3^{2-} = CaCO_{3(s)}$, that is, the crystallizations of calcite and aragonite. Near-surface waters are supersaturated and deep waters are undersaturated.

Reactions that reach equilibrium, such as the dissociation constant of bicarbonate, $HCO_3^- = H^+ + HCO_3^-$, do not depend on transport and biological as well as geological processes. Note that the total dissolved inorganic carbon, $TCO_2 = [CO_2] + [H_2CO_3] + [HCO_3^-] + [CO_3^{2-}]$, is at a steady state that depends on fluxes in and out of the oceans and their evolution with time, but that the components are in dissociation and hydration equilibria. It is interesting to observe that $[CO_3^{2-}]$ is not at solubility but is at dissociation equilibrium.

Nonequilibrium states may be steady states, in which the concentrations of the reactants and products depend on the set $(T, P, S\permil)$ or transient states that vary with $(T, P, S\permil, t)$, as in the case of several trace metals added by human activities.

A somewhat different elaboration of the consequences of the relative speeds of chemical reactions, transport processes, and biological activity can be found in Morgan (1967).

4.6   CHEMICAL THERMODYNAMICS, PARTIAL MOLAL QUANTITIES, AND MULTICOMPONENT SYSTEMS

4.6.1   Equilibrium Conditions in Multicomponent Systems

This section yields the key equations that will be used later for the derivation of the practical relationship between the chemical equilibrium constant and the thermodynamic functions of state.

In the chemical thermodynamics of solutions we are concerned with the energetics and the maximum extent of chemical reactions such as the dissociation of acids, redox processes, ion associations, complexes, and phase transformations which include the dissolution of gases and salts. The thermodynamics and equilibria of multielectrolyte solutions will be treated later in this chapter as well as in Chapter 5.

We often deal with systems containing several components and phases which may be open or closed depending on how we define the borders between systems and their surroundings. We may, for example, consider a solution of NaCl as an open phase to which we add solid NaCl removed from the surroundings. Alternatively, we may consider the NaCl solution in contact with solid NaCl as a closed system with two phases. The first approach can be used when studying the thermodynamics of salt solutions, whereas the second one is relevant to solubility work.

We define the number of components as the number of substances that can be varied independently within a phase or in a system containing several phases. Thus, in the case of the reaction

$$CaSiO_3 + CO_2 \rightleftarrows CaCO_3 + SiO_2 \quad (4.91)$$

there are three components. They are the total number of substances minus the number of restrictive conditions, such as those of mass balance and electrical neutrality. We may select any three among the four chemicals as the primary components.

If a system contains a number of components that may vary either in total amount (returning a cup of coffee to the pot) or in the proportions of its constituents then we must, in addition to properties such as P, V, and T, specify the amounts (moles, g, etc.) of the components. Extensive energy properties can then be expressed by relations such as $F = F(T, P, n_i)$, $E = E(S, V, n_i)$, and $A_H = A_H(T, V, n_i)$, where n_i is the number of moles of component i and $i = 1, \ldots, c$.

The total differential of E, as an example, becomes

$$dE = \left(\frac{\partial E}{\partial V}\right)_{S,n_i} dV + \left(\frac{\partial E}{\partial S}\right)_{V,n_i} dS$$

$$+ \sum_i \left(\frac{\partial E}{\partial n_i}\right)_{S,V,n_j} dn_i \quad (4.92)$$

with $n_i = n_1, \ldots, n_c$ and with $n_j \neq n_i$. The first two derivatives are taken with n_i constant and describe dE for a closed system, whereas the full three-term equation with n_i variable corresponds to an open system. Equation (4.92) can be rewritten as

$$dE = -P \, dV + T \, dS + \sum_i \mu_i \, dn_i \quad (4.93)$$

where

$$\mu_i = \left(\frac{\partial E}{\partial n_i}\right)_{S,V,n_j} \quad (4.94)$$

is known as the chemical potential. Spontaneous processes at constant S and V go from higher to lower values of E since internal energy is used to do work. In this vein dE may be thought of as a thermodynamic potential energy change and $\partial E/\partial n_i$ as a chemical potential change.

Counterparts of Equation (4.93) are

$$dH = T \, dS + V \, dP + \sum_i \mu_i \, dn_i \quad (4.95)$$

$$dA = -S \, dT - P \, dV + \sum_i \mu_i \, dn_i \quad (4.96)$$

$$dF = -S \, dT + V \, dP + \sum_i \mu_i \, dn_i \quad (4.97)$$

and one finds that

$$\mu_i = \left(\frac{\partial E}{\partial n_i}\right)_{S,V,n_j} = \left(\frac{\partial H}{\partial n_i}\right)_{S,P,n_j} = \left(\frac{\partial A_H}{dn_i}\right)_{T,V,n_j}$$

$$= \left(\frac{\partial F}{\partial n_i}\right)_{T,P,n_j} \quad (4.98)$$

Equations such as those above apply to open phases with compositions that can be changed by adding or by removing n_i through the phase walls. For closed phases

$$(dE)_{V,S} = (dH)_{S,P} = (dA)_{T,V} = (dF)_{T,P} \quad (4.99)$$

as, for example, $(dE)_{V,S} = (dE)_{V,S,n_j}$ since the n_j are fixed.

If the single-phase system without internal reactions or with such reactions at equilibrium is now considered closed, then the dn_i become zero and

$$\sum_i \mu_i \, dn_i = 0 \quad (4.100)$$

This is the general condition of equilibrium in a closed system.

Let us consider a closed system consisting of several open phases, at thermal and pressure equilibrium with each other. The equilibrium condition at constant T and P is that dF for the overall system, which is the sum of the terms $\Sigma \mu_i \, dn_i$ for all phases, be equal to zero. The transfer of components between phases may be diffusional, may correspond to a change of state such as the melting of ice, or may imply a chemical reaction between phases.

An example of the latter is the reaction of $CaCO_3$ with $MgCO_3$ to form the more stable dolomite. The system can consist of the solid phases $CaCO_{3(s)}$, $MgCO_{3(s)}$, $CaMg(CO_3)_{2(s)}$, $CO_{2(g)}$ and the aqueous solution in a thermally insulated container with a piston. Then, T and P remain constant in spite of changes in $p\,CO_2$ when the pH changes. If T and P did change, then, for dF as an example, Equation (4.97) would have to be used instead of (4.100).

4.6.2 Partial Molal Quantities and a Graphical Explanation

The fundamentals of partial molal quantities are developed here and will be used later to study the effects of chemical potentials, partial molal volumes, and partial molal enthalpies on chemical equilibria versus temperature and pressure.

The chemical potential μ_i is an example of a partial molal quantity. Let us consider a general extensive property G of a system on which no external force other than $P \times A$ acts. A is the area. If the system has c components and is large enough so that its surface properties do not affect significantly the bulk properties, then we may write

$$G = G(T, P, n_1, \ldots, n_c) \qquad (4.101)$$

to indicate the functional dependence of G on T, P, and the numbers of moles of the components. The general partial molal quantity is defined by

$$\bar{G}_i = \left(\frac{\partial G}{\partial n_i}\right)_{T,P,n_j} \qquad j \neq 1 \qquad (4.102)$$

The total differential of G is

$$dG = \left(\frac{\partial G}{\partial T}\right)_{P,n_i} dT + \left(\frac{\partial G}{\partial P}\right)_{T,n_i} dP$$

$$+ \sum_i \left(\frac{\partial G}{\partial n_i}\right)_{T,P,n_j} dn_i \qquad (4.103)$$

and, at constant T and P,

$$dG = \sum_i \left(\frac{\partial G}{\partial n_i}\right)_{T,P,n_j} dn_i = \sum_i \bar{G}_i dn_i \qquad (4.104)$$

$\bar{G}_i$ is the general partial molal quantity. This equation means that G changes by dG when an infinitesimal amount of i is added to the solution, whereas the numbers of moles of the components $j \neq i$, as well as P and T, are held constant. Another way to look at the process is to state that 1 mole of i is added to an infinite amount of solute. When I use the term partial molal, I am referring to a mass of solvent but one can also define a partial molar quantity on a volume of solution basis.

Integration of Equation (4.104), while holding $\bar{G}_i$ constant, results in

$$G = \sum_i n_i \bar{G}_i \qquad (T,P) \text{ constant} \qquad (4.105)$$

This corresponds to an increase in the amount of the solution while keeping its composition constant because otherwise $\bar{G}_i$ would change. This relation can also be obtained from Euler's theorem because experience has shown that extensive properties are linear homogeneous functions, that is, that they obey the relation

$$G(T, P, \lambda n_i) = \lambda G(T, P, n_i) \qquad (4.106)$$

Partial molal properties are quite valuable for the study of chemical processes in laboratory solutions and natural waters. They provide information on the effects of changes in the amounts of the species of interest on the properties of the system. As examples, chemical potentials yield information on chemical equilibria as we shall see later, partial molal volumes can be used to calculate the effects of pressure, while the partial molal enthalpy leads to the effects of temperature on solutions.

Many solutions contain two or more components in addition to water. Therefore, let us see how the general solution property G, mixing, and partial molal properties are interrelated. We shall do this with the binary version of Equation (4.105);

$$G = n_1 \bar{G}_1 + n_2 \bar{G}_2 \qquad (4.107)$$

Let us use the superscript (1) to indicate a given composition of a system, that is, $(n_1^{(1)}, n_2^{(1)})$. The superscript differs from the subscript which refers only to component 1.

In an actual process one may, for example, take $n_1^{(1)}$ moles of component 1 and place them in a container. The value of G at this time is shown by the line segment $n_1^{(1)}A_1$ in Figure 4.4. It has the value $n_1^{(1)}G_1$ where G_1 is the molal value of G for the pure component 1. Then, component 2 is added and G follows some curve which connects A_1 to B in the plane $n_1 = n_1^{(1)}$. The value of G when B is reached is given by

$$G^{(1)} \equiv G(n_1^{(1)}, n_2^{(1)}) = n_1^{(1)}G_1 + \int_0^{n_2^{(1)}} \bar{G}_2 \, dn_2$$

(4.108)

$G(n_1^{(1)}, n_2^{(1)})$ indicates a functional dependence. The value $\bar{G}^{(1)}$ could equally well have been reached by means of

$$G^{(1)} \equiv G(n_1^{(1)}, n_2^{(1)}) = n_2^{(1)}G_2 + \int_0^{n_1^{(1)}} \bar{G}_1 \, dn_1$$

(4.109)

If components (1) and (2) were added in constant proportions to build up the system, then the physical process would be described by Equation (4.107) in the form

$$G^{(1)} \equiv G(n_1^{(1)} + n_2^{(1)}) = n_1^{(1)}\bar{G}_1 + n_2^{(1)}\bar{G}_2 \quad (4.110)$$

with

$$\bar{G}_1 = \left(\frac{\partial G}{\partial n_1}\right)_{T,P,n_2^{(1)}}$$

(4.111)

and

$$\bar{G}_2 = \left(\frac{\partial G}{\partial n_2}\right)_{T,P,n_1^{(1)}}$$

(4.112)

Equation (4.110) is always valid, no matter how the system was built, but in general it does not represent an actual path. In Figure 4.4 it would correspond to a line connecting 0 to B. Combining, for example, Equations (4.110) and (4.111) one obtains

$$\int_0^{n_2^{(1)}} \bar{G}_2 \, dn_2 = n_1^{(1)}(\bar{G}_1 - G_1) + n_2^{(1)}\bar{G}_2 \quad (4.113)$$

which illustrates the fact that $\bar{G}_1$ and $\bar{G}_2$ are intensive quantities which depend on the relative composition and cannot be varied independently of each other. When we follow the path given by the integral in Equation (4.113) at $n_1 = n_1^{(1)} = $ constant, we find that $n_1^{(1)}(\bar{G}_1 - G_1)$ is intrinsic to it, that is, it is a variable in the integrand as is to be expected from the interdependence of $\bar{G}_1$ and $\bar{G}_2$. These considerations will be useful when we study electrolyte solutions.

4.6.3 Gibbs–Duhem Equation

An important equation for the thermodynamics of homogeneous mixtures such as solutions will be derived next. For i components

$$dG = \sum_i n_i \, d\bar{G}_i + \sum_i \bar{G}_i \, dn_i \quad (P, T \text{ constant})$$

(4.114)

as $G = \Sigma_i n_i G_i$. Comparing this relation to Equation (4.104) we find that

$$\sum_i n_i \, d\bar{G}_i = 0 \quad (P, T \text{ constant}) \quad (4.115)$$

In the special case of the free energy

$$\sum_i n_i \, d\mu_i = 0 \quad (P, T \text{ constant}) \quad (4.116)$$

This relation is known as the Gibbs–Duhem equation. It shows that the variations of chemical potentials with composition are mutually dependent. Chemical potentials drive chemical reactions. It is of value to have an equation that relates them for the constituents of a chemical reaction because it makes their determination easier.

It may seem that the determination of μ for a component would require a knowledge of the μ's for all the other components. Darken (1950) has shown that this is not the case if water containing the other components is considered to be the solvent. In this case, the opposite is true and the μ's of the other components can be calculated from that of one of them, provided it was measured over the full range of compositions. This and other methods are discussed in Lewis and Randall (1961).

Anticipating future results, the value of the Gibbs–Duhem equation resides in the relation $\mu_i = \mu_i^0 + RT \ln a_i = \mu_i^0 + RT \ln \gamma_i m_i$ with μ_i^0 being the value of μ_i^0 at an arbitrary standard state. μ_i applies to each component i of the solution. The determination of μ_i yields γ_i, a measure of the departures of the components i from ideality [see Equation (4.119) and Chapter 5].

All the methods to determine chemical potentials in binary and multicomponent solutions arise from the property of exact differentials shown below for a ternary solution:

$$\left(\frac{\partial \mu_1}{\partial n_2}\right)_{n_1, n_3} = \left(\frac{\partial \mu_2}{\partial n_1}\right)_{n_2, n_3} \qquad (4.117)$$

Terms for $\partial \mu_1 / \partial \mu_3$ and $\partial \mu_2 / \partial \mu_3$ must, of course, also be used.

Equation (4.117) becomes, in terms of measurable γ_i's (Stokes, 1979),

$$\nu_1 \left(\frac{\partial \ln \gamma_1}{\partial m_2}\right)_{m_1, m_3} = \nu_2 \left(\frac{\partial \ln \gamma_2}{\partial m_1}\right)_{m_2, m_3} \qquad (4.118)$$

ν_1 and ν_2 in the case of electrolytes correspond to the numbers of moles of ions formed by electrolytes 1 and 2. As an example, $\nu = 3$ for $H_2SO_4 \rightarrow 2H^+ + SO_4^{2-}$.

The Gibbs–Duhem equation becomes

$$\sum_i \nu_i m_i \, d \ln \gamma_i = 0 \qquad (4.119)$$

This important equation relates the activity coefficients of the components and permits the calculation of all the γ_i's from that of a given one. Free energies can be determined in principle from (4.116) but the procedure is difficult and one usually approaches F through γ (see Chapter 5).

4.6.4 Partial Molal Volume

Partial molal quantities can be derived for any extensive state function and, as was mentioned earlier, are intensive factors because they represent extensive (capacity) factors per mole. Besides the chemical potential, a partial molal quantity that is used quite often is the partial molal volume. This quantity is important in the study of activity coefficients and equilibria as functions of pressure (e.g., Owen and Brinkley, 1941).

The partial molal volume is defined by

$$\bar{V}_i = \left(\frac{\partial V}{\partial n_i}\right)_{T, P, n_j} \qquad n_j \neq n_i \qquad (4.120)$$

with the total volume given by

$$V = \sum_i n_i \bar{V}_i \qquad (4.121)$$

$\bar{V}_i$ is useful not only for pressure studies but also in the interpretation of the effect of adding component i on the structure of a solvent and the properties of a solution. Thus, one often finds that the addition of some salts to water yields a final volume that is smaller than the combined volumes of water and salt. This indicates electrostriction resulting from the compaction of water of hydration relative to the bulk water.

Some characteristic examples of standard partial molar volumes obtained from the data of Owen and Brinkley (1941) are, in cm³/mole at 25°C,

Ion	Na^+	K^+	Ca^{2+}	Mg^{2+}	Cl^-	SO_4^{2-}	CO_3^{2-}
$\bar{V}^0$	-1.5	8.7	1.9	9	18.1	14.5	-3.7

These results are used here only to illustrate $\bar{V}_i$ when used for ions. Further results for electrolyte solutions are presented in Chapter 5. It is not possible to measure partial molal volumes for single ions but only for molecules. Thus, a convention is needed. The scale used above is based on the convention $\bar{V}^0 = 0$ for H^+ and is obtained from solutions of single electrolytes (NaCl or CaCO$_3$ or MgSO$_4$, etc.).

Thus, $\bar{V}^0_{Na} = \bar{V}^0_{NaCl} - \bar{V}^0_{HCl}$. The experimental $\bar{V}^0_{NaCl}$ can be obtained again from $\bar{V}^0_{NaCl} = -1.5 + 18.1 = 16.6$ cm³/mole at 25°C. These values apply at the standard states of the electrolytes. As an example, the standard state of NaCl is a hypothetical one molal solution which behaves ideally, that is, has $\gamma_{NaCl} = 1$ (see Chapter 5).

In general, ions with large positive values of $\bar{V}^0$ either hydrate little so that the electrostrictive effect is slight or tend to break up the structure of water. The hydrogen-bonded structure of water is compact, whereas an assemblage of individual molecules occupies a larger volume (e.g., Horne, 1969). The impact of $\bar{V}^0$ on chemical equilibria occurs because $\partial \ln K / \partial \ln P = -\Delta \bar{V}^0 / RT$. In Chapter 5 I shall examine this effect for a single electrolyte and for several electrolytes present in solution. Emphasis will be placed on the solubility product of CaCO$_3$ at pressure in seawater in Chapter 5, Volume II.

EXAMPLE 4.21. EFFECT OF PRESSURE ON THE SOLUBILITY PRODUCT OF CaCO$_{3(s)}$ IN DISTILLED WATER AT 25°C (298.16°K). $v_{CaCO_3} = 36.94$ cm³/mole $=$

molal volume of $CaCO_{3(s)}$, $\bar{V}^0_{Ca} = -17.7$ cm^3/mole, $\bar{V}^0_{CO_3} = -3.7$ cm^3/mole in the $\bar{V}^0_H = 0$ scale. Thus,

$$\Delta \bar{V}^0 = -17.7 - 3.7 - (36.94) = -58.34 \text{ cm}^3/\text{mole}$$

(i)

for the reaction $CaCO_{3(s)} = Ca^{2+} + CO_3^{2-}$.

Let us assume for simplicity that $\Delta \bar{V}^0$ is constant with pressure. Integrating $\partial \ln K_{sO}/\partial P = -\Delta \bar{V}^0/RT$,

$$\log \left[\frac{(K_{sO})_P}{(K_{sO})_1} \right] = \frac{-\Delta \bar{V}^0 (P - 1)}{2.3026RT}$$

(ii)

$$R = (82.0597 \text{ cm}^3/\text{atm} \cdot \text{deg} \cdot \text{mole})(\text{deg})$$

$$= 82.0597 \text{ cm}^3/\text{atm} \cdot \text{mole}$$

At 1000 atm

$$\log \left[\frac{(K_{sO})_{1000}}{(K_{sO})_1} \right]$$

$$= \frac{-999 \times (-58.34)}{2.3026 \times 82.0597 \times 298.16} = 1.0345 \quad \text{(iii)}$$

$$\frac{(K_{sO})_{1000}}{(K_{sO})_1} = 10.83$$

(iv)

The solubility product increases with a pressure of 1000 atm, which corresponds to a depth of 1000 m, by a factor of 10.83 over the surface value.

Note that, if the convention $\bar{V}^0_H = 0$ is wrong, then $\bar{V}^0_{Na}$ and $\bar{V}^0_{Cl}$ are incorrect. The errors cancel out, however, when $\bar{V}^0_{Na} + \bar{V}^0_{Cl}$ is used. Let us say that $\bar{V}^0_H$ actually is $\bar{V}^0_H = e$. By the principle of the additivity of molal volumes $\bar{V}^0_{NaCl} = \bar{V}^0_{Na} + \bar{V}^0_{Cl}$. Then,

$$\bar{V}^0_{NaCl} - \bar{V}^0_{HCl} = \bar{V}^0_{Na} + \bar{V}^0_{Cl} - e$$

$$- \bar{V}^0_{Cl} = \bar{V}^0_{Na} - e$$

(4.122)

However,

$$\bar{V}^0_{HCl} = e + \bar{V}^0_{Cl}$$

(4.123)

so that, from (4.122) and (4.123),

$$\bar{V}^0_{Na} + \bar{V}^0_{Cl} = \bar{V}^0_{NaCl} + \bar{V}^0_{HCl}$$

(4.124)

Thus, the sum of the ionic partial molal volumes can be obtained from the sum of the experimental volumes for the components regardless of the convention used.

If our interest is in $\bar{V}^0$ for single ions then we should consider the value $\bar{V}_H$ (for H$^+$) of -4.5 cm^3/mole suggested by Murkejee (1961). This could be the case when the effect of single ions on the structure of water is under consideration.

Readers without a specific interest in $\bar{V}_i$ may wish to skip the rest of this subsection. On the other hand an examination of methods used to determine $\bar{V}_i$ can enhance our understanding of the subject.

Two main methods are available for the determination of partial molal quantities; the method of the intercept and the apparent molal property method. The direct determination of slopes, such as that of V versus n_i [see Equation (4.120)], is not precise enough. The method of intercepts for the determination of partial molal volumes of two-component systems is described here to add a sense of reality to the concept of partial properties.

Equation (4.121) can be rewritten as

$$v = \frac{V}{n} = \sum_i \frac{n_i}{n} \bar{V}_i = \sum_i X_i \bar{V}_i \quad (4.125)$$

where $n = \Sigma_i n_i$ and X_i is the mole fraction of component i. The term v is the mean molal volume. For two components, differentiation with respect to X_2 yields

$$\left(\frac{\partial v}{\partial X_2} \right)_{T,P} = \bar{V}_2 - \bar{V}_1 \quad (4.126)$$

because

$$X_1 + X_2 = 1 \quad (4.127)$$

v is plotted against X_2 in Figure 4.5.

The equation of the tangent to the curve v versus X_2 at $X_2 = X_2^{(1)}$ is

$$v = \bar{V}_1(X_2^{(1)}) + [\bar{V}_2(X_2^{(1)}) - \bar{V}_1(X_2^{(1)})]X_2 \quad (4.128)$$

The tangent intercepts the ordinate on the left at $v = \bar{V}_1(X_2^{(1)})$ when $X_2 = 0$ and the ordinate on the right at $v = \bar{V}_2(X_2^{(1)})$ when $X_2 = 1$. Thus, $\overline{AB} = \bar{V}_1(X_2^{(1)})$ and $\overline{CD} = \bar{V}_2(X_2^{(1)})$ become known and the $\bar{V}$'s can be determined.

It is of interest to observe the behavior of $\bar{V}$ in an actual case. A plot of the volume of solutions con-

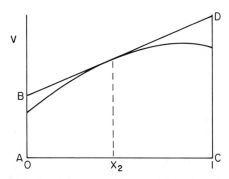

FIGURE 4.5 Plot of the mean volume v against the mole fraction X_2 of component 2. The solid line is the tangent to the curve at $X_2 = X_2^{(1)}$.

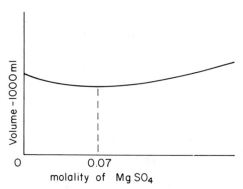

FIGURE 4.6 Plot of the volume of aqueous solutions of $MgSO_4$ against the salt molality at 18°C.

taining $MgSO_4$ against the molality at 18°C, is shown in Figure 4.6. It is based on Kohlrausch and Hallwachs (1894). The partial molal volume of $MgSO_4$, which may be thought of as the change in volume when 1 mole of salt is added to an infinite volume of solution of molality m, is the slope of the tangent to the curve at m. The partial molal volume is negative below about $m = 0.07$ molal, showing that electrostriction dominates over the added volumes of the Mg^{2+} and SO_4^{2-} ions. Thus, the volume of a solution containing 1000 g of water actually shrinks on addition of an infinitesimal amount of $MgSO_4$ at $m < 0.07$. The term $\bar{V}(MgSO_4)$ becomes positive and increases at higher concentrations.

The apparent molal (or molar) volume is often used and, in the molal case, refers to

$$v_a = \frac{V - 55.51 v_w^0}{m} \qquad (4.129)$$

where V is the volume of the solution containing 55.51 moles of water and m moles of salts and v_w^0 is the molal volume of pure water. Thus, v_a is the mean change per mole of salt in the volume of 1000 g of water when m moles of salt are added to the pure water. It serves as a tool to determine the partial molal volume of the salt because v_a tends to $\bar{V}_s$ for an infinitesimal change in m. Furthermore, an empirical fit of v_a versus m data by some function $f(m)$, when introduced in (4.129), yields after rearrangement

$$V = 55.51 \, v_w^0 + mf(m) \qquad (4.130)$$

Then $\bar{V}_s$ can be obtained from $(\partial V / \partial m) = \bar{V}_s$.

Mean and partial specific properties, expressed per unit mass, are sometimes used. We saw their application to hydrothermodynamics earlier. They are defined for a generic extensive property G as

$$g' = \frac{G}{M} = \frac{G}{\sum_i M_i} \qquad (4.131)$$

where M is the mass of the system. The primes distinguish $g' = G/M$ from $g = G/n$. The partial specific property is then

$$\bar{g}_i' = \left(\frac{\partial g}{\partial M_i} \right)_{T,P,M_j} \qquad i \neq j \qquad (4.132)$$

For a one-component system

$$G = n\bar{G}, \quad g = \bar{g}, \quad \text{and} \quad g' = \bar{g}' \quad (4.133)$$

since g and g' are intensive properties. $\bar{G}$ for such a system is simply G/n as in the case of $\bar{V} = V/n$.

Values of the thermodynamic properties as related to chemical equilibria shall be applied in several chapters while their use in the determination of the physicochemical properties of strong electrolytes will be considered in Chapter 5.

4.7 ACTIVITIES

4.7.1 Activities of Gases

The activities of gases are presented here primarily as a background for the study of the activities in the more complex aqueous solutions of electrolytes found in natural waters. Of course, gases are also of interest in such systems.

In this section I shall examine the relationships between concentration, activity, and energy in chemical systems because such relationships help forecast the probable course of reactions. I shall cover rigorously and in greater detail some aspects of this topic which were presented earlier.

One can show that

$$\left(\frac{\partial \mu}{\partial P}\right)_T = v \qquad (4.134)$$

where v is the molar volume of a gas or of any other single-component system. For an ideal gas

$$v = \frac{RT}{P} \qquad (4.135)$$

Introducing (4.135) into (4.134) and integrating from state A to state B,

$$\Delta \mu = \mu_B - \mu_A = RT \ln \frac{P_B}{P_A} \qquad (4.136)$$

If state A is a standard state for which $\mu_A = \mu^0$ and B is a variable state for which we omit the subscript, then

$$\mu = \mu^0 + RT \ln P \qquad (4.137)$$

μ corresponds to μ_B and μ^0 to μ_A^0.

For a real gas Equation (4.137) is only valid at all pressures if we replace the actual pressure by the fugacity P^*. The fugacity is the pressure that the gas would exert if it behaved ideally, in terms of Equation (4.137), at all temperatures and pressures. Let us look at this topic in greater detail because it provides a conceptual door to all manners of nonideal solutions.

During the historical development of the subject of effective concentrations, the concepts of fugacity and chemical potential were introduced independently. Fugacity and activity are redundant and one could use only one of them. Both quantities are presented in the literature, however, so it is necessary to establish relationships between them.

The activity of a gas is defined by

$$a = \frac{P^*}{P^{*0}} \qquad (4.138)$$

and is the fugacity relative to that at a standard state. Numerically, $a = P^*$ because the standard state is set at

$$P^{*0} = 1 \text{ atm} = P^0 \qquad (4.139)$$

$P^{*0} = P^0$ is valid because most gases behave ideally at ordinary temperatures and pressures of up to a few atmospheres. Under these conditions $P = P^*$ also holds true.

If an equation such as (4.137) is to be valid at all pressures and temperatures, then P must be replaced by the fugacity and

$$\mu = \mu^0 + RT \ln P^* \qquad (4.140)$$

I remind the reader, because fugacity is a difficult concept, that P^* is a fictional quantity which corresponds to P at all temperatures and pressures if the gas behaves ideally in the sense of Equation (4.137) and equals P^* when the gas is indeed ideal. The latter case occurs essentially at ordinary temperatures and pressures. The standard chemical potential μ^0 occurs when $P^* = P = 1$ atm regardless of T, since a low pressure means negligible interactions between the gas atoms or molecules. Usually though, T is set as 25°C. The noble gases such as He are present as atoms while O_2, N_2, and so on, are molecular.

At high pressures, P behaves nonideally, whereas P^* still obeys Equation (4.140). The activity a can be defined by

$$\mu = \mu^0 + RT \ln a \qquad (4.141)$$

The activity coefficient f can be introduced in the expression $a = f \times P$ which, together with Equation (4.138), yields

$$f = \frac{P^*}{P^{*0}P} \quad \text{or} \quad f = \frac{P^*}{P} \qquad (4.142)$$

The last equality is only numerical and not dimensional. It results from $P^{*0} = 1$ atm.

The ratio P^*/P of nitrogen is illustrated in Table 4.2.

At low pressures P^* usually equals P. Since $P^{*0} = 1$ by definition, the activity coefficient is $f = 1$. At high pressures (see Table 4.2) $P^* \neq P$ because the gas particles are closer to each other and can interact.

TABLE 4.2 The ratio of the fugacity to the pressure versus the pressure of nitrogen at 0°C (Otto et al., 1934; Sage and Lacey, 1950; Lewis and Randall, 1961).

P (atm)	P^*/P	P (atm)	P^*/P
1	0.99955	300	1.1353
50	0.9950	400	1.2566
100	0.9854	600	1.5242
150	1.0030	800	1.7964
200	1.00363	1000	2.070

The activity a is

$$a = fP = \frac{P^*P}{P^{*0}P} = \frac{P^*}{P^{*0}} \qquad (4.143)$$

and, numerically

$$a = P^* \qquad (4.144)$$

If the interactions result from attractive forces, $a = P^*$ required to make thermodynamics work [e.g., Equation (4.141)] is smaller than P and $f = P^*/P < 1$. The term $P^* = a$ is the thermodynamic pressure just as, in the case of solutes in aqueous solutions, $a = \gamma m$ is the thermodynamic concentration and γ is the practical (molar or molal) activity coefficient. The term a is numerically a ficticious molality which the solute of actual molality m would have if it behaved ideally. Attractive forces act as if the number of particles decreased and P^* as well as a represent the ideal behavior of this smaller number rather than the nonideal behavior of the true number. These concepts permit us to apply changes in thermodynamic quantities, such as ΔF, ΔH, ΔS, ΔA and ΔE, to chemical processes in solutions, and to changes of states of gases by means of activities, since actual pressures and molalities do not necessarily conform to thermodynamics. See Table 4.2.

EXAMPLE 4.22. ACTIVITY AND FREE ENERGY IN AN AQUEOUS SOLUTION. Assume that a solute has $m_s = 1.2$ and $\gamma_s = 0.80$. Then, $a_s = 0.96$. The activity may be obtained, for example, from vapor pressure data. Therefore,

$$\mu_s - \mu_s^0 = RT \ln 0.96 \qquad (i)$$

Since $\mu_s = (\partial F/\partial n_s)_{P,T,n_w}$, it can be used in principle to calculate the free energy change ΔF on dissolution of the solute s, by means of

$$\Delta F = \int_0^{m_s} \mu_s \, dm_s + \int_0^{m_s} \mu_w \, dm_w \qquad (ii)$$

m_s is used instead of n_s. The term ΔF can also be used as a measure of the spontaneity of the dissolution.

On the other hand,

$$\mu_s - \mu_s^0 \neq 1.2 \qquad (iii)$$

and has no thermodynamic significance.

Note that a in Equation (4.138) is dimensionless as it is the ratio of two pressures and f in (4.142) has the dimension of inverse pressure, that is, atm^{-1}. This is thermodynamically rigorous as is shown below. Let us define an absolute activity such that

$$\mu = RT \ln \lambda \qquad (4.145)$$

When we are concerned with a $\Delta\mu$, let us say, relative to a standard state, then

$$\mu - \mu^0 = RT \ln \frac{\lambda}{\lambda^0} = RT \ln a \qquad (4.146)$$

and the relative activity $a = \lambda/\lambda^0$ is indeed dimensionless.

The reader will find equations of state of nonideal gases, including the virial expansion, in texts such as Lewis and Randall (1961).

4.7.2 Mixtures of Gases

For a mixture of ideal gases Dalton's law yields

$$PV = RTn = RT \sum_i n_i \qquad (4.147)$$

As

$$X_i PV = RTn_i = P_i V \qquad (4.148)$$

we obtain

$$P_i = PX_i \qquad (4.149)$$

which represents the following important property of gases. The partial pressure of each component of a mixture of ideal gases at a given total pressure is proportional to its mole fraction. Equation (4.149) applies also to real gases at ordinary temperatures, around room temperature, and P below a few atmospheres. Furthermore, it is valid by definition at all pressures when expressed in terms of fugacities, that is,

$$P_i^* = P^*X_i \qquad P^* = \sum_i P_i^* \qquad (4.150)$$

For a single gas we saw that $a = P^*/P^{*0}$ is numerically equal to P^* because $P^{*0} = 1$. Generalizing this convention to a mixture of gases,

$$a_i = \frac{P^*X_i}{P^{*0}X_i^0} = \frac{P_i^*}{P_i^{*0}} = P_i^* \qquad (4.151)$$

The last equality is valid if the standard state is set as pure i when it exerts a pressure of 1 atm. By definition $a_i = f_iP_i$ and, from (4.150),

$$a_i = f_iPX_i \qquad (4.152)$$

at ordinary temperatures and pressures.

We are now in a position to examine the important concept of standard state in some detail. The standard state is defined by $a_i = 1$ so that $\mu_i = \mu_i^0$. For each component of a gas mixture

$$\mu_i = \mu_i^0 + RT \ln a_i \qquad (4.153)$$

with $\mu_i = \mu_i^0$ when $a_i = 1$. The standard state is realized in practice when the component is at a partial fugacity of 1 atm. For any given temperature the corresponding standard state is realized when $P_i^* = 1$ atm [see Equation (4.151)].

Under usual conditions $P_i^* = P_i$ and the standard state is achieved in practice when P_i is also 1 atm. Then $PX_i = P_i = 1$, the activity coefficient f_i is 1 by definition, and $a_i = 1$ [see Equation (4.152)]. This is not true at very low temperatures and at pressures above 10 atm (Garrels and Christ, 1965).

Solutions can be gaseous, liquid, or solid but I will use the terms gas mixture, solution, or aqueous solution when the solvent is water or a solution in water, and solid solution when the solvent is solid. Next, I shall extend the chemical potential–activity relation to solutions.

The concept of standard state probably arose from its mechanical counterpart in which some altitude, such as the mean sea level, is used as the standard level for potential energy. In the mechanical case as well as for aqueous or solid solutions, the standard level or state is arbitrary; we may use

the top of a mountain instead of the sea surface to determine Mgh for an object. The term h is the difference between the height of the object and the standard level. We shall examine the case of solutions soon.

The case of an ideal gas is more restricted because $\mu = \mu^0 + RT \ln P$ means that the standard state μ^0 is only achieved when $P = 1$ atm. The same is true if we use other units, such as mm Hg or psi. The arbitrariness enters when we consider an imperfect (nonideal) gas. Then, $\mu = \mu^0 + RT \ln a$ with $a = fP$ and the standard is usually reached by definition when $f = 1$ at $P = 1$, that is, when the real gas is assumed to behave ideally at 1 atm. This definition holds even if the gas is not ideal at 1 atm and can be used because the standard state is arbitrary and we are free to set it as we wish. We could just as well have set $\mu = \mu^0$ for $f = 0.5$ at $P = 2$ instead. In either case $P^{*0} = 1$ at the standard state but it is the change in the real quantity P from 1 to 2 which shows that there was an actual change in convention. Thus, the definitions of the activities $a = fP$ and the activity coefficient f are conventional for imperfect gases. This is similar to the shift from sea level to a mountain top and, as in this case, makes no difference in practice.

In summary:

	Standard State
Ideal gas	$P = 1$ (atm, mm Hg, psi, etc.)
Real (imperfect nonideal) gas	$P^* = a = fP = 1$. Usually $f = 1$ at $P = 1$ by convention. Defined as a gas that behaves ideally at 1 atm.

EXAMPLE 4.2.3.   THE STANDARD STATE OF NITROGEN AT 0°C (SEE TABLE 4.2).   In this case $a = P^{*0} = 0.99955$, that is, $f = 0.99955$ at $P = 1$ atm. The standard state is a fictitious one for which $f = 1$ when $P = 1$ atm.

The standard state is an artifice used because we cannot measure $\bar{V}$, μ, F, A, E, H, and S but only their changes either relative to the standard state or to another state, that is, $F_A - F^0$, $F_B - F^0$, and $F_B - F_A$ as an example.

4.7.3 Solutions and the Standard State of Solvents

I shall use the concepts of standard state, reference state, chemical potential, activity, and activity coefficient extensively in Chapter 5 when we shall examine electrolyte solutions. These are the solutions of acids, bases, and salts present in natural waters, for example, H_2CO_3, $NaCl$, $CaCO_3$, and so on. The above concepts are used to understand the behavior and determine the properties of such solutions as well as of nonelectrolyte ones (e.g., many organic molecules).

The standard state $a_i = 1$ occurs by definition when all components $j \neq i$ tend toward infinite dilution. This is the standard state of the solvent because i is the major constituent. If we are considering a solution in which i is a volatile component at equilibrium with its vapor, then

$$\mu_{i(\text{vapor})} = \mu_{i(\text{soln})} \qquad (4.154)$$

since no work is done when a small amount of i is transferred between the phases. This is analogous to the case of the minimum in potential energy discussed earlier.

The definitions of the activities and standard states of the gaseous components of the vapor, and of the same components in solution, together with (4.152), yield

$$\mu_{i(\text{vapor})}^0 + RT \ln a_{i(\text{vapor})} = \mu_{i(\text{soln})}^0$$
$$+ RT \ln a_{i(\text{soln})} \qquad (4.155)$$

Note that the standard states and, hence, the activities in the two phases are not in the same scale. In the vapor $\mu_i = \mu_i^0$ when the vapor exerts a partial pressure of 1 atm but in the solution $\mu_i = \mu_i^0$ when there is a pure solvent. The latter condition is less restrictive since there is no limitation imposed on the value of the pressure P to which the solution is submitted when $a_{i(\text{soln})} = 1$. Equation (4.155) is valid but

$$\mu_{i(\text{vapor})}^0 \neq \mu_{i(\text{soln})}^0 \qquad a_{i(\text{vapor})} \neq a_{i(\text{soln})} \qquad (4.156)$$

even when the two phases are at equilibrium.

The convention that the standard state of a solvent is the pure solvent has been generalized to nonvolatile solvents. The chemical potential expression for solvents is used in the form

$$\mu_{(\text{solv})} = \mu_{(\text{solv})}^0 + RT \ln a_{(\text{solv})}$$
$$= \mu_{(\text{solv})}^0 + RT \ln f_{(\text{solv})} X_{(\text{solv})} \qquad (4.157)$$

f now is the rational activity coefficient, that is, the coefficient defined in a mole fraction scale and is not the gaseous f. This is convenient because we can set $f_{(\text{solv})} = 1$ when $X_{(\text{solv})} = 1$, that is, for a pure solvent.

4.7.4 Ideal and Regular Solutions

Let us next examine the concepts of ideal (perfect) and regular solutions. These concepts help us classify solutions in terms of changes in properties when the components are mixed and understand those changes (e.g., Lewis and Randall, 1961).

An ideal solution is defined as one for which

$$P_i^* = X_i P_i^{*0} \quad \text{or} \quad P_i = X_i P_i^0 \qquad (4.158)$$

where P_i^0 is the equilibrium vapor pressure of pure i at some given pressure. I am excluding high pressures and low temperatures from the second equality. This definition implies that the only effect of mixing the components on their properties is due to the dilution of each of them by the others. It requires that the forces between like and unlike molecules be nearly equal. The heat and volume changes of mixing are then essentially zero. Nearly ideal solutions over the full mixing range are formed, for example, when benzene and toluene are mixed and when dilute solutions of nitrobenzene and aniline in each other are prepared.

Equation (4.151) may be written as

$$a_i = \frac{P_i^*}{P_i^{*0}} \cong \frac{P_i}{P_i^0} \qquad (4.159)$$

since the standard state is taken to be the pure component. Equation (4.158) applies to volatile components such as water. In solutions, the standard fugacities P_i^{*0} are those of the pure components at whatever T and P are being considered. Thus, P_i^{*0} is not fixed as unity at $P_i = 1$ atm in contrast to the case of gases. Under normal conditions at the standard state P_i^0 is the vapor pressure of pure i and P_i is its partial pressure when the concentrations of the

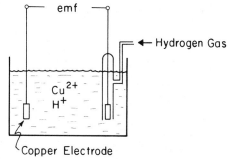

FIGURE 4.11 The copper–hydrogen cell.

Figure 4.11. A copper and a hydrogen electrode are immersed in an acid solution which contains a cupric salt. If $a_{Cu} = a_H = 1$ in a solution saturated with hydrogen gas at unit pressure, then the initial potential will be the standard one obtainable from tables. It is

$$E^0 = \frac{RT}{nF'} \ln K = 0.337 \text{ volts (V)} \quad (4.226)$$

and one can obtain K for the reaction

$$Cu^{2+} + H_2 \text{ (gas)} = Cu_{(s)} + 2H^+ \text{ (aqueous)} \quad (4.227)$$

because $E^0_{H_2}$ for $H_2 = 2H^+ + 2e^-$ is set equal to zero by convention. The reaction proceeds to completion because cupric ions are removed from solution by being deposited on the copper electrode while H_2 gas is evolved. ΔF^0 can be calculated from $nF'E^0$.

EXAMPLE 4.30. CALCULATION PROCEDURUE FOR ΔF^0 IN REACTION (4.227)

$$\Delta F^0 = -2F'E^0 = -2 \times F' \times 0.337 \quad (i)$$

where $F' = 96{,}493$ C/equiv or J/V-equiv $= 23{,}062$ cal/V-equiv, E^0 is in V, and 2 converts moles to equivalents for the specific reaction under study. Using 23,062 cal/V-equiv

$$\Delta F^0 = -2 \times 23{,}062 \times 0.337 = -15.54 \text{ kcal/equiv} \quad (ii)$$

while with $F' = 96{,}493$ J/V-equiv

$$\Delta F^0 = -2 \times 96{,}493 \times 0.337 = -65.04 \text{ J/equiv} \quad (iii)$$

For K

$$\log K = \frac{nF'}{2.3026RT} E^0 = \frac{2E^0}{0.05915} \text{ (at 25°C)} \quad (iv)$$

and

$$K = 2.482 \times 10^{11} \quad (v)$$

I shall use the following cell conventions:

1. | separates different phases. For example, $Pt|H_{2(aq)}$.

2. ; separates different components of a phase. For example, $Cu^{2+};Zn^{2+}$.

3. : separates different phases in the same state (solid, liquid, or gas) in contact but not mixed together. For example, $Ag:AgCl$.

The cell in Figure 4.11 may be represented by

$$Pt|H_{2(aq)};H^+;Cu^{2+}|Cu_{(s)} \quad (4.228)$$

The two half-cell reactions that occur at the two electrodes are

$$Cu^{2+} + 2e^- \rightarrow Cu_{(s)} \quad (4.229)$$

$$H_2 \rightarrow 2H^+ + 2e^- \quad (4.230)$$

The gain of electrons by the copper ions is known as reduction and the loss by the hydrogen gas as oxidation so that (4.227) is called a redox reaction. The electrode at which oxidation occurs (the hydrogen electrode in this case) is the anode and the electrode where reduction takes place is the cathode.

It is convenient in chemistry to have standard potentials for half-cells. These potentials obviously cannot be measured directly and the convention is used that the half-cell standard potential for reaction (4.230) is zero. Thus, half-cells with positive potentials as written and proceeding to the right will react spontaneously when coupled with the hydrogen electrode. The coupling of the half-cells $Pt|H_{2(aq)};H^+$ and $Cu^{2+}|Cu$ leads to reaction (4.227) which occurs spontaneously. The copper half-cell has a positive potential when its reaction is written as in Equation (4.229).

Zinc has a higher oxidizing potential than copper so that the spontaneous reaction is

$$Zn + Cu^{2+} = Zn^{2+} + Cu \quad (4.231)$$

solutes are finite. Note that the solvent in the oceans can be set as a matter of convenience as pure water or seawater to which further solutes are added. From (4.157) and (4.158) we derive the important property of ideal solutions

$$a_i \cong X_i \qquad (4.160)$$

Pure components (solvents) in infinitely dilute solutions obey Raoult's law for the special case $X_i = 1$ which leads to $P_i^* = P_i^{*0}$ and $P_i = P_i^0$. The physical reason for this is that when aqueous solutions of nonelectrolyes are very dilute, water behaves as a component of an ideal solution because there are not enough solute particles to alter the interactions between water molecules. This is a different concept than that of ideal gases since unperturbed interactions are part of the standard state. Then, $f_w = 1$ and this provides a rationale for selecting pure water (or any pure solvent in liquid or solid solutions) as the standard state because then

$$\mu_w = \mu_w^0 + RT \ln X_w \qquad (4.161)$$

leads to $\mu_w = \mu_w^0$ when $X_w = 1$.

The activity of water in any aqueous solution, whether ideal or not, is found from (4.156) and (4.160), to be

$$\mu_w = \mu_w^0 + RT \ln a_w \qquad (4.162)$$

with $f_w X_w = a_w$ and with $f_w = 1$ when $X_w = 1$.

The subscript w can be replaced by (solv) since (4.161) is quite general. The solvent, defined as the major component of the solution, can be either volatile or not and, by definition, $a_{\text{solv}} = 1$ for the pure solvent which leads to $\mu_{(\text{solv})} = \mu_{(\text{solv})}^0$. The solvent need not be a pure substance because the standard state can be defined arbitrarily. Thus, one can set the standard state as a seawater made of its major constituents if, for example, one wishes to study the equilibrium or the kinetic behavior of gas exchange across the sea surface, or equilibrium or kinetic behavior of a heavy metal such as lead introduced by the use of leaded gasoline.

In general $\mu_i = \mu_i^0 + RT \ln a_i$ are extended to pure substances or solutions, such as seawater, even if they are nonvolatile.

4.7.5 Standard and Reference States of Solutes

For solutes, whether they are volatile substances or not, again the standard state is $a_i = 1$. A solute is the minor component of a solution so that we cannot use $X_i = 1$ as the standard state. The use of m or c is more convenient than that of X. Furthermore, as $a_i = \gamma_i m_i$ we cannot use infinite dilution state for which $\gamma_i = 1$ because then $a_i = 1 \times 0 = 0$ and $\mu_i = \mu_i^0 - \infty$. Instead, an imaginary standard state consisting of a solution of unit concentration ($m = 1$) that behaves ideally ($\gamma = 1$) is defined. It has $a = 1 \times 1 = 1$ so that $\mu = \mu^0$. Here $\gamma = 1$ by convention.

By definition, the state at which γ is truly unity is called the reference state, in contrast to the standard state for which $a = 1$. The two coincide for pure substances and solvents but do not for solutes. As we saw in the preceding paragraph the reference state is defined as $m = 0$.

A regular solution is one that obeys Henry's law

$$P_i^* = k_H' X_i \qquad (4.163)$$

which may be written as

$$P_i = k_H' X_i \qquad (4.164)$$

or

$$P_i = k_H m \qquad (4.165)$$

at ordinary temperatures and pressures. k_H' and k_H are the Henry's law constants in mole fraction and in molal units. Most nonelectrolyte solutes obey these relations at low concentrations. The difference between ideal and regular solutions is illustrated in Figure 4.7 at a given temperature and pressure. Thus, solvents behave ideally, whereas solutes behave as components of regular solutions

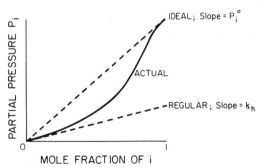

FIGURE 4.7   Ideal and regular solutions.

when $m \rightarrow 0$. The solute standard state, however, is the point on the ideal line which corresponds to $m_i = 1$.

Conventions for reference and standard states are as follows:

	Reference State (f or $\gamma = 1$)	Standard State ($a = 1$)
Pure gas	$P \cong P^* = 1$ atm	$P \cong P^* = 1$ atm
Pure solid	$X = 1$	$X = 1$
Pure liquid	$X = 1$	$X = 1$
Solvent	$X_i = 1$	$X_i = 1$
Solute	$X_i \rightarrow 0$ for $m_i \rightarrow 0$	$m = 1$ with $\gamma = 1$ (fictitious)

The concept of ideal behavior is different for solvents and solutes. In the first case it corresponds to a pure substance and Raoult's is obeyed. For solutes, however, it refers to infinite dilution and implies that there are no interactions between the solute particles, although there are interactions between solvent and solute. Ideal solutes are ideal in the sense of Henry's law, sometimes over the full range of concentrations and always when $m_i \rightarrow 0$, that is, in the reference state. In ideal gases there are no interactions of any kind besides elastic collisions.

Again, an expression of the type (4.153) applies to the chemical potential–activity relation for solutes, with $a_i = f_i X_i$ or $a_i = \gamma_i m_i$. The use of f_i versus γ_i depends on which concentration units are employed, X_i or m_i. I remind the reader that, strictly speaking, activities are dimensionless [see Equation (4.146)] and activity coefficients have units of (concentration)$^{-1}$. In practice, however, one can make activity coefficients dimensionless and assign concentration units to activities. It should also be noticed that the standard states and activity coefficients are usually defined at individual temperatures and pressures.

In a nutshell, activity coefficients, activities, and chemical potentials are related to and provide insights into interactions between particles in solution (see Chapter 5).

4.7.6 Effects of the Gravitational Field, Temperature, and Pressure in the Oceans

In this section, some interesting broad oceanographic applications of activities are examined. More conventional examples in solution chemistry will be viewed later.

Generalized Activity. The effect of temperature on activity coefficients is

$$\left(\frac{\partial \ln \gamma}{\partial T} \right)_P = -\frac{(\bar{H} - \bar{H}^0)}{RT^2} \qquad (4.166)$$

whereas that of pressure is

$$\left(\frac{\partial \ln \gamma}{\mu P} \right)_T = \frac{\bar{V} - \bar{V}^0}{RT} \qquad (4.167)$$

$\bar{V}^0$ and $\bar{H}^0$ are the partial molal volume and the partial molal enthalpy at the standard state. Note that upon integration of these equations the specific heat and the compressibility appear if the enthalpy and the partial molal volume are not constant over the ranges of temperature and pressure used.

In the oceans, one finds variations in chemical potential and activity coefficients not only because of changes in temperature and pressure but also because of depth, since the gravitational potential energy affects the free energy of solutes.

It can be shown by extending the results of Pytkowicz (1968) to temperature changes, that the variation in the chemical potential of a solute with depth z in the ocean is

$$d\mu = \frac{\partial \mu}{\partial z} dz + \frac{\partial \mu}{\partial P} dP + \frac{\partial \mu}{\partial T} dT + \frac{\partial \mu}{\partial X} dN \qquad (4.168)$$

This equation can be expressed as

$$d\mu = -(MW)g\,dz + \bar{V}\rho g\,dz + \bar{S}\,dT + RT\,d\ln a_z \qquad (4.169)$$

(MW) is the gram-molecular weight, g the acceleration of gravity, $\bar{S}$ the partial molal entropy, and a_z the activity of the solute at depth z referred to μ_z^0, the standard chemical potential defined at z.

The activity a_z is determined by

$$\mu_z = \mu_z^0 + RT \ln a_z \qquad (4.170)$$

where μ_z^0 corresponds to a unit concentration of the solute at depth z which is assumed to behave ideally. This is the conventional approach. One could, as an alternative, have defined a standard chemical potential μ_0^0 once and for all at the sea surface.

Then,

$$\mu_z = \mu_0^0 + RT \ln A_z \qquad (4.171)$$

A_z is a generalized activity which reflects not only the change in activity due to the concentration of the salt but also the change in activity which results from the pressure and the position in the gravitational field, with reference to the sea surface.

A_z and a_z are related by (Pytkowicz, 1968)

$$A_z = a_z \exp\left\{-\frac{1}{RT}\int_0^z [(MW) - \bar{V}\rho]g\,dz\right\} \qquad (4.172)$$

If the effect of temperature is neglected for the sake of simplicity, μ_0^0 and μ_z^0 are related by

$$\mu_0^0 - \mu_z^0 = \int_0^z [(MW) - \bar{V}\rho]g\,dz \qquad (4.173)$$

If the vertical distribution of a dissolved salt is at equilibrium, then $d\mu = 0$ and one obtains

$$a_z = a_0 \exp\left\{\frac{1}{RT}\int_0^z [(MW) - \bar{V}\rho]g\,dz\right\} \qquad (4.174)$$

$(MW)g\,dz$ reflects the settling effect due to the weight of the salt while $\bar{V}\rho g\,dz$ is akin to a buoyancy correction. Equation (4.174) yields roughly an 18% increase in the equilibrium concentration of NaCl/km of increasing depth (Pykowicz, 1962).

Gradients in concentration, such as the one above, are not observed in oceans or deep lakes because turbulent mixing is several orders of magnitude faster than the molecular diffusion of salts which would lead to vertical equilibration. This means, however, that the generalized activity A_z is not constant with depth and properties that depend on it also do vary. In the case of a uniform vertical distribution of a salt

$$A_z = A_0 \exp\left\{\frac{1}{RT}\int_0^z [(MW) - \bar{V}\rho]g\,dz\right\} \qquad (4.175)$$

The effect of the change in activity given by (4.175) could be detected if two galvanic cells connected by a wire were held at different depths in the ocean. Des Coudres (1893) performed an equivalent experiment, described in MacInnes (1961), by connecting two reversible calomel electrodes held at different heights through a salt bridge (Figure 4.8).

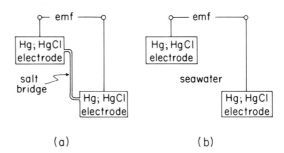

FIGURE 4.8 The experiment of Des Coudres (1893) and the oceanic experiment.

He was able to measure an emf resulting from the gravity and pressure effects.

Two interesting consequences of the effects of gravity, pressure, and temperature on distributions of water and ions are the osmotic pump (Levenspiel and de Nevers, 1974) and fossil brines (Mangelsdorf et al., 1970).

Osmotic Pump. First let us review the osmotic pressure. If we place saline and pure water on the two sides of a membrane that is permeable to water but not to salts, such as a dialysis bag, we observe that the pure water diffuses into the saline water for a while (Figure 4.9). This happens because the vapor pressure of water in the presence of salt is lower than that in distilled water. The difference is known as the osmotic pressure. The salt ions, by becoming hydrated and decreasing the free concentration of water in the solution, decrease its escaping tendency. The freshwater will, therefore, diffuse into the saltwater until the pressure exerted by the water column AB equals the pressure difference between the two sides of the membrane. This difference is known as the osmotic pressure.

The osmoregulation across cell membranes is very important in living organisms and osmotic imbalance can occur, for example, in the case of a heat stroke.

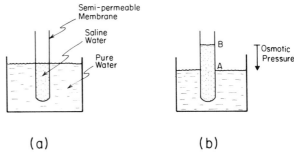

FIGURE 4.9 (a) Start of the osmotic experiment. (b) Final state of osmotic equilibrium.

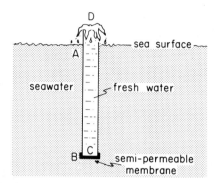

FIGURE 4.10 The osmotic fountain according to Levenspiel and de Nevers (1974).

The argument of Levenspiel and de Nevers (1974) regarding the osmotic pump goes as follows: Suppose that we stick an empty pipe, covered at the bottom by a semipermeable membrane, vertically in an isothermal ocean (Figure 4.10). The osmotic pressure difference between fresh and seawater is about 23 atm and corresponds to the pressure exerted by a column of water roughly 230 m high. The sequence of events will be:

1. Nothing happens at first as the column is lowered to 230 m.

2. From then on freshwater will flow into the pipe, at first up to a height 230 m below the sea surface, by reverse osmosis.

3. As the bottom of the pipe reaches great depths, the level of the freshwater in it will approach the sea surface. This will occur because seawater is about 2.8% denser than freshwater so that a difference in level smaller than 230 m is enough to compensate for an osmotic pressure difference of 23 atm.

4. When the pipe reaches a depth of about 8300 m, the freshwater will start to overflow at the top of the pipe because the higher density of the seawater will push the freshwater up 0.028×8300 m = 232 m.

This rough calculation indicates the general behavior that one may expect but is not exact since one assumes a uniform isothermal ocean and an ideal membrane.

The formal calculation is as follows. The increment in the chemical potential $d\mu$ is (Pytkowicz, 1962)

$$d\mu = \frac{\partial\mu}{\partial z}\, dz + \frac{\partial\mu}{\partial P}\, dP + \frac{\partial\mu}{\partial X}\, dX \quad (4.176)$$

μ and X refer to the water in seawater. X, the mole fraction, is the number of moles of water divided by the total number of moles of water plus salts. In a uniform ocean, $(\partial\mu/\partial X)\, dX$ is zero. The term in dz is not needed because the gravitational field is the same at the two sides of the membrane. Thus, we are left with the term in dP which, from hydrostatics, yields

$$d\mu = \frac{\partial\mu}{\partial P}\, dP = \bar{V}\rho g\, dz \quad (4.177)$$

Integrating (4.177) we obtain

$$\mu_B - \mu_A = g \int_0^{z_B} \bar{V}_{S(W)}\rho_{SW}\, dz \quad (4.178)$$

$$\mu_D - \mu_C = g \int_{z_C}^{z_D} V_{FW}\rho_{FW}\, dz \quad (4.179)$$

$S(W)$ refers to the water in seawater, SW to seawater, and FW to freshwater. V_{FW} is the molal volume of freshwater. At osmotic equilibrium $\mu_B = \mu_C$ and

$$\frac{\mu_D - \mu_A}{g} = \int_0^{z_B} \bar{V}_{S(W)}\rho_{SW}\, dz$$

$$+ \int_{z_C}^{z_D} V_{FW}\rho_{FW}\, dz \quad (4.180)$$

Introducing appropriate quantities in (4.180) one finds that the depth of the pipe needed to bring freshwater just to the surface is 8300 m. Note that I used a seawater density of 1.028 instead of 1.03, which was adopted by the original authors.

Soret Effect and Pore Waters. Mangelsdorf et al. (1970) reopened the question of the effects of gravity plus pressure and temperature (Soret effect) from a fresh viewpoint. They reasoned that molecular diffusion can lead toward an equilibrium distribution of salts in the pore waters of very old sediments (fossil brines) in which there is no turbulence. Departures from the known equilibrium in waters squeezed from cores, that is, samples captured in long pipes lowered into the sediments, can be interpreted in terms of specific chemical effects such as ion exchange with the sediments. Ion exchange occurs when, for example, sodium ions adsorbed on a solid phase are exchanged by potassium ions. This is an important process in the control of the chemical nature and the fertility of soils, in the purification of water, in chemical separations in the laboratory, in the oceans, and in pore waters.

TABLE 4.3 Percentage of change in the concentrations of ions with depth in seawater or pore waters of sediments if gravitational and pressure equilibria are reached [after Mangelsdorf et al. (1970)].

Depth (m)	Na$^+$	K$^+$	Mg^{2+}	Ca^{2+}	Cl$^-$	SO$_4^{2-}$	HCO$_3^-$	Br$^-$
				At 10°C				
100	0.88	1.10	1.64	2.15	0.81	3.42	1.64	2.42
500	4.50	5.63	8.54	11.27	4.09	18.27	8.43	12.67
1000	9.26	11.63	17.92	23.94	8.29	39.74	17.52	26.89
				At 25°C				
100	0.81	1.03	1.56	2.02	0.74	3.23	1.49	2.25
500	4.12	5.27	8.06	10.54	3.75	17.16	7.67	11.73
1000	8.45	10.86	16.86	22.31	7.60	37.15	15.87	24.77

Mangelsdorf et al. used better partial molal volume data than that available to Pytkowicz (1962, 1963), examined gradients over narrow depth ranges, and considered individual ions by means of the expression

$$[-(\text{IW})_i + \bar{V}_i\rho]g\,dz + RT\nu_i\,d\ln a_i = 0 \quad (4.181)$$

which applies at equilibrium when $d\mu = 0$. The term IW_i is the gram-ionic weight of the ion i and ν_i is the number of ions i produced by each mole of salt. IW is essentially equal to the atomic weight AW since the weight of electrons is negligible. Some typical results are shown in Table 4.3. It can be seen that an increase in temperature decreases the equilibrium gradients. The 15×10^6 yr refer to the diffusion of NaCl as an example. This occurs because an increase in the thermal energy of particles increases their kinetic energy and, consequently, their random motions.

The authors also examined the perturbation of the Soret effect due to the presence of sediments and the diffusional process. They concluded that times of the order of 15 million yr may be required to obtain vertical equilibrium distributions over a 1-km depth range, within the pore water.

In a chemical reaction solid → ions the effect of gravity is present on both sides and is canceled out. The effects of pressure and the chemical potential, however, are still present. If solid is present at all depths and the dissolution rate is fast, then the ions can reach and remain at equilibrium at all depths in the sediments. In this case equilibration is caused by the reaction instead of by molecular diffusion.

4.8 EQUILIBRIUM AND CHEMICAL REACTIONS

4.8.1 Derivation of Equilibrium Constants

Consider a chemical reaction

$$\nu_A{}^A + \nu_B{}^B = \nu_C{}^C + \nu_D{}^D \quad (4.182)$$

such as, for example,

$$H_2CO_3 + CO_3^{2-} = 2HCO_3^- \quad (4.183)$$

for which $\nu_A = 1$, $\nu_B = 1$, $\nu_C = 2$, and $\nu_D = 0$. The term ν_i represents the stoichiometric number of moles of i. Reaction (4.183) is a key reaction for the behavior of carbonates in seawater as we shall see later.

The condition for reaction (4.182) to be at equilibrium in a closed system at constant T and P is

$$\nu_A\mu_A + \nu_B\mu_B = \nu_C\mu_C + \nu_D\mu_D \quad (4.184)$$

because then no work is needed for a slight displacement from equilibrium.

Such a small change is somewhat analogous to a slight displacement in the position of a spring that was at mechanical equilibrium (see Chapter 1). In this case, the forces are balanced so that the net force $\mathbf{F}$ is zero. Therefore,

$$dW = \mathbf{F} \times d\mathbf{z} = 0 \quad (4.185)$$

where dW is the work done (the energy decrease) for the displacement of $d\mathbf{z}$.

To prove Equation (4.184) let us consider a

closed homogeneous system in which a generic chemical reaction

$$\sum_{i=1}^{r} \nu_i R_i = 0 \qquad (4.186)$$

occurs. R_i represents a reactant or a product, that is, A through D in (4.182) and 1 mole of a pure substance. In this notation, $\nu_i > 0$ for products and $\nu_i < 0$ for reactants. In the case of

$$H_2SO_4 + 2NaOH \rightarrow Na_2SO_4 + 2H_2O \quad (4.187)$$

$\nu(H_2SO_4)$ is -1, $\nu(NaOH) = -2$, $\nu(Na_2SO_4) = 1$, and $\nu(H_2O) = 2$.

We define the extent of a reaction by λ such that

$$d\lambda = \frac{dn_i}{\nu_i} \quad \text{or} \quad dn_i = \nu_i \, d\lambda \qquad (4.188)$$

For reaction (4.187) this yields

$$d\lambda = -\frac{dn_{H_2SO_4}}{1} = -\frac{dn_{NaOH}}{2}$$

$$= \frac{dn_{Na_2SO_4}}{1} = \frac{dn_{H_2O}}{2} \qquad (4.189)$$

as 2 moles of NaOH react with each mole of H_2SO_4.

For a closed system at constant T and P the equilibrium condition is

$$(\delta F)_{T,P} - \sum_{i=1}^{r} \mu_i \, \delta n_i = \sum_{i=1}^{r} \mu_i \nu_i \, \delta\lambda = 0 \qquad (4.190)$$

where δ now represents a virtual displacement. By virtual is meant a mathematically conceivable displacement that does not necessarily correspond to a physical one is meant. At equilibrium F must be a minimum for any displacement. Therefore, since $\delta\lambda$ can be negative or positive, the equilibrium condition (4.190) requires

$$\sum_{i=1}^{r} \nu_i \mu_i = 0 \qquad (4.191)$$

Equation (4.191) then is the condition for chemical equilibrium in a closed phase. For the reaction HAc $= H^+ + Ac^-$, it can be written as

$$-\mu_{HA} + \mu_H + \mu_{Ac} = 0 \qquad (4.192)$$

Note that for a phase at constant entropy and volume, the equilibrium condition is

$$(\delta E)_{S,V} = \sum_{i=1}^{r} \mu_i \, \delta n_i = 0 \qquad (4.193)$$

and Equation (4.191) is again obtained. Later in this section we shall see how μ_i is used for equilibrium calculations.

For a group of s reactions in a closed system, which may or may not have components in common, the equilibrium conditions are

$$\sum_{i=1}^{r} \nu_i^{(\sigma)} \mu_i = 0 \qquad (4.194)$$

where σ is the reaction index $\sigma = 1, \ldots, s$. If one reaction occurs in all the open phases within the closed system, then Equation (4.191) applies individually to the reaction in every phase. An example of this is a reaction $A + B = C + D$ which occurs in two immiscible organic solvents within a closed container. In the case of the simultaneous reactions

$$HAc \rightleftarrows H^+ + Ac^- \qquad (4.195)$$

$$H_2CO_3 \rightleftarrows 2H^+ + CO_3^{2-} \qquad (4.196)$$

Equation (4.194) yields

$$-\mu_{HAc} + \mu_H + \mu_{AC} = 0 \qquad (4.197)$$

$$-\mu_{H_2CO_3} + 2\mu_H + \mu_{CO_3} = 0 \qquad (4.198)$$

HAc represents acetic acid and Equation (4.196) is the sum of the two dissociation steps of carbonic acid.

Combining the equations for μ and K and rearranging, one obtains for the reaction (4.182)

$$\frac{a_C^{\nu_C} \times a_D^{\nu_D}}{a_A^{\nu_A} \times a_B^{\nu_B}}$$

$$= \exp\left[\frac{1}{RT}(\nu_A\mu_A^0 + \nu_B\mu_B^0 - \nu_C\mu_C^0 - \nu_D\mu_D^0)\right] = K$$

$$(4.199)$$

K is the thermodynamic equilibrium constant now obtained from an energy standpoint instead of from kinetic considerations. The symbols ν were not used earlier for the sake of simplicity but will be needed in the study of activity coefficients. It can be seen from the standard chemical potential terms in (4.199) that K is independent of the concentrations of the reactants and products. K depends, however, on the temperature and the pressure because of the μ^0 terms.

4.8.2 Temperature and Pressure Effects on Equilibrium Constants

Changes in temperature and pressure affect the kinetics and the chemical equilibria in the oceans, lakes, solid crust, and magma. In the oceans, for example, the depth ranges from 0 to about 10,000 m if deep trenches are excluded. This depth range corresponds to a variation in pressure from 1 to 1000 atm, with an average value of 380 atm. P and T effects must, therefore, be taken into consideration if we are to obtain equilibrium constants that are representative of the above media.

The temperature dependence of K is given by

$$\left(\frac{\partial \ln K}{\partial T}\right)_P = \frac{\Delta \bar{H}^0}{RT^2} \quad (4.200)$$

while the pressure dependence is

$$\left(\frac{\partial \ln K}{\partial P}\right)_T = -\frac{\Delta \bar{V}^0}{RT} \quad (4.201)$$

ΔH^0 is the sum of the partial molal enthalpies (partial molal heat contents) of the substances produced minus the sum for the substances consumed, when all of them are in their standard states. Each term must be multiplied by the number of moles of the constituent involved in the reaction. $\Delta \bar{V}^0$ is the corresponding quantity in terms of partial molal volumes.

It is worthwhile to notice two facts. First, Equations (4.200) and (4.201) reflect the Le Chatelier principle (1885) as was mentioned earlier. Thus, if the sum of the partial molal volumes of the reactants exceeds that of the products, then the equilibrium constant will increase with pressure and will push reaction (4.186) to the right. If the temperature is raised, then the equilibrium will shift in the direction in which heat is absorbed.

As

$$\Delta H^0 = \Delta H_0^0 + \frac{1}{RT^2} \int_0^T \Delta C_p \, dT \quad (4.202)$$

where the subscript zero refers to the hypothetical value of 0°K but in practice in an integration constant, the introduction of this equation into (4.200) yields

$$\left(\frac{\partial \ln K}{\partial T}\right)_P = \frac{\Delta H_0^0}{RT^2} + \frac{1}{RT^2} \int_0^T \Delta C_p \, dT \quad (4.203)$$

If ΔC_p is represented as a power series in T,

$$\Delta C_p = \alpha + \beta T + \gamma T^2 + \cdots \quad (4.204)$$

then integration between temperatures T_1 and T_2 yields (Glasstone, 1946)

$$\ln \frac{K_2}{K_1} = -\frac{\Delta H_0^0}{R}\left(\frac{1}{T_2} - \frac{1}{T_1}\right)$$

$$+ \frac{\alpha}{R} \ln \frac{T_2}{T_1} + \frac{\beta}{2R}(T_2 - T_1)$$

$$+ \frac{\gamma}{6R}(T_2^2 - T_1^2) + \cdots \quad (4.205)$$

When ΔH can be assumed to be nearly constant then

$$\ln \frac{K_2}{K_1} \cong -\frac{\Delta H^0}{R}\left(\frac{1}{T_2} - \frac{1}{T_1}\right)$$

$$\cong -\frac{\Delta H}{R}\left(\frac{1}{T_2} - \frac{1}{T_1}\right) \quad (4.206)$$

since then ΔH^0 is not very different from ΔH. The term ΔH_0^0 is obtainable graphically as an integration constant in the indefinite integral form of Equation (4.205).

In a similar way, the standard partial molal compressibility

$$\bar{K}_i^0 = -\left(\frac{\partial \bar{V}_i^0}{\partial P}\right)_{T,n_i} \quad (4.207)$$

must be taken into consideration if $\bar{V}_i^0$ varies with pressure. The pressure effect is given by (Owen and Brinkley, 1941)

$$RT \ln \frac{K_P}{K_1} = -\Delta \bar{V}^0 (P - 1)$$

$$+ \Delta \bar{K}^0 \left[(B + 1)(P - 1) - (B + 1)^2 \ln\left(\frac{B + P}{B + 1}\right)\right]$$

$$\quad (4.208)$$

where the subscripts 1 and P refer to the pressure and B is a constant characteristic of water. If $\Delta \bar{V}^0$ is relatively invariant then

$$RT \ln \frac{K_P}{K_1} \cong -\Delta \bar{V}^0 (P - 1) \quad (4.209)$$

EXAMPLE 4.24. PRESSURE EFFECTS (OWEN AND BRINKLEY, 1941). At infinite dilution and 25°C, with the pressure in atmospheres,

$$\underset{\text{calcite}}{CaCO_3} = Ca^{2+} + CO_3^{2-} \quad \Delta \bar{V}^0 = -58.3$$

$$\Delta \bar{K}^0 \times 10^4 = -157 \quad (i)*$$

$$\frac{(K_{sO})_{1000}}{(K_{sO})_1} = 8.10 \quad (ii)$$

$$\underset{\text{(anhydrate)}}{CaSO_4} = Ca^{2+} + SO_4^{2-} \quad \Delta \bar{V}^0 = -49.3$$

$$\bar{K}^0 \times 10^4 = -142 \quad (iii)$$

$$\frac{(K_{sO})_{1000}}{(K_{sO})_1} = 5.80 \quad (iv)$$

K_{sO} is the solubility product $\{Ca^{2+}\}\{CO_3^{2-}\}$ or $\{Ca^{2+}\}\{SO_4^{2-}\}$ in principle and also the concentration product since the γ's are unity at infinite dilution.

The effect of pressure is illustrated in Example 4.25.

EXAMPLE 4.25. CALCULATION OF K_{sO} FOR $CaSO_4$ FROM FREE ENERGIES OF FORMATION PLUS THE EFFECT OF PRESSURE ON K_{sO} FROM LEWIS AND RANDALL (1961)

$$\Delta F^0_{f(CaSO_4)(s)} = -315.9 \text{ kcal/mole} \quad (i)$$

$$\Delta F^0_{f(SO_4)(aq)} = -177.34 \text{ kcal/mole} \quad (ii)$$

$$\Delta F^0_{f(Ca)(aq)} = -132.18 \text{ kcal/mole} \quad (iii)$$

$$\Delta F^0 = +314.9 - 177.34 - 132.18$$

$$= 6.38 \text{ kcal/mole} = -RT \ln K_{sO} \therefore K_{sO} \quad (iv)$$

$$= 4.75 \times 10^{-4}$$

*This example was discussed earlier but without mention of $\Delta \bar{K}^0$.

From Equation (4.208), the values of $\Delta \bar{V}^0$ and $\Delta \bar{K}^0$ from Example (4.25), and $B = 2996$ bars (Owen and Brinkley, 1941)

Pressure (atm)	0	200	400
K_{sO}	4.75×10^{-4}	6.98×10^{-4}	1.01×10^{-3}
Pressure (atm)	600	800	1000
K_{sO}	1.43×10^{-3}	2.00×10^{-3}	2.76×10^{-3}

EXAMPLE 4.26. VERTICAL DISTRIBUTION OF DISSOLVED $CaCO_3$ IN A HYPOTHETICAL DEEP WATER BODY. This example illustrates the effects that pressure and temperature can have upon aqueous $CaCO_3$ in the absence of uptake by calcareous organisms.

Let T decrease and P increase with depth. Then, K_{sO} will increase roughly in an exponential manner with depth.

1. The waters are and have been at complete rest for millions of years. Then the ion-activity product IAP = $\{Ca^{2+}\}\{CO_3^{2-}\}$ equals K_{sO}.

2. The waters are stirred very vigorously. Then IAP is constant with depth while K_{sO} increases. The stirring causes IAP that was equal to K_{sO} to become larger near the surface and smaller than K_{sO} at depth.

3. The actual case in the oceans, if we neglect biological effects, falls between 1 and 2, with some supersaturation in the upper reaches and undersaturation in the deep waters. This, of course, requires slow kinetics of precipitation and dissolution relative to the mixing of waters and is the case in the sea (see Chapter 5, Volume II).

4.8.3 Thermal Properties of Solutions and Solids

In the previous section, we saw that the temperature behavior of $\Delta \bar{H}^0$ was required to obtain the effect of temperature on equilibrium constants. Actually, the thermal behavior of solutions has wider applications as can be seen in Chapter 25 of Lewis and Randall (1961). In this section, I shall limit myself to one remark.

First, the temperature dependence of activity coefficients [see Equation (4.163)] can be represented by

$$\left(\frac{\partial \ln \gamma}{\partial T} \right)_{P,n_i} = -\frac{\bar{L}}{RT^2} \quad \text{with} \quad \bar{L} = \bar{H} - \bar{H}^0$$

$$(4.210)$$

where $\bar{L}$ is the relative enthalpy and $\bar{H}^0$ represents the enthalpy of the solute in the reference state. Note that in Equation (4.163), when I was not differentiating solvents from solutes, $\bar{H}^0$ was referred to as the partial molal enthalpy in the standard state. When a substance is definitely a solute, the standard state of an ideally behaving one molal solution is clumsy for thermal data although it works for potentiometric data, as we shall see later. This is why the reference is used instead of the standard state in Equation (4.209).

Integration of Equation (4.210) yields

$$R \ln \gamma = \frac{\bar{L}}{T} + \text{constant} \qquad (4.211)$$

for a solute such as a dissolved salt.

If a small amount of salt is added to a large amount of pure water at any given temperature, then

$$\frac{\partial Q}{\partial n} = \bar{L} - \bar{L}^0 \qquad (4.212)$$

and $\bar{L}$ becomes known because $\bar{L} = \bar{H} - \bar{H}^0$ and, therefore, $\bar{L}^0 = \bar{H}^0 - \bar{H}^0 = 0$. Then, the constant in Equation (4.211) is determined and γ can be calculated versus T. In NaCl solutions (Lewis and Randall, 1961)

| m | 0 | 0.793 | 1.110 |
| $\bar{L}$ (cal/mole) | 0 | -153 | -248 |

4.8.4 Free Energy of Formation

The concept of the free energy of formation was examined briefly earlier in this work. ΔF_f^0 represents ΔF^0 in a conventional scale in which F^0 of pure elements in their most stable state are set as zero. Then, values of ΔF^0 for reactions can be calculated from the values of ΔF_f^0. As an example, in the reaction (Berner, 1971)

$$2H^+ + FeS = Fe^{2+} + H_2S \text{ (gas)} \qquad (4.213)$$

the standard energy change at 25°C and 1 atm is

$$\Delta F^0 = \Delta F_f^0[Fe^{2+}] + \Delta F_f^0[H_2S] - 2\Delta F_f^0[H^+] +$$

$$\Delta F_f^0[FeS] = -20.30 - 6.54 - (0 - 22.3)$$

$$= -4.54 \text{ kcal/mole} \qquad (4.214)$$

if the FeS is present as mackinawite. The equilibrium constant for reaction (4.213) can then be calculated by means of the equation relating ΔF^0 and K which is presented in the next section.

EXAMPLE 4.27. FREE ENERGIES OF FORMATION OF WATER AND HCl

Water vapor $\qquad H_{2(g)} + \frac{1}{2}O_{2(g)} = H_2O_{(g)}$

$$\Delta F_f^0 = -54,650 \text{ cal/mole} \qquad (i)$$

Liquid water $\qquad H_{2(g)} + \frac{1}{2}O_{2(g)} = H_2O_{(l)}$

$$\Delta F_f^0 = -56,700 \text{ cal/mole} \qquad (ii)$$

HCl vapor $\qquad \frac{1}{2}H_{2(g)} + \frac{1}{2}Cl_{2(g)} = HCl_{(g)}$

$$\Delta F_f^0 = -22,700 \text{ cal/mole} \qquad (iii)$$

One always has the number of gram-atoms which yields 1 mole on the right. These values are for 25°C. The gases are at unit activity, that is, essentially at unit pressure. The standard free energies of the elements are zero by definition.

Note that, if $H_2O_{(g)}$ is at its more usual pressure of 23.7 cm Hg, then we do not have the defined energy of formation but rather

$$\Delta F = \Delta F_f^0 - RT \ln \frac{23.7}{760} \qquad (iv)$$

EXAMPLE 4.28. DIFFERENCE BETWEEN THE STANDARD FREE ENERGY OF A REACTION AND THE STANDARD FREE ENERGY OF FORMATION (*Glasstone*, 1969). The reaction

$$CH_{4(g)} + 2O_2 = CO_{2(g)} + H_2O_{(l)} \qquad (i)$$

has the standard energies of formation, from the elements in their standard states to the compounds also in their standard states, equal to

$CH_{4(g)}$	$-12,200$ cal/mole	$CO_{2(g)}$	$-94,450$ cal/mole
O_2	0 cal/mole	$H_2O_{(l)}$	$-54,650$ cal/mole

The standard free energy of the reaction, with all substances in their standard states, is

$$\Delta F^0 = -113,400 + (-94,450) - (-12,200)$$

$$= -195,650 \text{ cal/mole} \qquad (ii)$$

4.8.5 Free Energy and the Equilibrium Constant

I mentioned earlier that ΔF, the change in free energy, indicates the tendency for a reaction to occur.

This is analogous to having the potential energy as a measure of the tendency of an object to fall. Let us elaborate on ΔF.

In a reversible process within a closed system (with no exchange of matter), and at constant P and T, the free energy can be transferred back and forth between the system and its environment in the direct and reverse reactions. There is no loss of F because there is no dissipation (degradation) of energy (entropy formation). Then

$$dF = dH = W \qquad (4.215)$$

Since $T\, ds = 0$, the term W is the reversible non $P\, dV$ work done by the system. If the system is at equilibrium, however, no work is done, and

$$dF = 0 \qquad (4.216)$$

This is the necessary condition for equilibrium.

Spontaneous processes occur at finite rates, with degradation of energy, so that the free energy (capacity to do non $P\, dV$ work) decreases. Thus,

$$dF < 0 \qquad (4.217)$$

is the condition for spontaneous processes, as was shown earlier.

Another way to determine the equilibrium condition is in terms of v_i and μ_i. Consider a chemical reaction that occurs within a closed system at constant temperature and pressure. For it

$$dF = dH - T\, dS + \sum_i \left(\frac{\partial F}{\partial n_R}\right)_{T,P} dn_R \qquad (4.218)$$

where, by definition, $(\partial F/\partial n_R)_{T,P} = \Sigma v_R \mu_R$ summed over all the components. R is the generic reactant or product. Thus, the necessary condition for equilibrium at constant T and P is Equation (4.191). Equations (4.191) and (4.216) yield, with F in a finite increment form,

$$\Delta F = \Sigma v_R \mu_R = 0 \qquad (2.219)$$

at equilibrium. As $\mu_R = \mu_R^0 + RT \ln a_R$,

$$\Delta F = \Delta F^0 + RT \ln Q_R \qquad (4.220)$$

with

$$\Delta F^0 = -RT \ln K \qquad (4.221)$$

K is the equilibrium constant, ΔF^0 is the change in free energy for a reaction when the reactants and products are in their standard states, and Q_r is the reaction quotient when the activities are not necessarily in their equilibrium configuration. When $Q_r = K$, that is, when the system is at equilibrium, then $\Delta F^0 = RT \ln Q_r = -RT \ln K$ and $\Delta F = 0$ because $Q_r = K$.

Some values of the $pK = -\log K$ for solutes are presented in Table 4.4.

If $Q_r > K$ then $\Delta F > 0$ and

$$v_A A + v_B B \rightarrow v_C C + v_D D \qquad (4.222)$$

is not a spontaneous reaction. It would have to be driven by an external source of energy and its reverse, with $\Delta F < 0$, would proceed until Q_r became equal to K.

EXAMPLE 4.29. RELATION BETWEEN K AND Q_r. Consider a generic reaction

$$A + B = C + D \qquad (i)$$

with the initial activities $(a_A)_i = 1.0$, $(a_B)_i = 0.2$, $(a_C)_i = 1.0$, and $(a_D)_i = 0.5$. Then, $(Q_r)_i = (1 \times 0.5)/(1 \times 0.2) = 2.5$.

If $K = 2.8$ then the reaction will proceed as written [left to right since a_C and a_D appear as products in the numerator of $(Q_r)_i$]. Equilibrium will occur when Q_r becomes equal to K.

If $K = 1.5$ then reaction i will proceed to the left since, in this case, a reaction to the right would imply $Q_r > K$ and $\Delta F > 0$ and would not be spontaneous.

The free energy of chemical reactions can be determined by a variety of methods, such as solubility data, emf (electromotive force) determinations, and spectroscopic and vapor pressure data. Let us look at the emf method because it reveals concepts employed in oceanographic practice.

4.8.6 Relationship Between the emf, the Free Energy, and the Equilibrium Constant

The emf method is a very accurate procedure for the determination of the ΔF of reactions at constant temperature and pressure, provided that the reaction can be carried out under nearly reversible conditions by drawing very little current. The subject is introduced briefly here and will be fully developed

TABLE 4.4 Values of the $pK = -\log K$ at 25°C for solutes [based in part on Stumm and Morgan (1970) and on Smith and Martell (1976)].

Solute	Reaction	pK
Acids		
$HClO_4$	$HClO_4 \rightleftharpoons H^+ + ClO_4^-$	-7
HCl	$HCl = H^+ + Cl^-$	-3
HNO_3	$HNO_3 = H^+ + NO_3^-$	-1
H_3PO_4	$H_3PO_4 = H^+ + H_2PO_4^-$	2.1
H_2CO_3	$H_2CO_3 = H^+ + HCO_3^-$	6.3
H_2S	$H_2S = H^+ + HS^-$	7.1
$H_2PO_4^-$	$H_2PO_4 = H^+ + HPO_4^{2-}$	7.2
H_3BO_3	$B(OH)_4^- = H^+ + B(OH)_3$	9.3
H_4SiO_4	$H_4SiH_4 = H^+ + H_3SiO_4$	9.5
HCO_3^-	$HCO_3^- = H^+ + CO_3^{2-}$	10.3
H_2O	$H_2O = H^+ + OH^-$	14
Bases		
ClO_4^-	$ClO_4^- + H_2O = HClO_4 + OH^-$	21
Cl^-	$Cl^- + H_2O = HCl + OH^-$	17
NO_3^-	$NO_3^- + H_2O = HNO_3 + OH^-$	15
$H_2PO_4^-$	$H_2PO_4^- + H_2O = H_3PO_4 + OH^-$	11.9
HCO_3^-	$HCO_3^- + H_2O = H_2CO_3 + OH^-$	7.7
NH_3	$NH_3 + H_2O = NH_4^+ + OH^-$	4.7
Solubilities		
$Mg(OH)_{2(s)}$	$Mg(OH)_{2(s)} = Mg^{2+} + 2OH^-$	11.15
$Ca(OH)_{2(s)}$	$Ca(OH)_{2(s)} = Ca^{2+} + 2OH^-$	5.19
$Al(OH)_3$	$Al(OH)_{3(s)} + 3H^+ = Al^{3+} + 3H_2O$	8.2
$MgCO_3$	$MgCO_{3(s)} = Mg^{2+} + CO_3^{2-}$	7.46
$CaCO_3$ (calcite)	$CaCO_{3(s)} = Ca^{2+} + CO_3^{2-}$	8.35
$CaCO_3$ (aragonite)	$CaCO_{3(s)} = Ca^{2+} + CO_3^{2-}$	8.22
$MnCO_3$	$MnCO_{3(s)} = Mn^{2+} + CO_3^{2-}$	9.30
$CuCl$	$CuCl_{(s)} = Cu^+ + Cl^-$	7.38
$AgCl$	$AgCl_{(s)} = Ag^+ + Cl^-$	7.94

in Chapter 7, Vol. II for readers with a special interest in redox reactions.

The work done in an electrochemical cell is $nF'E$ where F' is the Faraday equivalent, that is, the coulombs of electricity associated with the production or removal of 1 g-equiv of any ion in an electrochemical reaction. It is 96,493.5 C/g-equiv. E is the emf. A positive E is associated with a spontaneous cell reaction occurring spontaneously from left to right.

If the work for a reaction is reversible (in practice nearly reversible) then

$$F = -nF'E \qquad (4.223)$$

and n g-equiv of reactants are consumed. The minus sign appears because the free energy of the system decreases when it does the electrical work. The reaction is spontaneous, therefore, when $E > 0$. K is related to E^0, the standard potential, and ΔF^0 by

$$-\Delta F^0 = nF'E^0 = RT \ln K \qquad (4.224)$$

for the constituents in their standard states. Also,

$$E = E^0 - \frac{RT}{nF'} \ln Q_r = \frac{RT}{nF'} \ln \frac{K}{Q_r} \qquad (4.225)$$

where E^0 is the standard potential. Reactions are spontaneous to the right as written if $K > Q_r$. Thus, the oxidation $Zn_{(s)} = Zn^{2+} + 2e^-$ and the reduction $Cu^{2+} + 2e^- = Cu_{(s)}$ are spontaneous but $Zn^{2+} + 2e^- = Zn_{(s)}$ has to be driven.

Let us suppose that we have the cell shown in

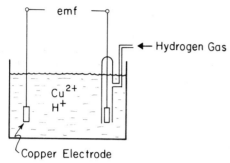

FIGURE 4.11 The copper–hydrogen cell.

Figure 4.11. A copper and a hydrogen electrode are immersed in an acid solution which contains a cupric salt. If $a_{Cu} = a_H = 1$ in a solution saturated with hydrogen gas at unit pressure, then the initial potential will be the standard one obtainable from tables. It is

$$E^0 = \frac{RT}{nF'} \ln K = 0.337 \text{ volts (V)} \quad (4.226)$$

and one can obtain K for the reaction

$$Cu^{2+} + H_2 \text{ (gas)} = Cu_{(s)} + 2H^+ \text{ (aqueous)} \quad (4.227)$$

because $E^0_{H_2}$ for $H_2 = 2H^+ + 2e^-$ is set equal to zero by convention. The reaction proceeds to completion because cupric ions are removed from solution by being deposited on the copper electrode while H_2 gas is evolved. ΔF^0 can be calculated from $nF'E^0$.

EXAMPLE 4.30.   CALCULATION PROCEDURUE FOR ΔF^0 IN REACTION (4.227)

$$\Delta F^0 = -2F'E^0 = -2 \times F' \times 0.337 \quad (i)$$

where $F' = 96,493$ C/equiv or J/V-equiv = 23,062 cal/V-equiv, E^0 is in V, and 2 converts moles to equivalents for the specific reaction under study. Using 23,062 cal/V-equiv

$$\Delta F^0 = -2 \times 23,062 \times 0.337 = -15.54 \text{ kcal/equiv} \quad (ii)$$

while with $F' = 96,493$ J/V-equiv

$$\Delta F^0 = -2 \times 96,493 \times 0.337 = -65.04 \text{ J/equiv} \quad (iii)$$

For K

$$\log K = \frac{nF'}{2.3026RT} E^0 = \frac{2E^0}{0.05915} \text{ (at 25°C)} \quad (iv)$$

and

$$K = 2.482 \times 10^{11} \quad (v)$$

I shall use the following cell conventions:

1. | separates different phases. For example, $Pt \mid H_{2(aq)}$.
2.   ; separates different components of a phase. For example, $Cu^{2+} ; Zn^{2+}$.
3.   : separates different phases in the same state (solid, liquid, or gas) in contact but not mixed together. For example, $Ag : AgCl$.

The cell in Figure 4.11 may be represented by

$$Pt \mid H_{2(aq)} ; H^+ ; Cu^{2+} \mid Cu_{(s)} \quad (4.228)$$

The two half-cell reactions that occur at the two electrodes are

$$Cu^{2+} + 2e^- \rightarrow Cu_{(s)} \quad (4.229)$$

$$H_2 \rightarrow 2H^+ + 2e^- \quad (4.230)$$

The gain of electrons by the copper ions is known as reduction and the loss by the hydrogen gas as oxidation so that (4.227) is called a redox reaction. The electrode at which oxidation occurs (the hydrogen electrode in this case) is the anode and the electrode where reduction takes place is the cathode.

It is convenient in chemistry to have standard potentials for half-cells. These potentials obviously cannot be measured directly and the convention is used that the half-cell standard potential for reaction (4.230) is zero. Thus, half-cells with positive potentials as written and proceeding to the right will react spontaneously when coupled with the hydrogen electrode. The coupling of the half-cells $Pt \mid H_{2(aq)} ; H^+$ and $Cu^{2+} \mid Cu$ leads to reaction (4.227) which occurs spontaneously. The copper half-cell has a positive potential when its reaction is written as in Equation (4.229).

Zinc has a higher oxidizing potential than copper so that the spontaneous reaction is

$$Zn + Cu^{2+} = Zn^{2+} + Cu \quad (4.231)$$

Copper will thus be deposited on the copper electrode and zinc will dissolve. This cell acts as a battery and its recharging consists of passing current in the opposite direction of that of the spontaneous process, to regenerate the zinc electrode and cupric ions. Standard free energy changes and equilibrium constants can be obtained from tables such as Table 4.3 where E^0 is relative to $E^0_{H_2} = 0$ for the reaction $H_2 = 2H^+ + 2e^-$.

If reactants and products are not removed from solution during a redox reaction, then one measures the standard potential E^0 in principle when we start the reaction with all the components in their standard state. The more usual procedure in practice is an extrapolation of E as a function of $m \to 0$. Then, the oxidation potential (the half-cell standard oxidation potential) for the oxidation of copper is -0.337 V. The emf obtained between the copper and the hydrogen electrodes was $+0.337$ V but it corresponded to Equation (4.229) which indicates reduction of copper from left to right.

The negative value of $E^0(\text{Cu}, \text{Cu}^{2+})$ indicates, from Equation (4.226), that ΔF^0 for the oxidation of copper to cupric ions is positive so that this oxidation would not occur spontaneously in the presence of H_2 and H^+. Instead, reduction of cupric ions to copper metal would be expected.

One can construct a table of oxidation potentials such as Table 4.5 for half-cell potentials relative to the reduction of hydrogen for which the half-cell potential is set as zero. This table tells us, for example, that reaction (4.231), which corresponds to the galvanic cell

$$\text{Zn} \,|\, \text{ZnCl}_{2(aq)} ; \text{CuCl}_{2(aq)} \,|\, \text{Cu} \qquad (4.232)$$

will have a standard potential $0.763 + 0.0337 = 0.110$ V when the activities of zinc and copper ions are unity. It will gradually tend to zero as the reaction nears equilibrium because Q_r will become equal to K in Equation (4.225).

$$-E = \frac{RT}{nF'} \ln \frac{Q_r}{K} \qquad (4.233)$$

Equation (4.233) is obtainable from (4.224) and (4.225). These results will be developed and be applied further in Chapter 7, Volume II. An illustration is provided next.

EXAMPLE 4.31.   USE OF E^0 AND THE MEASURED E TO CALCULATE THE DISSOCIATION CONSTANT K_{HA} OF A WEAK ACID.   The following cell can be used

TABLE 4.5 Some standard oxidation potentials (in V) at 25°C.

$\text{Zn} = \text{Zn}^{2+} + 2e^-$	0.763
$\text{Fe} = \text{Fe}^{2+} + 2e^-$	0.440
$\text{Pb} = \text{Pb}^{2+} + 2e^-$	0.126
$\text{H}_2 = 2\text{H}^+ + 2e^-$	0
$\text{Ag} + \text{Cl}^- = \text{AgCl} + e^-$	-0.222
$2\text{Hg} + \text{Cl}^- = \text{Hg}_2\text{Cl}_2 + 2e^-$	-0.268
$\text{Cu} = \text{Cu}^{2+} + 2e^-$	-0.337
$\text{Fe}^{2+} = \text{Fe}^{3+} + e^-$	-0.771
$2\text{Hg} = \text{Hg}_2^{2+} + 2e^-$	-0.789
$\text{Ag} = \text{Ag}^+ + e^-$	-0.920
$2\text{Cl}^- = \text{Cl}_2 + 2e^-$	-1.360

$$\text{Pt} \,|\, \text{H}_{2(aq)} ; \text{HA}(m_1) ; \text{NaA}(m_2) ; \text{NaCl}(m_3) \,|\, \text{Ag} : \text{AgCl}$$

$$\text{(i)}$$

The subscript aq or the symbol (m_i) indicate substances in solution.

For HA

$$K_{HA} = \frac{a_H a_A}{a_{HA}} = \frac{\gamma_H \gamma_A}{} \frac{m_H m_A}{m_{HA}} \qquad \text{(ii)}$$

The cell emf is

$$E = E^0 - RT \ln a_{HCl} \qquad \text{(iii)}$$

since the electrodes only sense H^+ and Cl^-. Applying (ii) to (iii) and remembering that $a_{HCl} = \gamma_H \gamma_{Cl} m_H m_{Cl}$,

$$E = E^0 - \frac{RT}{F'} \ln \frac{m_{Cl} m_{HA}}{m_A}$$

$$- \frac{RT}{F} \ln \frac{\gamma_{Cl} \gamma_{HA}}{\gamma_A} \frac{RT}{F} \ln K_{HA} \qquad \text{(iv)}$$

E^0 is obtained with the cell

$$\text{Pt} \,|\, \text{H}_{2(aq)} ; \text{HCl}_{(aq)} \,|\, \text{AgCl} : \text{Ag} \qquad \text{(v)}$$

In addition,

$$\frac{m_{Cl} m_{HA}}{m_A} = \frac{m_3 (m_1 - m_H)}{m_2 + m_H} \qquad \text{(vi)}$$

The term in the γ's in Equation (iv) tends to zero because $\gamma_{Cl}/\gamma_A \to 1$ when the solution concentrations tend to zero ($\gamma_{Cl} = \gamma_A$ for the limiting Debye–

Hückel equation) and $\gamma_{HA} \cong 1$ since HA is a neutral molecule. With this in mind, and introducing (vi) into (iv),

$$E = E^0 - \frac{RT}{F} \ln \frac{m_3(m_1 - m_H)}{(m_2 - m_H)} - \frac{RT}{F} \ln K_{HA}$$

(vii)

This derivation, made for a 1-1 electrolyte, was used by Harned and Ehlers (1932) to determine $K_{HAc} = 0.729 \times 10^{-5}$ at 25°C through the measurement of E for known E^0, m_H, m_1, m_2, and m_3.

4.8.7 A Brief Note on the Thermodynamics of Multicomponent Systems

The study of multicomponent solutions is a large field of research because it helps us understand and, at times, predict, important properties such as the free energy and γ's of complex solutions in terms of the behavior of simpler solutions. Results are useful in the study of laboratory solutions and natural waters such as the oceans (Millero, 1974; Whitfield, 1975). This subject will be elaborated upon in the next chapter, but some basic properties are examined at this point.

The basic thermodynamic relations are straightforward extensions of those of binary solutions, as long as the mixing is ideal (Kirkwood and Oppenheim, 1961). In the more frequent case of nonideality, excess properties are defined as is shown next for the case of the free energy.

Excess properties of solutions, such as $\Delta F_{mix}^{(e)}$, represent the excess values over that of mixing ideal solutions. It can be shown that (Lewis and Randall, 1961)

$$\Delta F_{mix} = 2RT\left(- m\phi + m_2\phi_2 + m_3\phi_2 \right.$$
$$\left. + m_2 \ln \frac{m_2\gamma_2}{m\gamma_{2(0)}} \; m_3 \ln \frac{m_3\gamma_3}{m\gamma_{3(0)}} \right) \quad (4.234)$$

ΔF_{mix} Free energy of mixing
ϕ_i Practical osmotic coefficient of i
γ_i Stoichiometric activity coefficient
$\gamma_{i(0)}$ Activity coefficient of pure i
m_2, m_3 Molalities of two solutes

The practical osmotic coefficient, in molal units, is $\phi_i = -1000 \ln \gamma_i/v_i m_i(MW)_i$. It originates from the rational osmotic coefficient g_i given by

$$\ln a_i = g_i \ln X_i = \ln f_i X_i \quad (4.235)$$

where X_i is the mole fraction. The term g represents a convenient alternative to f and γ when the data are obtained from vapor pressure and osmotic pressure measurements.

The ΔF for mixing of ideal solutions is given by

$$\Delta F_{mix}^{(i)} = 2RT(m_2 \ln f_2' + m_3 \ln f_3') \quad (4.236)$$

where $f_i' = m_i/m$. The term f' represents the number of moles of one solute divided by the sum of those of all the solutes and should not be confused with the rational activity coefficient f. Subtracting (4.236) from (4.234),

$$\Delta F_{mix}^{(e)} = 2mRT(-\phi + f_2'\phi_2 + f_3'\phi_3)$$
$$+ f_2 \ln \frac{\gamma_2}{\gamma_{2(0)}} + f_3 \ln \frac{\gamma_3}{\gamma_{3(0)}} \quad (4.237)$$

$\Delta F_{mix}^{(e)}$, the excess free energy of mixing, corresponds to the need to use activity coefficients to reflect deviations from ideality.

Another valuable, although empirical, property of multicomponent solutions is Young's rule

$$G = \sum_i X_i G_i \quad (4.238)$$

It lies at the opposite extreme from (4.237) in that it is applicable only to weakly interacting systems. G is a generic thermodynamic property, X_i is the mole fraction in terms of equivalents if we are dealing with ionic solutions, and G_i is the apparent equivalent property (volume, enthalpy, etc). Apparent equivalent properties are observables from which partial equivalent properties can be derived.

G can be obtained from Equation (4.233) for a mixture of electrolytes which interact very weakly. For the case of electrolytes AB and CD (Millero, 1974)

$$G = X_{AB}G_{AB} + X_{CD}G_{CD}$$
$$+ X_{CB}G_{CB} + X_{AD}G_{AD} \quad (4.239)$$

Other rules, such as that of Harned which can be used for interacting systems, will be discussed in the next section.

4.8.8 Homogeneous Equilibria

This is an important section in that it contrasts equilibrium within a single phase (a uniform or continuous region) such as a solution (salt in water or seawater) with concurrent equilibria within and between phases. An example of the latter is gaseous CO_2 at equilibrium with CO_2 in water which in turn is at equilibrium with aqueous H_2CO_3, HCO_3^-, and CO_3^{2-}.

Quartz pebbles, although there is a discontinuity among them, constitute a phase if each pebble has the same composition and is at the same temperature and pressure as the others. The same is true of KCl crystals at the bottom of a container with KCl-saturated water (see Chapter 1).

As we have seen earlier in Equation (4.218), the criterion for a reaction equilibrium, now expressed in terms of reactants R and products P, is

$$\sum_R \nu_R \mu_R = \sum_P \nu_P \mu_P \qquad (4.240)$$

For any physical or chemical process the general condition is, under isothermal and isobaric conditions (see Table 4.1),

$$(\Delta F)_{T,P} = 0 \qquad (4.241)$$

If T and V are constant, then

$$(\Delta A)_{T,V} = 0 \qquad (4.242)$$

Care should be taken not to confuse ΔF^0 and ΔF. The symbol ΔF^0 refers to all components of the reaction in their standard states, whereas ΔF indicates any concentration. For the reaction

$$CH_{4(g)} + 2O_{2(g)} = CO_{2(g)} + 2H_2O_{(l)} \qquad (4.243)$$

the standard free energies of formation of the reactants and products yield $\Delta F^0 = -196$ kcal/mole. The value of ΔF, however, will initially be negative if the reaction is spontaneous to the right and will drop to zero when equilibrium is reached. Then, there is no available energy to continue the reaction. $\Delta F^0 = -196$ kcal/mole is not the equilibrium ΔF which tends to zero. It does, however, yield $\Delta F^0 = -RT \ln K$ through thermodynamic algebra.

The spontaneous reaction conditions $(\Delta F)_{T,P} \leq 0$, with the equal sign for equilibrium, has its counterpart

$$(\Delta S)_{E,V} \geq 0 \qquad (4.244)$$

at constant internal energy and volume. These conditions apply to reactions in a single phase which is either closed, has reactions that are fast relative to transfer from adjacent phases, or is at equilibrium with those phases.

Note that uniformity can be an unnecessarily restrictive condition and that a phase may more generally be defined as a region in which properties vary continuously (see Chapter 1). Thus, for example, in a tall water column at rest which contains NaCl, the salt will have a slight but continuous gradient downward due to gravitation (Pytkowicz, 1962). The gradient of the concentration will be such that a_{NaCl} will be constant with depth (Pytkowicz, 1963). In general, however, scientists find it useful to define a phase as having constant values of (T, P), (T, V), (E, V), and so on, and this leads to the restricted definition of a phase as a uniform region.

4.8.9 Heterogeneous Equilibria and the Phase Rule

The phase rule is the key to an understanding of equilibria in multiphase systems such as $CaO_{(s)}$ + $MgO_{(s)}$ + $CO_{2(g)}$ + solution$_{(aq)}$. It deals in terms of numbers of components, phases, and degrees of freedom.

The number of components is defined as the minimum number of chemicals required to define the chemical system. As an example of the number of components, let us consider CO_2 gas at equilibrium with its solution in water. As a shortcut, we could guess right away that there are two components, CO_2 and H_2O. Let us proceed the long way around to clarify the concept of component.

The equations for the reactions at equilibrium are

$$CO_2 \text{ (gas)} = CO_2 \text{ (in solution)} \qquad (4.245)$$

$$CO_2 + H_2O = H_2CO_3 \qquad (4.246)$$

$$H_2CO_3 = H^+ + HCO_3^- \qquad (4.247)$$

$$HCO_3^- = H^+ + CO_3^{2-} \qquad (4.248)$$

$$H_2O = H^+ + OH^- \qquad (4.249)$$

The condition of electrical neutrality yields

$$[HCO_3^-] + 2[CO_3^{2-}] + [OH^-] - [H^+] = 0 \tag{4.250}$$

if the only solutes are CO_2 and its derivatives, plus H^+ and OH^-. Two times $[CO_3^{2-}]$ appears because each mole (the brackets indicate molar or molal concentrations) of CO_3^{2-} has two equivalents of charge. In addition we have the conservation of mass equations

$$TCO_2 = [CO_2] + [H_2CO_3]$$
$$+ [HCO_3^-] + [CO_3^{2-}] \tag{4.251}$$

$$TH_2O = [H_2O] + [H^+] + [OH^-] \tag{4.252}$$

and

$$pH_2O_{(g)} = pH_2O_{(soln)} \tag{4.253}$$

where pH_2O represents the partial pressure of H_2O.

Equations (4.251) and (4.252) can be combined so that one equation and one unknown are lost, as TCO_2 and TH_2O become the unknown $TH_2O + TCO_2$. Another equation is

$$P = pCO_2 + pH_2O + \Sigma P_i \tag{4.254}$$

where ΣP_i is the sum of the partial pressures of other gases present which do not participate in the reactions of interest. ΣP_i is assumed known.

Thus, there are 10 equations in the 11 unknowns pCO_2, $[CO_2]$, $[H_2CO_3]$, $[HCO_3^-]$, $[CO_3^{2-}]$, TCO_2, pH_2O, $[H_2O]$, $[H^+]$, $[OH^-]$, and TH_2O when the equations are written in terms of equilibrium constants and charge plus mass balances. Several equations and unknowns can be neglected in practical calculations but are needed to define the number of components and for use in the phase rule.

There are 11 unknowns minus 10 equations so that one compositional variable at each P and T must be fixed a priori or measured to determine all the concentrations of the components of the CO_2 and the H_2O systems.

The phase rule states that the number of degrees of freedom at equilibrium is $f = c - p + 2$, where c is the number of components and p is the number of phases. In the shorthand notation we had $c = 2$ (the components are CO_2 and H_2O), $p = 2$ (gas and solution phases), and $f = c - p + 2 = 2$. It would appear, therefore, that P and T would define the concentrations of all species at equilibrium. This is indeed the case if we are dealing with mole fractions because these are the units used in the derivation of the phase rule.

In the longhand version, however, we found that one compositional variable had to be added to P and T, so that we have $c = 3$ and $f_m = 3 - 2 + 2 = 3$. This happens because now we are dealing with molalities and to obtain them from mole fractions it is necessary to specify one additional variable, the total contents $TCO_2 + TH_2O$ of the aqueous phase. Thus, $f_m = 3$ is the number of molal degrees of freedom while $f = 2$ is the mole fraction number of degrees of freedom (RMP).

In general,

$$f_m = f + 1 \tag{4.255}$$

To show the need for the total contents of the solution when molal units are wanted rather than mole fractions,

$$X_i = \frac{n_i}{n_w + \Sigma_i n_i} = \frac{n_i/n_w}{1 + \Sigma_i(n_i/n_w)} \tag{4.256}$$

As

$$m_i = \frac{1000 n_i}{n_w (MW)_w} \tag{4.257}$$

$$X_i = \frac{m_i}{1000/MW_w + \Sigma m_i} = \frac{m_i}{m_w + \Sigma m_i} \tag{4.258}$$

which results from multiplying and dividing Equation (4.256) by $1000/MW_w$. Thus, to obtain m_i we must apply the transformation

$$m_i = X_i(m_w + \Sigma m_i) \tag{4.259}$$

In other words, we must know $m_w + \Sigma m_i$ in the solution.

The number of degrees of freedom for the simple case of O_2 and H_2O is $f = 2$ and

$$P = pO_2 + pH_2O \tag{4.260}$$

$$[O_2] = s_{O_2} pO_2 \tag{4.261}$$

s_{O_2} is the solubility of O_2. Earlier, in the case of CO_2, I used $pCO_2 = k_H^{(CO_2)} (CO_2)$, that is, Henry's law. This law, however, is only valid for dilute solutions at which $s_{CO_2} = 1/k_H^{(CO_2)}$ and, in general, it is better to use solubilities. Another equation for dilute solutions arises from the ideal behavior of water, in the sense of Raoult's law, and is

$$p\text{H}_2\text{O} = (p\text{H}_2\text{O})^0 X_w \qquad (4.262)$$

To replace it by an equation in terms of molalities, as $X_w = m_w/(m_w = \Sigma m_i)$, we obtain

$$p\text{H}_2\text{O} = (p\text{H}_2\text{O})^0 \frac{m_w}{m_w + \Sigma m_i} \qquad (4.263)$$

where $(p\text{H}_2\text{O})^0$ is the partial pressure of pure water.

It is not necessary in this case to use the condition $\text{H}_2\text{O} = \text{H}^+ + \text{OH}^-$ because there is no acid–base reaction but the mass balance $TO_2 + TH_2O = [\text{O}_2] + [\text{H}_2\text{O}]$ is required in principle. Thus, there are four equations in the unknowns $p\text{O}_2$, $[\text{O}_2]$, $p\text{H}_2\text{O}$, $[\text{H}_2\text{O}]$, and $TO_2 + TH_2O$. This last quantity is needed to convert X_i into m_i. Therefore, we again have one compositional degree of freedom plus P and T so that $f_m = 3$.

Note that the total contents of the solutions can be replaced in speciation calculations by any other relevant variable such as, for example, $[\text{H}^+]$ in the CO_2 case.

The actual equations that would be used for calculations based on (4.245) through (4.253) would be, if we neglect activity coefficients for the time being,

$$p\text{CO}_2 = k_H^{(CO_2)} [\text{CO}_2] \qquad (4.264)$$

$$K_h'' = \frac{[\text{H}_2\text{CO}_3]}{[\text{CO}_2][\text{H}_2\text{O}]} \qquad (4.265)$$

$$K_1'' = \frac{[\text{H}^+][\text{HCO}_3^-]}{[\text{H}_2\text{CO}_3]} \qquad (4.266)$$

$$K_2'' = \frac{[\text{H}^+][\text{CO}_3^{2-}]}{[\text{HCO}_3^-]} \qquad (4.267)$$

$$K_w'' = \frac{[\text{H}^+][\text{OH}^-]}{[\text{H}_2\text{O}]} \qquad (4.268)$$

The double prime indicates stoichiometric constants, k_H is Henry's law constant, K_h'' is the hydration constant of CO_2, and K_w'' is the dissociation constant of water. $p\text{H}_2\text{O}$ is a function of the concentrations and the nature of the solutes at a given T and P.

Let us turn now to another simple application of the phase rule before proceeding to its derivation. The phase rule, as was mentioned earlier, states that

$$f = c - p + 2 \qquad (4.269)$$

Let us see how it works in the case of a single component which could, for example, be water (Figure 4.12).

At points A or C there is only one phase but at B and D there are equilibria between two phases; solid–liquid at B and liquid–gas at D. Then, $c = 1, p = 2$, and, therefore, $f = 1 - 2 + 2 = 1$. This means that there is only one independent variable, T or P, if the system is to remain at equilibrium. This means that the liquid–gas system can evolve along the curve EF and remain at equilibrium if only P or T is varied arbitrarily. T, for example, is fixed by each arbitrary P. The phase rule does not tell us which specific equilibrium will occur (D or G, for example) but that, given certain constraints, a state of equilibrium will occur. E, the triple point, corresponds to $f = 1 - 3 + 2 = 0$ and occurs only at the one set (T, P). There are no degrees of freedom.

Let us derive the phase rule next, in a manner akin to that presented by Stumm and Morgan (1970). The number of degrees of freedom is given by

$$f = v - e \qquad (4.270)$$

v is the number of variables (components, temperature, pressure and, in some cases, others such as electric fields) while e is the number of independent equations within the phases. It can be easily shown that, if phases have different numbers of equations, then the largest e enters into Equation (4.270).

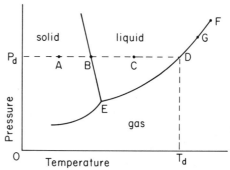

FIGURE 4.12 Phase diagram for a single component and three phases.

In a given phase there are $c - 1$ compositional variables because we include the condition of conservation of mass

$$\sum_{i=1}^{n} X_i = 1 \qquad (4.271)$$

in addition to equilibrium and charge balance relations used to fix c. Equation (4.271) shows that any $n - 1$ values of the X_i's fix the remaining one. Thus, in the phases there are

$$v = p(c - 1) + 2 \qquad (4.272)$$

variables.

The μ_i's must be the same at the interphase equilibria for a given component i. Thus,

$$\mu_{i1} = \mu_{i2} = \cdots = \mu_{ip} \qquad (4.273)$$

and there are $p - 1$ relations per component. Thus,

$$e = c(p - 1) \qquad (4.274)$$

Entering (4.272) and (4.274) into (4.270) we arrive at the phase rule, that is, at Equation (4.269).

4.8.10 Triangular Diagrams

Stability diagrams are illustrated in this section by means of the general features of triangular diagrams. Such diagrams are very useful in geochemistry and will be applied later in this book to the geochemistry of three component systems. The main application will be to the system $CaCO_{3(s)}$–$MgCO_{3(s)}$–H_2O and will include the solid solutions referred to as magnesium calcites. Note that triangular diagrams apply only at a given temperature and pressure and are in reality projections of systems with five variables; three compositional plus T and P, into the plane for which $(P, T) = $ constant. In the system $CaCO_{3(s)}$–$MgCO_{3(s)}$–H_2O it is also necessary that pCO_2 be constant.

Consider the three components A, B, and C in Figure 4.13. The state denoted by X corresponds to the fractional amounts of A, B, C given by the length of the lines $\overline{Xa}$, $\overline{Xb}$, and $\overline{Xc}$. Thus, the distance from X to a given side, drawn parallel to one of the others, gives the percentage of the component at the opposite corner. An example may be $A = H_2O$, $B = NaCl$, $C = Na_2SO_4$ if we neglect for

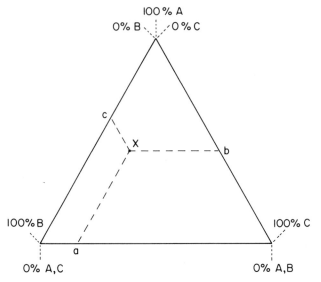

FIGURE 4.13   The principle of triangular diagrams.

the time being the hydrated form of Na_2SO_4. There is no component A when X is a point on the BC line.

A straight line such as AM in Figure 4.14 indicates a constant ratio B/C with A varying while NO represents a constant proportion of A with B/C varying. If P and Q are two mixtures then $\overline{PR}$ and $\overline{RQ}$ are the proportions of Q and P in R.

Consider next one pair of partly miscible components B and C while A and B as well as A and C are completely miscible in each other as in the case of acetic acid, chloroform, and water. Acetic acid and chloroform are only partly miscible in each other. The general triangular diagram is shown in Figure 4.15.

Note that the general classes of behaviors de-

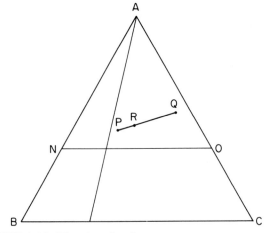

FIGURE 4.14   The triangular diagram.

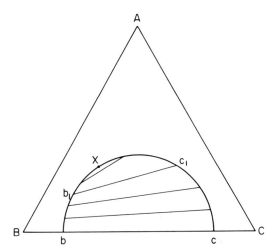

FIGURE 4.15 Partly miscible liquids.

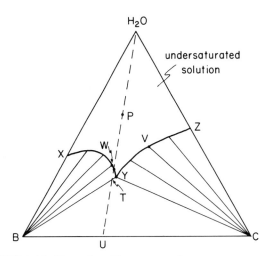

FIGURE 4.16 Two salts with a common ion.

scribed here can be applied to the earth sciences in terms of natural waters exposed to organic solutes and mineral assemblages. Furthermore, they pertain to assemblages of solid and magmatic minerals, including their solid solutions. An example of minerals which enter into three phase diagrams is provided by $A = SiO_2$, $B = NaAlSiO_4$, $C = KAlSiO_4$ which appears in the study of the crystallization of silicate melts (Mason, 1966).

The points b and c in Figure 4.15 indicate conjugate mixtures, a solution of C saturated with B and vice versa, with A absent. I remind you that A represents a component and not a point on the plane. The tie lines between b_1 and c_1 give the compositions of the conjugate solutions which vary with A. X, the Plait point, is the one for which the conjugate solutions reach the same composition. b is indeed a solution of B in C because, when it is moved to the right along a horizontal line, the ratio B/C decreases.

Let us consider two salts with a common ion (Figure 4.16). This figure corresponds to two salts that do not form a compound or solid solutions. The curves XY and YZ represent the saturation states of B and C. The point P is in the region of a solution undersaturated with respect to B and C. Points within the tie lines in the zones BXY and CYZ correspond to solutions supersaturated with regard to B and C, respectively, whereas points above XYZ indicate undersaturation. The tie lines connect the composition of the solution to that of the solid phase at equilibrium with it. Thus, VC connects a saturated solution of composition given by V to the saturating solid C. If C were an hydrated salt such

as $Na_2SO_4 \cdot 10H_2O$, the point at which the tie lines intercepted the water–C axis would be above C, as is shown in Figure 4.17.

Note that a solution of composition W in Figure 4.16 is at equilibrium with B while V is at equilibrium with the solid C. If a solution of composition P were evaporated gradually, the solution would reach point W. Further evaporation would cause precipitation of B and would increase the concentration of C until point Y was reached. Now C as well as B would crystallize with continued evaporation and, when the water was gone, the system would reach point U. This would occur because the ratios B/C at P and at U must be the same due to the conservation of mass. An example of such a system is $NH_4Cl–(NH_4)_2SO_4–H_2O$. Note that the solubility,

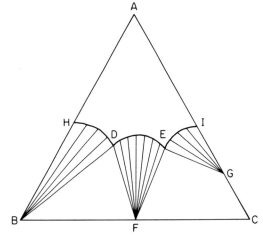

FIGURE 4.17 Triangular diagram for two salts, one of which is hydrated, which can form a compound (double-salt).

but not the solubility product of each component, is affected by the presence of the other salt. This is the common ion effect; an increase in the amount of NH_4^+ due to a further addition of $(NH_4)_2SO_4$ to the solution would depress the amount of NH_4Cl which can dissolve.

In Figure 4.17 point G may represent $Na_2SO_4 \cdot 10H_2O$ while C indicates Na_2SO_4. The solubility curve of the compound is $\overset{\frown}{DE}$. The various parts of the diagram show the following:

$AHDEI$ Unsaturated solution

HBD B + saturated solution of B

DEF BC + saturated solution of BC

EIG C + saturated solution of C

BDF B + BC + saturated solution of D

The number of degrees of freedom varies in the usual manner. For BDF there are three components H_2O, B, and BC and the three phases H_2O, B, and BC so that $f = 2$. Thus, at fixed T, P, the system has no degrees of freedom left.

If we add water to a double salt BC, such as $(NH_4)_2SO_4 \cdot 2NH_4NO_3$, congruent dissolution occurs if $\overline{AF}$ crosses DEF. If, however, $\overline{AF}$ crosses HD (DEF slanted to the right) then BC dissolves since $B + BC$ and the dissolution is incongruent.

Finally, let us consider the case of solid solutions (Figure 4.18). An example in question would be $Ca_xMg_{1-x}CO_3$ although magnesian calcites have a limited range of values of x while Figure 4.18 represents a full range of mixtures. Each solid solution of composition such as M is at equilibrium with a liquid solution. The composition M depends on the composition of the solution at equilibrium with it,

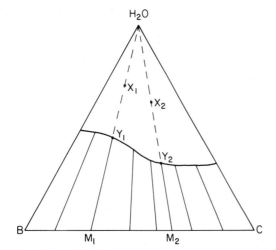

FIGURE 4.18 Solid solutions.

that is, on the initial composition of the liquid solution. Thus, evaporation of X_1 leads to M_1, whereas that of X_2 forms M_2.

In the case of $CaCO_3$, $MgCO_3$, and H_2O (Figure 4.19) the system at equilibrium may consist of solutions saturated with magnesian calcites, that is, solid solutions of composition $Ca_xMg_{(1-x)}CO_3$. This is indicated by $CaCO_3$–A–B–F, with the compound dolomite $CaMg(CO_3)_2$ in the zone B–E–C, and with $MgCO_3$ in C–D–$MgCO_3$. At point C, for example, the solution is saturated with dolomite and $MgCO_3$. Pure calcite is present only on the line A–$CaCO_3$.

Note that I mentioned equilibrium. This often is not the case in nature because rates of reactions may lead to the formation of phases which appear stable over a short period, but which are changed to the truly stable forms only over long periods. Thus, aragonite precipitates from seawater under normal conditions although magnesian calcites appear to be the stable phases. We shall return to this point later.

If a solid solution of composition C_i is exposed to a solution, it dissolves congruently at first (line C_i–H_2O) but then follows a tie line C_f–D_f which yields the final compositions of the liquid solution (D_f) and the solid solution (C_f). The specific tie line depends on the initial compositions in and the amounts of the phases. The concept of final states, which depend on the amounts of the initial phases in solid solutions in contact with liquid solutions, was introduced by Wollast and Reinhard-Derie (1977).

In Figure 4.19 I show a case where the final phase is richer in magnesium than the initial phase. This can happen in seawater if the equilibrium phase is to the right of the initial phase C_i. In freshwater, however, the final phase is always richer in

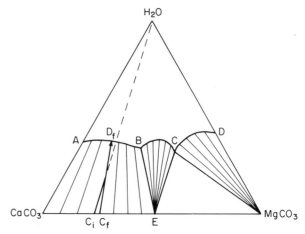

FIGURE 4.19 The system $CaCO_3$–$MgCO_3$–H_2O.

calcium so that a pure calcite can only be precipitated by repeated crystallizations if magnesium is present. What happens then is a congruent dissolution of the magnesian calcite of composition C_i followed by an incongruent dissolution, in which there is more calcite on the surface of the initial solid because of a faster dissolution of $MgCO_3$ (Land, 1967; Plummer and Mackenzie, 1974). I shall return to the $CaCO_3$–$MgCO_3$–H_2O in Chapters 4 and 5 in Volume II.

It should be kept in mind in the above discussion that tie lines connect the compositional (H_2O–BC) point Y to M, the solid at equilibrium with Y (see Figure 4.19).

4.8.11 Other Types of Stability Diagrams

I shall present further types of stability diagrams in this subsection and in others throughout this book. Magnesium minerals will be used to illustrate several types of stability plots which let us know what types of minerals are stable under varying conditions.

The main magnesium minerals are

Brucite	$Mg(OH)_2$
Magnesite	$MgCO_3$
Nesquehonite	$MgCO_3 \cdot 3H_2O$
Hydromagnesite	$Mg_4(CO_3)_3(OH)_2 \cdot 3H_2O$
Biotite	$K(Mg, Fe)_3Si_3O_{10}(OH)_2$
Chlorite	$(Mg, Fe)_5(Al, Fe)_2Si_3O_{10}(OH)_8$
Sepiolite	$Mg_2Si_3O_8 \cdot 2H_2O$
Talc	$Mg_3Si_4O_{10}(OH)_2$
Vermiculite	$Mg_3Si_4O_{10}(OH)_2$

The dissolution reactions of magnesium compounds in natural waters often occur together with those of Mg calcites ($Ca_xMg_{1-x}CO_3$). I shall at first neglect calcium for simplicity. Stumm and Morgan (1970) used the first four minerals listed above to illustrate the stability relations in the system Mg–CO_2–H_2O although such a simple case is not usual in nature. Still, the example developed by these authors reveals very well the algebraic procedures used to construct various types of diagrams. It will be presented in a simpler version and with additional clarifying material next.

Note, in this example, that thermodynamic solubility products and activities were used in the early equations of Stumm and Morgan, but that there was a shift to concentrations in midstream, due to the

deletion of activity coefficients for the CO_2 system. This means that the values of γ were set equal to 1 for simplicity and/or for use in very dilute solutions. The same is true of $[Mg^{2+}]$ through Equation (xvi) below.

EXAMPLE 4.32. STABILITY RELATIONS IN PART OF THE SYSTEM Mg–CO_2–H_2O IN FRESHWATER AND SEAWATER (BASED ON STUMM AND MORGAN, 1970). The reactions to be considered and the thermodynamic solubility products are

$$Mg(OH)_2 = Mg^{2+} + 2OH^- \quad \frac{\log K_{sO}}{-411.6} \quad \text{(i)}$$
$$\underset{\text{brucite}}{}$$

$$Mg(CO_3) = Mg^{2+} + CO_3^{2-} \quad -4.9 \quad \text{(ii)}$$
$$\underset{\text{magnesite}}{}$$

$$MgCO_3 \cdot 3H_2O = Mg^{2+} + CO_3^{2-} + 3H_2O - 5.4$$
$$\underset{\text{nesquehonite}}{} \quad \text{(iii)}$$

Since I wish to plot various quantities against the pH, the term H^+ is made explicit in the following equilibrium equations when possible:

$$\frac{[Mg(OH)_2]}{[Mg^{2+}]} = \frac{[OH^-]^2}{(K_{sO})_b} = \frac{[H_2O]^2}{[H^+]^2(K_{sO})_b} \quad \text{(iv)}$$

$$\frac{[MgCO_3]}{[Mg^{2+}]} = \frac{[CO_3^{2-}]}{(K_{sO})_m} = \frac{[TMg]\alpha_2}{(K_{sO})_m} \quad \text{(v)}$$

$$\frac{[MgCO_3 \cdot 3H_2O]}{[Mg^{2+}]} = \frac{[CO_3^{2-}]}{(K_{sO})_n} = \frac{(TMg)\alpha_2}{(K_{sO})_n} \quad \text{(vi)}$$

α_2 is the relative distribution $[CO_3^{2-}]/TCO_2$, with $TCO_2 = [CO_2] + [H_2CO_3] + [HCO_3^-] + [CO_3^{2-}]$, or (see Chapter 2), Volume II

$$\alpha_2 = \frac{K_1'' K_2''}{[H^+]^2 + K_1''[H^+] + K_1'' K_2''} \quad \text{(vii)}$$

K_1'' and K_2'' are the stoichiometric dissociation constants of H_2CO_3. Thus, a transition is made from activities to concentrations with the assumption $\gamma = 1$ and a link is made between Equations (v), (vi), and $[H^+]$.

Equations (iv) through (vi) become, in logarithm form and with the numerical values of the constants,

$$\log \frac{[Mg(OH)_2]}{[Mg^{2+}]} = 16.4 + 2pH \quad \text{(viii)}$$

$$\log \frac{[MgCO_3]}{[Mg^{2+}]} = 4.9 + \log TCO_2 + \log \alpha_2 \quad \text{(ix)}$$

$$\log \frac{[MgCO_3 \cdot 3H_2O]}{[Mg^{2+}]} = 5.4 + \log TCO_2 + \log \alpha_2 \quad \text{(x)}$$

Next let us consider three types of plots versus the pH:

1. Activity ratio diagram, $\log \{[\text{solid}]/[Mg^{2+}]\}$.
2. Solubility diagram, $\log [Mg^{2+}]$.
3. Predominance diagram, pCO_2.

If we assume that we have pure solids then {solids} = 1. Let us also set $K_1'' = K_1$ and $K_2'' = K_2$ for simplicity although this is only valid in very dilute solutions. Equations (viii) through (ix) are ready for the computation of the first two types of plots but CO_2 has to be introduced for the third. This is done, for example, through the procedure illustrated for nesquehonite for which

$$MgCO_3 \cdot 3H_2O = Mg^{2+} + CO_3^{2-} + 3H_2O$$
$$\log K_{sO} = -5.4 \quad \text{(xi)}$$

$$CO_3^{2-} + 2H^2 = H_2CO_3 \quad -\log K_1K_2 = 16.6 \quad \text{(xii)}$$

$$H_2CO_3 = CO_2 + H_2O \quad -\log K_h = 1.5 \quad \text{(xiii)}$$

$$MgCO_3 \cdot 3H_2O + 2H^+ = Mg^{2+} + CO_{2(g)}$$
$$+ 3H_2O \log K_{sO} = 12.7 \quad \text{(xiv)}$$

with an analogous procedure for other phases.

Note that $[Mg^{2+}]$ and the CO_2 system are not independent of one another. The condition of electroneutrality requires that

$$[Mg^{2+}] = [HCO_3^-] + 2[CO_3^{2-}] + [OH^-] - [H^+] \quad \text{(xv)}$$

or

$$[Mg^{2+}] = \alpha_1 T(CO_2) + 2\alpha_2 T(CO_2)$$
$$+ \frac{K_w}{[H^+]} - [H^+] \quad \text{(xvi)}$$

Thus, in a numerical example we have the choice of fixing $[Mg^{2+}]$ or $T(CO_2)$. We can assume that $TCO_2 = 0.01$ as an example.

Stumm and Morgan recommended that in an activity ratio diagram the plotting be started at high pH's where $\log \alpha_2 = 0$. In the region $pK_1 < pH < pK_2$, the slope $d \log \alpha_2/d$ pH is $+1$.

Elaborating on this statement by Stumm and Morgan (1970), we have $\log \alpha_2 = \log [CO_3^{2-}] - \log TMg$ and, as $TMg = $ constant, $d \log \alpha_2 = d[CO_3^{2-}]$. Furthermore, from K_2 we obtain $pK_2 = pH - \log [CO_3^{2-}] + \log [HCO_3^-]$. The term pK_2 is constant and $\log [HCO_3^-]$ is almost constant in the pK_1 to pK_2 range. Thus, $d \log[CO_3^{2-}]/d$ pH $= d \log \alpha_2/d$ pH $= 1$. The term TMg is the sum of the concentrations of all dissolved Mg species.

I should caution the reader that realistic results should be worked out for the actual system under consideration. Furthermore, if concentrations are used, then stoichiometric equilibrium constants should be employed while activities would be used to match thermodynamic equilibrium constants.

4.8.12 Solid Solutions

Typical solid solutions are rhodochrosite, where the solvent (major component) is $MnCO_{3(s)}$, and which usually contains solutes such as Fe^{2+}, Mg^{2+}, and Zn^{2+}; ferromanganese minerals (nodules or coatings on sediments), and calcites where Mg^{2+} can replace some of the Ca^{2+}. The formula for a solid solution is usually written as $(M_1)_x(M_2)_{1-x}A$, where $1 - x$ is the number of g-atom of M_2 which replace x g-atom of X_1 and A is the M_1. Examples are $Mn_xFe_{1-x}CO_3$ in rhodochrosite and $Ca_xMg_{1-x}CO_3$ in calcite.

The criterion for a solid solution versus adsorption is the substitution of ions in lattice sites either in the surface or in the bulk of a crystal. Thus, the interlayer uptake of ions by clays does not yield such a solution. Ion exchange is more difficult to interpret. Let it represent, for example, the exchange of H^+ by metal ions in silenol groups on the surface of amorphous silica. The fundamental compositional structure of the silica does not change. By this I mean that the fundamental structural element is still $SiO_{2(s)}$ when the surface changes from $SiO_2 \cdot 2H_2O_{(s)}$ to $SiO_2 \cdot 2M$. The overall reaction is

$$SiO_2 \begin{array}{c} OH-H \\ \diagup \\ \diagdown \\ OH-H \end{array} + 2M^+ = SiO_2 \begin{array}{c} OH-M \\ \diagup \\ \diagdown \\ OH-M \end{array} + 2H^+$$

$$\text{(4.275)}$$

The replacement of Ca^{2+} by Mg^{2+}, even if only at the surface of calcite, is an actual compositional change of the lattice. The reaction is

$$CaCO_{3(s)} + (1 - x)Mg^{2+}$$
$$= Ca_xMg_{1-x}CO_{3(s)} + (1 - x)Ca^{2+} \quad (4.276)$$

and there is a solid solution formed.

The extent and the stability of a solid solution depends on how alike the radii and the charges of the two ions are, and on the type and the spacing of the lattice. Thus, Mg^{2+} is taken up by calcite but Sr^{2+} substitutes for calcium in aragonite.

For a pure solid such as $MnCO_{3(s)}$ or $CaCO_{3(s)}$ the activity is one but this is not the case for solid solutions, as we shall see for the case of $Ca_xMg_{1-x}CO_{3(s)}$ later in this book.

If $CaCO_3$ behaved ideally in magnesian calcites, we would have

$$a_{CaCO_3(ss)} = X_{CaCO_3} \quad (4.277)$$

but this is not the case and

$$a_{CaCO_3(ss)} = \lambda_{CaCO_3}X_{CaCO_3} \quad (4.278)$$

$$a_{MgCO_3(ss)} = \lambda_{MgCO_3}X_{MgCO_3} \quad (4.279)$$

λ is the rational activity in the solid solution. The standard states are the pure components for which, by definition, the a_i's are unity. Since the values of X_i are also unity, this means that $\lambda_i = 1$ at the standard state. The subscript i represents, for example, $CaCO_3$ or $MgCO_3$.

In the solid case, the reference states ($\lambda_i = 1$) are also the single components as, when $X_i \rightarrow 0$, we find that λ_i diverges from 1 and cannot be used [see Garrels and Christ (1965)]. Thus, in very dilute solid solutions the solvent behaves ideally ($a_i \cong \lambda_i \cong X_i \cong 1$) but the solute is far from ideal.

EXAMPLE 4.33. SOLUBILITIES AND ACTIVITIES FOR SOLID SOLUTIONS IN THE CASE OF $Ca_{0.9}Mn_{0.1}CO_3$. X_{Ca} = 0.9 and it has been found that $\lambda_{CaCO_3} \cong 1.2$ (Garrels and Christ, 1965).

Ideal solution $a_{CaCO_3} = 0.9$
Nonideal solution $a_{CaCO_3} = 1.2 \times 0.9 = 1.08$

As $a_{Ca}a_{CO_3} = K_{so}a_{CaCO_3} = 4.571 \times 10^{-8} a_{CaCO_3}$, we have

Pure calcite $a_{Ca}a_{CO_3} = 4.571 \times 10^{-9}$
Ideal solution $a_{Ca}a_{CO_3} = 4.114 \times 10^{-9}$
Nonideal solution $a_{Ca}a_{CO_3} = 4.936 \times 10^{-9}$

4.9 THE PHYSICOCHEMICAL PROPERTIES OF SEAWATER

These properties fall into two groups. The thermodynamic properties of interest are the enthalpy, the free energy, the specific volumes, the partial molal volumes, the molal compressibilities, the coefficients of thermal expansion, the osmotic and the vapor pressure, and the lowering of the freezing point. The transport properties are the viscosity, the thermal conductance, and the electrical conductance.

Early results on the above properties are reviewed by Sverdrup et al. (1942) and Fofonoff (1962). Pytkowicz and Kester (1971) examined some of the problems associated with some of these quantities.

Recent data on some of the physicochemical properties of seawater can be found in:

Doherty and Kester (1974)	Freezing point
Millero et al. (1974)	Heat capacity
Millero et al. (1973)	Enthalpy
Robinson (1954)	Osmotic coefficient
Emmet and Millero (1975)	Density of seawater relative to standard mean oceanic water
Cox et al. (1970)	Density at 1 atm
Chen and Millero (1977)	Density versus P, T, and $S\%_0$
Horne (1969)	Viscosity

4.10 STATISTICAL THERMODYNAMICS OF SOLUTIONS

This is an enormous field and I shall limit myself in this and in the following sections to an examination of fundamental concepts and key results which can provide a basis for further study by the readers and which will appear in Chapter 5.

The tenet of statistical mechanics and thermodynamics is that a real system, such as a gas or a solute, can be represented by an ensemble of points. Each point represents a possible state of the system and is characterized in a phase space by the

$3N$ coordinates and $3N$ momenta of all N particles in the system. Thus, when the molecules of a gas move randomly within a container, for each instant there is a point in the phase space representative of the configuration represented by the $6N$ coordinates.

The term "can be represented" used in the previous paragraph applies to the concept that the ensemble average of a thermodynamic property such as the energy equals the time average measured for the system.

If the system is isolated then it is described by the microcanonical ensemble for which N, V, and E of the members (points in the phase space) are the same. E is the internal energy.

Let us consider next the canonical ensemble in terms of discrete energy levels for the particles. The members have the same (N, V, T) which are the independent thermodynamic variables. Each member is enclosed by a diathermal rigid wall (porous to energy but not to mass) and we let it equilibrate with a large heat reservoir of temperature T. Then, we isolate the ensemble of systems by an adiabatic rigid wall. This is canonical ensemble and, as it is isolated, it has a constant (T, V, N).

If the allowed energy levels are $E_1, \ldots, E_j$ then there are n_j members in energy level E_j. Let k be 1, $\ldots, j$. The term n_j is the occupation number. The following conditions hold:

$$\Sigma n_j E_j = E_t \qquad (4.280)$$

$$\Sigma n_j = N_c \qquad (4.281)$$

E_t is the energy of the ensemble which is fixed because the ensemble has an adiabatic wall. N_c is the total number of configurations since there are n_j members in each energy state.

The following a priori probability can be applied since the ensemble is isolated (like a number of the microcanonical ensembles); that every distribution of occupation numbers $\mathbf{n}$, that is, $(n_1, \ldots, n_j, \ldots)$ compatible with the preceding two equations is equally probable. A distribution is a set $(n_1, \ldots, n_j, \ldots)$.

The number of ways in which any one distribution of the n_j's can be realized is

$$W(n) = \frac{N_c!}{n_1! \ldots n_j!} = \frac{N_c!}{\prod_j n_j!} \qquad (4.282)$$

$W(n)$ is the number of ways that N_c distinguishable objects can be arranged in groups with n_k members in the E_k levels. There are many distributions compatible with the above and in any one distribution n_j/N_c is the fraction of members with energy E_j. The overall probability that a system is in an energy level E_j can be obtained by averaging n_j/N_c over all allowed distributions, while weighting each allowed distribution equally. The result is

$$P_j = \frac{n_j}{N_c} = \frac{1}{N_c} \frac{\Sigma_j W(n) n_j}{\Sigma_j W(n)} \qquad (4.283)$$

Thus P_j is the probability that a lattice configuration picked at random has an energy E_j.

The canonical ensemble of any mechanical property M_j is

$$M = \Sigma_j M_j P_j \qquad (4.284)$$

It can be proven that, when $w/N_c \to \infty$,

$$P_j = \frac{n_j^*}{N_c} = \frac{e^{-\beta E_j(N,V)}}{\Sigma_j e^{-\beta E_j(N,V)}} \qquad (4.285)$$

where n^* is that distribution which maximizes $W(n_k)$ under the constraints imposed on the system. Thus, this equation gives the probable distribution. The statistical average internal energy $\bar{E}$, which is equal to the thermodynamic E, is given by

$$\bar{E} = \bar{E}(\underset{\uparrow}{N}, V, \beta) = \frac{\overset{\downarrow}{\Sigma_j E_j(N, V)} e^{-\beta E_j(N,V)}}{\Sigma_j e^{-\beta E_j(N,V)}} \qquad (4.286)$$

N is the number of ions of each kind and E_j is a mechanical or thermodynamic quantity. β equals kT.

$$\beta = kT \qquad (4.287)$$

$Q(N, V, T)$, known as the partition function

$$Q(N, V, T) = \sum_j e^{-\beta E_j(N,V)} \qquad (4.288)$$

provides the connection with thermodynamic functions such as the relations

$$E = kT^2 \left(\frac{\partial \ln Q}{\partial T} \right)_{N,V} \qquad (4.289)$$

$$S = kT\left(\frac{\partial \ln Q}{\partial T}\right)_{N,V} + k \ln Q \qquad (4.290)$$

$$A(N, V, T) = -kT \ln Q(N, V, T) \qquad (4.291)$$

If we have n similar, distinguishable, noninteracting molecules with n_i of them having energy E_i, then the equilibrium configuration is

$$n_i = n_0 g_i e^{-E_i/kT} \qquad (4.292)$$

where n_0 = number of molecules in $E_0 = 0$ energy state

g_i = statistical weight (there may be g_i energy levels with almost equal energy E_i)

The above equation is an expression of Maxwell–Boltzmann law. The total number of molecules is

$$n = g_0 n_0 + g_1 n_0 e^{-E_1/kT} + g_2 n_0 e^{-E_2/kT} + \cdots$$

$$= n_0 \sum_i g_i e^{-E_i/kT} \qquad (4.293)$$

The total energy, referred to zero-point energy, set equal to zero by appropriate choice of scale, is

$$E = E_0 g_0 n_0 + E_0 g_0 n_0 e^{-E_1/kT} + \cdots$$

$$= n_0 \sum_i E_i g_i e^{-E_i/kT} \qquad (4.294)$$

The equilibrium energy is obtained from the last two equations and is

$$E = E - E_0 = \Delta E = n \frac{\sum_i E_i g_i e^{-E_i/kT}}{\sum_i g_i e^{-E_k/kT}}$$

$$= RT^2 \frac{d \ln Q}{dT} \qquad (4.295)$$

E is the configurational internal energy per mole at equilibrium where Q, the canonical partition function, is

$$Q = \sum_i g_i e^{-E_i/kT} \qquad (4.296)$$

Note also that

$$\frac{n_i}{n} = \frac{g_i e^{-E_i/kT}}{\sum_i g_i e^{-E_i/kT}} = \frac{g_i e^{-E_i/kT}}{Q} \qquad (4.297)$$

The continuum energy counterpart of this last equation is given by

$$Q = \int \cdots \int e^{-H/kT} dp_i dx_i \qquad (4.298)$$

H is the Hamiltonian

$$H = \sum_{i=1}^{3} p_i \dot{x}_i - L \qquad (4.299)$$

where p_i are the momenta, $\dot{x}_i$ are the time derivatives of the coordinates x_i, and L is the Lagrangian. L is defined by

$$L \equiv P_K - P_E \qquad (4.300)$$

where P_E is the potential energy and P_K is the kinetic energy $0.5M \sum \dot{x}_i^2$.

In the integral above, $i = 1, \ldots, s$ and the dp_i and dx_i are the number of coordinates and momenta needed to completely specify the position or the motion of the particle.

The relations that permit thermodynamic quantities to be calculated once Q is known for the canonical ensemble are

$$A = -kT \ln Q \qquad (4.301)$$

$$S = k \ln Q + kT\left(\frac{\partial \ln Q}{\partial T}\right)_{N,V} \qquad (4.302)$$

where N = total number of particles

$$E = kT^2\left(\frac{\partial \ln Q}{\partial T}\right)_{N,V} \qquad (4.303)$$

$$\mu = -kT\left(\frac{\ln Q}{\partial N}\right)_{V,T} \qquad (4.304)$$

Thus, the statistical thermodynamic method consists in replacing theoretically a real system in time by simultaneous ensembles that represent the possible states of the system, and then determining Q. Then, the thermodynamic quantities for the system can be calculated from the equations above.

4.11   IRREVERSIBLE THERMODYNAMICS

This topic will only be outlined in terms of concepts because, although important for geochemical cycles, it has not yet been adapted for the quantitative study of their energetics.

An important theorem from the early work of Prigogine (1955) is that the most stable nonequilibrium states are steady states. They are time-invariant but not equilibrium states. It was demonstrated that steady states generate a minimum rate of increase of entropy in the surroundings of the system. This is a result for systems that are not far from equilibrium and may apply to some geochemical systems such as that of CO_2.

The relationship between a flux J_i and a generalized force X_j is given by

$$J_i = L_{ij}X_j \qquad (4.305)$$

J_i may be the diffusion flux while X_j is the gradient of concentration. L_{ij} is a proportionality constant.

Onsager (1931) derived the important reciprocity relation $L_{ij} = L_{ji}$ which, by being general rather than depending on specific models, put irreversible thermodynamics on a firm foundation.

In the thermomolecular effect (Hemley, 1964), that is, the thermomolecular pressure difference,

$$J_1 = L_{11}X_1 + L_{12}X_2 \qquad (4.306)$$

$$J_2 = L_{21}X_1 + L_{22}X_2 \qquad (4.307)$$

J_1 and J_2 are the flows of heat and matter, whereas X_1 and X_2 are the forces that produce these flows. X_1 and X_2 are known as conjugate forces and are shown for linear processes. Aspects of nonequilibrium thermodynamics in biophysics, including membrane and diffusive phenomena, have been examined by Katchalsky and Curran (1965).

Later Glansdorff and Prigogine (1971) extended nonequilibrium thermodynamics to the concepts of stability, structure, and fluctuations.

It can be seen in Prigogine (1978) that structure can be generated from fluctuations in hydrodynamic, chemical, and other systems that are far from equilibrium. This leads to an important distinction in the laws of systems near and far from equilibrium and permits a thermodynamic study of the latter. Thus, the restriction of irreversible thermodynamics to the study of equilibria and near-

equilibria is removed. Such a result will eventually have a strong impact on geochemical systems.

EXAMPLE 4.34.   A CONCEPTUAL VIEW OF GEOCHEMICAL CYCLES AND IRREVERSIBLE THERMODYNAMICS.   Let us again consider reservoirs a and b with contents A and B of a given element and fluxes J_{AB} and J_{BC} (see Chapter 2).

At first b is empty and then it is gradually filled up by the flux $J_{AB} = k_{AB}A_0$. I assume again that A_0 is constant due to its size or to replenishment.

Eventually a steady state for which $J_{AB} = J_{BC} = k_{BC}B_0$ is reached, with B_0 representing the steady-state content of b. For a given $k_{AB}A_0$, the value of B_0 increases for decreasing values of the rate constant k_{BC}. This trend results from the fact that J_{AB} acts as an accumulator for b for a longer time before J_{BC} can catch up with it.

Irreversible thermodynamics of near-equilibrium systems tells us that, for given values of $k_{AB}A_0$ and k_{BC}, the steady state $J_{AB} = J_{BC}$ generates less entropy in the surroundings than the transient buildup of B. It is also logical to conclude that smaller steady-state fluxes generate less entropy than large ones since at rest dS/dt would be zero (RMP). Therefore, a large $k_{AB}A_0$ corresponds to a large dS/dt.

At this point I may ask the question: Is the conclusion of the previous paragraph as far as irreversible thermodynamics goes or does it also have a controlling role on the values of k_{AB} and k_{BC}? This is equivalent to asking whether irreversible thermodynamic controls forces that drive material fluxes in nature. In the example above I represented such forces by reservoir contents instead of chemical potentials for simplicity.

There is no general theorem that states that irreversible thermodynamics can act on $J_{AB} = k_{AB}A_0$ so as to minimize the rate of entropy production, that is, to minimize the steady-state flux of matter within the constraints imposed on the system. These constraints are the applied forces such as A_0, gradients of chemical potential or temperature, and so on. If the above theorem existed, irreversible thermodynamics could operate through the many processes that enter into k_{AB}.

It is doubtful that such a theorem will ever be proven because fluxes can be controlled simultaneously by equilibrium and steady states. An example of this is the saturation, in some instances, of running water with $CaCO_3$. The water content is at equilibrium with the mineral but its motion is an irreversible process.

Let us, as another aspect of this problem, consider the heat budget of the earth once again. The radiation of heat follows the Stephan–Boltzmann law

$$\frac{1}{A}\frac{dQ_{rad}}{dt} = k_{SB}T^4 \qquad (i)$$

with $k_{SB} = 1.36 \times 10^{-12}$ cal/cm$^2 \cdot$ sec $\cdot$ deg^4. At the outer part of the atmosphere the radiation on a surface perpendicular to it is 2 cal/cm$^2 \cdot$ min and is known as the solar constant.

From the standpoint of light, Wien's displacement law is

$$\lambda_{max}\, T = 2.9 \text{ mm/deg} \qquad (ii)$$

In other words, the wavelength of maximum radiation decreases when the temperature of the radiating surface increases. $T = 6000°K$ for the sun and, therefore $\lambda_{max} = 5 \times 10^{-4}$ mm $= 0.5$ µm. Most of the radiation is in the visible and ultraviolet bands.

The earth reflects 35% of the incoming radiation on top of the atmosphere and absorbs the remaining 65%. The 65% corresponds to 0.0054 cal/cm$^2 \cdot$ sec. The earth is at a thermal steady state so that the energy absorbed must be radiated into space. The average actual radiation temperature of the earth is 288°K and the radiation (back radiation) is mostly in the infrared.

The input radiation from the sun is controlled by a physical law, that of Stephan–Boltzmann, which reflects atomic processes, plus the orbit and geometry of the earth. I do not see my way clearly to a possible, more general thermodynamic principle that can control this input, that is, k_{SB}. The same appears to be true of the back radiation.

It is interesting to realize that, if k_{SB} for the back radiation had been smaller than that for the solar radiation, the temperature of the earth would have reached a steady-state value higher than 15°C. Actually, this is what we are doing to the effective k_{SB} when we burn fossil fuels.

SUMMARY

Thermodynamics deals with energy–matter interactions. It yields the direction of chemical reactions and their extent at equilibrium.

The first law of thermodynamics states that the total energy of the universe is constant. The second law postulates that the part of the total energy which is available to do work decreases (the entropy increases) in spontaneous processes. Engines operate in cycles to remain in operation. Reversible processes are idealizations for which the entropy is constant.

Some of the fundamental quantities and equations of themodynamics are as follows:

E = internal (total) energy

H = enthalpy or heat content

A = Helmholz free (available to do work) energy

F = Gibbs free (available to do non $P\,\Delta V$ work) energy

S = entropy

C = heat capacity

$\bar{X}$ = partial molal quantity

T = absolute temperature

P = pressure

V = volume

n = number of moles

W = work

Q = heat

Q_r = reaction quotient

K = equilibrium constant

μ = chemical potential

X^0 = quantity at the standard state

ΔX_f = quantity for the formation of 1 mole of a substance

$dE = \delta Q - \delta W$

$H = E + PV$

$A = E - TS$

$F = H - TS$

$\Delta F^0 = -RT \ln K$

$\Delta F = -RT \ln(K/Q_r)$

$\partial \ln K / \partial P = -\Delta \bar{v}^0 / RT$

$\partial \ln K / \partial T = \Delta H^0 / RT^2$

$\mu = \partial F / \partial n$

$\bar{v} = \partial V / \partial n$

Spontaneous reactions occur when $(dA)_{V,T} < 0$, $(dS)_{E,V} > 0$, and $(dS)_{P,T} < 0$. These equations relate thermodynamic properties of solutions to spontaneity and equilibrium. The latter occurs when the symbols $<$ and $>$ are replaced by an equality.

An important relation is $\Delta F = RT \ln K/Q_r$ where Q_r is the reaction quotient. $\Delta F < 0$ when $Q_r > K$ and the reactions as shown proceeds spontaneously to the right.

The applications of general thermodynamics to hydrothermodynamics are demonstrated.

It is shown that systems such as CO_2–CO_3^{2-} are not in general equilibrium in large water bodies although localized dissociation equilibria can occur.

The thermodynamics of multicomponent systems, the Gibbs–Dunhem equation, and partial molal quantities are examined in some detail because they correspond to many processes in natural waters.

The concepts of activities and activity coefficients, needed to study the properties and equilibria in real systems, are defined for gases, pure liquids and solids, and solutions. The chemical potential μ is related to the activity a by $\mu = \mu^0 + RT \ln a$. In aqueous solutions $a = \gamma m$ where γ is the activity coefficient and m is the molality. Standard reference states for various types of systems are defined and tabulated. These states are needed because no absolute values of μ and other related thermodynamic properties can be determined.

The concepts of ideal and regular solutions as well as Henry's law are introduced for use with dilute solutions. A generalized activity is defined and applied to the effect of gravity in deep systems such as the oceans.

Equilibrium conditions for chemical reactions, including the effects of temperature and pressure, are examined. Topics such as thermal properties, the free energy of formation, equilibrium constants, the free energy of reactions, and the reaction quotient are studied. They pertain to the direction and extent of chemical reactions.

The electromotive forces of redox reactions are related to the equilibrium constants and free energies. This topic is vital for the study of trace metals in solution.

The thermodynamics of multicomponent systems, heterogeneous equilibria, and the phase rule are reviewed since they apply to natural waters.

Triangular diagrams are examined because of their usefulness in the study of stability relations for three component systems such as $CaCO_{3(s)}$–$MgCO_{3(s)}$–$H_2O_{(l)}$. Other types of diagrams are also studied and special attention is paid to the effects of solid solution formation on stable mineral assemblages.

References for the study of seawater as a water–sea salt system are presented.

Concepts of statistical and irreversible thermodynamics are discussed because of their potential application to natural waters.

The ideas and tools presented in this chapter are illustrated by seawater processes but do apply as well to freshwaters.

REFERENCES

Berner, R. A. (1971). *Principles of Chemical Sedimentology,* McGraw-Hill, New York.

Broecker, W. S., and V. M. Oversby (1971). *Chemical Equilibria in the Earth,* McGraw-Hill, New York.

Cox, R. A., M. J. McCartney, and F. Culkin (1970). *Deep-Sea Res.* **17,** 689 (1950).

Darken, L. S. (1950). *J. Am. Chem. Soc.* **72,** 2909.

Des Coudres (1893). *Th. Ann. der Physik, u. Chemie* **49,** 284.

Doherty, B. R., and D. R. Kester (1974). *J. Mar. Res.* **32,** 285.

Eckart, C. (1940). *Phys. Rev.* **58,** 269.

Emmet, R. T., and F. J. Millero (1975). *J. Geophys. Res.* **79,** 3463.

Fofonoff, N. P. (1962). In *The Sea,* Vol. 1, M. N. Hill, Ed., Interscience, New York, p. 3.

Garrels, R. M., and C. L. Christ (1965). *Solutions, Minerals and Equilibria,* Harper & Row, New York.

Glansdorff, P., and I. Prigogine (1971). *Thermodynamic Theory of Structure, Stability, and Fluctuations,* Interscience, New York.

Glasstone, S. (1969). *Textbook of Physical Chemistry,* 2nd Ed., MacMillan, London.

Harned, H. S., and R. W. Ehlers (1932). *J. Am. Chem. Soc.* **54,** 1350.

Hemley, H. J. M. (1964). *J. Chem. Educ.* **41,** 647.

Holland, H. D. (1978). *The Chemistry of the Atmosphere and Oceans,* Wiley, New York.

Horne, R. A. (1969). *Marine Chemistry: The Structure of Water and the Chemistry of the Hydrosphere,* Interscience, New York.

Katchalsky, A., and P. F. Curran (1965). *Nonequilibrium Thermodynamics in Biophysics,* Harvard University Press, Cambridge.

Kirkwood, J. G., and I. Oppenheim (1961). *Chemical Thermodynamics,* McGraw-Hill, New York.

Kohlrausch, T., and W. Hallwachs (1894). *Ann. Physik* **53,** 14.

Krauskopf, K. B. (1967). *Introduction to Geochemistry,* McGraw-Hill, New York.

Land, L. S. (1967). *J. Sedim.* **37** 914.

Levenspiel, O., and N. de Nevers (1974). *Science* **183,** 157.

Lewis, G. N., and M. Randall (1961). *Thermodynamics,* revised by K. S. Pitzer and L. Brewer, 2nd ed., McGraw-Hill, New York.

Leyendekkers, J. V. (1976). *Thermodynamics of Seawater,* Marcel Dekker, New York.

Mangelsdorf, Jr., P. C., F. T. Manheim, and J. M. T. M. Gieskes (1970). *Am. Assoc. Petrol. Geol. Bull.* **54,** 617.

Millero, F. J. (1974). In *The Sea,* Vol. 5. E. D. Goldberg, Ed., Interscience, New York, p. 3.

Millero, F. J., G. Perron, and J. E. Desnoyers (1973). *J. Geophys. Res.* **78,** 4499.

Millero, F. J., L. D. Hansen, and E. V. Hoff (1974). *J. Mar. Res.* **31,** 21.

Morgan, J. J. (1967). *Equilibrium Concepts in Natural Water Systems,* W. Stumm and J. J. Morgan, Eds., *Am. Chem. Soc. Publ. Adv. in Chem. Ser.* **67,** Washington, p. 1.

Murkejee, P. (1966). *J. Phys. Chem.* **70** 2708.

Murray, C. N., and J. P. Riley (1971). *Deep-Sea Res.* **18,** 533.

National Bureau of Standards (1952). *Selected Values of Chemical Thermodynamic Properties,* Circular 500, National Bureau of Standards, Washington.

Onsager, L. (1931). *Phys. Rev.* **37,** 405.

Otto, J., A. Michels, and H. Wouters (1934). *Physik. Z.* **35,** 97.

Owen, B. B., and S. R. Brinkley, Jr. (1941). *Chem. Rev.* **29,** 461.

Plummer, N. L., and F. T. Mackenzie (1974). *Am. J. Sci.* **274,** 61.

Prigogine, I. (1955). *Thermodynamics of Irreversible Processes,* Interscience, New York.

Prigogine, I. (1978). *Science* **201,** 777.

Pytkowicz, R. M. (1962). *Limnol. Oceanogr.* **7,** 434.

Pytkowicz, R. M. (1963). *Limnol. Oceanogr.* **8,** 286.

Pytkowicz, R. M. (1968). In *Oceanography, Marine Biology Annual Review,* Vol. 6, H. Barnes, Ed., Allen and Unwin, London, p. 83.

Pytkowicz, R. M., and D. R. Kester (1971). In *Oceanography, Marine Biology Annual Review,* Vol. 9, H. Barnes, Ed., Allen and Unwin, London, p. 11.

Reid, R. O. (1964). *Hidrotermodinamica,* Secretaria de Marina, Buenos Aires.

Robinson, R. A. (1954). *J. Mar. Biol. Assoc. U.K.* **33,** 449.

Sage, B. H., and W. N. Lacey (1950). *Thermodynamic Properties of Lighter Paraffin Hydrocarbons and Nitrogen,* American Petroleum Institute, New York.

Siever, R. (1962). *J. Geol.* **70,** 127.

Smith, R. M., and A. E. Martell (1976). *Critical Stability Constants,* Vol. 4, Plenum Press, New York.

Stokes, R. (1979). In *Activity Coefficients in Electrolyte Solutions,* R. M. Pytkowicz, Ed., CRC Press, Boca Raton, p. 1.

Stumm, W., and J. J. Morgan (1970). *Aquatic Chemistry: An Introduction Emphasizing Chemical Equilibria in Natural Waters,* Interscience, New York.

Sverdrup, H. V., M. W. Johnson, and R. H. Fleming (1942). *The Oceans: Their Physics, Chemistry, and General Biology,* Prentice-Hall, Englewood Cliffs, N.J.

Takahashi, T. (1961). *J. Geophys. Res.* **66,** 477.

Whitfield, M. (1975). *Mar. Chem.* **3,** 197.

Wollast, R., and D. Reinhard, Derie (1977). In *the Fate of Fossil Fuel Co₂ in the Oceans,* N. R. Andersen and A. Malahoff, Eds., Plenum Press, New York, p. 479.

SUGGESTED READINGS

Daniels, F., and R. A. Alberty, *Physical Chemistry,* Wiley, New York, 1975.

Eggers, Jr., D. F., N. W. Gregory, G. D. Halsey, Jr., and B. S. Rabinovitch, *Physical Chemistry,* Wiley, New York, 1964.

Fowler, R. H., and E. A. Guggenheim, *Statistical Thermodynamics,* Cambridge University Press, Cambridge, 1939.

Kelley, K. K., *U.S. Bur. Mines. Bull.* **477,** 1950. Summary of heat capacity data versus temperature.

Kirkwood J. G., and I. Oppenheim, *Chemical Thermodynamics,* McGraw-Hill, New York, 1961.

Klotz, I. M., *Chemical Thermodynamics,* Prentice-Hall, Englewood Cliffs, N.J., 1950.

Lewis, G. N., and M. Randall, *Thermodynamics,* McGraw-Hill, New York, 1961.

McQuarrie, D. A., *Statistical Thermodynamics,* Harper & Row, New York, 1973.

Moore, W. J., *Physical Chemistry,* Prentice-Hall, Englewood Cliffs, N.J., 1972.

Plank, M., *Treatise on Thermodynamics,* Longmans Green, New York, 1927.

CHAPTER FIVE

ELECTROLYTE SOLUTIONS

5.1 AN OVERVIEW AND THE USEFULNESS OF ACTIVITY COEFFICIENTS

5.1.1 General Considerations

The purposes of this chapter are to provide the reader with a critical knowledge of electrolyte solutions and a solid foundation for research on this subject. This precludes the tidy use of a textbook approach because the presentation of the subject must be searching.

Such an approach reveals a wealth of theories, hypotheses, assumptions, results, and conclusions which do not fit in a single mold but which are highly individualistic. The development of this chapter, therefore, is not tidy but is logical. This logic will be perceived by readers as they progress through the various topics.

Those readers who only wish for a brief encounter with the subjects of this chapter may find Section 5.12 and the tables in the text to be of use. I must caution such readers about the fairly frequent blind use of methods and results which has resulted from a superficial knowledge and which has led to errors and misconceptions.

5.1.2 The Topics in This Chapter

Electrolytes, for the purposes of this chapter, are substances that separate into ions when they are dissolved in water. The resulting ion–ion and ion–water interactions influence the properties of the solvent and the solute. Examples of such properties are activity coefficients, which reflect the nonideal behavior due to such interactions, dissociation constants of weak acids and bases, solubilities of minerals, the vapor pressure, and the electrical conductance.

The major part of the chapter is aimed at understanding the above interactions as well as explaining their effects on the properties of solutions. The results are explanatory and, at times, predictive.

The topics in this chapter are:

1. General concepts such as types of and conventions for electrolytes. The conventions are needed due to the arbitrariness of standard states and the various definitions of activity coefficients which are available.

2. The Debye–Hückel theory which first explained the nonideality of dilute electrolyte solutions. Results can still be used in simple systems.

3. Modifications and extensions of the Debye–Hückel equations which allow them to be used in mixed electrolyte solutions but which are still limited to ionic strengths below 0.1.

4. Our lattice model of solutions based on flickering quasilattices which results from coulombic interactions and causes partial ordering in the sense of time-average thermodynamic properties.

5. The hydration theory which accounts for changes in concentrations due to the removal of the water of hydration.

6. Activity coefficients of single ions. They cannot be measured directly and, therefore, their accuracy is unknown. Still, these coefficients are useful at times. The pioneering model of Garrels and Thompson (1962) is examined exhaustively because it is convenient under certain circumstances, is still in use by some authors, and it requires a large number of assumptions.

The experimental approaches used in my laboratory, which reduce the number of required assumptions, as well as the methods of a number of workers are as follows.

7. Thermodynamics of electrolyte solutions and the determination of excess functions.

8. Theories of mean activity coefficients for electrolytes. These theories are generally developed for electrolytes such as NaCl, $CaCO_3$, and so on, but can often be adapted to single ions. Other approaches are geared directly to single ions. It is better, when possible, to deal with electrolytes because single-ion activity coefficients require a non-thermodynamic assumption and cannot be tested experimentally. Still, there are times when the single-ion approach is the more convenient one.

9. Concepts from statistical thermodynamics.

5.1.3 Importance of Activity Coefficients

In this chapter the thermodynamic properties of electrolyte solutions are examined, with emphasis on their activity coefficients. Electrolytes are substances that dissociate into ions (see Chapters 1, 3, and 4). They occur in lakes, rivers, oceans, soil waters, body fluids, and so on, and affect the properties of natural as well as laboratory waters. Some conceptual cases that illustrate the importance of activity coefficients in such solutions are presented next. These examples require some knowledge of activities and can be bypassed for the time being and returned to later by newcomers to this field.

EXAMPLE 5.1. SCHEMES TO DETERMINE THE DEGREE OF SATURATION ILLUSTRATED FOR A SYMMETRIC ELECTROLYTE: FIRST CASE

1. *Thermodynamic.* K_{sO}, the thermodynamic solubility product $a_{Ca}a_A$, is known. Measure $(m_C)_T$ and $(m_A)_T$ in natural waters and, from these quantities plus $(\gamma_{\pm CA})_T$ calculate the ion activity product IAP. The degree of saturation is IAP/K_{sO}.

This scheme requires a knowledge of the measured K_{sO} at a given (T, P) and of $(\gamma_{\pm CA})_T$ measured or calculated theoretically, which is not always an accurate procedure, for a given $(T, S‰, P)$. The use of $S‰$ to represent the ionic strength and the composition is only applicable to ionic media. The concept of ionic medium will be studied later in this chapter.

2. *Stoichiometric.* K_s, the stoichiometric solubility product, is known. Measure the concentrations of the two ions and calculate IP, the concentration ion product. The degree of saturation is IP/K_s.

If $K_{s,F}$ was tabulated then the free concentrations must be used to determine IP_F.

Note that the symbols F for free ions and T for total (free ions plus ion-paired ones) will not be confused with temperature and free energy within the context of the subject matter.

The quantities K_{sO} and K_s are solubility products obtained when the solution is at equilibrium (saturation) with the mineral $CaCO_3$. We shall see in this chapter that there are two classes of γ's and m's; the total ones (subscript T) such as $[m_{Ca}]_T$ and the free quantities [e.g., $(m_{Ca})_F = (m_{Ca})_T - (m_{CaCl}) - m_{CaSO_4} - m_{HCO_3} - \cdots$]. In this case m_{CaCl}, for example, represents the molality of $CaCl^+$ ion pairs. The term m_{CaSO_4} can be used to represent $CaSO_4^0$ ion pairs but can also indicate the stoichiometric molality of $CaSO_4$. The distinction between these two meanings will be made in the text or by the use of $[CaSO_4^0]$ and $[CaSO_4]_T$ instead of the common symbol m_{CaSO_4}. The symbol $\gamma_\pm$ represents mean activity coefficients which, for $CaCO_3$, result in $\gamma_{\pm CaCO_3} = (\gamma_{Ca}\gamma_{CO_3})^{0.5}$.

5.1.4 Notations

Alternative notations are used in this chapter depending on which provide greater clarity. Thus a_i is a clearer symbol for activities than $\{i\}$ but the latter is the better if the charge of i is shown. Similarly, m_i, c_i, and $[i]$ are used on the basis of convenience.

The notations used in solubility work (Example 5.1) are presented next in a more systematic way.

The relationships among the quantities used in Example 5.1 are

$$K_{sO} = (\gamma_{Ca})_{T,e}(\gamma_{CO_3})_{T,e}(m_{CO_3})_{T,e}$$
$$= (\gamma_{\pm CaCO_3})^2_{T,e}(m_{Ca})_{T,e}(m_{CO_3})_{T,e} \quad (5.1)$$

$$K_s = (m_{Ca})_{T,e}(m_{CO_3})_{T,e} \quad (5.2)$$

$$IAP = (\gamma_{\pm CaCO_3})^2(m_{Ca})_T(m_{CO_3})T \quad (5.3)$$

$$IP = (m_{Ca})_T(m_{CO_3})_T \quad (5.4)$$

$$\Omega_{sO} = \frac{IAP}{K_{sO}} \quad (5.5)$$

$$\Omega_s = \frac{IP}{K_s} \quad (5.6)$$

TABLE 5.13 Molar mean activity coefficient of NaCl at 25°C calculated from the Debye–Hückel equation with a constant dielectric constant equal to that of pure water (78.30) and with a linearly decreasing dielectric constant (see text).

c (moles/L)	D_e	A_D	B_D	$-\log \gamma_\pm{}^a$	$-\log \gamma_\pm^{(0)b}$	$\gamma_\pm$ (molar)	$\gamma_\pm^{(0)}$ (molar)	$\gamma_\pm/\gamma_\pm^{(0)}$
0	78.30	0.5116	0.3292	0	0	1.000	1.000	1.000
0.01	78.19	0.5127	0.3294	0.0452	0.0451	0.901	0.904	0.997
0.1	77.20	0.5525	0.3315	0.1155	0.1133	0.766	0.770	0.995
1.0	67.30	0.6420	0.3551	0.2612	0.2175	0.548	0.606	0.904
2.0	56.30	0.8391	0.3882	0.3646	0.2484	0.432	0.564	0.766
3.0								

$^a\gamma_\pm$ in the case of a decreasing dielectric constant.
$^b\gamma_\pm^{(0)}$ for $D_{e,0} = 78.30$.

We see from Table 5.13 that a decrease in dielectric constant leads to values of $\gamma_\pm$ which become progressively smaller relative to $\gamma_\pm^{(0)}$ which in itself decreases. Thus, a decrease in dielectric constant leads to a larger salting-in, that is in the decrease of γ's with the increasing ionic strength.

Experiments show that the activity coefficients decrease with increasing concentration up to a certain value and then start to increase. This latter behavior is referred to as salting-out. The reason for the term salting-out is easy to see. Let us assume that we have a sparingly soluble salt CA. Its solubility product is

$$K_{sO} = \{C^+\}\{A^-\} = \gamma_C\gamma_A[C^+][A^-] \quad (5.178)$$

with $\gamma_C\gamma_A = (\gamma_{\pm CA})^{0.5}$. Large values of $\gamma_{\pm CA}$ at equilibrium imply low values of the concentration ion product $[C^+][A^-]$ in the saturated solution. Thus, an increase in $\gamma_\pm$ at high concentrations implies smaller concentration solubilities (salting-out) than if $\gamma_\pm$ had continued to decrease, provided that K_{sO} was the same in the two cases. Of course, this is only a conceptual exercise to illustrate the meaning of salting-out since experience shows that K_{sO} differs from salt to salt. An example of salting-out is shown in Table 5.14.

The transition occurs at lower values of I for higher valence electrolytes. The decrease in the macroscopic dielectric constant does not help us explain salting-out, since it leads to a decrease in $\gamma_\pm$. This effect can also be obtained by decreasing $D_{e,0}$ in Equation (5.153).

The local dielectric constant is that due to the orientation (rotational polarization) of water molecules in the fields of the ions. This dielectric

constant also decreases with increasing concentration since additional ions have fewer water molecules to orient and because of orientation saturation. The strong interaction between ions and water at high concentrations can be seen from the small number of interionic waters (see Table 5.15).

The next equation does not solve the problem of the salting-in–salting-out transition. Still, it is useful at low to moderate concentrations and serves as a transition to the more sophisticated approaches that will be presented later.

5.4.10 The Equation of Hückel

Hückel (1925) proposed the equation

$$\log \gamma_{\pm(H)} = -\frac{|z_C z_A| A_D I^{0.5}}{1 + B_D a_D I^{0.5}} + C_H I \quad (5.179)$$

where the term $C_H I$ was introduced presumably to take care of changes in dielectric constant. It does

TABLE 5.14 Salting-in and salting-out for KCl at 25°C (Robinson and Stokes, 1968)

m	$\gamma_\pm$	m	$\gamma_\pm$	
0.1	0.770	1.2	0.593	
0.2	0.718	1.4	0.586	
0.3	0.688	1.6	0.580	
0.4	0.666	1.8	0.576	Salting-in
0.5	0.649	2.0	0.573	
0.6	0.637	2.5	0.569	
0.7	0.626	3.0	0.569	
0.8	0.618	3.5	0.572	Salting-out
0.9	0.610	4.0	0.577	
1.0	0.604	4.5	0.583	

TABLE 5.15 Interionic distance versus the molality for NaCl at 25°C. The molality is represented by m, ρ is the density, r is the distance, and h_r is the number of water molecules between the two ions.

m	ρ	r (Å)	h_r
1.0	1.037	9.4633	2.429
2.0	1.075	7.5556	1.738
3.0	1.112	6.6383	1.405
4.0	1.148	6.0649	1.197
5.0	1.184	5.6591	1.050

not accomplish this. In actual practice it is used empirically, with C_H as a second curve-fitting parameter in addition to a_D and is useful up to $I = 0.1$. The problem of interpreting a_D in mixed electrolyte solutions was carried over from the Debye–Hückel equation to that of Hückel.

The term $C_H I$ does not represent the effect of changes in D_e. $C_H I$ increases with increasing concentration and increases $\gamma_{\pm}$. As an example, if the Debye–Hückel term is -0.5, it yields $\gamma = 0.316$ but in conjunction with $C_H I = 0.1$ the result is $\gamma = 0.398$. It was shown above, however, that the effect of an increase in concentration is to decrease D_e and that the effect of this decrease would lower $\gamma_{\pm}$. Thus, $C_H I$ does not represent the effect of the dielectric constant. Brown and MacInnes (1935) and Shedlovsky and MacInnes (1936) showed that the Hückel equation can be made to fit the activity coefficients of NaCl and HCl quite well up to $c = 0.1$ mole/L when a_D and C_H were set as 4.00 and 0.047 for NaCl and 4.65 and 0.105 for HCl. The term a_D is quite different from that used in conjunction with the Debye–Hückel equation alone and adds further to the uncertainty in the meaning of this distance or radius.

5.4.11 Effects To Be Considered in Activity Coefficient Studies

Let us examine at this point, at which the reader has acquired some relevant background, factors that can affect γ's so that the reader can observe whether they are or are not taken into consideration in the various available models. Actually, most of the following processes are not considered explicitly in equations for activity coefficients. This means that, in some cases, the form of the equations is such that they cannot be fitted to data over wide ranges of concentrations. An example of this is the Debye–Hückel equation for $\gamma_{\pm}$ since it can-

not be used in the salting-out region. By fitting I mean that experimental values of $\gamma_{\pm}$ are used in conjunction with quasitheoretical equations to obtain their parameters. An example of this is the earlier determination of a_D. Other equations can be made to fit experimental γ's over wider ranges of concentrations as in the cases of the hydration, specific interaction, and ion-pair relations.

The reader may wonder what is the point of having equation parameters obtained by force fitting to data if all they can do is regurgitate the data to which they were fitted. There are two points to the exercise. First, it helps one to better understand the behavior of electrolyte solutions within the framework of individual theories even though it does not confirm such theories. Second, as will be shown later in this chapter, the determination of the equations in simpler electrolyte solutions permits one to calculate activity coefficients and related properties in multicomponent natural waters.

Let us examine some processes that are not always presented in equations for activity coefficients. The absence of terms for these processes plus the limitations of the Debye–Hückel theory mentioned earlier limit the theoretical significance of equations for activity coefficients.

Hydration of the Ions. Ions have a primary sheath of water molecules around them. This sheath, and perhaps some molecules in the next layers, are ordered toward the ion. It is thought that next to this inner layer there is a random region in which there occurs a transition from the radial ordering of hydration to the hydrogen-bonded one of the bulk water and in which hydrogen bonding forms one or more types of structures. Hydration has an effect on the chemical potential and, therefore, on activity coefficients as we shall see later (Robinson and Stokes, 1968; Pytkowicz and Johnson, 1979). Hydration provides the energy for the dissolution of a crystal and is, therefore, a spontaneous process. It yields a larger negative change in the dissolution of a crystal while a conceptual evaporation of the crystal yields a positive energy change.

All processes that occur in the formation of a solution affect its free energy. In the case of activity coefficients, however, we start from hydrated ions and then calculate ion–ion interactions. Thus, hydration only appears in terms for its effects on the concentrations of free waters and ions. Thus, the free energy of the actual hydration is absorbed into the standard and the reference states.

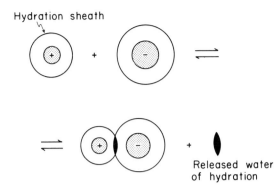

FIGURE 5.9 Gurney cosphere overlap.

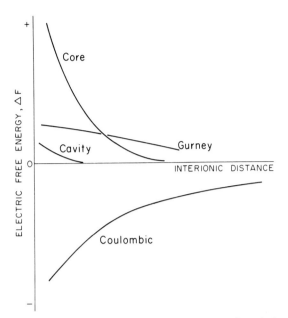

FIGURE 5.10 Energy due to various processes in solution, based on Ramanathan and Friedman (1971).

Ion Pairing. The formation of ion pairs changes the distribution of charges in the solution as in the case of the ions Na^+ and Cl^- which form the electrically neutral dipole $NaCl^0$.

The formation of ion pairs decreases the field intensity on the free ions, since the ion pairs are dipoles. Such entities exert smaller forces than those produced by ions. Thus, as the total electric forces on the free ions are lower than if ion pairs did not exist, the deviations from ideality are smaller. Thus, the values of $(\gamma_\pm)_F$ are larger above than if ion pairs were not formed. This field effect is part of the general decrease in $(\gamma_\pm)_F$ with increasing I.

I shall show later that $(\gamma_\pm)_F > (\gamma_\pm)_T$, where T indicates free ions plus ion pairs, and the inequality is due to decreases in the concentrations of free ions because of the formation of ion pairs.

Gurney Cosphere Overlap (Gurney, 1953). This occurs when the hydration spheres of two ions overlap to some extent during ionic association as is shown in Figure 5.9. According to Ramanathan and Friedman (1971), the electrical energy (free energy) for some of the effects considered here follow curves of the types shown in Figure 5.10.

It can be seen, as is to be expected, that the Gurney cosphere overlap increases with decreasing distance between ions. Furthermore, the contribution to the free energy is positive because water of hydration, which stabilizes the ions in solution and, hence, contributes a negative free energy, is lost (see Figure 5.9).

Cavity Effects. Polarized charge is developed on the surface of a dielectric sphere if it is placed in an electric field. The same is true of the surface of a dielectric medium surrounding a cavity, as is shown in Figure 5.11. These cavities, with their induced

dipole moments, are formed and work is done when ions interact with the cavities of other ions. These cavities are actually occupied by ions but they are superimposed to a dielectric vacuum which was formed when water molecules were pushed away by the ions.

Core Repulsion. This is a repulsion term of quantum mechanical origin for which the potential is of the form (Ramanathan and Friedman, 1971)

$$\phi_{ij} = B_{ij}\left(\frac{r_i^* \rightarrow r_j^*}{r}\right)^n \tag{5.180}$$

r_i^* and r_j^* are the crystal radii for the two ions, r is the interionic distance, and n is the Pauling (1948) repulsion exponent. B_{ij} is a constant which includes Madelung's term (see the lattice theory section of this chapter) and, therefore, averages ϕ over all (i, j) couples.

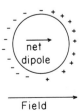

FIGURE 5.11 Cavity in a dielectric medium.

TABLE 5.16 Comparison of $\gamma_\pm$ from Davies' equation with experimental values for NaCl.

$I = m$	$\gamma_\pm \gamma_{NaCl(exp)}$	$\gamma_{\pm(D)}$
0.001	0.9649	0.9657
0.01	0.9024	0.9037
0.05	0.8205	0.8241
0.1	0.7813	0.7854
0.2	0.735	0.7516
0.5	0.632	0.7379
1.0	0.657	0.7943

Dielectric Constant Change with m. This effect is not usually taken into consideration except for concentrated solutions.

5.4.12 Guggenheim Equation

Guggenheim (1935) proposed the use of

$$\log \gamma_{\pm(GG)} = -\frac{A_D|z_C z_A|I^{0.5}}{1 + I^{0.5}} + C_{(GG)}I \quad (5.181)$$

which is an extension of the equation of Guntelberg and avoids the need to explain a_D by fixing it as $1/B_D$. The fit between γ's obtained experimentally, by means of the Guntelberg equation, and by Equation (5.181) is shown in Table 5.16. It can be seen that considerable improvement is obtained by means of the term $C_G I$. Equation (5.181) is useful up to nearly $I = 0.5$ for NaCl at least. It is important to realize that the fit of an equation to γ_T's for strong electrolytes gives us an idea of its validity for weak electrolytes and, therefore, for equilibrium calculations.

5.4.13 Davies Equation: Free and Total Activity Coefficients

Davies (1938) proposed an empirical equation which he later revised to Davies (1962)

$$\log_{\pm(D)} = -0.50|z_C z_A| \left(\frac{I^{0.5}}{1 + I^{0.5}} - 0.30I \right)$$

$$(5.182)$$

The revision consisted in the replacement of a coefficient 0.20 by 0.30. It can be seen from Table 5.16 that the agreement between calculated and the measured γ's becomes poor at $I > 0.1$. In this table I accepted provisionally the assumption of Davies

to the effect that Cl^- is not associated, even though we now know that this is not the case (Johnson and Pytkowicz, 1979b).

Davies obtained the coefficient 0.30 by trying different values until calculations based on Equation (5.182) agreed with experimental data for a number of electrolytes. Furthermore, he applied a correction to the calculated values if specific electrolytes were known to be associated.

The correcting procedure was as follows. Let α be the degree of dissociation with the molecule or ion pair splitting into ν ions ($\nu = 2$ for ion pairs). The symbol α represents the fraction of the molecules or ion pairs present in solution which are dissociated at any one time. If there can be a maximum of n molecules or pairs, then there will be $n(1 - \alpha + \nu\alpha)$ particles of which $n(1 - \alpha)$ are not dissociated and $\nu\alpha n$ are present as individual ions.

The relations of the previous paragraph lead to

$$K_{d'}^* \text{ or } K_d' = \frac{n(1 - \alpha)}{(n\alpha)^2} = \frac{m(1 - \alpha)}{(m\alpha)^2} = \frac{(1 - \alpha)}{m\alpha^2}$$

$$(5.183)$$

where K^* and K' apply respectively to ion pairs and weak electrolytes (held by covalent bonds). The thermodynamic counterparts are

$$K^* \text{ or } K \frac{(1 - \alpha)}{m\alpha^2} \frac{1}{\gamma_{\pm(D)F}^2} \quad (5.184)$$

$(\gamma_\pm)_F^2$ is the mean free activity coefficient. Note that Davies uses dissociation constants, whereas I present association constants so that the two values of K^* are inverted relative to one another.

If K^* is known, α can be calculated from Equation (5.184). Then $\gamma_{\pm(D)}$ can be obtained from the relation

$$\gamma_{\pm(D)} = \gamma_{\pm(exp)}\alpha \quad (5.185)$$

In principle $\gamma_{\pm(D)} = (\gamma_\pm)_F$ if there is ion pairing and $\gamma_{\pm(exp)} = (\gamma_\pm)_F = (\gamma_\pm)_T$ if the electrolyte is completely dissociated. Therefore, $\gamma_{\pm(D)}/\alpha$ should be used to equate $\gamma_{\pm(D)}$ to $(\gamma_\pm)_T$ or to $\gamma_{\pm(exp)}$. Some examples from Davies (1962) are presented in Table 5.17.

Equation (5.185) results from the fact that the activity a_i of a given species is given by

$$a_i = (\gamma_i)_F(m_i)_F = (\gamma_i)_T(m_i)_T \quad (5.186)$$

TABLE 5.17 Results from experimental data (column 2), the data corrected for association (column 5), and the deviation Δ from the Davies equation [selected from Davies (1962)] at $m = 0.1$.

Electrolyte	$(\gamma_\pm)_{exp} = (\gamma_\pm)_T$	pK	$1 - \alpha$	$(\gamma_\pm)_F$	Δ
$MgCl_2$	0.528			0.528	-0.017
$MgAc_2$	0.450	0.78	0.240	0.537	-0.014
$CaCl_2$	0.518			0.518	-0.027
$Ca(NO_3)_2$	0.488	0.28	0.098	0.522	-0.024
$SrCl_2$	0.515			0.515	-0.030
$Sr(NO_3)_2$	0.478	0.82	0.256	0.578	$+0.036$
$Ba(NO_3)_2$	0.428	0.92	0.295	0.549	-0.018
$Ba(Ac)_2$	0.450	0.41	0.126	0.492	-0.055
$CuCl_2$	0.510	0.40	0.124	0.557	$+0.010$
Na_2SO_4	0.452	0.70	0.106	0.527	-0.022
K_2SO_4	0.436	0.90	0.143	0.542	-0.010
K_2CrO_4	0.456	0.60	0.089	0.519	-0.029

(Pytkowicz et al., 1966) and because $\alpha = (m_i)_F/(m_i)_T$ by definition. Therefore, one obtains the relation $\alpha = (\gamma_i)_T/(\gamma_i)_F$ which in turn yields $\alpha(\gamma_i)_F = (\gamma_i)_T$.

Equation (5.184) is an important result because, when $(\gamma_\pm)_T$ is measured and $m = m_T$ is known, K or K^* can be determined from α. This can be done, for example, from conductance measurements. If, on the other hand, K or K^* is tabulated, $(\gamma_\pm)_T$ is measured, and m is known, then α can be obtained.

It is important to realize, as I mentioned earlier, that Davies tabulated values of $(\gamma_\pm)_F = (\gamma_i)_T/\alpha$ and used them to obtain his $0.30I$ term where association was known to occur. In the absence of association he used $(\gamma_\pm)_T = (\gamma_\pm)_{exp} = (\gamma_\pm)_F$. He was not aware, however, of the Cl^- association with alkaline and alkaline–earth cations. Thus, Davies found that the factor 0.30 best fitted what he thought to be $(\gamma_\pm)_F = (\gamma_\pm)_T$ for chloride salts.

The equations of Debye–Hückel, Guntelberg, Guggenheim, and Culberson, on the other hand, pertain to $(\gamma_\pm)_T$ as they are fitted to yield $\gamma_{\pm(exp)}$ instead of $\gamma_{\pm(exp)}\alpha$ for all salts. The application of the Davies equation would probably improve if C_D was different for different classes of compounds. As it now stands, it becomes a poorer predictor of $(\gamma_\pm)_F$ as the concentration increases. This can be seen by comparing Table 5.17 with the values of $(\gamma_\pm)_T = 0.681$ and $(\gamma_\pm)_F = 0.805$ obtained by Johnson and Pytkowicz (1978) for 0.60 m NaCl.

In general, simple equations such as the ones presented so far provide a better fit for 1-1 electrolytes than for electrolytes of types 2-1, 2-2, 3-1, and so on, because the onset of salting-out, the increase in γ with m, occurs at significantly lower molalities

for higher valence electrolytes (see Figure 5.7). The simple equations are not powerful enough to represent γ's well beyond the transition from salting-in to salting-out and, often, even when this transition is approached.

The Guggenheim equation appears to be a better empirical equation than that of Davies, with a judicious choice of C_G, as is shown in the last two tables. Furthermore, it can be refined because it contains the first two terms of the Culberson equation, which provides a fine fit for up to $m = 1.0$ for NaCl, KCl, $CaCl_2$, and $MgCl_2$ (see Table 5.6). Thus, as the concentration increases, the last two terms of the Culberson equation, presented earlier in this chapter, become increasingly important. The main advantage of the Guggenheim and the Culberson equations over the Davies equation arises because they contain adjustable parameters while in Davies C_D is fixed at 0.30. This last equation is best used when there is no a priori $\gamma_{\pm(exp)}$ data to be fit for the calculation of coefficients of equations. This occurs mainly in mixed electrolyte solutions. In the case of mixed electrolytes, however, there are more powerful, although more laborious, methods than that of the Davies equation for the determination of γ's. We shall examine this case later on in this chapter.

5.5 HYDRATION

5.5.1 Notes on Hydration

In this section some aspects of hydration, namely, its effects on standard states, activity coefficients, and salting-out are examined. I will not attempt to cover the structural aspects of hydration since they transcend the scope of this book. The discussion that follows is akin to that based on Figure 4.4 but is expanded and has a different emphasis. In essence, I show that the integration of $\gamma_{\pm(s)}$, when s represents an electrolyte, yields the free energy for s as well as for water. This is followed by an application of the concept to an equation for $\gamma_{\pm(s)}$ which is used in practice.

Relationships among some of the subscripts used in this work are as follows. Let X be a general property that can represent ΔF, F, μ, γ, and so on.

1. X_w refers to water.
2. X_s indicates an electrolyte.
3. X_{el} represents the electrical (coulombic) interaction of ions plus the effect of hydration on the concentrations of ions and the effect of water on X. This is a general definition independent of any theory.
4. X_{th} implies a theoretical representation of X_{el} as it requires assumptions for the derivation of the microscopic equation.
5. X_{exp} indicates the experimental value of X.

The last three quantities are exchangeable as follows:

1. $X_{el} \equiv X_{th}$ when it is assumed that X_{th} truly represents ion–ion interactions and changes in concentrations due to hydration.
2. $X_{el} \equiv X_{exp}$ and $X_{th} \equiv X_{exp}$ when X_{el} and X_{th} are obtained by equating them to the experimental data. This is done for X_{th} by the use of one or more adjustable parameters. If the experimental data does always result from the effects present in X_{el} and nothing more, then $\gamma_{el} \equiv \gamma_{exp}$ is always valid.

In the specific case of activity coefficients, $\gamma_T = \gamma_{exp}$ since measurements yield total quantities while γ_F has to be calculated. Exceptions, which are pointed out, occur when γ_T is calculated indirectly from γ_F. Subscripts such as DH, H, G, GG, D correspond to γ_{th} but become equal to γ_{exp} when the curve fitting is done to determine equation parameters.

First, I shall look at how the activities of water and salt are related and are taken into account in solution studies (Pytkowicz and Johnson, 1979). The total free energy F when $n_w = 55.51$ moles of

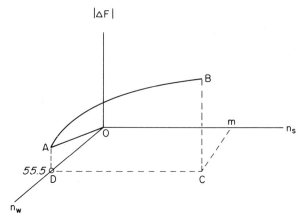

FIGURE 5.12 Free energy of formation, F, of an m molal solution. The symbols n_w and n_{sl} represent the number of moles of water and salt.

water (1 kg) are mixed with n_s moles of salt is given by

$$F = 55.51\mu_w + m\mu_{sl} = F_w + F_{sl} \quad (5.187)$$

where μ_w and μ_{sl} are the chemical potentials of the water and the electrolyte. F_w and F_s are their free energies. The quantity F corresponds to the length of the line segment $\overline{BC}$ (Figure 5.12). Physically, Equation (5.187) corresponds to the addition of small increments of a premixed solution of molality m to a container until 1 kg of water and m moles of salt are present. The path is represented by the straight line $\overline{OB}$.

The preparation of most solutions is best simulated, however, by $\overline{OA}$, the addition of 1 kg of water to the container, followed by the arc $\overbrace{AB}$ when m moles of salt are subsequently added to the water.

The length $\overline{AD}$ is simply

$$F_w^0 = 55.51\mu_w^0 \quad (5.188)$$

F_w^0 is the free energy of 55.51 moles of water at its standard state, the pure solvent.

The path AB corresponds to an increase in free energy $\Delta F_{s,w}$ given by

$$\Delta F_{s,w} = \int_0^m \left(\frac{\partial F}{\partial n_{sl}}\right)_{P,T,n_w} dn_s + 55.51RT \ln a_w$$

$$= \int_0^m dF_w + \int_0^m dF_s \quad (5.189)$$

at constant P, T, and n_w. The subscript sl, w refers to changes in the free energies of the water and the salt. If the salt behaved ideally then the segment of curve AB would be replaced by a segment of a line tangent to AB at A. A change dF_w in the free energy of water appears even though $n_w = 55.51$ is constant. This happens because when salt is added the composition changes, and, therefore, the intensive quantities μ_w and μ_{sl}, which are not independent of each other, also change.

$dF_w = 55.51RT\, d \ln a_w = 0$ when $m = 0$ because $a_w = 1$ in this case. For $m > 0$,

$$\int_0^m dF_w = 55.51RT \ln a_w < 0 \qquad (5.190)$$

which is a negative term because $a_w < 1$ for $m > 0$. Equation (5.189) also yields

$$\int_0^m dF_s = F_{sl} = m\mu_{sl} = m(\mu_{sl}^0 + 2RT \ln \gamma_{\pm(sl)}$$
$$+ 2RT \ln m) \qquad (5.191)$$

Equation (5.190) corresponds to the change in the fugacity ratio P_w^*/P_w^{*0} of the water in the solution of molality m to that in pure water (or water at an infinite dilution of salt). We have $P_w^* < P_w^{*0}$ because of the actual binding by the salt of a finite number of water molecules. These molecules have a lower fugacity than that of bulk water when $m > 0$ and, furthermore, X_w is decreased.

From Equations (5.188), (5.190), and (5.191),

$$F = 55.51(\mu_w^{0(X)} + RT \ln a_w) + m(\mu_s^0 + 2RT \ln \gamma_{\pm(sl)}$$
$$+ 2RT \ln m) = 55.51\mu_w + m\mu_{sl} \qquad (5.192)$$

μ_w is independent of the concentration units used for water by definition, and the standard state μ_w^0 absorbs the differences in concentration units which may appear in a_w. We can, therefore, write

$$a_w = f_w X_w \qquad (5.193)$$

with the understanding that $\mu_w^{0(X)}$ corresponds to mole fraction units ($\mu_w = \mu_w^0$ when $X_w = 1$). The symbol X_w represents the mole fraction of water. Introducing Equation (5.193) into (5.192) and regrouping,

$$F = [55.51(\mu_w^{0(X)} + RT \ln X_w) + m(\mu_{sl}^0 + 2RT \ln m)]$$
$$+ [55.51RT \ln f_w + 2RTm \ln \gamma_{\pm(sl)}] \qquad (5.194)$$

The first pair of brackets contains the ideal free energy $F^{(i)}$, that is, the free energy of a m molal solution, in which the ions do not interact with each other and do not affect the activity of water. The second set of brackets contains the electrical work.

$$\Delta F_{el} = 55.51RT \ln f_w + 2RTm \ln \gamma_{\pm(sl)} \qquad (5.195)$$

with

$$\Delta F_{el(w)} = 55.51\,\Delta\mu_{el(w)} = 55.51RT \ln f_w \qquad (5.196)$$

and

$$\Delta F_{el(sl)} = m\,\Delta\mu_{el(sl)} = 2RTm \ln \gamma_{\pm(sl)} \qquad (5.197)$$

Care should be exercised in interpreting Equations (5.196) and (5.197) because the changes in the nonideal free energies of water, $\Delta F_{el(w)}$, and of salt, $\Delta F_{el(s)}$, are not independent of one another since f_w as well as $\gamma_{\pm(sl)}$ are affected by electrolyte–water interactions. The value of $\gamma_{\pm exp}$ contains both ion–ion and ion–water interaction. ΔF_{el} is also known as the excess free energy.

From the above equations,

$$\Delta F_{el} = \Delta F_{el(sl)} + 55.51RT \ln f_w \qquad (5.198)$$

From Equation (5.183),

$$\Delta\mu_{el(sl)} = \frac{\partial F_{el}}{\partial m} = 2RT \ln \gamma_{\pm sl} \qquad (5.199)$$

Thus,

$$\Delta F_{el} = 2RT \int_0^m \ln \gamma_{\pm sl}\, dm = \Delta F_{exp} \qquad (5.200)$$

where the subscript (exp) refers to experimental values.

Thus, we see that the integral in Equation (5.200) contains effects of hydration as well as of ion–ion interactions [see Equation (5.198)] if $\gamma_{\pm sl}$ is the experimental value.

Note that the standard and the reference states for $\Delta F_{el} = E_{el}$ were chosen earlier with ions hydrated a priori. Thus, ΔF_{el} does not include the free energy to break water molecules away from the hydrogen-bonded bulk water and to attach them to ions. Therefore, the water part of ΔF_{el} only includes effects of ions on the solvent such as changes in D_e and in the concentration of water.

The interpretation presented above shows that the integration of $\gamma_{\pm(sl)}$ yields a ΔF_{el} which depends not only on the free energy change of the electrolyte but also on that of the water. This leads us to the practical application that follows.

5.5.2 Hydration Theory

Several forms of the equation for hydration will be derived. Robinson and Stokes (1968) presented an extension of the Debye–Hückel theory in which concentration effects of hydration were derived explicitly. A different but very clear derivation can be found in Bockris and Reddy (1970). They make the calculation in two steps; the effect of the change in the concentration of water and the change in the concentration of ions that result from hydration.

The electrical free energy F_{el} of a given amount of solution, with n_w moles of water and 1 mole of solute that dissociates into v_C moles of cations and v_A moles of anions, can be expressed in two equivalent ways:

$$F_{el} = n_w\mu_w + v_C\mu_C + v_A\mu_A \quad (5.201)$$

and

$$F_{el} = (n_w - h)\mu_w + v_C\mu'_C + v_A\mu'_A \quad (5.202)$$

Since the chemical potentials are taken to be the electrical potentials. h is the hydration number, that is, the number of moles of water which solvate each mole of salt and μ'_i is the chemical potential of the hydrated form of ion i.

Equations (5.201) and (5.202) are identical because F_{el} is independent of where the water of hydration is assigned. Thus, setting these equations equal to each other and developing the chemical potentials by means of

$$\mu_i = \mu_i^0 + RT \ln \gamma_i m_i \quad (5.203)$$

one obtains the expression

$$\left(\frac{h\mu_w^0}{RT} + h \ln a_w\right) + \frac{v_C}{RT}(\mu_C^0 - \mu_C^{0'})$$

$$+ \frac{v_A}{RT}(\mu_A^0 - \mu_A^{0'}) + v_C \ln \frac{m_C}{m'_C} + v_A \ln \frac{m_A}{m'_A}$$

$$= v_C \ln \gamma'_C - v_C \ln \gamma_C + v_A \ln \gamma'_A - v_A \ln \gamma_A \quad (5.204)$$

Since

$$m = \frac{1000}{18}(n_w + v) \quad \text{and} \quad m' = \frac{1000}{18}(n_w - h + v) \quad (5.205)$$

$$\frac{m}{m'} = \frac{m_C}{m'_C} = \frac{m_A}{m'_A} = \frac{n_w - h}{n_w} \quad (5.206)$$

since there are n_w moles of water and v moles of ions per mole of salt in the solution.

Introducing Equation (5.206) into (5.204)

$$\frac{h\mu_w^0}{RT} + h \ln a_w + \frac{v_C}{RT}(\mu_C^0 - \mu_C^{0'})$$

$$+ \frac{v_A}{RT}(\mu_A^0 - \mu_A^{0'}) + v \ln \frac{n_w - h}{n_w}$$

$$= v_C \ln \gamma'_C - v_C \ln \gamma_C \, v_A \ln \gamma'_A - v_Z \ln \gamma_A \quad (5.207)$$

with $v = v_C + v_A$. When $n_w \to \infty$ all activity coefficients and a_w tend to unity, their logarithms tend to zero, and

$$v \ln \frac{n_w - h}{n_w} = v \ln\left(1 - \frac{h}{n_w}\right) \to 0 \quad (5.208)$$

Therefore, all the terms that include standard chemical potentials must add up to zero because these terms do not depend on n_w. Equation (5.205), therefore, is reduced to

$$v_C \ln \gamma'_C - v_C \ln \gamma_C + v_A \ln \gamma'_A - v_A \ln \gamma_A$$

$$= v \ln \frac{n_w - h}{n_w} + h \ln a_w \quad (5.209)$$

Then, remembering that

$$\gamma_\pm = (\gamma_C^{v_C} \gamma_A^{v_A})^{1/2} \quad (5.210)$$

one obtains

$$\ln \gamma'_\pm = \ln \gamma_\pm + \frac{h}{v} \ln a_w + \ln \frac{n_w - h}{n_w}$$

$$= \ln \gamma_\pm + \frac{h}{v} \ln a_w + \ln(1 - 0.018hm) \quad (5.211)$$

as $n_w = 1000/18m$ leads to $(n_w - h)/n_w = 1 - 0.018hm$.

Equation (5.211) becomes, in terms of n_s moles of electrolyte per n_w moles of water,

$$\ln \gamma'_{\pm} = \ln \gamma_{\pm} + \frac{hn_{sl}}{\nu} + \ln \frac{n_w - hn_{sl}}{n_w} \quad (5.212)$$

In terms of mole fractions and rational activity coefficient, Equation (5.211) becomes

$$\ln f'_{\pm} = \ln f_{\pm} + \frac{h}{\nu} \ln a_w + \ln \frac{n_w + \nu - h}{n_w + h} \quad (5.213)$$

Robinson and Stokes (1968) and Bockris and Reddy (1970) used $f'_{\pm}$ although I prefer $\gamma'_{\pm}$ because molal activity coefficients are used more often than rational ones in solution work. For n_s moles instead of 1 mole of electrolyte in n_w moles of water, Equation (5.213) becomes

$$\ln f'_{\pm} = \ln f_{\pm} + \ln \frac{n_w + n_{sl} - hn_{sl}}{n_w + hn_{sl}} \quad (5.214)$$

This equation is akin to that of Bockris and Reddy (1970) except for the fact that these authors used n_h to represent hn_s.

Bockris and Reddy calculated the hydrated mean activity coefficient from the change in free energy due to the changes in the concentrations of free water and ions dissolved in it, as the results of hydration. They obtained the same results as Robinson and Stokes (1968) who approached the problem through electrical potential calculations.

For the practical osmotic coefficient

$$\phi = -\frac{1000}{18\nu m} \ln a_w \quad (5.215)$$

Equation (5.211) becomes

$$\ln \gamma'_{\pm} = \ln \gamma_{\pm} - 0.018hm\phi + \ln[1 + 0.018(\nu - h)m] \quad (5.216)$$

The rational osmotic coefficient g is defined by

$$\ln a_w = g \ln X_w = \ln f_w X_w \quad (5.217)$$

It is useful at high concentrations where f_w is insensitive to departures from ideality. In a 2 M KCl solution $f_w = 1.004$ while $g = 0.944$ (Robinson and Stokes, 1968). The practical (molal) osmotic coefficient given by Equation (5.215) is related to the osmotic pressure by

$$\pi = \frac{\nu RT \times 18}{1000 \bar{V}_w} \phi m \quad (5.218)$$

a simpler relation than that between π and g.

I shall change the meaning of the symbols slightly in that n_w will contain n_s instead of 1 mole of salt. This will permit me to obtain the corrected and the uncorrected molalities. The true concentrations are as follows:

1. For X'_w

$$X_w = \frac{n_w}{n_w + n_{sl}} X'_w$$

$$= \frac{n_w - \nu hn_{sl}}{(n_w - \nu hn_{sl}) + \nu hn_{sl} + n_{sl}} \quad (5.219)$$

so that

$$X'_w = X_w \frac{n_w - \nu hn_{sl}}{n_w} \quad (5.220)$$

2. For c

$$c' = c \quad (5.221)$$

since c is defined per liter of solution regardless of how the water is distributed.

3. For m

$$m' = m \frac{n_w}{n_w - hn_{sl}} \quad (5.222)$$

since the units are moles/kg-free H_2O for m'_s and moles/kg-total H_2O for m_s. w represents weight except in the subscript.

The terms n_w, h, and n_s are:

1. n_w = number of moles of water in a given amount of solution (weight or volume).
2. n_s = number of moles of electrolytes in n_w.
3. h = number of moles of water of hydration per moles of electrolyte.

Thus, Equation (5.221) may be rewritten, per kg-total H_2O, as

$$m' = m \frac{1000}{1000 - hm(MW)_w} \quad (5.223)$$

The hydration equation has merit since it goes one step beyond the Debye-Hückel equation. Furthermore, it introduces a reasonable treatment of salting-out, that is, for the transition from a decreasing to an increasing $\gamma_\pm$ when the concentration goes up. It is thought that salting-out results in part from the decreased availability of water of hydration for new ions due to the hydration of those ions already in solution and because less water is available to supply the hydration energy to break up solid lattices. Other reasons are changes in the structure of water and in the dielectric constant (the Hückel effect).

Note that $\gamma'_\pm$ and $f'_\pm$ are defined in an m or X scale. An alternative procedure is to set an m' or X' scale based on Equations (5.220), (5.221), or (5.222) and to make measurements in these scales. The standard state for m' would be $\gamma' = 1$ when $m' = 1$.

Robinson and Stokes (1968) and Bockris and Reddy (1970) showed that, in terms of my notation,

$$\ln \gamma_{\pm(exp)} = \ln \gamma_{\pm(DH)} - \frac{hn_{sl}}{\nu} \ln a_w$$

$$- \ln \frac{n_w + n_{sl} - hn_{sl}}{n_w + n_{sl}} \quad (5.224)$$

Robinson and Stokes (1968) concluded that $f_{\pm DH}$ is a hydrated $f'_\pm$ because the electrical potential and free energy were calculated excluding spheres of radius $r_i + a_D$.

The correct interpretation of this equation, based on the approach of Bockris and Reddy (1970), is that the last two terms represent $- 1/RT$ times the free energy required to change the concentrations of the electrolyte and the water. We have $\gamma_{\pm(DH)} < \gamma_{\pm(exp)}$ as is to be expected because $\gamma_{\pm(DH)}$ contains the additional nonideality due to hydration. This reasoning is equivalent to that of Robinson and Stokes (1968) based on a_D and presented above.

In Equation (5.224), n_w is the number of moles of water which contain 1 mole of salt. This equation was obtained by calculating the $RT \ln(c_2/c_1)$ type work and the free energy change (minus the work) necessary to change the concentrations of water and ions due to the removal and fixing of the water of hydration.

Bockris and Reddy obtained their terms as follows. When ions are introduced, the activity of water goes from 1 to a_w causing a work $RT \ln a_w$ and $\Delta F = RT \ln a_w$. Then, the free energy when h moles of water are taken up by hydration is $-(h/\nu) RT \ln a_w$ per mole of electrolyte.

The free energy change when ions are concentrated due to there being less free water is

$$RT \ln \frac{n_w + \nu}{n_w - h + \nu} \quad (5.225)$$

In the Bockris and Reddy (1970) treatment the work results from the changes in concentrations and is easier to understand than the approach of Robinson and Stokes (1968). Note that $f_{\pm DH}$ is not necessarily the best choice for reasons discussed earlier, especially since Bates et al. (1970) pushed the hydration equation up to $m = 6.0$. Furthermore, the use of hydration and charging terms alone is not sufficient to describe the complex processes which affect activity coefficients and are masked by the curve fitting by means of a_D and h.

$\gamma_{\pm(DH)}$ is a hydrated $\gamma'_\pm$ because spheres of radius $r_J + a_D$ were excluded from the calculations of the electrical free energy. $\gamma_{\pm(exp)}$, measured in an m scale, does not reflect hydration. Thus, no hydration correction is needed when theories that do not take hydration effects into consideration are matched against $\gamma_\pm$. If theories include h, then an m' or X' scale may be desirable for measurements unless one wishes to relate $\gamma'_\pm$ to $\gamma_\pm$ as was done above.

Robinson and Stokes (1968) recommended the values for a_D and h shown in Table 5.18 and Bates et al. (1970) obtained the fit shown in Table 5.19. This fit has no fundamental significance since it was obtained by calculating the values of a_D and h which yielded agreement between experimental results and Equation (5.171). Still, the fact that an equation

TABLE 5.18 Selected parameters of the hydration equation [from Robinson and Stokes (1968)].

Electrolyte	h	a_D (Å)
HCl	8.0	4.47
LiCl	7.1	4.32
NaCl	3.5	3.97
KCl	1.9	3.63
$MgCl_2$	13.7	5.02
$CaCl_2$	12.0	4.73
$HClO_4$	7.4	5.09
$NaClO_4$	2.1	4.04

TABLE 5.19 Forced fit between the hydration equation with a Debye–Hückel term and experimental results. Selected results for $\gamma_\pm$.

	NaCl		CaCl$_2$	
I	Calculated	Experimental	Calculated	Experimental
0.1	0.778	0.775	0.614	0.616
0.2	0.735	0.731	0.552	0.550
0.5	0.681	0.678	0.482	0.480
1.0	0.657	0.656	0.453	0.449
2.0	0.668	0.670	0.447	0.450
3.0	0.714	0.715	0.500	0.498
6.0	0.986	0.981	0.792	0.814

Source: Reprinted with permission from R. G. Bates, B. R. Staples, and R. A. Robinson, *Anal. Chemistry* **42**, 867–871. Copyright © 1970 American Chemical Society.

with two adjustable parameters can reproduce the shape of the experimental $\gamma_\pm$ versus I curve, including the transition from salting-in to salting-out, and common sense do indicate that hydration plays a role. It cannot be considered the only effect, however, since it must coexist with nonspecific coulombic terms, ion pairing, repulsive core potential, ion cavity, and Gurney overlap effects.

The values of a_D and the hydration numbers for a number of electrolytes are shown in Table 5.20 while h versus crystallographic radii are plotted in Figure 5.13. It must be remembered that these numbers depend on the use of the extended Debye–Hückel equation with an adjusted a_D. We saw earlier that the equation of Robinson and Stokes (1968) is an empirical equation which does not have a fundamental meaning, especially at moderate and high concentrations. Thus, in practice, h obtained by curve fitting is of dubious validity. Still, the trend of h with ionic radius (rather than its true values which cannot be calculated by the present method) is interesting.

The extrapolation of h for CsCl done on the basis of the Cs radius is not entirely convincing since the circles for KCl and RbCl should be given greater weight than the circle of CsCl because the former two were placed from calculated values of h. Still, the results of Bates et al., if approximately correct, suggest that the hydration number of CsCl is nearly zero and we shall see later that this can perhaps help us with rough estimates of the activity coefficients of single ions. There is, however, an uncertainty in h as shown in Table 5.21.

In practice there is little difference between the use of $f_\pm$ or $\gamma_\pm$ for the determination of a_D at the low concentrations for which the Debye–Hückel equation can be made to fit experimental data. In effect, the coefficient A_D differs only by the square root of the density of the solvent (nearly 1) between the two scales (Lewis and Randall, 1961) and B_D is the same. Thus, if we plot $f_{\pm\,exp}$ versus X and $\gamma_{\pm\,exp}$ versus m, then essentially the same value of a_D will

TABLE 5.20 Hydration numbers h and distances of closest approach a_D (in Å) (Bates et al., 1970).

	h	a_D
HCl	8.0	4.47
LiCl	7.1	4.32
NaCl	3.5	3.97
KCl	1.9	3.63
RbCl	1.2	3.49
NH$_4$Cl	1.6	3.75
MgCl$_2$	13.7	5.02
CaCl$_2$	12.0	4.73
SrCl$_2$	10.7	4.61
BaCl$_2$	7.7	4.45

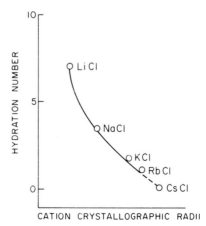

FIGURE 5.13 Hydration numbers versus crystallographic radii (Bates et al., 1970).

TABLE 5.21 Hydration numbers of Bates et al. (1970) and average values from compilations by Bockris and Reddy (1970). Quantities in parentheses represent number of independent methods used for the averages.

	Bates	Bockris
LiCl	7.1	6 (>3)
NaCl	3.5	6 (>3)
KCl	1.9	5 (>3)
RbCl	1.2	4 (>3)

be obtained when we fit $f_{\pm DH}$ to the first curve and $\gamma_{\pm DH}$ to the second one.

Robinson and Stokes (1968) used $f_\pm$ instead of $\gamma_\pm$ because these authors maintained that the Debye–Hückel equation must be represented in terms of rational activity coefficients and mole fractions. This is not necessarily so as was shown by Lewis and Randall (1961) and shown earlier in this section, and molalities can indeed be used.

5.5.3 Debye–Hückel and Related Equations: A Summary

1. *Debye–Hückel extended laws*

$$\log \gamma_{\pm(DH)} = -\frac{A_D|z_1 z_2|I^{0.5}}{1 + B_D a_D I^{0.5}} \quad \text{for an electrolyte}$$

$$\text{(i)}$$

$$\log \gamma_{i(DH)} = -\frac{A_D z_i^2 I^{0.5}}{1 + B_D a_D I^{0.5}} \quad \text{for ion } i \quad \text{(ii)}$$

These are valid up to about $I = 0.1$ in single-electrolyte solutions.

2. *Debye–Hückel limiting laws*

$$\log \gamma_{\pm(DH)} = -A_D|z_1 z_2|I^{0.5} \quad \text{(iii)}$$

$$\log \gamma_{i(DH)} = -A_D z_i^2 I^{0.5} \quad \text{(iv)}$$

The upper limit of usefulness is $I \leqslant 10^{-3}$ in single-electrolyte solutions.

3. *Guntelberg equations*

$$\log \gamma_{\pm(G)} = -\frac{A_D z_1 z_2 I^{0.5}}{1 + I^{0.5}} \quad \text{(v)}$$

$$\log \gamma_{i(G)} = -\frac{A_D z_i^2 I^{0.5}}{1 + I^{0.5}} \quad \text{(vi)}$$

applicable up to about $I = 0.005$ in mixed electrolyte solutions.

4. *Culberson equation*

$$\log \gamma_{\pm(C)} = \frac{A_C|z_C z_A|}{1 + B_C I^{0.5}} = C_C I + D_C I^{1.5} + E_C I^2$$

$$\text{(vii)}$$

applicable from $I = 0.1$ to 1.0.

5. *Hückel equation*

$$\log \gamma_{\pm(H)} = \log \gamma_{\pm(DH)} + C_H I \quad \text{(viii)}$$

applicable up to $I = 0.1$ and sometimes beyond the salting-in–salting-out transition.

6. *Guggenheim equation*

$$\log \gamma_{\pm(GG)} = -\frac{A_D z_C z_A I^{0.5}}{1 + I^{0.5}} + C_G I \quad \text{(ix)}$$

The upper useful limit is somewhere in the range $I = 0.1$ to 0.5 depending on the electrolyte.

7. *Davies equation*

$$\log \gamma_{\pm(D)} = -0.50|z_C z_A|\left(\frac{I^{0.5}}{1 + I^{0.35}} - 0.30I\right) \quad \text{(x)}$$

good to $I = 0.1$ in single-electrolyte solutions.

8. *Hydration equation of Robinson and Stokes*

$$\ln \gamma'_\pm = \ln \gamma'_\pm + \frac{h}{v} \ln a_w + \ln \frac{n_w - h}{n_w} \quad \text{(xi)}$$

for molal activity coefficients, of value for curve fitting at all concentrations.

None of the above equations are especially useful in mixtures of electrolytes. In such cases it is best to use experimental results, obtained by potentiometry, vapor pressure, and so on, or the theories of ion pairs and specific interactions which shall be used further along.

In single-salt solutions:

1. Only Equations (iii) and (iv) for $I < 10^{-3}$, (v) and (vi) for $I < 0.005$, and (vii) for $I < 10^{-1}$ can be used directly to calculate $\gamma_\pm$. In the case of Equation (viii) association constants must be known from another source, for example, conductance.

2. The remaining equations require the determination of unknown parameters which are

obtained by curve fitting of $\gamma_{\pm\,\text{(th)}}$ to $\gamma_{\pm\,\text{(exp)}}$. They are not, therefore, predictive.

In mixed electrolytes the limits of usefulness of the above equations decrease but alternative methods are available.

5.5.4 A Further Word on Salting-In and Salting-Out

The causes of salting-in are the energy of hydration, the interionic attraction, and the decrease of D_e with increasing m. This last effect increases the energy of interaction between ions. When $\gamma_{\pm}$ decreases, $K_s = K_{s0}a_{(s)}/\gamma_{\pm}$ increases so there is less tendency for escape into a solid phase.

Salting-out, on the other hand, is caused primarily by the decrease in the availability of waters of hydration for further added ions when the concentration increases, and by the changes in the effective concentrations of water and electrolytes.

Salting-out also occurs when nonelectrolytes are added to electrolyte solutions. The effect of the latter on the activity coefficient is, according to Debye and McCauley (1925) who used only electrostatic considerations,

$$\ln f = \frac{\beta}{2k_B T D_e} \sum_j \frac{n_j e_j^2}{b_j} \quad (5.226)$$

where β is a proportionality constant and b_j is the radius of the ion j of type j, which is assumed to be spherical (j goes from 1 to s). An alternative approach due to Setchenov (1892) yields an equation which, for seawater, has the empirical form

$$\ln B_T = b_1 + b_2\,S\text{\textperthousand} \quad (5.227)$$

where B_T is the Bunsen coefficient at the temperature T. B_T is the volume of gas (STP) which can dissolve in a unit volume of solution if the partial pressure of the gas is 1 atm. In practice the volume of gas after dissolution must be converted to the initial one and the total measured pressure minus the saturation value of $p\text{H}_2\text{O}$ must equal 1 atm. A more general expression (Whitfield, 1979) is

$$\ln \frac{m_0}{m} = km_{sl} \quad (5.228)$$

where m_0 and m are the molal solubilities of the nonelectrolyte when m_{sl}, the molality of the electrolyte, is zero. Equation (5.228) yields

$$\ln m = \ln m_0 - km_{sl} \quad (5.229)$$

This equation is similar in form to Equation (5.226).

Solubilities of gases in seawater, in terms of μmoles-gas/kg-SW at $S\text{\textperthousand} = 35.0$ are presented in Table 5.22 (Kester, 1975). Kester also presents compilations versus T and $S\text{\textperthousand}$.

The Debye and McCauley (1925) equation for the self-salting-out of electrolytes

$$\ln f_{\pm} = \frac{(D_{e,0} - D_e)e^2}{\nu k_B T D_{e,0}^2} \sum_j \frac{\nu_j z_j^2}{b_j} \quad (5.230)$$

does not appear correct to me because it indicates that the effect of a decrease in D_e increases $f_{\pm}$ although the interionic field should increase.

Another approach is based on the effect of ions on the structure of water. Some ions, such as Na^+, are known as structure formers, whereas others, including K^+, are structure breakers. These effects are observed, for example, in the decreased viscosity of water when salts of structure breakers (e.g., KCl) are dissolved in it. NaCl, on the other hand, increases the viscosity of water (Horne, 1969). This view is supported by proton NMR and infrared work.

Frank and Wen (1957) interpreted structure making and breaking in terms of a double hydration layer around ions in which the inner layer has the primary waters of hydration oriented toward the ion, whereas the outer layer has random water molecules. This randomness results from the disruption of the normal structure of water by the field of the ion. The outer shell predominates in structure breakers or, according to Gurney (1962), penetrates into and decreases the inner shell.

The ions, whether of one or several electrolytes present concurrently, affect the structure of water

TABLE 5.22 Solubilities of gases in μmoles-gas/kg-SW at $S\text{\textperthousand} = 35.0$ and 760 mm Hg water vapor saturated air.

Temperature (°C)	N_2	O_2	Ar
0	616.4	349.5	16.98
5	549.6	308.1	15.01
10	495.6	274.8	13.42
15	451.3	247.7	12.11
20	414.4	225.2	11.03
25	383.4	206.3	10.11

Source: Reprinted with permission from D. R. Kester, *Chemical Oceanography*, Vol. 1, 2nd Ed., 1975, pp. 497–556. Copyright © 1975 Academic Press Inc., London, Ltd.

and, therefore, the ionic free energies of hydration. Salting-out results if, as a result of this effect, it becomes more difficult to hydrate the ions of the electrolyte of interest. Salting-out tends to occur at higher concentrations at which the ions sense the effects of the hydration of other ions that are present.

5.6 LATTICE MODEL

5.6.1 Introduction

The main features of lattice models were examined in detail by Pytkowicz et al. (1976) and by Pytkowicz and Johnson (1979) who proposed a refined new model. In this work I concentrate on the conceptual highlights of our theory. Earlier approaches as well as our work are examined extensively in the above references.

The main difficulty with earlier models is that they were expressed in terms of perfect long-range order (PLRO). Consequently, a $c^{1/3}$ dependence was expected which did not match $\gamma_{\pm\,exp}$ over most of the concentration ranges that were examined.

5.6.2 Postulates

Our work was based on the following postulates:

1. There is a flickering PLRO resulting from the balance of strong coulombic forces between ions and their thermal energy. I mean, by flickering, that ions move around. Furthermore, the time-average properties of the solution correspond to those of a partial long-range ionic order. The concept of partial order permits us to introduce the effects of T as well as of the concentration, in contrast to earlier work. At any instant an ionic configuration has a degree of long-range order L induced by strong coulombic forces. Its average over a period of time is $\bar{L}$ and is due to ion–ion interactions.

2. The use of the time average for a system (an ionic solution in this case) is equivalent, by a tenet of statistical thermodynamics, to an ensemble average. Thus, there is, in principle, a tie between our lattice model and the statistical thermodynamic models discussed elsewhere in this chapter.

3. We introduced a new concept beyond the PLRO, namely, that the degree of order not only decreases with an increasing temperature but that it also increases with an increase in concentration.

The reason for the latter is that the ions become closer together and that this increases the coulombic energy and its ordering influence.

4. Our lattice model is explanatory rather than predictive because, like the equations of Debye–Hückel (1923), Hückel (1925), Guggenheim (1935), Robinson and Stokes (1968), Pitzer (1973), Ramanathan and Friedman (1971), and so on, it has a curve-fitting parameter.

5. The lattice model contains terms not only for the average value of thermodynamic properties but also for the contribution of ionic fluctuations.

6. The partial ordering of ions is fast relative to the motion and mixing of water parcels because the ordering process only requires exchanges between nearest neighbor cations and anions (Pytkowicz et al., 1976). The partial order need only extend within all the moving parcels of the solution and is not required to be coherent for the whole solution for the correct time-average thermodynamic properties to occur. This is equivalent to the case of crystals of a solid, discussed in Chapter 4.

5.6.3 Meaning of the PLRO in Aqueous Solution and the Configurational Free Energy

Consider a solid solution. Its degree of order is, according to the traditional concept of long-range order in solids (e.g., Kittel, 1959)

$$L = 1 - 2w \qquad (5.231)$$

w is the fraction of cations C on sublattice a and of anions A on sublattice c. The symbol c refers to the sites (sublattice) that contain 50% or more of cations C and a applies in the same way to A. For perfect order $w = 0$ and $L = 1$, while for complete randomness $w = 0.5$ and $L = 0$. Cations on A can be called "wrong ions" for short.

In terms of a time average in an aqueous solution, at a given (T, m)

$$\bar{L} = 1 - 2w \qquad (5.232)$$

Our problem consists of determining the electrical free energy $\Delta \bar{F}_L$ due to ionic interactions in terms of $\bar{L}$ because then $\gamma_\pm$ can be calculated from

$$\frac{\partial \Delta \bar{F}_L}{\partial m} = 2RT \ln \gamma_{\pm(L)} \qquad (5.233)$$

$\bar{F}'_L$ will express that part of the free energy $\bar{F}_L$ which does not include the configurational entropy. The subscript L refers to the lattice (electrical) free energy. Thus,

$$\Delta \bar{F}_L = \Delta \bar{F}'_L - T \Delta \bar{S}_L \qquad (5.234)$$

The horizontal bars indicate time averages. The configurational entropy corresponds to the degree of order and will be presented soon. It is zero for a perfectly ordered lattice and a maximum for a random one.

In our work on ionic lattices in aqueous solutions we adapted the concept of the Madelung constant A_M from solid ionic solutions. For solids, the attractive term (the electrical interaction energy) is

$$E_L = - \frac{N_A z^2 e^2 A_M}{r_0} \qquad (5.235)$$

where N_A is Avogadro's number, z is the valence, e is the elementary charge, and r_0 is the distance between nearest neighbors. A_M is the term that accounts for the fact that each ion interacts with all the other ones, whether they be nearest neighbors or not.

5.6.4 The Lattice Equation

The Coulombic Free Energy. In the case of aqueous ionic solutions, Pytkowicz et al. (1976) and Pytkowicz and Johnson (1979) found

$$\Delta \bar{F}'_L = - \frac{N_A (\bar{L} z e)^2 A_M}{D_e r_0} m + \Delta \bar{F}'_R \qquad (5.236)$$

where now $\pm \bar{L}$ is the effective charge of each ion. $\Delta \bar{G}_R$ is the noncoulombic repulsion term while the coulombic one for the whole lattice is taken into account by A_M.

The time-average behavior of the L lattice is shown in Figures 5.14a and 5.14b. The top left ion in Figure 5.23b has 75% cationic character ($\bar{w} = 0.25$, $\bar{L} = 0.50$). This ion has an effective charge of 0.50 because it has a 0.75 cationic minus a 0.25 anionic character. The ion in Figure 5.23a is a pure cation ($w = 0$, $L = 1$) and there are no ions on the wrong sites, that is, no C on a or A on c.

It is exceedingly important to understand the basis upon which the order concept is proposed. In reality there is no site in space in which cations are present 75% of the time and anions 25%. At any one

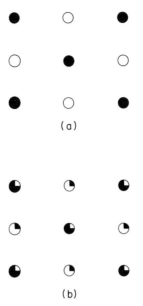

FIGURE 5.14 Representations of ionic solutions with properties that correspond to (a) perfect order and (b) partial order.

instant there is a degree of long-range order L induced by the strong coulombic forces. It corresponds to some flickering configuration and the values of L over a period of time lead to $\bar{L}$ and $\Delta \bar{G}'_L$.

The Effect of Fluctuations. Let us consider next the effect of ionic fluctuations, which was not entered in Equation (5.236). We found it to be

$$\frac{\bar{x}}{r_0} = \left(1 - \frac{\bar{r}_x}{r_0} \right)^{0.5} \qquad (5.237)$$

where $\bar{x}$ is the time-average departure of the L ions from the ideal lattice-geometry sites (equal interionic spacing r_0) and $\bar{r}_x$ is the effective interionic distance that replaces r_0 in Equation (5.236). Then, this equation becomes

$$\Delta \bar{F}'_L = \frac{N_A (\bar{L} z e)^2 A_M \bar{v}}{D_e r_0} m + \Delta \bar{F}'_R \qquad (5.238)$$

$\Delta \bar{F}'_R$ can be neglected except at very high concentrations because the quantum mechanical repulsion is negligible when ions are not nearly in contact.

The term $\bar{v}$ is

$$\bar{v} = \frac{r_0}{\bar{r}_x} = \frac{r_0^2}{r_0^2 - x^2} \qquad (5.239)$$

This is an approximate equation based on a calculation with ion triplets but $\bar{v}$ will be determined empirically and, therefore, independently of its actual form in terms of r_0 and x.

Configurational Entropy. The entropy term is, according to solid solution theory (Kittel, 1959) and adapted to aqueous solutions,

$$\Delta\bar{S}_L = -Rm[(1 + \bar{L})\ln(1 + \bar{L}) \\ + (1 - \bar{L})\ln(1 - \bar{L})] \qquad (5.240)$$

where R is the gas constant. The correspondence between the solid and the aqueous solutions arises from the time-average behavior of the latter which leads to properties equivalent to those of a partial degree of long-range order.

Lattice Equation. The distance r_0 can be expressed in terms of molalities as

$$r_0 = \frac{[1000 + m(\text{MW})_{sl}]^{1/3}}{2N_A m\rho} \qquad (5.241)$$

where ρ is the density of the solution. sl indicates the solute. Introducing (5.241) into (5.238) and adding the entropy $-T\Delta\bar{S}_L$ we obtain the lattice equation, with $\Delta\bar{F}_L = \Delta\bar{F}'_L - T\Delta\bar{S}_L$,

$$\Delta F_L = -\frac{2^{1/3}N_A^{4/3}(\bar{L}ze)^2 A_M \bar{v}\rho^{1/3}}{D_{e,0}[1000 + m(\text{MW})_{sl}]^{1/3}} m^{4/3} \\ + RTMm[(1 + \bar{L})\ln(1 + \bar{L}) + (1 - \bar{L})\ln(1 - \bar{L})] \qquad (5.242)$$

5.6.5 Results

The stability condition for the lattice requires that $d\Delta\bar{F}_L/d\bar{L}$, in which $\Delta\bar{F}_L$ is given by (5.242), be zero. This condition, when $\Delta\bar{F}_L$ is differentiated, leads to

$$\frac{1}{\bar{L}\bar{v}}\ln\frac{1 + \bar{L}}{1 - \bar{L}} = k \qquad (5.243)$$

where k represents the terms in Equation (5.242) which do not depend on $\bar{L}$. One sees, from Equation (5.243), that for each value of $\bar{v}$ there corresponds only one value of $\bar{L}$ which causes $\Delta\bar{F}_L$ to be a minimum for a given (P, T, m) set.

Another condition that must be met by $(\bar{L}, \bar{v})$ pairs is $\Delta\bar{F}_L = \Delta F_{\text{exp}}$ where ΔF_{exp} is the experimen-

tal free energy. As $\gamma_\pm$ is the measured mean activity coefficient, it is necessary that $(\bar{L}, \bar{v})$ yield

$$\Delta\bar{F}_L = \Delta F_{\text{exp}} = 2RT\int_0^m \ln\gamma_\pm \, dm \qquad (5.244)$$

because $\partial\Delta F_{\pm(\text{exp})}/\partial m = 2RT\ln\gamma_{\pm(\text{exp})}$. At a given (T, P, m) only one among the $(\bar{L}, \bar{v})$ sets that obey Equation (5.243) will also yield the above equality (5.244).

A computer may search for values of these parameters which yield minimum values of $\Delta\bar{F}_L$ and then select that set $(\bar{L}, \bar{v})$ which causes one of the minima to yield $\Delta\bar{F}_L$ to equal $\Delta F_{(\text{exp})}$. As an alternative the computer may search for $(\bar{L}, \bar{v})$ sets that yield

$$\bar{F}_L = \Delta F_{(\text{exp})} = 2RT\int_0^m \ln\gamma_{\pm(\text{exp})} \, dm \qquad (5.245)$$

Once $\bar{L}$ is known, $\Delta\bar{S}_L$ can be calculated from Equation (5.240) and $\Delta\bar{F}'_L$ from $\Delta\bar{F}_L + T\Delta\bar{S}_L$.

In order to obtain results for single-salt solutions of 1-1 electrolytes, the integration in Equation (5.245) was done by means of the limiting Debye–Hückel equation at concentrations below those for which measurements are available, and from the Culberson equation from $m = 10^{-4}$ to $m = 1$.

Lattice parameters are presented for NaCl and KCl at 25°C in Tables 5.23 and 5.24. Plots of $\bar{L}$ and $\bar{v}$ versus m, using NaCl as an example, are shown in Figures 5.15 and 5.16. An increase in $\bar{L}$ with m is to be expected as is demonstrated below and in Figure 5.17.

Figure 5.17 shows that the values of $\bar{L}$ which yield minimum free energies increase with increasing concentration m. The reason for this is that the thermal energy that drives the system toward randomness becomes relatively less important when the lattice energy increases with increasing m. At a given $\bar{L} = \bar{L}'$ in Figure 5.17, we see that the slopes of the curves, $\partial\Delta F_L/\partial L$, become steeper at higher m's. Thus, for a given $\partial L > 0$ at L' the value of $\partial\Delta\bar{F}_L$ becomes a larger negative number and the ordering process is more spontaneous at higher m's, whereas a disordering process, for which $\partial L < 0$, becomes more difficult.

Calculations based on our simple model show that, although $r_0 - \bar{x}$ and $\bar{x}/r_0$ decrease slightly with increasing concentration, $r_0/\bar{r}_x$ and $\bar{v}$ decrease markedly. This probably occurs because the noncoulombic repulsion is stronger over the distance r_0

TABLE 5.23   Lattice parameters for NaCl.

m	$-\Delta\bar{F}_L$	$-\Delta\bar{F}'_L$	$-\Delta\bar{S}_L$	$-T\,\Delta\bar{S}_L$	$\bar{L}$	$\bar{v}$
10^{-10}	3.400×10^{-5}	2.239×10^{-2}	7.499×10^{-5}	2.236×10^{-2}	0.0949	1.636×10^3
10^{-8}	3.870×10^{-2}	7.565×10^{0}	2.524×10^{-2}	7.526×10^{0}	0.1738	3.549×10^2
10^{-6}	3.890×10^{1}	2.390×10^{3}	7.885×10^{0}	2.351×10^{3}	0.3055	7.818×10^1
10^{-4}	3.890×10^{4}	7.456×10^{5}	2.370×10^{3}	7.067×10^{5}	0.5208	1.808×10^1
10^{-3}	1.188×10^{6}	1.285×10^{7}	3.912×10^{4}	1.167×10^{7}	0.6575	9.077
10^{-2}	3.508×10^{7}	2.169×10^{8}	6.100×10^{5}	1.818×10^{8}	0.7988	4.817
0.1	9.116×10^{8}	3.432×10^{9}	8.454×10^{6}	2.521×10^{9}	0.9092	2.731
0.5	7.560×10^{9}	2.200×10^{10}	4.833×10^{7}	1.441×10^{10}	0.9527	1.862
1.0	1.762×10^{10}	4.742×10^{10}	9.994×10^{7}	2.980×10^{10}	0.9630	1.561

TABLE 5.24   Lattice parameters for KCl.

m	$-\Delta\bar{F}_L$	$-\Delta\bar{F}'_L$	$-\Delta\bar{S}_L$	$-T\,\Delta\bar{S}_L$	$\bar{L}$	$\bar{v}$
10^{-10}	3.412×10^{-5}	2.233×10^{-2}	7.499×10^{-5}	2.236×10^{-2}	0.0947	1.636×10^3
10^{-8}	1.570×10^{-2}	7.542×10^{0}	2.524×10^{-2}	7.526×10^{0}	0.1738	3.549×10^2
10^{-6}	3.163×10^{0}	2.383×10^{3}	7.885×10^{0}	2.351×10^{3}	0.3055	7.818×10^1
10^{-4}	3.663×10^{4}	7.433×10^{5}	2.370×10^{3}	7.067×10^{5}	0.5208	1.808×10^1
10^{-3}	1.192×10^{6}	1.287×10^{7}	3.923×10^{4}	1.170×10^{7}	0.6583	9.108
10^{-2}	3.544×10^{7}	2.180×10^{8}	6.124×10^{5}	1.826×10^{8}	0.8002	4.840
0.1	9.426×10^{8}	3.488×10^{9}	8.536×10^{6}	2.545×10^{9}	0.9124	2.764
0.5	8.214×10^{9}	2.290×10^{10}	4.924×10^{7}	1.468×10^{10}	0.9585	1.923
1.0	2.117×10^{10}	5.156×10^{10}	1.019×10^{8}	3.038×10^{10}	0.969	1.682

when this distance decreases so that fluctuating ions spend relatively less time in the neighborhood of their counterions.

Further considerations such as the limiting behaviors of the model at low and high concentrations are presented by Pytkowicz and Johnson (1979) who also discuss the adequacy of earlier models.

5.7   ACTIVITY COEFFICIENTS OF SINGLE IONS

The following symbols are used in this book:

$(\gamma)_F$ and γ_T or $(\gamma_i)_F$ and $(\gamma_i)_T$ Activity coefficient of the free ion (F) and of the free plus the ion-paired ion

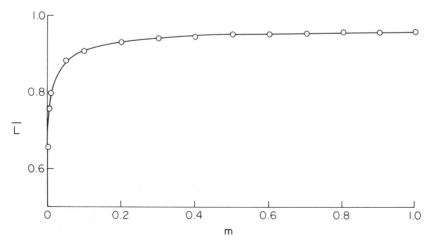

FIGURE 5.15   Values of I versus m.

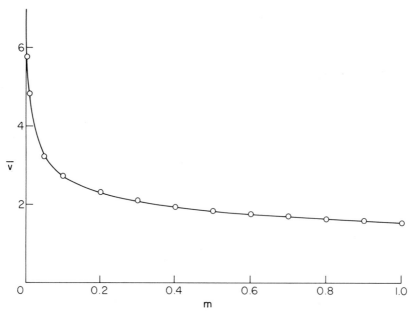

FIGURE 5.16 Values of $\bar{v}$ versus m.

$(\gamma_{\pm})_F$ and $(\gamma_{\pm})_T$ or $(\gamma_{\pm CA})$ and $(\gamma_{\pm CA})_T$	Corresponding mean activity coefficients
a_i, a_{CA}, $\{i\}$, $\{Ca^0\}$, $\{CA\}$	Activities of the ion i, of the ion pair CA^0, and of the electrolyte CA. The symbol a_{CA} can refer to the ion pair or to the electrolyte and is used where it is convenient and there is no chance of confusion. A symmetric electrolyte is used for simplicity.
m_F and m_T	Molalities of an ion, an ion pair, or an electrolyte

$(m_i)_F$ and $(m_i)_T$	Ionic molalities
$(m_{CA})_F$ and $(m_{CA})_T$	Molalities of an ion pair
m_{ip}	Molality of an ion pair.
$[i]_F$ and $[i]_T$, $[CA^0]$, $[CA]_F$ and $[CA]_T$	Molalities of an ion, an ion pair, and an electrolyte

It is useful to employ all these symbols because they appear interchangeably in the literature and because some are clearer than others for various topics.

In this section I consider γ_i's for single ions and models that yield them. Later on I shall examine mean activity coefficient approaches.

Single ions do exist in solutions but their individual properties cannot be measured individually. As an example, the Eh_A of NaCl in an aqueous solution is

$$Eh_A = Eh_A^0 - \frac{RT}{F'} \ln a_{Na}a_{Cl} \qquad (5.246)$$

This equation yields $\Delta F = -mF'E$ and is, therefore, thermodynamically valid if a reversible measurement is made, that is, if very little current is drawn. The cell can consist of a sodium-sensitive and an Ag/AgCl electrode. Thus $a_{Na}a_{Cl}$ has thermodynamic significance. The separation of a_{Na} from a_{Cl}, however, requires the use of a nonthermo-

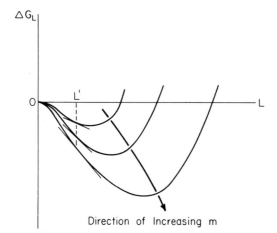

FIGURE 5.17 Values of L which yield minimum free energies.

dynamic assumption. A similar problem arises with any other relevant measurements.

In this section methods and results obtained before we knew that Cl^- also forms ion pairs such as $NaCl^0$, KCl^0, $MgCl^+$, and $CaCl^+$ are presented. An exception to this are the results based on the work of Johnson (1979) and of Johnson and Pytkowicz (1978, 1979a, 1979b).

The main reasons for presenting data that are partly obsolete is that the concepts and the methods can be readily extended to the presence of Cl^- ion pairs and that trends in the results are often valid. Furthermore, not all solutions contain appreciable amounts of chloride. Readers may, at times, wish to add their own chloride association results to the earlier work. The main purpose of this book is to present concepts. Actual data can become rapidly obsolete but concepts do not. In any event, the results present in this section are often the most up-to-date ones even though they do not yet include the effects of Cl^- association.

Single-ion activities can be bypassed in many problems but can be convenient in some instances, such as in the determination of the speciation of ions in multicomponent solutions. Even then, mean coefficients can be used as in the case of solutions containing Na^+, Ca^{2+}, Mg^{2+}, Cl^-, SO_4^{2-}, and their ion pairs (Johnson and Pytkowicz, 1979a).

Another case in which γ's for single ions are of value is in the determination of (CO_3^{2-}) by means of

$$[CO_3^{2-}]_T = \frac{K_2\{HCO_3^-\}}{\{H^+\}_T(\gamma_{CO_3})_T} \qquad (5.247)$$

K_2 is the second thermodynamic dissociation of carbonic acid and T represents total (free plus ion-paired) quantities.

5.7.1 Ion Pairs: Concepts

The subject of ion pairs will be approached at first through the properties of single ions, which require a nonthermodynamic assumption, and later on from mean properties of electrolytes.

I should state at the beginning that, if one accepts the coulomb law and the thermal motion of particles, then one must accept the fact that a fraction of the ions in solution will be in contact at each moment. This fraction depends on m and T. This conclusion is perfectly general and can be used for our sophisticated models as well as for the original model of Bjerrum (1926). The starting points for the

various types of models will be seen to be quite different, however.

Confirmation for the existence of ion pairs was obtained from Raman spectroscopy, sound attenuation, theory, and so on. The concept is quite elegant in that our recent method yields free activity coefficients that reflect all processes, such as hard-core effects, ion-cavity interactions, changes in dielectric constants, electrostatic interactions of ions, and hydration, with the exception of ion pairing. There is no need to use empirical equations such as the Debye–Hückel one.

The Bjerrum Approach. Bjerrum (1926) showed that the distribution of ions in solution is bimodal; that is, that there are high probabilities of ions being close together or of their being far apart (see Figure 5.18a). The probability is proportional to the density of ions found in a shell of volume $4\pi r^2 \, dr$. It reflects two factors: the coulombic attraction between oppositely charged ions (Figure 5.18b) and the increasing size of the spherical shells when r

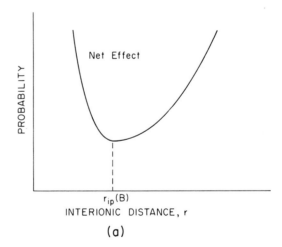

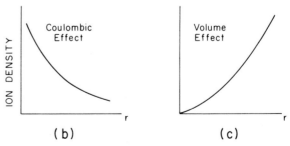

FIGURE 5.18 Probability distribution of single ions and ion pairs. (a) Net effect, (b) coulombic effect, (c) volume effect.

increases (Figure 5.18c). According to Bjerrum, ion pairs occur when $r < r_{ip(B)}$ with

$$r_{ip(B)} = \frac{z_+ z_- e^2}{D_e k_B T} \qquad (5.248)$$

and $r_{ip(B)} > a_D$. This theory is an oversimplification in that approximations had to be made in the calculation of the potential. Furthermore, $r_{ip(B)} > a_D$ means that ions that are not in contact form ion pairs while this term is used to indicate new species with a different charge than those of free ions with which they coexist. As examples, three species exist in the solutions that contain the salts such as those shown below:

their water of hydration. In this case we may refer to pairs of ions. Thus, Ca^{2+} and Cl^- form the ion pair $CaCl^+$ while Pb^{2+} and Cl^- produce the complex $PbCl^+$. The difference arises from the different electronegativity of the atoms in question. Na has a low electronegativity and denotes electrons easily to Cl so that the ions Na^+ and Cl^- are stable, whereas Pb^{2+} and Cl^- have a tendency to share their electrons in a coordinate covalent manner when in contact with each other. In the latter case most or all of the water of hydration is squeezed out by the strength of the Pb^{2+}–Cl^- attraction.

Let us combine the concept of contact ion pairs with that of Bjerrum (1926). We then have three major types of ions; true ion pairs, near-ions which

Salt	Cation	Anion	Ion Pair	Salt	Cation	Anion	Ion Pair
NaCl	Na^+	Cl^-	$NaCl^0$	$MgCl_2$	Mg^{2+}	Cl^-	$MgCl^+$
Na_2SO_4	Na^+	SO_4^{2-}	$NaSO_4^-$	$CaSO_4$	Ca^{2+}	SO_4^{2-}	$CaSO_4^0$

In terms of the Bjerrum theory, the Debye–Hückel equation can only be valid for values of the interionic distance $r > r_{ip(B)} > a_D$ because these authors did not take into account ion pairs as new entities. Thus, only such values of r yield free ions. This is not a definitive conclusion because $r < r_{ip(B)}$ does not yield only ion pairs and such pairs were included in the Debye–Hückel potential when the potential was integrated from $r = a_D$. Still, the Bjerrum approach brought a new concept to solution theory. Note that the ion pairs intrinsic to the Debye–Hückel theory are based on spherical ions and are subject to the other limitations listed earlier for this theory.

A further development of ion-pair theories is due to Fuoss (1958) and was used to compare experimental with theoretical values in seawater by Kester and Pytkowicz (1975). Direct evidence for the existence of such pairs was obtained, for example, by Raman spectroscopy (Daly et al., 1972) and sound attenuation (Fisher, 1967).

The concept of ion pairs as used in this work corresponds to contact resulting from a coulombic attraction between ions that retain all or most of their water of hydration. This is known as outer sphere association even if there is a small amount of cosphere overlap. The term "complexing" is applied by me to association between ions which involves covalent bonding and the loss of most of

are below the Bjerrum critical distance but are not in contact, and ions which are far apart. True ion pairs have an interionic distance a_D which is not known because a_D is determined by curve fitting for a theory that loses its meaning at $m > 10^{-3}$ (Frank and Thompson, 1959a). Near-ions have a significantly lower probability of existence than ion pairs (see Figure 5.18) and are sensed approximately as ion pairs by far-ions. I shall, therefore, include near-ions into the concept of ion pairs and will consider far-ions to be free ones. The ambiguity between ion pairs and near-ions can introduce small differences in properties measured by different techniques.

As I develop the subject of ion pairs, we shall see that we must choose between models with and without Cl^- association with K^+, Na^+, Mg^{2+}, and Ca^{2+}. In principle ion pairs with Cl^- must exist because, statistically, a fraction of the cations and anions will be in contact at any one time, under the influence of their coulombic attraction. The corresponding association constants must be small, especially in the case of K^+ and Na^+, since we are dealing with large hydrated cations with low charges (Kester and Pytkowicz, 1971). Their e/r ratios, where r is their radius, are small. The low values of the association constants are offset in media such as seawater by the large concentrations of the ions of interest.

Free and Total Quantities. Now, we are in a position to present two distinct definitions of terms related to a_i, the activity of a cation or of an anion (Pytkowicz et al., 1966). The use of single ions is rigorous conceptually; also the results must be combined to yield mean quantities when the theoretical results are compared to experimental ones.

$$a_i = (\gamma_i)_T (m_i)_T = (\gamma_i)_F (m_i)_F \qquad (5.249)$$

T refers to the stoichiometric (total) quantities while F indicates only those that correspond to the free ions.

If an ion i is 20% paired with a counterion then Equation (5.249) becomes

$$a_i(\gamma_i)_T [0.8(m_i)_F] = a_i = [0.8(\gamma_i)_F](m_i)_T$$
$$= (\gamma_i)_F [0.8(m_i)_T] \qquad (5.250)$$

with $(\gamma_i)_T = 0.8(\gamma_i)_F$ and $(m_i)_F = (m_i)_T$. In other words, the 20% decrease in a_i due to ion pairing can be formally ascribed either to the total activity coefficient or to the free concentration.

Let us return for a while to $\gamma_\pm$ since γ_i cannot be determined by experiments alone. In general, activity coefficients are obtained in a c_T or m_T scale where c and m represent the molarity and the molality. Therefore, $\gamma_{\pm(\exp)}$ is a total coefficient. The experimental mean activity coefficients correspond conceptually to total ones because they are obtained in a total m or c scale. Thus, any theories for activity coefficients which use adjustable parameters to fit calculated to measured γ's, for example, the extended Debye–Hückel theory, yield $(\gamma_\pm)_T = \gamma_{\pm(\exp)}$. It is clear, therefore, that $(\gamma_\pm)_T$ has physical meaning because it is calculated from measured values of m_T and the activity. The expression that is used is

$$(\gamma_\pm)_T = \left(\frac{a}{m_T}\right)^{0.5} \qquad (5.251)$$

Let us consider next the significance of the term $(\gamma_\pm)_F$. Ion pairs are real according to measurements by Raman spectra (Daly et al., 1972) and sound attenuation (Fisher, 1967). The spectroscopic measurements showed that $NaSO_4^-$ competes with HSO_4^- for SO_4^{2-} ions, whereas the acoustic work demonstrated the absorption of sound by $MgSO_4^0$. Theoretical computation based on the ion-pairing model to Fuoss (1958) yielded association constants

that were in rough agreement with constants determined potentiometrically (Kester and Pytkowicz, 1975). Stoichiometric solubility products, such as that of $CaCO_3$ in seawater, are depressed relative to thermodynamic products to a greater extent than that which is expected from the ionic strength alone and suggest a further decrease due to ion association (Pytkowicz, 1975; Pytkowicz and Gates, 1968). Furthermore, these different techniques yielded roughly similar results.

The activity a of an electrolyte is decreased by ion pairing and this effect can obviously be assigned to $(\gamma_\pm)_T$ when we operate in an m_T scale. The determination of $(\gamma_\pm)_F$ in an m_F scale is not as straightforward except in the cases of Raman spectra and sound attenuation. In general, it is necessary to obtain $(\gamma_\pm)_F$ and m_F from a model and from the measured activity, as I shall show later in this chapter. This model must obey

$$a = (\gamma_\pm)_F^2 m_F = (\gamma_\pm)_T^2 m_T \qquad (5.252)$$

It is interesting to consider whether experiments, besides Raman and sound work, respond to free or to total concentrations. This question becomes, in the determination of activities by ion-sensitive electrodes, whether only free ions are sensed or if the electrode responds also to ion pairs. Thus, an electrode may respond only to $[Ca^{2+}]_F$ or it may also sense $[CaSO_4^0]$, $[CaCO_3^0]$, and so on. There are two types of indications that electrodes may sense only free ions. First, there is the fact that different techniques such as solubilities (Pytkowicz and Gates, 1968), potentiometry (Thompson and Ross, 1966), and sound attenuation (Fisher, 1967) yielded similar values for $[Mg^{2+}]_F$ in seawater. Second, potentiometric (electrode) results in binary solutions lead to the correct $(\gamma_\pm)_F$ in more complex solutions in which new types of ion pairs and a reapportionment of the previous pairs occur (Johnson and Pytkowicz, 1979a).

$(\gamma_\pm)_F$ and m_F would still be useful conceptual constructs even if they were not sensed by electrodes. The term $(\gamma_\pm)_F$, which can be calculated from measured activities and models, represents the effects of nonspecific interactions of the Debye–Hückel type, the ion-cavity and hard-core effects to be discussed later in this chapter, and the effects of hydration plus changes in the dielectric constant. The total $(\gamma_\pm)_T$ would only differ from $(\gamma_\pm)_F$ because it also represents the results of ion pairing with some cosphere overlap. The strength of the ion-pair approach depends on the model used. The most realistic one is that of Johnson and Pytkowicz (1979a).

Effective Ionic Strength. A new concept, the effective ionic strength I_e, must be introduced for the self-consistency of the ion-pairing model (Pytkowicz and Kester, 1969).

The conventional ionic strength I is

$$I = 0.5 \sum_i m_i z_i^2 \qquad (5.253)$$

It was introduced by Lewis and Randall (1921) because they observed that in very dilute solutions log

Ion	m_T	Free Metal	$CaSO_4^0$	$CaHCO_3^+$	$CaCO_3^0$	$MgCaCO_3$	CaF
Ca^{2+}	0.01063	88.35	10.87	0.29	0.41	0.07	0.02

$\gamma_{\pm \exp}$ was proportional to $I^{0.5}$. The proportionality constant was independent of the specific ions and depended only on the charge type and the temperature. Later, Debye and Hückel (1923) provided a rationale for this observation through their limiting law. I shall term I the stoichiometric or total activity coefficient.

EXAMPLE 5.5. THE TOTAL IONIC STRENGTH. In the case of a 10^{-4} m Na_2SO_4 solution:

$$\begin{aligned} I &= 0.5([Na^+] + 4[SO_4^{2-}]) \\ &= 0.5\,[2 \times 10^{-4} + 4 \times 10^{-4}] \qquad \text{(i)} \\ &= 3 \times 10^{-4}\ m \end{aligned}$$

For a 10^{-4} m $CaCl_2$ solution I is also 3×10^{-4}. For a mixture of 10^{-3} m $NaCl$ and 10^{-4} m Na_2SO_4

$$\begin{aligned} I &= 0.5[1.2 \times 10^{-3} + 10^{-3} + 4 \times 10^{-4}] \qquad \text{(ii)} \\ &= 1.3 \times 10^{-3}\ m \end{aligned}$$

In the presence of ion pairs the effective ionic strength (Pytkowicz and Kester, 1969) is

$$I_e = 0.5\left[\sum_i (m_i)_F z_i^2 + \sum_{ip} m_{ip} z_{ip}^2\right] \qquad \text{(iii)}$$

ip represents the ion pairs that occur in the solution. In the case of Na_2SO_4

$$I_e = 0.5[(Na^+)_F + 4(SO_4^{2-})_F + (NaSO_4^-)] \quad \text{(iv)}$$

I shall select, among several available ion-pairing models, those that best illustrate quantitatively the concepts of γ_F, γ_T, m_F, m_T, and I_e.

EXAMPLE 5.6. PARAMETERS OF THE ION-PAIRING MODEL. For a seawater of 34.8‰ salinity ($I = 0.72$) Pytkowicz and Hawley (1974) reported the percent concentrations of the calcium species for calcium, among other results for the major cations of seawater:

These results were obtained potentiometrically and in conjunction with the earlier results of Kester and Pytkowicz (1969). Thus, $m_F = 0.01063 \times 0.8835 = 0.00939$. As $(\gamma_{Ca})_F = 0.255$ according to Pytkowicz et al. (1977), $(\gamma_{Ca})_T = 0.255 \times 0.8835 = 0.225$.

I_e, taking the other major constituents of seawater studied into consideration, was found to be 0.668 versus $I = 0.718$ at 34.8‰ salinity.

Association Constants and I_e. The concept of the effective ionic strength leads to an extension of higher concentrations of the original definition of stoichiometric ionic strength to higher concentrations. Consider the relation

$$K^{*\prime} = K^* \frac{(\gamma_C)_F (\gamma_A)_F}{(\gamma_{CA})_{ip}} \qquad (5.254)$$

where $K^{*\prime}$ is the stoichiometric association constant. K^*, the thermodynamic constant, depends by definition only upon P and T and not upon the ionic strength and the composition of the solution.

Pytkowicz and Kester (1969), Pytkowicz and Hawley (1974), and Johnson and Pytkowicz (1979a) showed that $K^{*\prime}$ was only a function of P, T, and the effective ionic strength I_e and did not vary with composition of solutions.

The results that support this conclusion will be discussed fully later and will only be mentioned briefly here. Pytkowicz and Kester (1969) found that $K^{*\prime} = K^*(\gamma_{Na})_F(\gamma_{SO_4})_F/\gamma_{NaSO_4}$ was independent of composition in $NaCl$–Na_2SO_4 solutions at a given effective ionic strength. It is improbable that the γ terms varied in such a way as to cancel each other with compositional changes. These authors did not take chloride association into consideration but Johnson (1979) showed that the above $K^{*\prime}$ is also insensitive to chloride effects.

Johnson and Pytkowicz (1978, 1979a, 1979b) found that the correct total activity coefficients can be calculated for complex solutions based on results in binary ones, if γ_F is assumed to depend only on I_e at a given temperature and pressure. Similar results were obtained in bicarbonate solutions by Pytkowicz and Hawley (1974).

The extended ionic strength principle can be stated as follows: Free activity coefficients for ions of electrolytes are relatively invariant with composition at fixed effective ionic strengths. This result applies to single-ion as well as to mean free activity coefficients.

Chemical Potential of Ion Pairs. It is interesting to clarify the chemical potential of ion pairs in terms of the association equilibrium in order to avoid the pitfall of assuming that it can be disregarded at the standard state when there are no ionic interactions.

For the ion pairs CA^0,

$$\mu_{ip} = \mu^0_{ip} + RT \ln a_{ip} \tag{5.255}$$

while for the cations and anions

$$\mu_C = \mu^0_C + RT \ln a_C \tag{5.256}$$

$$\mu_A = \mu^0_A + RT \ln a_A \tag{5.257}$$

At equilibrium $\mu_{ip} = \mu_C + \mu_A$ by definition and $\mu^0_{ip} = \mu^0_C + \mu^0_A \neq 0$. Consequently,

$$K^* = \frac{a_{ip}}{a_C a_A}$$

$$= \exp\left[\frac{1}{RT}(\mu^0_C + \mu^0_A - \mu^0_{ip})\right] \tag{5.258}$$

5.7.2 Interconversions of Cl‰, S‰, I, and I_e

These interconversions do not take into account Cl^- ion pairs. The equation to convert I into S‰ is (Lyman and Fleming, 1940)

$$I = 0.00147 + 0.019885\,S‰$$

$$+ 0.000038\,(S‰)^2 \tag{5.259}$$

It can be used to determine γ_i versus S‰. The values of γ_i are obtained from the Culberson equation and the mean-salt method at given ionic strengths. Values of S‰ and Cl‰ versus I are presented in Table 5.25. The conversion of the salinity into the chlorinity and vice versa, based on S‰ = 1.80655 Cl‰, is shown in Table 5.26. Values of I and I_e (Courant et al., manuscript) are compared in Table 5.27. Values of $(\gamma_i)_F$ versus S‰, I, and I_e are shown in Table 5.28.

Atlas and Pytkowicz (unpublished results) calculated values of I versus Cl‰ using the chlorinity ratios given by Pytkowicz and Kester (1971) and the 1969 atomic weights. They compared the results to those obtained from the Lyman and Fleming (1940) equation. Several values of the pH (7.5, 8.1, 8.5) were used to calculate I at a specific alkalinity of 0.123 and 20°C.

TABLE 5.25 S‰ and Cl‰ versus the total ionic strength based on Equation (5.259) and S‰ = 1.80655 Cl‰.

S (‰)	I (molal)	S (‰)	I (molal)
20	0.4075		
21	0.4282	31	0.6379
22	0.4490	32	0.6591
23	0.4698	33	0.6803
24	0.4907	34	0.7016
25	0.5116	35	0.7229
26	0.5325	36	0.7442
27	0.5535	37	0.7657
28	0.5745	38	0.7871
29	0.5956	39	0.8086
30	0.6167	40	0.8301

Cl (‰)	I (molal)	Cl (‰)	I (molal)
10	0.3675	18	0.6701
11	0.4048	19	0.7085
12	0.4423	20	0.7471
13	0.4799	21	0.7858
14	0.5177	22	0.8246
15	0.5556	23	0.8636
16	0.5936	24	0.9027
17	0.6318	25	0.9420

TABLE 5.26 Interconversion of $S‰$ and $Cl‰$ based on $S‰ = 1.80655\ Cl‰$.

S	Cl	S	Cl
20	11.070	32	17.713
21	11.624	33	18.266
22	12.177	34	18.820
23	12.731	35	19.373
24	13.284	36	19.927
25	13.838	37	20.481
26	14.392	38	21.034
27	14.945	39	21.588
28	15.499	40	22.141
29	16.052	41	22.695
30	16.606	42	23.248
31	17.159		

Cl	S	Cl	S
10	18.065	18	32.517
11	19.872	19	34.324
12	21.678	20	36.131
13	23.485	21	37.937
14	25.291	22	39.744
15	27.098	23	41.550
16	28.904	24	43.357
17	30.711	25	45.163

TABLE 5.27 Values of I_e (Courant et al., manuscript) versus T and $S‰$ and an intercomparison of I, I_e, and $S‰$

	T				
S	5	10	15	20	25
30	0.5687	0.5701	0.5715	0.5730	0.5744
31	0.5876	0.5890	0.5904	0.5918	0.5932
32	0.6068	0.6081	0.6096	0.6110	0.6124
33	0.6263	0.6277	0.6291	0.6305	0.6319
34	0.6461	0.6475	0.6489	0.6503	0.6518
35	0.6663	0.6677	0.6691	0.6705	0.6720
36	0.6868	0.6882	0.6896	0.6910	0.6924
37	0.7076	0.7090	0.7104	0.7118	0.7132

Comparison of I_e with I_T

S‰	I_e	I_T
30	0.5744	0.6167
31	0.5932	0.6379
32	0.6124	0.6591
33	0.6319	0.6803
34	0.6518	0.7016
35	0.6720	0.7229
36	0.6924	0.7442
37	0.7132	0.7657

TABLE 5.28 Values of $(\gamma_i)_F$, the free activity coefficients for various salinities obtained from I, the total ionic strength, and I_E, the effective ionic strength.

	$(\gamma_i)_F$			
	K^+	Na^+	Ca^{2+}	Mg^{2+}
30‰ S				
$I_E = 0.579$	0.640	0.713	0.261	0.287
$I_T = 0.617$	0.636	0.712	0.259	0.286
34.8‰ S				
$I_E = 0.668$	0.629	0.711	0.257	0.285
$I_T = 0.718$	0.625	0.711	0.255	0.285
37‰ S				
$I_E = 0.713$	0.625	0.711	0.256	0.286
$I_T = 0.766$	0.620	0.711	0.254	0.285

The differences between the Atlas–Pytkowicz and Lyman–Fleming calculations are:

Cl‰	16	19
	-0.0001 ± 0.0026	$+0.0003 \pm 0.0031$

	22
	$+0.0006 \pm 0.0036$

These differences include the variation of I versus pH but are negligible in spite of the more recent data used by us.

The equation of Courant et al. is a preliminary one, based in part on the results of Kester and Pytkowicz (1969), and does not take into account chloride ion pairs. The equation, which contains considerable extrapolation, is

$$I_e = 0.16085 + 8.876 \times 10^{-3}\ S‰ + 1.636 \times 10^{-4}$$
$$(S‰)^2 - 2.826 \times 10^{-4}(25 - T°C) \quad (5.260)$$

5.7.3 The Mean-Salt Method

The mean-salt method has been used extensively by the school of Garrels (e.g., Garrels and Thompson, 1962; Lafon, 1969, Berner, 1971; van Breemen, 1973). This method is based on an early hypothesis of MacInnes (1919) to the effect that the activity of chloride ions in any 1-1 chloride solution is the same as that in a KCl solution of the same concentration.

Derivation of $\gamma_K = \gamma_{Cl}$

$$a_{Cl(MCl)} = a_{Cl(KCl)} \quad (5.261)$$

if $m_{MCl} = m_{KCl}$. The subscript in parentheses indicates the electrolyte. This hypothesis was based on transference numbers and conductance measurements as well as on the similarity in the electronic structures of K^+ and Cl^-. As $a_{Cl(KCl)} = (a_K a_{Cl})^{0.5}$,

$$a^2_{Cl(KCl)} = a_K a_{Cl} \quad \text{and} \quad a_K = a_{Cl} \quad (5.262)$$

This result leads to the fundamental relationship of the mean-salt method because $m_K = m_{Cl}$ yields, in conjunction with Equation (5.262),

$$\gamma_K = \gamma_{Cl} \quad (5.263)$$

The subscripts F and T are not needed because chlorides are assumed to be completely dissociated.

Another way to obtain this result is to write

$$\gamma_{Cl} m_{Cl} = \gamma_{\pm KCl} m_{\pm KCl} \quad (5.264)$$

or

$$\gamma_{Cl} = \gamma_{\pm KCl} \quad (5.265)$$

because $m_{\pm KCl} = m_{KCl} = m_{Cl}$ for a strong electrolyte. As

$$\gamma_{\pm KCl} = (\gamma_K \gamma_{Cl})^{0.5} = \gamma_{Cl} \quad (5.266)$$

then $\gamma_K \gamma_{Cl} = \gamma^2_{Cl}$ and

$$\gamma_{\pm KCl} = \gamma_K = \gamma_{Cl} \quad (5.267)$$

$\gamma_K = \gamma_{Cl}$ even if $m_K \neq m_{Cl}$ in a mixture of chloride because γ is independent of the composition of MCl–KCl in the absence of chloride association at a given ionic strength. Still, this equality is usually applied to a pure KCl solution.

Preliminary Discussion. Equation (5.267) is the MacInnes convention. At first I shall discuss it in its original form, proposed by MacInnes (1919) for 1-1 chlorides of concentration below 0.1 m. It was later extended by Garrels and Thompson (1962) to higher order electrolytes and to higher molalities. The convention cannot be proven but it is likely that it can provide useful approximate data since it received some support from conductance data and from the similar electronic structures of K^+ and Cl^-.

Let us examine MacInnes' (1919) argument in more detail. MacInnes observed that the product of

the transference number and the equivalent conductance of chloride ions is a constant at a given concentration for 1-1 chlorides. He concluded that the activity of chloride ions in these solutions is independent of the nature of the cations.

MacInnes assumed that univalent ions of nearly the same weight and size, such as K^+ and Cl^-, have the same activity and, therefore, that when K^+ and Cl^- are equimolar, then $\gamma_K = \gamma_{Cl}$. Based on this assumption, he was able to calculate single-ion γ's from potentiometric measurements. MacInnes showed that the emf calculated from his values of a_H and a_{Cl} agreed with the measurements by Lewis et al. (1917) in the case of 0.1 m HCl.

Garrels and Thompson (1962) extended the MacInnes convention to higher order electrolytes such as $MgCl_2$ and to $[Cl^-] = 0.7$ although the approximate validity of the convention under these conditions has not been tested. It is based on an extension of the assumptions of MacInnes to the effect that γ_{Cl} for 1-1 electrolytes is independent of the solution composition when $I \leq 0.1$ and that no Cl^- ion pairs are formed.

Use of the Convention. The use of the MacInnes convention to estimate single-ion activity coefficients can be illustrated, for instance, for Ca^{2+}. If the mean activity coefficient of $CaCl_2$ is known, then

$$\gamma_{Ca} = \frac{\gamma^3_{\pm CaCl_2}}{\gamma^2_{\pm KCl}} = \frac{\gamma_{Ca} \gamma^2_{Cl}}{\gamma^2_{Cl}} \quad (5.268)$$

The $CaCl_2$ and the KCl solution have the same I. We shall see later that the formation of chloride ion pairs limits the validity of this approach.

The values of single-ion activity coefficients for some chlorides are shown in Tables 5.29 and 5.30. They were obtained from the mean activity coefficients calculated by the Culberson equation used in conjunction with the MacInnes convention $\gamma_K = \gamma_{Cl}$, and from the relation between I and S‰.

It is important to realize that the work on single ions described in this part of the book was done before the existence of chloride ion pairs was ascertained by us (Kester and Pytkowicz, 1968, 1969; Johnson and Pytkowicz, 1978, 1979a). The results are presented anyway for several reasons:

1. The results for chloride ion pairs are relatively difficult to obtain and some readers may wish to start with the procedures and

TABLE 5.29 Single-ion activity coefficients at 25°C based on the Culberson Equation (5.174) and the mean-salt method (Pytkowicz et al., 1977).

Ionic Strength I (Total Molal)	Activity Coefficients			
	K^+ and Cl^-	Na^+	Mg^{2+}	Ca^{2+}
0.00	1.000	1.000	1.000	1.000
0.01	0.901	0.904	0.670	0.676
0.02	0.869	0.874	0.588	0.596
0.03	0.847	0.854	0.537	0.547
0.04	0.830	0.838	0.501	0.513
0.05	0.816	0.826	0.473	0.486
0.06	0.804	0.816	0.451	0.564
0.07	0.794	0.807	0.433	0.447
0.08	0.784	0.800	0.417	0.432
0.09	0.776	0.793	0.404	0.419
0.10	0.769	0.787	0.392	0.408
0.11	0.762	0.782	0.382	0.398
0.12	0.755	0.777	0.373	0.389
0.13	0.750	0.772	0.365	0.381
0.14	0.744	0.768	0.357	0.374
0.15	0.739	0.765	0.351	0.368
0.16	0.734	0.761	0.344	0.362
0.17	0.730	0.758	0.339	0.356
0.18	0.726	0.755	0.334	0.351
0.19	0.721	0.752	0.329	0.347
0.20	0.718	0.750	0.325	0.343
0.21	0.714	0.748	0.320	0.339
0.22	0.711	0.745	0.317	0.335
0.23	0.707	0.743	0.313	0.332
0.24	0.704	0.741	0.310	0.328
0.25	0.701	0.739	0.308	0.326
0.26	0.698	0.738	0.304	0.323
0.27	0.695	0.736	0.301	0.320
0.28	0.692	0.734	0.298	0.318
0.29	0.690	0.733	0.296	0.316
0.30	0.687	0.732	0.294	0.313
0.31	0.685	0.730	0.292	0.311
0.32	0.682	0.729	0.289	0.310
0.33	0.680	0.728	0.288	0.308
0.34	0.678	0.727	0.286	0.306
0.35	0.676	0.726	0.284	0.305
0.36	0.674	0.725	0.282	0.303
0.37	0.672	0.724	0.281	0.302
0.38	0.670	0.723	0.279	0.300
0.39	0.668	0.722	0.278	0.299
0.40	0.666	0.721	0.277	0.298
0.41	0.664	0.721	0.275	0.297
0.42	0.662	0.720	0.274	0.296
0.43	0.661	0.719	0.273	0.295
0.44	0.659	0.719	0.272	0.294
0.45	0.657	0.718	0.271	0.293
0.46	0.656	0.718	0.270	0.293
0.47	0.654	0.717	0.269	0.292
0.48	0.653	0.716	0.268	0.291
0.49	0.651	0.716	0.267	0.290

TABLE 5.29 (*Continued*)

| Ionic Strength I (Total Molal) | Activity Coefficients | | | |
	K^+ and Cl^-	Na^+	Mg^{2+}	Ca^{2+}
0.50	0.650	0.716	0.266	0.290
0.51	0.648	0.715	0.265	0.289
0.52	0.647	0.715	0.265	0.289
0.53	0.646	0.714	0.264	0.288
0.54	0.644	0.714	0.263	0.288
0.55	0.643	0.147	0.263	0.288
0.56	0.642	0.713	0.262	0.287
0.57	0.641	0.713	0.261	0.287
0.58	0.639	0.713	0.261	0.287
0.59	0.638	0.713	0.260	0.286
0.60	0.637	0.712	0.260	0.286
0.61	0.636	0.712	0.259	0.286
0.62	0.635	0.712	0.259	0.286
0.63	0.634	0.712	0.258	0.286
0.64	0.633	0.712	0.258	0.285
0.65	0.631	0.712	0.258	0.285
0.66	0.630	0.711	0.257	0.285
0.67	0.629	0.711	0.257	0.285
0.68	0.628	0.711	0.257	0.285
0.69	0.627	0.711	0.256	0.285
0.70	0.626	0.711	0.256	0.285
0.71	0.625	0.711	0.256	0.285
0.72	0.625	0.711	0.255	0.285
0.73	0.624	0.711	0.255	0.285
0.74	0.623	0.711	0.255	0.285
0.75	0.622	0.711	0.255	0.286
0.76	0.621	0.711	0.255	0.286
0.77	0.620	0.711	0.254	0.286
0.78	0.619	0.711	0.254	0.286
0.79	0.618	0.711	0.254	0.286
0.80	0.618	0.711	0.254	0.286
0.81	0.617	0.711	0.254	0.287
0.82	0.616	0.711	0.254	0.287
0.83	0.615	0.711	0.254	0.287
0.84	0.614	0.711	0.254	0.287
0.85	0.614	0.711	0.253	0.288
0.86	0.613	0.711	0.253	0.288
0.87	0.612	0.711	0.253	0.288
0.88	0.611	0.712	0.253	0.288
0.89	0.611	0.712	0.253	0.289
0.90	0.610	0.712	0.253	0.289
0.91	0.609	0.712	0.253	0.289
0.92	0.609	0.712	0.253	0.290
0.93	0.608	0.712	0.253	0.290
0.94	0.607	0.712	0.253	0.291
0.95	0.607	0.712	0.253	0.291
0.96	0.606	0.713	0.253	0.291
0.97	0.605	0.713	0.253	0.292
0.98	0.605	0.713	0.253	0.292
0.99	0.604	0.713	0.253	0.293
1.00	0.603	0.713	0.253	0.293

TABLE 5.30 Free activity coefficients of single ions at 25°C and the salinity, calculated from Table 5.29 and the MacInnes assumption.

Salinity (‰)	K^+	Na^+	Ca^{2+}	Mg^{2+}	Cl^-
25	0.648	0.715	0.265	0.289	0.648
26	0.646	0.714	0.264	0.288	0.646
27	0.643	0.714	0.263	0.288	0.643
28	0.641	0.713	0.261	0.287	0.641
29	0.637	0.712	0.260	0.286	0.637
30	0.635	0.712	0.259	0.286	0.635
31	0.633	0.712	0.258	0.285	0.633
32	0.630	0.711	0.257	0.285	0.630
33	0.628	0.711	0.257	0.285	0.628
34	0.626	0.711	0.256	0.285	0.626
35	0.625	0.711	0.255	0.285	0.625
36	0.623	0.711	0.255	0.285	0.623
37	0.620	0.711	0.254	0.286	0.620
38	0.618	0.711	0.254	0.286	0.618
39	0.617	0.711	0.254	0.287	0.617
40	0.615	0.711	0.254	0.287	0.615

some results for SO_4^{2-}, HCO_3^-, and CO_3^{2-} pairs alone before extending their models to chlorides.

2. Not all solutions of interest, especially in the laboratory, contain significant amounts of chloride ions as I mentioned earlier.

3. The historical development of ion pairs will give the reader a sound grasp of the subject.

4. The comparison between models without and with chloride association will help us decide the type of results (e.g., the effects of T and P) not yet worked out in the latter case.

The Mean-Salt Method: A Further Discussion. Let us consider the mean-salt method in terms of γ_F, γ_T, and γ_e with Cl^- ion pairs taken into consideration.

MacInnes (1919), Garrels and Thompson (1962), and many following workers assumed that chloride ions do not associate with alkali and alkaline–earth ions. Therefore, the conditions

$$(\gamma_K)_{exp} = (\gamma_K)_T = (\gamma_K)_F = (\gamma_{Cl})_{exp}$$
$$= (\gamma_{Cl})_T = (\gamma_{Cl})_F \qquad (5.269)$$

were implicit in their work. It is likely that these conditions were approximately met in pure KCl at the highest dilutions used by MacInnes because the degree of association was slight so that $(\gamma_i)_T \cong (\gamma_i)_F$. This is not the case at the higher concentrations that he studied (0.1 m) and in the work of Garrels and Thompson (1962) and others at $m \cong 0.7$.

The problem that I shall consider next is whether the use of γ_F or γ_T is the correct one in terms of the original MacInnes equality $a_K = a_{Cl}$, and Cl^- ion pairs. This equality can be expanded to

$$(\gamma_K)_F(m_K)_F = (\gamma_{Cl})_F(m_{Cl})_F = (\gamma_K)_T(m_K)_T$$
$$= (\gamma_{Cl})_T(m_{Cl})_T \qquad (5.270)$$

with $\gamma_F \neq \gamma_T$ and $m_F \neq m_T$. These inequalities differ from the relations (5.269) due to chloride association.

Our results discussed earlier showed that it is γ_F rather than γ_T which depends only on the effective ionic strength I_e and not on the composition of the solution. From Equation (5.210) this means that

$$(\gamma_K)_F = (\gamma_{Cl})_F \qquad (5.271)$$

When $(m_K)_F = (m_{Cl})_F$ and that Equation (5.271) is also valid when the free molalities differ, if I_e is kept invariant. Thus, the modified MacInnes convention is that $(\gamma_K)_F$ and $(\gamma_{Cl})_F$ are equal at any given effective ionic strength, regardless of the solution composition, for example, KCl + NaCl versus KCl + LiCl. The extension of the rule to higher valence electrolytes such as $CaCl_2$, and to ionic strengths above 0.1 m is tentative but perhaps time will show that it is useful.

From Equation (5.270) we find that $(\gamma_K)_T = (\gamma_{Cl})_T$ only when $(m_K)_T = (m_{Cl})_T$ in a pure KCl solution, since γ_T, in contrast to γ_F, is a function of composition. This happens because in $a = \gamma_T m_T$ it is γ_T that absorbs the effects of ion pairing while in $a = \gamma_F m_F$ the molality m_F reflects the ionic association. Therefore, $(\gamma_K)_F = (\gamma_{Cl})_F$ should be used instead of $(\gamma_K)_T (\gamma_{Cl})_T$ if the MacInnes convention and chloride association are taken into consideration.

The resulting association constants for alkali chlorides should be small due to the low charges of the ions but the ionic concentrations are high in media such as seawater and offset the low constants.

5.7.4 The Mean-Salt Method Compared to the Debye–Hückel Equation

Before leaving this subject, it is of interest to compare results for single ions obtained from the mean-salt method, without including chloride association, and the Debye–Hückel equation. I sought the values of a_D which provided the best fit for this equation.

Note that the values of γ_C, where C is a univalent cation, are actually obtained by

$$(\gamma_C)_T \cong \frac{\gamma_{\pm CCl}^2{}_T}{(\gamma_{\pm KCl})_T} \qquad (5.272)$$

and are, therefore, rough approximations since γ_{Cl} in the numerator and in the denominator are not the same. They cancel out and $(\gamma_C)_T$ approaches the desired $(\gamma_C)_F$ only to the extent that CCl and KCl form similar fractions of ion pairs at a given I_e. This can be seen from

$$(\gamma_C)_F = \frac{(\gamma_{\pm CCl})_F^2}{(\gamma_{\pm KCl})_F} \frac{(m_{CCl})_F/(m_{CCl})_T}{(m_{KCl})_F/(m_{KCl})_F} \qquad (5.273)$$

which results from $a = \gamma_F m_F = \gamma_T m_T$ as we saw earlier. These considerations can be readily extended to 2-1 electrolytes.

EXAMPLE 5.7. A COMPARISON OF $(\gamma_{Na})_T$ AND $(\gamma_{Na})_F$ OBTAINED BY THE MEAN-SALT METHOD. According to the results of Johnson and Pytkowicz (1979a), $(\gamma_{\pm NaCl})_T = 0.6626$, $(\gamma_{\pm NaCl})_F = 0.7251$, $(\gamma_{\pm KCl})_T = 0.6264$, and $(\gamma_{\pm KCl})_F = 0.6264$ at $I = 0.7$. Thus, neglecting the differences between I and I_e,

$$(\gamma_{Na})_T \cong \frac{(0.6626)^2}{0.6264} = 0.7010 \qquad (i)$$

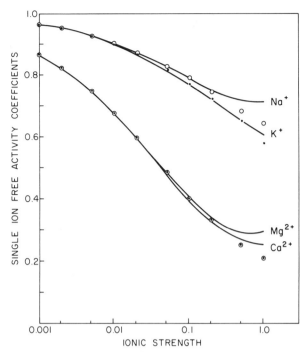

FIGURE 5.19 Fit between the Debye–Hückel equation (solid line) and the mean-salt method, without considering Cl^- ion pairs. The symbols for the MSM results are o for (Na^+, Mg^{2+}) and ⊙ for (K^+, Ca^{2+}). The values of a_D used were K^+ (4 Å), Na^+ (5 Å), Mg^{2+} (6 Å), Ca^{2+} (6 Å).

and

$$(\gamma_{Na})_F \cong \frac{(0.7251)^2}{0.6264} = 0.8393 \qquad (ii)$$

$I_e = 0.5892$ for NaCl and 0.5472 for KCl. The disagreement between γ_T and γ_F is even larger for 2-1 electrolytes. 0.8393 is the correct quantity if the method is assumed to be valid at $I = 0.7$. A better comparison would have been obtained if the γ's had been calculated at the same I_e.

The Debye–Hückel equation is empirical at concentrations above 10^{-3} m, whereas the mean-salt method is not valid because of Cl^- ion pairing. The values of a_D used to force fit γ_{DH} to γ_{MSM} differ slightly from those used to fit $\gamma_{\pm DH}$ for the chlorides of the cations to $\gamma_{\pm exp}$. The fit shown by the data in Figure 5.19 does not confirm that the mean-salt method is theoretically sound as has been inferred in the literature. Rather, it shows that we have two empirical compatible tools.

In Table 5.31 I contrast the use of I versus I_e for the calculation of γ_F of ions at given salinities and 25°C. The effect of using I versus I_e is negligible for

TABLE 5.31 A comparison of the use of I and I_e for the estimate of γ_F of single ions versus $S‰$. Values of I versus $S‰$ were obtained from Equation (5.259). The corresponding I_e versus $S‰$ were calculated by means of Equation (5.260). Values of γ were then read from Table 5.29. The results show the effect of neglecting the difference between I and I_e on γ. Note that Cl^- ion pairs were neglected.

			γ			
	K	Na	Ca	Mg	HCO₃	CO₃
30‰S						
$I_e = 0.5794$	0.640	0.713	0.261	0.287	0.684	0.220
$I = 0.6167$	0.636	0.712	0.259	0.286	0.680	0.216
37‰S						
$I_e = 0.7132$	0.625	0.711	0.256	0.285	0.672	0.206
$I = 0.7657$	0.620	0.711	0.254	0.286	0.668	0.201

K^+, HCO_3^-, and CO_3^{2-} and is sizable for the other ions. One should keep in mind, however, that the differences are minimal because Cl^- ion pairing was not taken into consideration.

5.7.5 Formal Relation Between γ_F and γ_T

At this point I shall derive an important equation needed for several chapters of this book. The association constant for ion pairs CA^0 in the simple case of a symmetric electrolyte

$$K^{*\prime} = \frac{[CA^0]}{[C]_F[A]_F} \qquad (5.274)$$

or

$$[CA^0] = K^{*\prime}\,[C]_F[C_A]_F \qquad (5.275)$$

Also,

$$[C]_T = [C]_F + [CA^0]$$

$$= C_F[1 + K^{*\prime}\,[A]_F] \qquad (7.275)$$

As $(\gamma_C)_F/(\gamma_C)_T = [C]_T/[C]_F$,

$$\frac{(\gamma_C)_F}{(\gamma_C)_T} = (1 + K^{*\prime}[A]_F) \qquad (5.277)$$

5.7.6 The Garrels and Thompson (1962) Method and Critical Comparisons

My scheme to present the model of Garrels and Thompson is as follows:

1. To show how these authors obtained γ_F and to discuss the results.

2. To examine critically their estimates of γ_{ip}.

3. To present the values of K^* and to compare them to those obtained by other workers.

4. Outline the overall method of Garrels and Thompson and compare the speciation to those of others.

The method of Garrels and Thompson (1962) was a pioneering effort which required a number of assumptions that are now unwarranted. Still, it will be presented critically and in greater detail than in the authors' presentation. The reason for this presentation is that it will permit readers to select knowingly those steps that are still acceptable and to modify the other ones for their specific problem.

Furthermore, Garrels and Thompson (1962) applied their method only at $I = 0.70$ in seawater but it can be extended to other solutions. The main value of the method is that it requires a minimum of measurements, in contrast to other ion-pairing procedures. It provides, therefore, fast rough estimates of the ion pairing in solutions of interest in which experimental data have not yet been obtained and cannot be obtained within the time available to the reader.

Activity Coefficients. Next the methods used by Garrels and Thompson to obtain the activity coefficients shown in Table 5.32 are presented. The coefficients were used in conjunction with thermodynamic association constants of the type shown below for a symmetric electrolyte.

$$K^*_{CA} = \frac{\gamma_{CA}}{(\gamma_C)_F(\gamma_A)_F}\,\frac{[CA^0]}{[C]_F[A]_F} \qquad (5.278)$$

TABLE 5.32 Activity coefficients used by Garrels and Thompson for seawater of $I = 0.70$ ($Cl‰ = 19$) at 25°C and 1 atm total pressure

Single Ions		Ion Pairs		
Ion	γ_F		Ion Pair	γ_{ip}
Na^+	0.76	Uncharged	$NaHCO_3^0$	1.13
K^+	0.64		$MgCO_3^0$	1.13
Mg^{2+}	0.36		$CaCO_3^0$	1.13
Ca^{2+}	0.28		$MgSO_4^0$	1.13
SO_4^{2-}	0.12		$CaSO_4^0$	1.13
HCO_3^-	0.68	Net charge $+1$	$MgHCO_3^+$	0.68
CO_3^{2+}	0.20		$CaHCO_3^+$	0.68
		Net charge -1	$NaCO_3^-$	0.68
			$NaSO_4^-$	0.68
			KSO_4^-	0.68

These constants will be examined later. Equation (5.278) requires the use of free activity coefficients but it will be shown below that some of the γ's used were actually total ones. Still, the general method is of interest because it can be corrected for this, for Cl^- ion pairs, and for other deficients that shall be pointed out.

The derivation will be presented critically and steps glossed over by the authors will be expanded for greater clarity.

Free Chloride Activity Coefficients. Expressions such as

$$(\gamma_{Na})_F = \frac{\gamma^2_{\pm NaCl}}{\gamma_{\pm KCl}} \qquad (5.279)$$

$$(\gamma_{Ca})_F = \frac{\gamma^2_{\pm CaCl_2}}{\gamma_{\pm KCl}} \qquad (5.280)$$

were used to calculate the single-ion γ's for 1-1, 1-2, and 2-1 electrolytes. These values were assumed to correspond to free coefficients although, as was discussed earlier, this is not correct.

It is interesting to compare values of $(\gamma_i)_F$ from values of $(\gamma_\pm)_F$ obtained by Johnson and Pytkowicz (1979a) and the mean-salt method with those of Garrels and Thompson (1962).

EXAMPLE 5.8. COMMON USE OF THE MEAN-SALT METHOD AND COMPARISON OF THE ACTIVITY COEFFICIENTS OF γ_T (GARRELS AND THOMPSON, 1962) WITH THOSE CALCULATED FROM JOHNSON AND PYTKOWICZ (1978, 1979a, 1979b)

1. Values of $\gamma_{\pm F}$ from Johnson and Pytkowicz at $I = 0.6$.

	KCl	NaCl	MgCl_2
$(\gamma_\pm)_F$	0.808	0.805	0.699

2. Calculation of $(\gamma_i)_F$ by the mean-salt method.

$$(\gamma_{Na})_F = \frac{(\gamma_{\pm NaCl})^2_F}{(\gamma_{\pm KCl})_F} = 0.802 \qquad (i)$$

$$(\gamma_{Mg})_F = \frac{(\gamma_{\pm MgCl_2})^3_F}{(\gamma_{\pm KCl})_F} = 0.423 \qquad (ii)$$

3. Comparison with Garrels and Thompson (1962) at $I = 0.7$. Note that Garrels and Thompson used total quantities obtained from measurements but assumed that they were free ones.

	$(\gamma_i)_F$	
	Johnson and Pytkowicz	Garrels and Thompson
Na^+	0.802	0.76
Mg^{2+}	0.423	0.36

It can be seen that the differences in γ between the two sets of results is not very large and would be a bit closer if they had been reported at the same total ionic strength. The effect of the differences is amplified, however, when ionic speciations are determined.

The meaning of the γ's obtained by Garrels and Thompson is not clear-cut because $(\gamma_{Cl})_T$ differs in different solutions due to ion pairing.

The ionic speciation, in contrast to the γ's, differs considerably when chloride ion pairs are neglected (Garrels and Thompson, 1962; Kester and Pytkowicz, 1969) and when they are taken into consideration (Johnson and Pytkowicz, 1978, 1979a, 1979b) as is shown in Table 5.28.

Note that the activities obtained by the two methods differ as we have

$$a = \gamma_F \frac{m_F}{m_T} = 0.802 \times 0.851 = 0.682$$

for Na^+ (Johnson and Pytkowicz, 1979a), and

$$a = \gamma_{GT} \frac{m_F}{m_T} = 0.76 \times 0.99 = 0.75$$

for Na^+ (Garrels and Thompson, 1962). This results from the fact that $\gamma_{GT} = \gamma_T$ which equals $(\gamma_{\pm NaCl})\exp$ while γ_F must be used in the above equations.

Free Sulfate Activity Coefficient. $(\gamma_{SO_4})_F$ could not be calculated by the mean-salt method because all known sulfate salts are associated. Therefore, Garrels and Thompson estimated $(\gamma_{SO_4})_F$ from:

γ_K obtained by the mean-salt method $(\gamma_K)_F = (\gamma_{Cl})_F = (\gamma_{\pm KCl})_{exp}$.

γ_{KSO_4} assumed to be the same as γ_{HCO_3} which was determined as is shown later on. This assumption is based on similar e/r ratios for the species KSO_4^- and HCO_3^-, and is a questionable one. KSO_4^- is larger even without considering its water of hydration. Note that KSO_4^- is an ion pair that retains its water of hydration while HCO_3^- is not. Furthermore, KSO_4^- is quite polar since K^+ and SO_4^{2-} retain their charges.

$K^*_{KSO_4} = \{KSO_4^-\}/\{K^*\}_F \{SO_4^{2-}\}_F$ obtained from the value in the literature.

$(\gamma_{\pm K_2SO_4})_T$ obtained from Harned and Owen (1958).

The procedure used to calculate $(\gamma_{SO_4})_F$ was not presented but probably was of the following type.

The unknowns, related to the given quantities, are $[KSO_4^-]$, $[K^+]_F$, $[SO_4^{2-}]_F$, $(\gamma_{SO_4})_F$, $(\gamma_K)_T$, and $(\gamma_{SO_4})_T$ and six equations are required to determine them. These may be

$$(\gamma_{\pm K_2SO_4})_T^3 = (\gamma_K)_T^2 (\gamma_{SO_4})_T \qquad (5.281)$$

$$K^*_{KSO_4} = \frac{\gamma_{KSO_4}[KSO_4^-]}{(\gamma_K)_F[K^+]_F[\gamma_{SO_4}]_F[SO_4^{2-}]_F} \qquad (5.282)$$

$$[K^+]_T = [K^+]_F + [KSO_4^-] \qquad (5.283)$$

$$[SO_4^{2-}]_T = [SO_4^{2-}]_F + [KSO_4^-] \qquad (5.284)$$

$$[K^+]_F = 2[SO_4^{2-}]_F + [KSO_4^-] \qquad (5.285)$$

$$\frac{(\gamma_K)_F}{[K^+]_T} = \frac{(\gamma_K)_T}{[K^+]_F} \qquad (5.286)$$

The term on the left in the last equation is known. Equation (5.285) is the condition of electrical neutrality.

EXAMPLE 5.9. CALCULATION OF $(\gamma_\pm)_F$ FOR TABLE 5.33. I used the following equations, in conjunction with the data of Johnson and Pytkowicz (1979a):

For a symmetric electrolyte (1-1, 2-2, etc.), represented by CA

$$(\gamma_{\pm CA})_F^2 = (\gamma_{\pm CA})_T^2 \frac{(m_C)_T^2 (m_A)_T}{(m_C)_F (m_A)_F} \qquad (i)$$

where $\gamma_{\pm CA} = (\gamma_C \gamma_A)^{0.5}$.

For a 2-1 electrolyte,

$$(\gamma_{\pm C_2A})_F^3 = (\gamma_{\pm C_2A})_T^3 \frac{(m_C)_T^2(m_A)_T}{(m_C)_F^2(m_A)_F} \qquad (ii)$$

This results from the following relationships:

$$a_{ca} = (a_C)^{v+} (a_A)^{v-} \qquad (iii)$$

$$a_C = \gamma_C m_C \qquad a_A m_A \qquad (iv)$$

and, therefore

$$a_{CA} = \gamma_C^{v+} \gamma_A^{v-} m_C^{v+} m_A^{v-} \qquad (v)$$

As

$$\gamma_C^{v+} \gamma_A^{v-} = (\gamma_{\pm C_2A})^v \qquad (vi)$$

we have

$$a = (\gamma_{\pm C_2A})^v (m_C)^{v+} (m_A)^{v-} \qquad (vii)$$

TABLE 5.33 Values of $(\gamma_\pm)_F$ based on the results of Johnson (1979) and Johnson and Pytkowicz (1978, 1979a) and on those of Garrels and Thompson (1962).

Electrolyte	$(\gamma_\pm)_T$[a]	$(\gamma_\pm)_F$	
		Johnson and Pytkowicz	Garrels and Thompson
NaCl	0.661	0.796	0.70
KCl	0.642	0.798	0.64
MgCl$_2$	0.467	0.676	0.53
CaCl$_2$	0.458	—	0.49
Na$_2$SO$_4$	0.354	0.739	0.41
K$_2$SO$_4$	0.351	0.763	0.37
MgSO$_4$	0.155	0.559	0.21
CaSO$_4$	0.151	—	0.18

[a] From Johnson (1979) at the effective ionic strength of seawater.

For the case of Garrels and Thompson, I simply used their value for $(\gamma_C)_F$ and $(\gamma_A)_F$ and obtained the mean free coefficients from the expressions

$$(\gamma_{\pm CA})_F = [(\gamma_C)_F (\gamma_A)_F]^{0.5} \qquad \text{(viii)}$$

$$\gamma_{\pm C_2 A} = [(\gamma_C)_F^2 (\gamma_A)_F]^{1/3} \qquad \text{(ix)}$$

The results are presented in Table 5.33 in which one observes the expected depression of γ_T due to the formation of ion pairs. The effect is much larger in our case than in that of Garrels and Thompson because we took into account not only the effect of SO_4^{2-} ion pairs but also that of Cl^- association. Ion pairs with HCO_3^- and CO_3^{2-} are included in the Garrels–Thompson column but not in the Johnson–Pytkowicz one. The corrections, however, are slight for the γ's of the cations. This is due to their high concentrations relative to those of SO_4^{2-} and of the CO_2 species.

HCO$_3^-$ and CO$_3^{2-}$ Activity Coefficients (The Method of Walker et al., 1927). Garrels and Thompson (1962) stated that they used values of $(\gamma_{\pm KHCO_3})_F$ and $(\gamma_{\pm K_2CO_3})_F$ obtained from Walker et al. (1927), in conjunction with the mean-salt method in the form

$$(\gamma_{HCO_3})_F = \frac{(\gamma_{\pm KHCO_3})_F^2}{\gamma_{\pm KCl}} \qquad (5.287)$$

$$(\gamma_{CO_3})_F = \frac{(\gamma_{\pm K_2CO_3})_F^2}{\gamma_{\pm KCl}^2} \qquad (5.288)$$

However, as it will be shown next, Walker et al. (1927) determined total activity coefficients. Their approach is developed fully because it may be useful for the determination of total coefficients in carbonate solutions.

Several problems appear in the procedure of Garrels and Thompson (1962) which should be avoided by workers who elect their method.

Chloride ion pairs were not considered in the mean-salt calculation.

Walker et al. (1927) determined $(\gamma_{HCO_3})_T$ and $(\gamma_{CO_3})_T$ although Garrels and Thompson (1962) assumed in Table 5.32 and in Equations (5.287) and (5.288) that they were values of the free activity coefficients.

Total activity coefficients depend on the composition of the medium and, therefore, the results of Walker et al. (1927) apply only to the KCl and to NaCl solutions used by them. Table 5.34 shows that $(\gamma_{HCO_3})_T$ and $(\gamma_{CO_3})_T$ are depressed more by Na^+ than by K^+. This result is surprising because K^+, due to its smaller hydrated radius, is expected to ion pair more than Na^+ when $m_K = m_{Na}$.

Two methods were used by Walker et al., titrations with a colorimetric end point and potentiometric measurements. The potentiometric determinations were carried out only up to $I = 0.1$ and the assumption was made in the calculations that the Debye–Hückel limiting ratio $d \log \beta / d \log \alpha = 4$ is valid up to this ionic strength. The two sets of results were in good agreement. α and β are $(\gamma_{HCO_3})_T$ and $(\gamma_{CO_3})_T$.

The nonthermodynamic assumption, required to obtain γ's for single ions in both methods, was that the single-ion a_H can be obtained from pH measurements.

TABLE 5.34 Colorimetric values at 25°C versus the ionic strength I for $(\gamma_{HCO_3})_T$ and $(\gamma_{CO_3})_T$.

	Potassium		Sodium	
I	HCO_3^-	CO_3^{2-}	HCO_3^-	CO_3^{2-}
0.00	1.000	1.000	1.000	1.000
0.01	0.904	0.667	0.901	0.658
0.02	0.876	0.589	0.871	0.575
0.04	0.843	0.503	0.835	0.485
0.06	0.821	0.453	0.810	0.431
0.08	0.803	0.415	0.790	0.389
0.10	0.790	0.389	0.773	0.356
0.20	0.746	0.309	0.710	0.255
0.40	0.704	0.246	0.638	0.165
0.60	0.682	0.216	0.591	0.122
0.80	0.667	0.198	0.557	0.096
1.00	0.654	0.183	0.529	0.079
1.50	0.636	0.164	0.495	0.055
2.00	0.627	0.154	0.452	0.042
2.50	0.619	0.147	0.428	0.033

Source: Reprinted from A. C. Walker, V. B. Bray, and J. Johnson, *American Chemical Society* **49**, 1235, published in 1927 by the American Chemical Society.

Walker et al. wrote the expressions

$$K_1 = \frac{a_H\{HCO_3^-\}}{\{H_2CO_3\}} \qquad K_2 = \frac{a_H\{CO_3^{2-}\}}{\{HCO_3^-\}} \qquad (5.289)$$

$$[H_2CO_3^*] = \gamma_{CO_2}\gamma_{H_2O}(TCO_2)(pCO_2) \quad (5.290)$$

from which

$$\frac{(1/pCO_2)\{HCO_3^-\}^2}{\{CO_3^{2-}\}} = \gamma_{CO_2}\gamma_{H_2O}TCO_2\frac{K_1}{K_2} \quad (5.291)$$

Note that a_H and $\{i\}$ represent activities.

The solutions contained Na_2CO_3 or K_2CO_3 and $Na_2CO_3 + NaCl$ or $K_2CO_3 + KCl$. They were equilibrated with $CO_{2(g)}$ of known pCO_2 and were analyzed for the alkalinity $[HCO_3^-[_T + 2[CO_3^{2-}]_T$, TCO_2, and $[HCO_3^-]$ so that $[CO_3^{2-}]$ could be calculated. Actually the measurements yielded $[MHCO_3]$ and $[M_2CO_3]$ and the activities were obtained from

$$\{HCO_3^-\} = \alpha[MHCO_3] = \alpha[HCO_3^-]_T \quad (5.292)$$

and

$$\{CO_3^{2-}\} = \beta[M_2CO_3] = \beta[CO_3^{2-}]_T \quad (5.293)$$

α and β are the ion-activity coefficients $(\gamma_{HCO_3})_T$ and $(\gamma_{CO_3})_T$ because $[HCO_3^-]_T$ and $[CO_3^{2-}]_T$ are total quantities.

Walker et al. (1927) then derived

$$g = \frac{[MHCO_3]^2}{[M_2CO_3]pCO_2} = \gamma_{CO_2}\gamma_{H_2O}TCO_2\frac{K_1}{K_2}\frac{\beta}{\alpha^2} \quad (5.294)$$

from the above equations. Values of g, for each pCO_2, were calculated from $[MHCO_3]$, $[M_2CO_3]$, and pCO_2. They were then extrapolated to infinite dilution ($M \rightarrow 0$). $TCO_2 = (TCO_2)^0$ at this limit is the solubility of CO_2 in water calculated for $pCO_2 = 1$. $\gamma_{CO_2} = \gamma^0_{CO_2}$ is known, and $\gamma_{H_2O} = 1$.

In addition, α and β tend to unity. Thus, the ratio K_1/K_2 can be calculated and, as the two thermodynamic dissociation constants are independent of the ionic strength, this ratio can be applied at the original ionic strength for the calculation of β/α^2.

Activity Coefficients of Ion Pairs. The activity coefficient of nonelectrolytes, 1.13, was chosen by Garrels and Thompson (1962) for pairs with no net charge despite the fact that ion pairs are dipolar entities (Table 5.32).

It is of interest to compare γ_{ip} assumed by these workers with those estimated by Kester (manuscript)

TABLE 5.35 Activity coefficients of dipolar ions with zero net charge in aqueous sodium chloride and in potassium chloride. The measured values are based on solubility, emf, vapor pressure, or freezing point measurements (compilation by C. H. Culberson). Calculated values of $\gamma_{\pm T}$ for HCl^0, $NaCl^0$, and KCl^0 according to Johnson and Pytkowicz (1978, 1979a, 1979b).[a]

| Ionic Strength (Molal) | Glycine | | | Asparagine | Alanine | Cystine | HCl^0 | $NaCl^0$ | KCl^0 |
| | NaCl | | KCl | | | | | | |
	25°C	0°C	25°C	NaCl (25°)	NaCl (25°)	NaCl (25°)	25°C	25°C	25°C
0.0	1.000	1.000	1.000	1.000	1.000	1.000	1.000	1.000	1.000
0.1	0.975	0.954	0.970	0.950	0.968	0.916	0.995	0.953	0.914
0.2	0.954	0.921	0.947	0.916	0.943	0.867	1.070	1.010	0.913
0.3	0.936	0.893	0.928	0.888	0.922	0.832			
0.4	0.919	0.870	0.912	0.863	0.904	0.803	1.360	1.240	1.000
0.5	0.903	0.850	0.901	0.841	0.889	0.779			
0.6	0.889	0.832	0.892	0.821	0.875	0.758	1.800	1.530	1.140
0.7	0.876	0.818	0.881	0.801	0.864	0.739			
0.8	0.863	0.806	0.871	0.783	0.854	0.721	2.400	1.880	1.300
0.9	0.852	0.796	0.862	0.765	0.846	0.704			
1.0	0.841	0.790	0.854	0.747	0.838	0.689			

[a]The dipole moments of these molecules are glycine = 14, asparagine = 15, alanine = 17, cystine = 27.

Charge type γ_{ip}	Zero spherical symmetric	Zero, dipolar	Univalent, dipolar
	1.13	0.8	0.4

and with the data in Tables 5.35 and 5.36.

The values of Kester are rule of thumb values based on data in the Ph.D. thesis of Kester (1970). Kester did not present the method used to obtain γ_{ip} but it probably was similar to that of Johnson (1979) which follows.

If K^* is known from the literature and $K^{*\prime}$ is measured, then

$$\gamma_{ip} = \gamma_{CA} = \frac{K^*}{K^{*\prime}} (\gamma_{\pm CA})_F^2 \qquad (5.295)$$

where

$$(\gamma_{\pm CA})_F^2 = (\gamma_{\pm CA})_T^2 \frac{[C]_T [A]_T}{[C]_F [A]_F} \qquad (5.296)$$

The molality ratios were obtained from Kester and Pytkowicz (1969).

In the case of sulfate, for seawater at $I_e = 0.67$ (Kester, 1970), and, therefore

	ΔF^*	$\Delta F^{*\prime}$	(Seawater, $I_e = 0.67$)
$NaSO_4^-$	−0.98	−0.42	
$MgSO_4^-$	−3.22	−1.38	
$CaSO_4^-$	−3.15	−1.41	

with

$$\Delta F_{MSO_4}^{*\prime} = -RT \ln K_{MSO_4}^{*\prime} \qquad (5.297)$$

Of course now chloride ion pairs must be considered in conjunction with sulfate ones.

C. H. Culberson (personal communication) assembled the data for the activity coefficients of the

TABLE 5.36 Activity coefficients (molal) of uncharged molecules and atoms (nonelectrolytes) in 35‰ seawater (compiled by C. H. Culberson, personal communication).

| Molecule | Temperature | |
	2°C	25°C
He	1.189	1.153
Ne	1.221	1.171
Ar	1.259	1.210
Kr		1.250 (24°C)
Xe		1.330
H_2	1.196	1.152
N_2	1.282	1.230
O_2	1.256	1.210
CO	1.259	1.242
CO_2	1.189	1.158
CH_4	1.250	1.200
Naphthalene ($C_{10}H_8$)		1.363
$Si(OH)_4$		1.320 (20°C)

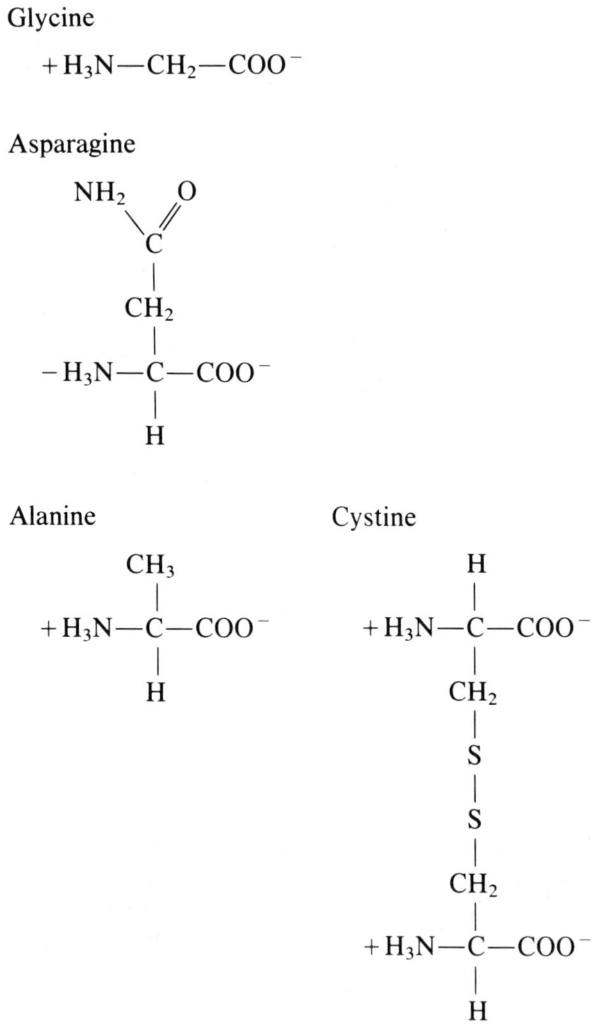

Glycine

$$+H_3N-CH_2-COO^-$$

Asparagine

$$
\begin{array}{c}
NH_2 \quad\ \ O \\
\diagdown \ \ \diagup\!\!\diagup \\
C \\
| \\
CH_2 \\
| \\
-H_3N-C-COO^- \\
| \\
H
\end{array}
$$

Alanine   Cystine

$$
\begin{array}{c}
CH_3 \\
| \\
+H_3N-C-COO^- \\
| \\
H
\end{array}
\qquad\qquad
\begin{array}{c}
H \\
| \\
+H_3N-C-COO^- \\
| \\
CH_2 \\
| \\
S \\
| \\
S \\
| \\
CH_2 \\
| \\
+H_3N-C-COO^- \\
| \\
H
\end{array}
$$

FIGURE 5.20 Amino acids.

dipolar amino acids shown in Figure 5.20. The values of γ are presented in Table 5.35. True nonelectrolytes behave as is shown in Table 5.36. We saw estimates of γ_{ip}. Let us see how γ_{ip} should behave.

1. Ion pairs of the 1-1 type such as HCl^0, $NaCl^0$, and KCl^0 have weak dipole moments when compared to amino acids with large intercharge space and to 2-2 ion pairs such as $MgSO_4^0$ and $CaSO_4^0$.

I may remind you at this point that the dipole moment is given by the interionic space multiplied by the absolute value of the charge of one of the ions, that is, $\mu = |ze|d$. The term z is large for amino acids and 2 for $MgSO_4^0$.

2. The potential decays with r^{-2} where r is the distance between the two entities, if we consider the electrostatic interactions between ions and dipoles in zero net charge ion pairs. In contrast, inter-

ionic potentials decay with r^{-1}. Thus, in very dilute solutions one may expect ions to see ion pairs primarily as nonelectrolytes. I am referring here to neutral ion pairs and not to $NaSO_4^-$, $CaCl^+$, and so on.

Thus, there should be the following progression in how ions perceive ion pairs.

Very dilute solutions	Dilute to moderate concentrations	High concentrations
Nonelectrolytes	Dipoles	Resolved ions

The last category simply means that an ion whose distance to the dipole is of the same order of magnitude as the length of the dipole should see the dipolar ions as separate entities.

3. The pattern indicated in (2) indicates that the ion pair should follow a Setchenov type behavior in very dilute solutions (salting-out of the ion pair) with γ_{ip} increasing from 1 on up as the concentrations of salts in the solution increase. Next, as ion dipole forces increase, there should be salting-in due to the ion–dipole attraction which leads to a decrease in γ_{ip} with the concentration. The decay of γ_{ip} should be faster than that for ions in ion–ion interactions due to the higher inverse power dependence of the former.

4. In the case of ion pairs with a net charge, interactions such as that of K^+ with $NaSO_4^-$ should be sensed by the ions as follows:

Very dilute solutions	Dilute to moderate concentrations	High concentrations
Singly charged ion	Single charge plus dipole	Resolved ions

At intermediate concentration the ion K^+ perceives one net charge from SO_4^{2-} plus the dipole moment $Na^+ - SO_4^-$.

5. It is not possible at this time to calculate the dipole moment of ion pairs from first principles because the extent of the water of hydration and its effect on the dipole moment are not known. Ion pairs, as coulombic entities, retain at least part of their water of hydration, depending on the extent of cosphere overlap.

6. The data of Johnson (1979) shown in Table 5.35 do not conform to the above pattern in two ways. First, as γ_{ip} must be 1 at infinite dilution, there is an unexplained dip of γ_{ip} at $I = 0.1$. Second,

the values of γ_{ip} increase monotonically with increasing concentration and do not reveal a salting-in expected from ion–dipole interactions. The second reservation may not be necessarily valid if because of the small dipole moments in question, the ion pairs are sensed primarily as nonelectrolytes throughout the concentration range of the measurements.

The first concern appears valid. The values of γ_{ip} were obtained by Kester (1970) and by Johnson (1979) from expressions of the type (5.295). Kester used literature values of K^* while Johnson obtained his values of the thermodynamic constants by extrapolation of data on $K^{*'}$ to $I = 0$. It could be that the dip in γ_{ip} at $I = 0.1$ represents the results of systematic errors in the values of K^* obtained by extrapolation.

The value of Kester, $\gamma = 0.8$, pertains to 2-2 ion pairs and is probably valid for seawater ($I \cong 0.7$). Unfortunately no data are available at other ionic strengths. The compilations by Culberson (personal communication), shown in Tables 5.35 and 5.36, are of considerable interest for two reasons. The results for amino acids represent an upper limit to those expected for seawater-type electrolytes, due to the large dipole moments of the former. Thus 1-1 type ion pairs of simple salts should be curves of γ_{ip} versus concentrations that lie above those for the amino acids. Second, a comparison of the values of γ_{ip} for amino acids and for truly neutral electrolytes (Tables 5.35 and 5.36) reveals the salting-in effect of dipole moments.

In general, a comparison of the values in Table 5.35 and Table 5.36 suggests that HCl^0, $NaCl^0$, and KCl^0 may act somewhat like nonelectrolytes, possibly because their dipole moment is small relative to those of amino acids. This can be due to the smaller intercharge distance (remember that the dipole moment is given by the charge of one of the ions times the distance between them) and to the decrease in dipole moment because of the waters of hydration. This is shown schematically by $-\boxed{+\ -}+$ where the rectangle represents one of the water molecules of hydration.

A comparison of the values of γ_{ip} at $I \cong 0.7$ yields for neutral ion pairs:

TABLE 5.37 Values of the thermodynamic association constants, K^*, used by Garrels and Thompson (1962) and of the stoichiometric association constants, $K^{*'}$, resulting from their work. Actually, these authors used the dissociation constants $1/K^*$.

Ion Pair	K^*	$K^{*'}$
$NaSo_4^-$	5.25	0.792
KSO_4^-	9.12	0.660
$MgSO_4^0$	229	8.26
$CaSO_4^0$	204	5.74
$NaHCO_3^0$	0.562	0.246
$MgHCO_3^+$	14.5	7.01
$CaHCO_3^+$	18.2	6.70
$NaCO_3^-$	18.6	4.01
$MgCO_3^0$	2512	142
$CaCO_3^0$	1585	90.8

A selection among these data cannot be made without a further knowledge of the dipole moments of hydrated ion pairs.

Thermodynamic Stability Constants. The values of K^* used by Garrels and Thompson (1962) are shown in Table 5.37 together with the values of $K^{*'}$ calculated from $K^{*'} = K^* (\gamma_C)_F (\gamma_A)_F / \gamma_{ip}$. It is interesting to note that small values of $(\gamma_C)_F (\gamma_A)_F / \gamma_{ip}$ lead to much smaller values of $K^{*'}$ and of $[CA^0]/[C]_F[A]_F$, that is, of the extent of association than one would expect from a cursory inspection of K^*. Speciation calculations are made at times with K^* and $\gamma_{ip} = (\gamma_C)_F = (\gamma_{FA}) = 1$ for valuable illustrative purposes (e.g., Stumm and Morgan, 1970). The numerical results of such calculations should not be accepted literally.

The values of K^* for sulfate ion pairs were obtained from Davies (in Hamer, 1959). Again, I should remind you that these data in the original papers as well as the other ones used or derived by Garrels and Thompson (1962) are dissociation constants but I present them in the form of association constants.

Thermodynamic Association and Stability Constants from Davies (1962) and Johnson (1979). A more extensive and recent compilation of values of

Garrels and Thompson (1962)	Kester (1970)	Culberson*	Johnson (1979)
1.13	0.8	~0.86	~1.3

*Personal communication.

TABLE 5.38   Values of K^*, the thermodynamic association constant for ion pairs, and of β, the thermodynamic stability constant for true complexes, selected from the compilation by Davies (1962). The values of K^* for chlorides are from Johnson (1979).

	$K^{*(t)}$				$\beta^{(t)}$					
	Na^+	K^+	Mg^{2+}	Ca^{2+}	Zn^{2+}	Cu^{2+}	Mn^{2+}	Cd^{2+}	Pb^{2+}	Fe^{3+}
OH^-	0.2	a	38.0	20.0	—	—	—	—	6.3×10^7	10^{12}
F^-	a	—	6.6	10	18.2	17.0	—	—	—	1.1×10^6
Cl^-	0.89	0.116	4.1	5.2	—	2.51	1.0	100	31.6	32
Br^-	—	—	a	a	—	1.0	—	158	100	4.0
NO_3^-	0.25^b	0.63	a	1.9	—	—	—	2.51	15.1	10
IO_3^-	0.32	0.50	5.25	7.8	—	—	—	—	—	—
ClO_4^-	—	0.32	—	—	—	—	—	—	—	1.7×10^4
SO_4^{2-}	5.01	10.0	169.8	190.5	204	229	190	224	—	1.6×10^4

^{a}No evidence for ion pairs or complexes found.
^{b}The second significant figure is uncertain for Na^+ and K^+ ion pairs except for the data of Johnson (1979).

K^*, the thermodynamic association constants for ion pairs, plus some values of β, the thermodynamic stability constants for true complexes, are presented in Table 5.38. The few values of β are shown to form the basis for the discussion below. Ion pairs are coulombic entities while complexes involve electron sharing.

There is at times a considerable discrepancy between data presented in various tables. This occurs for two reasons. First, there are experimental inaccuracies that are difficult to resolve. A useful criterion, therefore, is to select the most comprehensive data unless there are reasons to suspect them. The largest source of β values is the critical compilation by Smith and Martell (1976) which is presented in part in Chapter 6, Vol. 2 and is in good agreement with that of Davies (1962).

The second reason for discrepancies is that even Smith and Martell did not consider the competition for anions by ion pairs and complexes. If β for a given complex was measured in a solution of alkalies and/or alkaline–earths (Na^+, K^+, Ca^{2+}, Mg^{2+}, etc.) and halides (F^-, Br^-, Cl^-, I^-), then ion pairs such as $NaCl^0$ compete with $PbCl^+$ for Cl^-. The presence of ion pairs must be taken into consideration when β for $PbCl^+$ is determined (Sipos et al., 1980) (see Chapter 6, Volume II).

It can be seen in Table 5.38 that the association constants for OH^- ion pairs are larger than those for chlorides. They are not used for the calculations of ion pairing in most natural water in spite of this because the concentrations of chloride are much larger than those of hydroxide.

Values of β tend to be higher than those of K^*

because the β's reflect coordinate covalent (inner sphere) bonds which are stronger than the coulombic bonds across waters of hydration (outer sphere) present in ion pairs. Some ions form exceedingly strong complexes with OH^- as in the case of Fe^{3+} (see Table 5.38 and Chapter 6, Volume II).

Determination of Thermodynamic Association Constants of HCO_3^- and CO_3^{2-} Ion Pairs [According to Garrels and Thompson, (1962)].   The constants for $CaHCO_3^+$ and $MgHCO_3^+$ were obtained by modifying those of Greenwald (1941), which had been determined at $I \cong 0.15$. The "free" activity coefficients of Ca^{2+} and Mg^{2+}, calculated by the mean-salt method at $I = 0.15$, are 0.34 and 0.41. For HCO_3^- Garrels and Thompson used 0.76 for $(\gamma_{HCO_3})_F$ in seawater.

Greenwald's stoichiometric association constants in the case of $CaHCO_3^+$ was $10^{0.8}$ and, therefore,

$$K = 10^{0.8} \frac{\gamma_{CaHCO_3}}{(\gamma_{Ca})_F (\gamma_{HCO_3})_F}$$

$$= 10^{0.8} \frac{0.76}{0.34 \times 0.76} = 10^{-1.26} \quad (5.298)$$

The convention $(\gamma_{HCO_3})_F = \gamma_{CaHCO_3}$ was used.

The constants for $NaCO_3^-$ and $MgCO_3^0$ were obtained from earlier work (Garrels et al., 1960). Those for $CaCO_3^0$ and $NaHCO_3^0$ were determined later by the same method. As by-products, K_{sO} was measured for calcite ($10^{-8.35}$) and aragonite

$(10^{-8.22})$. The values of K_{sO} can also be calculated from the free energy of formation data in Garrels et al. (1960).

Let us examine how values of K^* were determined. The first step of the method used by Garrels et al. (1960) consisted essentially in the titrating HCO_3^- solutions with, for example, NaCl. The ratio

$$\frac{a_{HCO_3}}{[HCO_3^-]_T} = (\gamma_{HCO_3})_{(T)} \qquad (5.299)$$

was obtained from the reaction

$$CO_{2(g)} + H_2O_{(l)} = H^+ + HCO_3^- \qquad (5.300)$$

with

$$\frac{a_H a_{HCO_3}}{a_{CO_2} a_{H_2O}} = K_1 K_h = 10^{-7.82} \qquad (5.301)$$

calculated from free energy data in Latimer (1952). Equation (5.301) rewritten in terms of $a = \gamma m$ leads to

$$(\gamma_{HCO_3})_T = \frac{10^{-7.82} a_{CO_2} a_{H_2O}}{a_H (HCO_3^-)_T} \qquad (5.302)$$

In this equation $a_{CO_2} = 1$ by holding $pCO_2 + pH_2O \cong pCO_2 = 1$ atm. The value of a_{H_2O} in salt solutions is available in the International Critical Tables, a_H is measured, and $[HCO_3^-]$ is obtainable by a titration with acid if the pH is high enough for $[CO_3^{2-}]_T$ to be significant, or simply from the amount of $NaHCO_3$ added. Thus, $(\gamma_{HCO_3})_T$ can be calculated.

In solutions in which HCO_3^- and CO_3^{2-} are present the ratios $(\gamma_{CO_3})_T / (\gamma_{HCO_3})_T$ can be obtained from the reaction

$$HCO_3^- = H^+ + CO_3^{2-} \qquad (5.303)$$

which yields

$$\frac{a_H a_{CO_3}}{a_{HCO_3}} = K_2 = 10^{-10.33} \qquad (5.304)$$

These determinations were made in the following solutions: $NaHCO_3$–NaCl, Na_2CO_3–$NaHCO_3$–NaCl, $NaHCO_3$–$MgCl_2$, $NaHCO_3$–NaCl–$MgCl_2$, $NaHCO_3$–Na_2CO_3–$MgCl_2$, and $NaHCO_3$–Na_2Co_3–NaCl–$MgCl_2$.

Let us examine as an example of the above method how $K^*_{MgCO_3}$ was determined. Garrels et al.

(1960) assumed that $(\gamma_{CO_3})_T$ in K_2CO_3 solutions is equal to $(\gamma_{CO_3})_F$ because ion pairing between K^+ and CO_3^{2-} is negligible, and that this result also applies in the presence of Na^+ and Mg^{2+}. This is correct since γ_F depends only on I_e. Measurements were available in such solutions. Therefore,

$$\frac{{}^aCO_3(MgCl_2)}{(\gamma_{CO_3})_{T(K_2CO_3)}} = [CO_3^{2-}]_F \qquad (5.305)$$

$(\gamma_{Mg})_F$ was obtained by the mean-salt method

$$(\gamma_{Mg})_F = \frac{(\gamma_{\pm Mg_2})_F^3}{(\gamma_{\pm KCl})_F^2} \qquad (3.306)$$

since the authors were unaware of Cl^- ion pairing (Johnson and Pytkowicz, 1978, 1979a). The Debye–Hückel equation was used at $I < 0.1$. For $Mg^{2+} + CO_3^{2-} = MgCO_3^0$ we have

$$[Mg^{2+}]_T = [Mg^{2+}]_F + [MgCO_3^0] \qquad (5.307)$$

The association constant is

$$\begin{aligned} K^* &= \frac{a_{MgCO_3}}{a_{Mg} a_{CO_3}} \\ &= \frac{\gamma_{ip}[MgCO_3^0]}{(\gamma_{Mg})_F [Mg^{2+}]_F (\gamma_{CO_3})_F [CO_3^{2-}]_F} \\ &= \frac{\gamma_{ip}[MgCO_3^0]}{(\gamma_{Mg})_T [Mg^{2+}]_T (\gamma_{CO_3})_T [CO_3^{2-}]_T} \qquad (5.308) \end{aligned}$$

a_{CO_3} was calculated from the equation

$$\frac{a_H (\gamma_{CO_3})_T [CO_3^-]_T}{(\gamma_{HCO_3})_T [HCO_3^-]_T} = 10^{-10.33} \qquad (5.309)$$

which corresponds to the reaction (5.303). The term $(\gamma_{HCO_3})_T$ was determined in the presence of $MgCl_2$ by Equation (5.302), a_H was measured, and the total concentrations were determined by titrations.

The activity coefficient of the ion pair $MgCO_3^0$ was assumed to equal that of H_2CO_3 (Harned and Owen, 1958) as given by Markham and Kobe (1941). The value of $[MgCO_3^0]$ was obtained from Equation (5.307) so that $a_{MgCO_3} = \gamma_{ip}[MgCO_3^0]$. The term $[Mg^{2+}]_T$ is that added as $MgCl_2$. The activity coefficient $(\gamma_{Mg})_F$ is obtained by the mean-salt method. $[CO_3^{2-}]_F = [CO_3^{2-}]_T - [MgCO_3^0]$ was obtained from Equation (5.305).

Next, the extent of ion pairing was used to calcu-

TABLE 5.39 Values of $(\gamma_{CO_3})_F$ and γ_{ip} for the ion pair $NaCO_3^-$ versus the ionic strength [from Garrels et al. (1960)].

I	$(\gamma_{CO_3})_F$	γ_{ip}	I	$(\gamma_{CO_3})_F$	$\gamma_{ip} = (\gamma_{HCO_3})_F$
0.112	0.488	0.785	1.612	0.141	0.640
0.362	0.283	0.716	2.112	0.130	0.626
0.612	0.210	0.686	3.112	0.119	0.611
0.862	0.179	0.668	4.112	0.113	0.600
1.112	0.161	0.654	5.112	0.110	0.590

TABLE 5.40 Values of $(\gamma_{CO_3})_F$ and for the ion pair $MgCO_3^0$ versus the ionic strength (Garrels et al., 1960).

I	$(\gamma_{CO_3})_F$	γ_{MgCO_3}	I	$(\gamma_{CO_3})_F$	γ_{MgCO_3}
0.057	0.460	—	0.252	0.285	1.050
0.086	0.407	1.020	0.279	0.275	1.055
0.114	0.373	1.025	0.304	0.268	1.060
0.142	0.343	1.030	0.331	0.260	1.065
0.170	0.321	1.035	0.357	0.253	1.070
0.198	0.308	1.040	0.383	0.248	1.075
0.224	0.295	1.045	0.408	0.242	1.080

late a new value of I and the whole procedure was repeated until K^* converged to an invariant quantity. Values for $(\gamma_{CO_3})_F$ and for γ_{ip} in the case of $NaCO_3^-$ are presented in Table 5.39 while the corresponding values for $MgCO_3^0$ are shown in Table 5.40. The values of γ_{NaHCO_3} for the $NaHCO_3^0$ ion pair were assumed equal to those of $(\gamma_{HCO_3})_F = (\gamma_{HCO_3})_{T(MgCl_2)}$ at the same ionic strength, for lack of a better procedure.

The decision on whether the approximation $\gamma_{MgCO_3} = \gamma_{H_2CO_3}$ is better than that of Kester mentioned earlier or than the values presented by Culberson (see Table 5.35) depends on the dipole moments of the neutral ion pairs. These in turn depend on the unknown extent of hydration between the two ions of the pairs.

A Word on Greenwald (1941). Garrels and Thompson (1962) used the results of Greenwald for the stoichiometric association constants, $K^{*'}$, in the cases of $MgHCO_3^+$ and $CaHCO_3^+$ at $I \cong 0.15$. The growing evidence for association of Ca^{2+} and Mg^{2+} with organic acids led Greenwald to seek the presence of ion pairs in the case of the dissociation products of H_2CO_3, that is, HCO_3^- and CO_3^{2-}. The ionic strength used was aimed at making the results valid for physiological fluids.

Greenwald made two types of measurements, titrations, a method with which we are familiar in principle, and solubility measurements. The titrations indicated, for example, that the dissociation of H_2CO_3 was greater in the presence of $MgCl_2$ than in its absence. This is due to the removal of part of the free HCO_3^- and CO_3^{2-} from the equilibria

$$H_2CO_3 = HCO_3^- + H^+ = CO_3^{2-} + 2H^+ \quad (5.310)$$

so that the reaction is pushed to the right. The shift results from the reactions

$$Mg^{2+} + HCO_3^- = MgHCO_3^+$$
$$Mg^{2+} + CO_3^{2-} = MgCO_3^0 \quad (5.311)$$

and appears in the calculation of $m_{H_2CO_3}$ from the dissociation equilibria for $H_2CO_3^*$ (see Chapter 1, Volume II).

The principle of the method was as follows. The average solubility, expressed as $pK_s = -\log K_s$, was 8.007 in KCl solutions containing $CaCl_2$ or $MgCl_2$ but decreased (K_s increased) in the presence of $KHCO_3$. This indicated removal of part of the Mg^{2+} and the Ca^{2+} as $MgHCO_3^+$, $CaHCO_3^+$, $MgCO_3^0$, and $CaCo_3^0$ from the equilibria

$$Mg^{2+} + CO_3^{2-} = MgCO_{3(s)} \quad (5.312)$$

$$Ca^{2+} + CO_3^{2-} = CaCO_{3(s)} \quad (5.313)$$

Therefore, more HCO_3^- was converted to CO_3^{2-} and more $MgCO_3$ and $CaCO_3$ could dissolve before saturation occurred.

The Garrels and Thompson Model of Seawater. The determination of the chemical model of seawater was as follows: A mass balance was written for each constituent, in the form illustrated for sodium by

$$[Na^+]_T = [Na^+]_F + [NaHCO_3^0]$$
$$+ [NaCO_3^-] + [NaSO_4^-] \quad (5.314)$$

The corresponding association constant for, let us say $NaSO_4^-$, is

$$K^* = \frac{\gamma_{NaSO_4}[NaSO_4^-]}{(\gamma_{Na})_F[Na]_F(\gamma_{SO_4})_F[SO_4^{2-}]_F} \quad (5.315)$$

There are 17 equations of the types (5.314) and (5.315). The solution of the system was simplified by assuming, in the first pass, that $[C]_T = [C]_F$ for

TABLE 5.41 The Garrels and Thompson (1962) model of seawater at 25°C and 1 atm total pressure.

Ion	m_T	% Free	% CSO_4	% $CHCO_3$	% CCO_3
Na^+	0.4752	99	1.2	0.01	—
K^+	0.0100	99	1.0	—	—
Mg^{2+}	0.0540	87	11.0	1.0	0.3
Ca^{2+}	0.0104	91	8.0	1.0	0.2

Ion	m_T	% Free	% NaA	% KA	% MgA	% CaA
SO_4^{2-}	0.0284	54	21.0	0.5	21.5	3
HCO_3^-	0.00238	69	8.0	—	19.0	4
CO_3^{2-}	0.000259	9	17.0	—	67.0	7

the cations, calculating the speciation model, correcting the cations for ion pairs, and proceeding in an iterative manner to convergence. The final results are shown in Table 5.41.

EXAMPLE 5.10. A SIMPLE APPLICATION OF THE METHOD DEVELOPED BY GARRELS AND THOMPSON (1962). DETERMINATION OF THE SPECIATION FOR NaCl–Na$_2$SO$_4$. Consider a NaCl–Na$_2$SO$_4$ solution with [NaCl] = 0.4 m and [Na$_2$SO$_4$] = 0.1. Let us assume that there is no Cl^- ion pairing and that I can be used instead of I_e.

In reality the calculations shown below would have to be repeated until successive iteration yielded the speciation at I_e (Kester and Pytkowicz, 1969).

First,

$$I = 0.4 + 0.3 = 0.7 \quad \text{(i)}$$

Then,

$$\gamma NaSO_4 = 0.68 \quad \text{and} \quad (\gamma_{SO_4})_F = 0.12 \quad \text{(ii)}$$

According to Garrels and Thompson (1962)

$$\gamma_{\pm NaCl} = 0.667 \quad \text{and} \quad \gamma_{\pm KCl} = 0.626 \quad \text{(iii)}$$

at I = 0.7 (Robinson and Stokes, 1968).
The mean-salt method yields

$$(\gamma_{Na})_F = \frac{\gamma_{\pm NaCl}^2}{\gamma_{\pm KCl}} = 0.711 \quad \text{(iv)}$$

According to Garrels and Thompson

$$K^*_{NaSO_4} = \frac{[NaSO_4^-]}{[Na^+]_F[SO_4^{2-}]_F} \times \frac{\gamma_{NaSO_4}}{(\gamma_{Na})_F(\gamma_{SO_4})_F}$$

$$\text{(v)}$$

Furthermore

$$[Na^+]_T = [Na^+]_F + [NaSO_4^-] = 0.6 \quad \text{(vi)}$$

and

$$[SO_4^{2-}]_T = [SO_4^{2-}]_F + [NaSO_4^-] = 0.1 \quad \text{(vii)}$$

Let us combine the last three equations with $NaSO_4^- = x$. Then,

$$K^*_{Na_2SO_4} = \frac{x}{(0.6 - x)(0.1 - x)} = 5.248 \quad \text{(viii)}$$

The results are

$$[NaSO_4^-] = 0.0734 \ m \qquad [Na^+]_F = 0.5266 \ m$$

$$[SO_4^{2-}]_F = 0.02656$$

and

$$\frac{[Na^+]_F}{[Na^+]_T} = 0.878 \qquad \frac{[SO_4^{2-}]_F}{[SO_4^{2-}]_T} = 0.266$$

In an iteration procedure one would then correct the initial I by using $[Na^+]_T$ for the first pass at I_e by means of

$$I_e = 0.5 \{[Na^+]_F + 4[SO_4^{2-}]_F + [NaSO_4^-]\} \quad \text{(ix)}$$

I_e would then be used to obtain corrected values of $(\gamma_{Na})_F$, $(\gamma_{SO_4})_F$, γ_{NaSO_4}, and K^*, and so on.

The method of Pytkowicz and Kester (1969) is initially more time consuming since the stoichiometric $K^{*\prime}$ is actually measured versus I in the solutions of interest. Later on it is simpler and has fewer assumptions since the γ's need not be known.

Results of Johnson and Pytkowicz. Results from Johnson and Pytkowicz (1979a) are presented here for comparative purposes but their method will be developed later on. From the results of Johnson (1979) and Johnson and Pytkowicz (1979a) for seawater:

	$NaCl^0$	KCl^0	$MgCl^+$	$CaCl^+$
$(\gamma_{\pm CCl})_T$	0.661	0.642	0.467	0.458
$(m_C)_F/(m_C)_T$	0.83	0.78	0.48	0.44
$(m_{Cl})_F/(m_{Cl})_T$	0.83	0.83	0.83	0.83

Values for γ_F/γ_T for ions in seawater are presented below for the results of Garrels and Thompson (1962), in which chloride ion pairs are neglected, and for the data of Johnson (1979), in which they are taken into consideration. The γ's and the m's are related by

$$\frac{\gamma_F}{\gamma_T} = \frac{m_T}{m_F} \qquad \text{(ii)}$$

The values for m_F/m_T are those of the previous tabulation but with more significant figures.

		Johnson (1979)		Garrels and Thompson (1962)	
	Ion	m_F/m_T	γ_F/γ_T	m_F/m_T	γ_F/γ_T
KCl^0	K^+	0.7828	1.277	0.99	1.01
$NaCl^0$	Na^+	0.8294	1.205	0.99	1.08
$MgCl^+$	Mg^{2+}	0.4814	2.077	0.87	1.15
$CaCl^+$	Ca^{2+}	0.4354	2.297	0.99	1.01

In addition the values of $K^{*\prime}$ are 0.34 for $NaCl^0$, 0.52 for KCl^0, 1.75 for $MgCl^+$, and 2.13 for $CaCl^+$. Note that $(m_{Cl})_F/(m_{Cl})_T = (\gamma_{Cl})_T/(\gamma_{Cl})_F$ and $(m_C)_F/(m_C)_T = (\gamma_C)_T/(\gamma_C)_F$. Thus, the contention of Garrels and Thompson that $(\gamma_{Cl})_T = (\gamma_{Cl})_F$ and $(\gamma_C)_T = (\gamma_C)_F$ must be considered to be a rough approximation.

Next, we may compare the values of ion-pairing parameters for the electrolytes, based on the data of Garrels and Thompson (1962) and Johnson (1979).

Note that Johnson included Cl^- ion pairs and did not resort to the mean-salt method and single-ion γ's. One can apply this method to compare our results to those of single-ion approaches.

EXAMPLE 5.11. A COMPARISON OF THE VALUES OF m_F/m_T AND γ_F/γ_T FROM GARRELS AND THOMPSON (1962) AND FROM JOHNSON (1979). Johnson (1979) made his calculations for the speciation of a seawater for which $I = 0.720$ and $I_e = 0.5527$ when chloride ion pairing was taken into consideration. The ionic strength I was set a priori and I_e was calculated from the speciation by means of

$$I_e = 0.5 \sum_i [(m_i)_F z_i^2 + m_{ip} z_{ip}^2] \qquad \text{(i)}$$

where ip denotes ion pairs. The corresponding salinity is 35‰.

γ_T is smaller in the third column than in the last one because of the formation of $NaCl^0$, KCl^0, $MgCl^+$, and $CaCl^+$.

5.7.7 Work Related to That of Garrels and Thompson

Table 5.41 shows that free cationic cation concentrations change less than the free anionic ones. This is due to the fact that, with the assumption of no Cl^- ion pairing, the concentrations of associating anions are smaller than those of the associating cations. Thus, experimental tests of the Garrels and Thompson (1962) model, based on free cations and a nonthermodynamic assumption, are less sensitive than those that are based on anions. Unfortunately the former are easier to obtain.

Results from Early Potentiometric Methods. Potentiometric measurements with ion-selective electrodes were used by Thompson (1966) and Thompson and Ross (1966) to determine m_F/m_T for Mg^{2+}, and for Ca^{2+} in seawater by Hostetler (1963) to determine $K^{*\prime}$ for $MgHCO_3^+$ in chloride solutions at $I = 0.15$. The first two workers found good agreement with Garrels and Thompson (1962) which is not surprising, since the reference and the test solutions contained similar amounts of chloride, assumed to be free. Therefore, the chloride effect was canceled out. The results were 88% free Mg^{2+} and 82% free Ca^{2+}. Hostetler tested the Greenwald value, $K^{*'} = 0.17$. He avoided the presence of CO_3^{2-} by working at relatively low pH's to simplify the calculations, and found $K^{*'} = 0.37$.

The Solubility Method. Pytkowicz and Gates (1968) used a solubility method in which again Cl^- ion pairs were neglected. It was only after Kester and Pytkowicz (1969) applied the Fuoss theory (1958) to electrolyte solutions that the presence of these pairs was suspected.

The method of Pytkowicz and Gates (1968) was as follows. Two solutions of the same effective ionic strength were prepared. I_e was estimated from the results of Kester and Pytkowicz (1969). Extra Cl^- was added to solution A to compensate for the lack of SO_4^{2-}. The $[Mg^{2+}]_T$ values were the same in the solutions.

Solution	Ions
A	Na^+, K^+, Mg^{2+}, Ca^{2+}, Cl^-
B	Artificial seawater without HCO_3^- and CO_3^{2-} contains Na^+, K^+, Mg^{2+}, Ca^{2+}, Cl^-, SO_4^{2-}, Br^-, F^-

A and B were equilibrated with brucite, $Mg(OH)_2$. K_{sO} of brucite is the same in A and B because it depends only on T and P, provided that no surface coatings of different compositions are formed. Therefore,

$$K_{sO} = (a_{Mg})_A(a_{OH})_A^2 = (a_{Mg})_B(a_{OH})_B^2 \quad (5.316)$$

Since free activity coefficients depend only on I_e and $a_i = (\gamma_i)_F(m_i)_F = (\gamma_i)_T(m_i)_T$,

$$\frac{(a_{OH})_A^2}{(a_{OH})_B^2} = \frac{(\gamma_{Mg})_{FB}[Mg^{2+}]_{FB}}{(\gamma_{Mg})_{FA}[Mg^{2+}]_{FA}} = \frac{[Mg^{2+}]_{FB}}{[Mg^{2+}]_{FA}}$$

$$(5.317)$$

The subscript FA refers to the free ions in solution A. The "lack" of ion pairing in A means that $[Mg^{2+}]_{FA} = [Mg^{2+}]_{FB}$. Equation (5.317) can be rewritten as

$$[Mg^{2+}]_{FB} = [Mg^{2+}]_{TA} \frac{(a_H)_B^2}{(a_H)_A^2} \quad (5.318)$$

a_{OH} was replaced by a_H because K_w is independent of composition. The terms a_H refer to activities at

saturation with brucite. The value of $[Mg^{2+}]_{TA} = [Mg^{2+}]_{TB}$ is the known stoichiometric one while $(a_H)_A$ and $(a_H)_B$ are measured. Therefore, $[M^{2+}]_{FB}$ can be calculated.

Three runs were made and a typical set of results is

Steady-State pH				
A	B	$[Mg^{2+}]_{FB}$	$[Mg^{2+}]_{TB}$	% $MgSO_4^0$
9.314	9.335	0.0493	0.0055	10

The average for the runs was $[MgSO_4^0]/[Mg^{2+}]_{TB} \times 100 = 10\%$ in good agreement with the results of Garrels and Thompson (1962), Thompson (1966), and Thompson and Ross (1966).

A by-product of our results was $K_s = [Mg^{2+}]_T(a_{OH})^2 = 2.4 \times 10^{-11}$. $K_{sO} = (\gamma_{Mg})_F[Mg^{2+}]_F(a_{OH})^2 = 7.8 \times 10^{-12}$ is obtained from $(\gamma_{Mg})_F = 0.36$ from Garrels and Thompson (1962) and our $[(Mg^{2+})_F]$. This result is in good agreement with Hostetler's (1963) result, namely, $K_{sO} = 7.1 \times 10^{-12}$.

If Cl^- ion pairs had been taken into consideration the following would have happened:

1. No change in K_s since $[Mg^{2+}]_T = [MgSO_4]$ and a_{OH} are measured. $[MgSO_4]$ should not be confused with $[MgSO_4^0]$ where $MgSO_4^0$ is the ion pair.

2. $(a_H)_A$ and $(a_H)_B$ would not change since they are measured without regard for the model used.

3. $[Mg^{2+}]_{FA} \neq (Mg^{2+})_{TA}$ due to $MgCl^+$ ion pairs.

4. $(I_e)_A \neq (I_e)_B$ since there are only $MgCl^+$ ion pairs in A but $MgCl^+$ and $MgSO_4^0$ are present in B.

5. Therefore, Equations (5.316) and (5.317) would not be valid.

6. The trends would be $[Mg^{2+}]_{FA} < [MgSO_4]$, $[Mg^{2+}]_{FB} < [Mg^{2+}]_{FA}$, and $[Mg^{2+}]_{FB} < 0.9[MgSO_4]$. Correct results would require that A and B be prepared at a truly equal value of I_e.

Thus, our results represent a first cut at the problem but are shown at this point because of the value of the method.

Sound Attenuation. Fisher (1967) used ultrasonic adsorption determined in $MgSO_4$ solutions to calculate the extent of $MgSO_4^0$ formation in seawater. The adsorption of sound had been found to be proportional to the electrical conductivity. Therefore, the sound attenuation in a pure solution of $MgSO_4$ yielded, by means of the conductivity, the extent of ion pairing. Fisher found that 9.2% of the Mg^{2+} was associated to SO_4^{2-}. He then assumed that this value was applicable to seawater. This is a poor assumption because Mg^{2+} and SO_4^{2-} are also tied to a number of other ions and because other ion pairs absorb sound in seawater.

Extensions of the Method of Garrels and Thompson. There have been several extensions and applications of the Garrels and Thompson (1962) method and results. I shall mention a few at this point.

van Breemen (1973) was interested in waters in contact with acid sulfate soils derived from pyritic sediments. These waters had pH's between 1.8 and 8.2 and ionic strengths ranging from 10^{-3} to 1.1. The determination of the speciation in the soil solution needed to study the soil genesis required the determination of the ionic activities for 33 dissolved species. These were used in conjunction with thermodynamic equilibrium constants to obtain the stoichiometry of the waters. The procedure was essentially an expansion of that of Garrels and Thompson (1962).

Lafon (1969) extended the method of Garrels and Thompson to high pressures, a procedure which required several further assumptions and is somewhat shaky. Berner (1971) recalculated the speciation of seawater by the method of Garrels and Thompson (1962) using slightly different input data for $\gamma_\pm$'s of chlorides (Robinson and Stokes, 1968). pK values for the ion pairs were calculated from literature data on ΔF_f^0 and differ a little from those of Garrels and Thompson (1962).

Truesdell and Jones (1969) extended the mean-salt method approach to supersaline brines encountered in some lakes. The validity of this method at high ionic strengths is even more uncertain than for seawater. Berner (1971) recommended that the mean-salt method be corrected by Harned's rule in hypersaline brines. This rule, derived from a different concept, is also important for the work of Pytkowicz and Kester (1969) and for many other researches. It will, therefore, be discussed at this point.

Harned's Rule. This is an empirical rule which can be written in several forms. If there are two electrolytes A and B in a solution then

$$\log(\gamma_A)_T = \log(\gamma_{A,0})_T - \alpha_A(m_B)_T \quad (5.319)$$

See the meanings of the symbols listed below.

$$\log(\gamma_{0A})_T = \log(\gamma_{A,0})_T - \alpha_A(m)_T \quad (5.320)$$

Therefore, Equation (5.319) becomes

$$\log(\gamma_A)_T = \log(\gamma_{0,A})_T + \alpha_A(m_B)_T \quad (5.321)$$

For component B

$$\log(\gamma_B)_T = \log(\gamma_{B,0})_T - \gamma_B(m_A)_T \quad (5.322)$$

or

$$\log(\gamma_B)_T = \log(\gamma_{0,B})_T + \alpha_B(m_B)_T \quad (5.323)$$

The symbols mean:

$(\gamma_A)_T$, $(\gamma_B)_T$	Total mean activity coefficients of A and B
$(\gamma_{A,0})_T$, $(\gamma_{B,0})_T$	Values for the indicated electrolyte when the other one is absent
$(\gamma_{0A})_T$, $(\gamma_{0B})_T$	γ's for trace amounts of A and B
α_A, α_B	Harned's rule coefficients, valid at constant $(m)_T$

When $(m_B)_T = (m)_T$, that is, there is only a trace of A in the solution, then

$$(m)_T = (m_A)_T + (m_B)_T$$

An example, based on the results of Harned and coworkers and presented by Robinson and Stokes (1968), for the cell

$$Pt \mid H_2; \; HCl(m_A); \; KCl(m_B) \mid Ag:AgCl \quad (5.324)$$

is

$$(\gamma_{\pm HCl})_T = 0.00358 - 0.0580(m_{KCl})_T \quad (5.325)$$

where $0.00358 = \log \gamma_{\pm(HCl,0)}$ and $0.0580 = \alpha_{HCl}$ at $m_T = 2.0$. This equation can be used from known values of $\gamma_{\pm(HCl,0)}$ and α_{HCl} to calculate $(\gamma_{\pm HCl})_T$ versus $(m_{KCl})_T$ at $m_T = 2.0$.

One drawback of Harned's rule is that α varies rapidly with I at low ionic strengths. I stated earlier that Berner (1971) advocated the use of Harned's rule for hypersaline media. He referred to a demonstration by Åkerlöf (1937) to the effect that Harned's rule can be extended to multicomponent solutions by means of

$$\log(\gamma_A)_T = \log(\gamma_{A,0})_T + \Sigma \alpha_i I_i \quad (5.326)$$

where the α_i are calculated for pairs of electrolytes. I have not verified this result but it sounds plausible if the concentrations are not high enough to require quadratic cross terms. The plus sign in the last term in Equation (5.326) should be a minus. Berner recommended this approach when electrode data were not measured or available in the literature. I_i is the contribution of i to the ionic strength.

5.7.8 The Ion Pairing Model of Kester and Pytkowicz

In our work we reduced the number of assumptions used by Garrels and Thompson (1962) through an experimental approach, but did not include the ion pairing of chloride.

Let us first see how we derived Harned's rule from an ion-pairing model (Pytkowicz and Kester, 1969) to counteract the impression that only the Brønsted–Guggenheim model (Brønsted, 1923; Guggenheim, 1935, 1936), to be discussed later, can lead to this result (Robinson and Stokes, 1968).

Harned's Rule and Ion Pairs. The rule will be derived for NaCl–Na$_2$SO$_4$, with the assumption that $[Cl^-]_F = [Cl^-]_T$. First,

$$(\gamma_{\pm NaCl})_T = (\gamma_{\pm NaCl})_F \left(\frac{[Na^+]_F}{[Na^+]_T} \right)^{0.5} \quad (5.327)$$

and, by definition,

$$[NaSO_4^-]_F = K^{*'}_{NaSO_4}[Na^+]_F[SO_4^{2-}]_F \quad (5.328)$$

since

$$[Na^+]_T = (Na^+)_F + [NaSO_4^-] \quad (5.329)$$

Equation (5.328) is introduced in (5.329), which is rearranged to

$$\frac{[Na^+]_F}{[Na^+]_T} = \frac{1}{1 + K^{*'}_{NaSO_4}[SO_4^{2-}]_F} \quad (5.330)$$

It follows from Equations (5.274) and (5.277) that

$$\log[\gamma_{\pm NaCl}]_T = \log[\gamma_{\pm NaCl}]_F$$
$$- 0.5\log[1 + K^{*'}_{NaSO_4}[SO_4^{2-}]_F] \quad (5.331)$$

In effect,

$$(\gamma_{\pm NaCl})_T = (\gamma_{\pm NaCl})_F \frac{[Na^+]_F^{0.5}}{[Na^+]_T^{0.5}} \quad (5.332)$$
$$= (\gamma_{\pm NaCl})_F [1 + K^{*'}_{NaSO_4} [SO_4^{2+}]_F]^{0.5}$$

If $[NaSO_4^-] < [Na^+]_F$ then $K^{*'}_{NaSO_4}[SO_4^{2-}]_F < 1$. The last term of Equation (5.332) can then be expressed by the series

$$\log(1 + K^{*'}_{NaSO_4}[SO_4^{2-}]_F) = \frac{K^{*'}_{NaSO_4}[SO_4^{2-}]_F}{2.303}$$
$$- \frac{(K^{*'}_{NaSO_4}[SO_4^{2-}]_F)^2}{4.606} + \cdots \quad (5.333)$$

Equation (5.334),

$$[SO_4^{2-}]_F = \frac{[SO_4^{2-}]_T}{1 + K^{*}_{NaSO_4}[Na^+]_F} \quad (5.334)$$

can be obtained in the same way as (5.330).

Next we introduced Equation (5.334) into (5.333). We used the equality $[SO_4^{2-}]_T = [Na_2SO_4]$, where $[Na_2SO_4]$ is the molality of the Na$_2$SO$_4$, and entered Equation (5.333) into (5.331) with $\log(\gamma_{\pm NaCl,0})_T = \log(\gamma_{\pm NaCl})_F$. This equality is obtained because NaCl is present. The result was

$$\log(\gamma_{\pm NaCl})_T = \log(\gamma_{\pm NaCl})_F - k_1 [Na_2SO_4]$$
$$+ k_2[Na_2SO_4]^2 + \cdots \quad (5.335)$$

with

$$k_i = \frac{(K^{*'}_{NaSO_4})^i/(1 + K^{*'}_{NaSO_4}[Na^+]_F)^i}{2(2.303)^i} \quad (5.336)$$

Equation (5.335) is reduced to Harned's rule if $[NaSO_4^-] \ll [Na^+]_F$ or $K^{*\prime}_{NaSO_4}[SO_4^{2-}]_F \ll 1$, since this last term equals $[NaSO_4^0]/[Na^+]_F$. These conditions simply mean that the extent of association is small and that the second and the higher order terms can be neglected in Equation (5.333) and, therefore in (5.335).

Similar derivations can be made for the other electrolytes found in natural solutions such as seawater. If chloride ion pairs are to be introduced then one can examine, for example, the couple $HClO_4$–$NaCl$ and use the same type of derivation as for $NaCl$–Na_2SO_4. Pytkowicz and Kester (1969) further showed that the Harned rule derived in terms of I_e, the effective ionic strength

$$I_e = 0.5([Na^+]_T + 4[SO_4^{2-}]_F + [NaSO_4^-]) \quad (5.337)$$

The Method of Kester and Pytkowicz. Kester and Pytkowicz (1968, 1969) and Pytkowicz and Kester (1969) determined the stoichiometric association constants for $NaSO_4^-$, $MgSO_4^0$, and $CaSO_4^0$ in seawater at 25°C and 1 atm. Then, in conjunction with the values of $K^{*\prime}$ for HCO_3^- and CO_3^{2-}, ion pairs are calculated from Garrels and Thompson (1962) (see Table 5.36). The speciation was calculated by means of a series of equations. These equations were of the types

$$K^{*\prime}_{NaSO_4} = \frac{[NaSO_4^-]}{[Na^+]_F[SO_4^{2-}]_F} \quad (5.338)$$

$$[Na^+]_T = [Na^+]_F + [NaSO_4^-]$$
$$+ [NaHCO_3^0] + [NaCO_3^-] \quad (5.339)$$

plus the counterpart of (5.339) for anions, extended to all the ion pairs expected to occur at that time.

became the same in the two solutions since $E_{Na} = E_{Na}^0 + (RT/F')\ln a_{Na}$ if two assumptions, now open to question, are accepted:

1. There is no chloride association (Davies, 1962; Sillen and Martell, 1964).

2. There is no change in liquid junction potential at the salt bridge of the reference electrode–solution interface for NaCl and NaCl–Na_2SO_4 solutions.

Let us continue with the procedure. Since $(a_{Na})_S = (a_{Na})_X$, where S represents the standard and X the test solution,

$$(\gamma_{Na})_{FS}[Na^+]_{FS} = (\gamma_{Na})_{FX}[Na^+]_{FX} \quad (5.340)$$

The equation for I_e and relations of the type (5.340) for $[Na^+]_{TX}$ and $[SO_4^{2-}]_{TX}$ were added to the system of equations. The ratio $(\gamma_{Na})_{FS}/(\gamma_{Na})_{FX}$ was obtained by the mean-salt method with $(\gamma_{Na})_{FX}$ at I_e for the test solution and $(\gamma_{Na})_{FX}$ at I for the standard solution. An iterative procedure to a constant γ ratio was used to determine $K^{*\prime}_{NaSO_4}$. Equations (5.339), (5.340), its counterparts for $[SO_4^{2-}]_T$, and the mean-salt method relations of $(\gamma_{Na})_{FS}$ and $(\gamma_{Na})_{FX}$ yielded seven equations in the unknowns $K^{*\prime}_{NaSO_4}$, $[NaSO_4^-]$, $[Na^+]_{FX}$, $[SO_4^{2-}]_{FX}$, $(\gamma_{Na})_{FS}$, $(\gamma_{Na})_{FX}$, and I_e, that is, seven unknowns. Note that the mean-salt method entered into the ratio $(\gamma_{Na})_{FS}/(\gamma_{Na})_{FX}$ so that inaccuracies in the method were partly offset.

Results and Discussion. The values of $K^{*\prime}$ determined by Kester (1969), Kester and Pytkowicz (1968, 1969), and Pytkowicz and Kester (1969) were:

Ion pair	$NaSO_4^-$	KSO_4^-	$MgSO_4^0$	$CaSO_4^0$
$K^{*\prime}$	2.02 ± 0.03	1.03	10.2 ± 0.5	10.8 ± 0.7

An iteration procedure was required to convert I to I_e.

The procedure for the determination of the $K^{*\prime}$ values for sulfate will be illustrated for the case of $NaSO_4^-$. A standard NaCl solution was titrated with concentrated NaCl until the emf, obtained with a sodium-sensitive glass–calomel reference couple, became the same as that obtained with a NaCl–Na_2SO_4 solution. At this point the activity of Na^+

An important result was that $K^{*\prime}_{NaSO_4}$ is independent of composition at a given I_e (see Figure 5.21). This result is valid, even on the high $NaSO_4^-$ side in the presence of Cl^- ion pairing, because $K^{*\prime}_{NaSO_4}$ is small. The invariance of $K^{*\prime}_{NaSO_4}$ is expected for large mole fractions of Cl^- according to Johnson (1979) because $K^{*\prime}_{NaSO_4}$ increases in the presence of $NaCl^0$ pairs but $[Na^+]$ decreases due to the interaction with Cl^-.

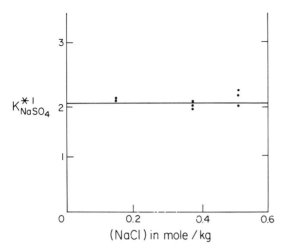

FIGURE 5.21 Stoichiometric association constant $K^{*'}_{NaSO_4}$ for $NaSO_4^-$ ion pairs versus composition of $NaCl-Na_2SO_4$ solutions at $I_e = 0.687$ at $25°C$.

TABLE 5.42 $K^{*'}_{Na_2SO_4}$ in $NaCl-Na_2SO_4$ solutions at $25°C$ and 1 atm at $I_e = 0.687$ (Kester, 1969).

(Na_2SO_4) (Molal)	$(NaCl)$ (Molal)	I_e (Molal)	$K^{*'}_{Na_2SO_4}$
0.34998	0.00000	0.6987	2.04 ± 0.02
0.28740	0.13293	0.6852	2.11 ± 0.01
0.12976	0.44425	0.6887	2.00 ± 0.04
0.06720	0.56167	0.6834	2.23 ± 0.16

the electrostatic effects with the exception of the pairing of ions.

Further Results and a Discussion of the Method of Kester and Pytkowicz. Results for $NaCl-Na_2SO_4$ solutions are shown in Table 5.42 and for our model in seawater in Table 5.43. Values of γ_T in seawater are obtainable from $\gamma_T = \gamma_F m_F/m_T$ where γ_F is available in Table 5.29, $I = 0.68$ while m_F/m_T data are obtainable from Table 5.43. Of course, the results for γ_T determined in this way hinge upon the validity of the mean-salt method. Cationic values of γ_T for seawater at about $S\%_0 = 34.7$ are shown below:

Ion	Na^+	K^+	Mg^{2+}	Ca^{2+}
γ_T	0.695	0.620	0.254	0.227

Comparisons of values obtained from the models of Kester and Pytkowicz (1969) and Johnson (1979) are made in Table 5.44 with measurements of $(\gamma_\pm)_T$. Again, other workers besides Johnson neglected Cl^- ion pairs. Interestingly enough, the results of Johnson (1979), which include Cl^- association, agree well with the experimental and theoretical values of others. The reason for this is that the change in $K^{*'}$ in the case of Cl^- ion pairs offsets the changes in the free ion concentrations. Another way to look at this is that one starts with $\gamma_{\pm(exp)}$ and then returns to $(\gamma_\pm)_T$ which must agree with the former. The speciations obtained by these methods are, however, quite different. This is equivalent to saying that changes in m compensate for changes in γ.

Algebraically, if the subscript Cl^- is used to denote inclusion of Cl^- ion pairs,

$$(\gamma_\pm)_{T,Cl} = (\gamma_\pm)_{F,Cl} \frac{(m_F)_{Cl}}{(m_T)_{Cl}} \quad (5.342)$$

$$(\gamma_\pm)_T = (\gamma_\pm)_F \frac{m_F}{m_T} \quad (5.343)$$

The result, at least in the high $[SO_4^{2-}]$ band, implies that, as

$$K^{*'}_{NaSO_4} = K^{*'}_{NaSO_4} \frac{(\gamma_{Na})_F (\gamma_{SO_4})_F}{\gamma_{NaSO_4}} \quad (5.341)$$

$$= constant$$

$(\gamma_{Na})_F$, $(\gamma_{SO_4})_F$, and γ_{NaSO_4} are independent of solution composition. The alternative is a sheer coincidence which would make their variations with composition cancel each other. This type of behavior was also observed indirectly by Pytkowicz and Hawley (1974) for bicarbonates and by Johnson (1979) and Johnson and Pytkowicz (1979a) for chlorides and sulfates since γ's in ternary solutions could be predicted from results in binary ones at the same I_e.

The values of $K^{*'}$ for $NaSO_4^-$, $MgSO_4^0$, and $CaSO_4^0$ measured by Kester (1969) were supplemented by values for the HCO_3^- and CO_3^{2-} ion pairs of the above cations calculated from the results of Garrels and Thompson (1962) for the calculation of the speciation of seawater.

The Extended Ionic Strength Principle. The above conclusions permit us to enumerate an "extended ionic strength principle," namely, that free activity coefficients depend on I_e, P, and T but are independent of the composition. The original principle (Lewis and Randall, 1921) did not distinguish between total and free activity coefficients because it was meant for dilute solutions. I should remind you at this time that γ_F is a construct which includes all

TABLE 5.43   The speciation model of Kester and Pytkowicz (1969) in seawater at 25°C and 1 atm.

	m_T	% Free	% MSO_4	% $MHCO_3$	% MCO_3
Na^+	0.48230	97.7	2.2	0.03	—[a]
K^+	0.01020	98.8	1.2	—	—
Mg^{2+}	0.05485	89.0	10.3	0.6	0.13
Ca^{2+}	0.01062	88.5	10.8	0.6	0.07

	m_T	% Free	% NaA	% KA	% MgA	% CaA
SO_4^{2-}	0.02909	39.0	37.2	0.4	19.4	4.0
HCO_3^-	0.00186	70.0	8.6	—	17.8	3.3
CO_3^{2-}	0.00011	9.1	17.3	—	67.3	6.4

[a] — indicates negligible amounts.

Since we observe that $(\gamma_\pm)_{T,Cl} \cong (\gamma_\pm)_T$,

$$(\gamma_\pm)_{F,Cl} \frac{(\gamma_\pm)_{F,Cl}}{(\gamma_\pm)_F} = \frac{m_F}{m_T} \times \frac{(m_T)_{Cl}}{(m_F)_{Cl}} \quad (5.344)$$

so that changes in m due to the inclusion of Cl^- ion pairing offset those in $(\gamma_\pm)_F$. The subscript Cl is only used to contrast values in the presence and in the assumed absence of Cl^- when it is needed for clarity.

In conclusion, the method of Kester and Pytkowicz (1969) represents an advance over that of Garrels and Thompson (1962) because the number of assumptions is reduced. It does not yield the true extent of ion pairing in seawater, but is a first approximation, because Cl^- ion pairs are not included. It has some uncertainty due to possible changes in liquid junction potentials when reference electrodes are transferred from standard to test solutions. Valuable concepts, such as I_e (Pytkowicz

and Kester, 1969), arose from the work. The method can be extended to natural waters in general if all the species are accounted for.

5.7.9 Pressure and Temperature Effects on Ion Pairing and Stoichiometric Thermodynamic Functions

We already examined the effects of pressure on chemical equilibria as presented by Owen and Brinkley (1941) in an earlier chapter, and will deal further with the subject later on when the CO_2–carbonate and the SiO_2 systems will be considered. At this point, I shall briefly present some theoretical considerations and results of Kester and Pytkowicz (1969) on ion pairing at high pressures and low temperatures. This involves the use of apparent stoichiometric thermodynamic functions which will be clarified next. Values for them can be found in the original paper.

Consider a reaction $A + B = C + D$ for which

$$K = K' \frac{(\gamma_C)_T(\gamma_D)_T}{(\gamma_A)_T(\gamma_B)_T} \quad (5.345)$$

If the reaction is one that involves solubilities or dissociations of acids, as examples, then the γ's are total reactions while for ion pairing one has

$$K^* = K^{*\prime} \frac{(\gamma_C)_F(\gamma_A)_F}{\gamma_{CA}} \quad (5.346)$$

The change in K with pressure is $\partial \ln K/\partial P = -\Delta \bar{v}^0/RT$ and a counterpart of this relation can be written for the temperature variation. The term $\Delta \bar{v}^0$ is $\bar{v}_C^0 + \bar{v}_D^0 - \bar{v}_A^0 - \bar{v}_B^0$ where the components are in their standard states.

TABLE 5.44   A comparison of total mean γ's obtained by several methods for seawater.

	$\gamma_{\pm(th)}$[a]	$\gamma_{\pm(th)}$[b]	$\gamma_{\pm(exp)}$[b]	$\gamma_{\pm(exp)}$
NaCl	0.66	0.661	0.667	0.672[c]
				0.668[d]
Na_2SO_4	0.34	0.354	0.385	
KCl		0.648		0.645[d]
K_2SO_4		0.347		0.352[e]
$CaSO_4$	0.164	0.151		0.136[f]

[a] Kester (1969).
[b] Johnson (1979).
[c] Platford (1965).
[d] Gieskes (1966).
[e] Whitfield (1975).
[f] Culberson et al. (1978).

Therefore, the variation in K' with pressure is

$$\frac{\partial \ln K'}{\partial P} = -\frac{\Delta \bar{v}^0}{RT} \frac{\partial \ln}{\partial P} \left[\frac{(\gamma_A)_T (\gamma_B)_T}{(\gamma_C)_T (\gamma_D)_T} \right] \quad (5.347)$$

if $\Delta \bar{v}^0$ is a constant over the range of the experiments. The equation becomes more complex if the compressibility is required and a number of cross terms appears. We can define a $\Delta \bar{v}'$, the change in partial molal volume in the medium of interest instead of in water, by means of

$$\frac{\partial \ln K'}{\partial P} = -\frac{\Delta \bar{v}'}{RT} \quad (5.348)$$

with

$$\Delta \bar{v}' = \Delta \bar{v}^0 \frac{\partial \ln}{\partial P} \left[\frac{(\gamma_A)_T (\gamma_B)_T}{(\gamma_C)_T (\gamma_D)_T} \right] \quad (5.349)$$

This equation is thermodynamically correct because $\Delta \bar{v}'$ is a function of the initial and final states as are $\Delta \bar{v}^0$ and the γ's. This result tells us that it is better to measure $\Delta \bar{v}'$ directly than to use $\Delta \bar{v}^0$ and have to estimate the variations of the γ's with pressure if one wishes to know, through $\partial \ln K'/\partial P$, the effect of pressure on the stoichiometry of a reaction.

We also have $\Delta F^0 = -RT \ln K$. Therefore,

$$\Delta F^0 = -RT \ln K' + RT \ln \frac{(\gamma_A)_T (\gamma_B)_T}{(\gamma_C)_T (\gamma_D)_T} \quad (5.350)$$

or

$$\Delta F' = -RT \ln K' \quad (5.351)$$

with $\Delta F'$ defined by

$$\Delta F' = \Delta F^0 + RT \ln \frac{(\gamma_C)_T (\gamma_D)_T}{(\gamma_A)_T (\gamma_B)_T} \quad (5.352)$$

$\Delta F'$ is also a function of state but one that represents the nonideal work as $\gamma \neq 1$. It is the work done by the system when it is brought from the standard state to the actual state under consideration, for example, seawater brought to a given state $(P, T, S‰)$. If the γ's become unity,

$$\Delta F' = \Delta F^0 \quad \text{and} \quad K' = K \quad (5.352)$$

Thus, it is seen that $\Delta F'$, the stoichiometric free energy, has thermodynamic meaning as do the other functions of state.

In general terms, for a reaction away from equilibrium $(Q_{r(a)} \neq K)$

$$\Delta F = -RT \ln K + RT \ln Q_{r(a)} \quad (5.353)$$

with

$$Q_{r(a)} = \frac{a_C a_D}{a_A a_B} = \frac{(\gamma_C)_T (\gamma_D)_T}{(\gamma_A)_T (\gamma_B)_T} Q_r \quad (5.354)$$

and $\Delta F^0 = -RT \ln K$. Q_r represents the stoichiometric ratio while $Q_{r(a)}$ is the activity ratio. Then, subtracting (5.353) from (5.352), $\Delta F'$ becomes

$$\Delta F' = \Delta F + RT \ln Q_r \quad (5.355)$$

$\Delta F' - \Delta F$ is the difference in the change of free energy needed to bring the reaction to equilibrium in terms of concentrations of activities.

In summary,

1. Work to bring the system from $Q_{r(a)}^0$ to K is ΔF^0. $Q_{r(a)}^0 = 1$ is the activity quotient in the standard state.
2. Work to bring $Q_{r(a)}$ to K is ΔF. $Q_{r(a)}$ is the activity quotient.
3. Work to bring Q_r to K' is $\Delta F'$. Q_r is the concentration quotient.

Similar expressions can be obtained for the other thermodynamic functions and the determination of $\Delta \bar{H}'$ will be shown next. $\Delta H'$ is the partial stoichiometric molal enthalpy of a reaction.

Kester (1969) demonstrated from thermodynamic principles, for the case of $NaSO_4^-$ as an example, that

$$\Delta \bar{H}' = \frac{R[\ln(K^*)_{T_2} - \ln(K^*)_{T_1}]}{1/T_1 - 1/T_2} \quad (5.356)$$

Temperature data were obtained for sulfate ion pairs (Kester and Pytkowicz, 1969) by the method used earlier at 25°C. The association constants are shown in Table 5.45. The constants decrease with increasing temperature indicating that, according to Le Chatelier's principle, the ion pairing is exothermic. This is to be expected because the electri-

TABLE 5.45 Stoichiometric association constants and thermodynamic functions of sulfates versus temperature in seawater (Kester and Pytkowicz, 1969). $\Delta \bar{F}'$ in kcal/mole, $\Delta \bar{H}'$ in kcal/mole, and $\Delta \bar{S}'$ in cal/deg · mole.

	$T(°C)$	$K^*_{NaSO_4}$	$\Delta \bar{F}'$	$\Delta \bar{H}'$	$\Delta \bar{S}'$
$NaSO_4^-$	2.4	3.42	−0.42	−3.8	−11
	25.0	2.00			
$MgSO_4^0$	1.7	15.0	−1.38	−2.7	−4
	25.0	10.2			

cal potential energy of the constituent ions decreases and is released as degraded kinetic energy when they approach each other.

The effects of pressure were also examined. The equation used was analogous to that for the pH (see Chapter 2, Volume II) for a cell

$$Ag : AgCl \mid reference\ soln \mid glass \mid test\ soln \mid AgCl : Ag$$

(5.357)

without liquid junction. Then,

$$\log \frac{\{Na^+\}_P}{\{Na^+\}_1} = -\frac{E_P - E_1 - \Delta E_{asym}}{s}$$ (5.358)

The effect of pressure on $\{Cl^-\}$ is canceled out between the reference solution and the test solution (0.4949 m NaCl). The change, $\Delta E_{asym} = (E_{asym})_P - (E_{asym})_1$, the asymmetry of the glass membrane, was estimated from pressure runs with 0.4949 m NaCl on both sides of the glass membrane.

The partial molal volume at 25°C, obtained from Equation 5.348 was 15.8 mL/mole. This quantity has a prime subscript because the association constant K^*' is determined in terms of concentrations. $\Delta \bar{v}^0$ is obtainable from K^* in the relation $\Delta \bar{v}^0 = -RT(\partial \ln K^*/\partial P)$. Thus, one obtains Equation (5.349) from (5.347) and (5.359) if $\Delta \bar{v}^0$ is constant over the pressure range of interest. This means that, if one uses $\Delta \bar{v}^0$ to calculate $\partial \ln K^*'/\partial P$, it is necessary to determine the effect of pressure on the γ's. The required experiments and/or calculations are difficult and uncertain. This can be done experimentally but the use of measured Δv's measured in the medium of interest is more straightforward. Our results for $\ln K^*_{NaSO_4}$ were roughly 1.25 at 1 atm and 0.55 at 1000 atm. The decrease in $K^*_{NaSO_4}$ with increasing pressure shows that the extent of association decreases because the change in electrostriction, randomness in the transition region, and the bulk water have a net positive $\bar{v}'$.

The ion-sensitive electrodes were not sensitive enough to determine K^*' for $CaSO_4^0$ versus T and P and K^*' for $MgSO_4^0$ versus P. At the time of the work $NaCl^0$, $MgCl^+$, and $CaCl^+$ were not taken into consideration.

Other approaches to $\Delta \bar{v}$ will be examined later on.

5.7.10 Activity Coefficients of HCO_3^- and CO_3^{2-}

The Method of Hawley and Pytkowicz

Equations and Results. Hawley (1973) and Pytkowicz and Hawley (1974) developed a potentiometric method to determine the ion pairing of HCO_3^- and CO_3^{2-} with the major cations of seawater.

We saw earlier and will examine further (Chapters 1 and 2, Volume II) the equation $K_1' = k_H' a_H [HCO_3^-]_T/[H_2CO_3^*]$ for the reactions $H_2O + CO_2 = H_2CO_3$ and $H_2O + CO_2 = HCO_3^- + H^+$. In the method of Pytkowicz and Hawley (1974) it was placed in the form $K_1' = k_H' a_H [HCO_3^-]_T/a_w[CO_2^*]$. In these equations $[CO_2^*] \equiv [H_2CO_3^*] \equiv [CO_2] + [H_2CO_3] \cong [CO_2]$. Note that CO_2^* and $H_2CO_3^*$ are the same quantity by definition and that $[H_2CO_3]$ is essentially negligible compared to $[CO_2]$. The term k_H' represents the effect of the change in liquid junction potential when electrodes are transferred from a buffer to seawater or some other complex concentrated solution.

The first definition of K_1' is of practical value and works when a_w is essentially constant, as happens in seawater (see Chapter 2, Volume II), while the one with a_w explicit is more useful when a_H changes.

Furthermore,

$$^a HCO_3 = (\gamma_{HCO_3})_F [HCO_3^-]_F$$

$$= (\gamma_{HCO_3})_T [HCO_3^-]_T$$ (5.359)

$$[HCO_3^-]_T = [HCO_3^-]_F + [NaHCO_3^0]$$
$$+ [MgHCO_3^+] + [CaHCO_3^+]$$ (5.360)

$$K^{*'}_{MHCO_3} = \frac{[MHCO_3]}{[M]_F[HCO_3^-]_F}$$

$$= K^* \frac{(\gamma_M)_F(\gamma_{HCO_3})_F}{\gamma_{MHCO_3}}$$ (5.361)

where the charges are omitted for simplicity.

Combining the above equations yields

$$\frac{K_1'}{a_w \gamma^*_{CO_2}} = K_1'' (1 + K^*_{NaHCO_3}[Na^+]_F$$

$$+ K^*_{MgHCO_3}[Mg^{2+}]_F + K^*_{CaHCO_3}[Ca^{2+}]_F) \quad (5.362)$$

where

$$K_1'' = \frac{k_H K_1}{(\gamma_{HCO_3})_F} \quad (5.363)$$

The association constants and K_1'' were obtained by the application of Equation (5.362) and the measurement of K_1' to a sufficient number of solutions of different composition at what was thought to be the same I_e.

The corresponding association constants for carbonates were based on $K_1' K_2'$. A complication arose due to the appearance of what may be triple ions. We cannot be certain of this as the triple ions may be an artifact of changes in liquid junction potentials, and a reference electrode with a salt bridge was used (see Chapters 1 and 2, Volume II). Another reason for the observed behavior may be Cl^- association which could change I_e significantly among the various solutions.

The seawater model that resulted from this work is shown in Table 5.45. It was obtained by the usual set of equations, namely, the association constants and the mass balance relations. The stoichiometric association constants used are presented in Table 5.46 and the speciation is shown in Table 5.47.

The results of Hawley (1973) and Pytkowicz and Hawley (1974) are close to those of Kester and Pytkowicz (1969) for cations but differ considerably for anions. This results from the fact that the anion concentrations, with the exception of $[Cl^-]_T$, are smaller than those of the cations so that they are more sensitive to ion pairing.

The value of $(\gamma_{HCO_3})_F$ can be calculated from

$(\gamma_{HCO_3})_F = (k_H' K_1 a_w \gamma_{CO_2}/K_1')(1 + K^*_{NaHCO_3}[Na^+]_F + \cdots)$, in which k_H' contains the nonthermodynamic assumption required for single-ion properties, and a similar method can be used for K_2'. The results are

Species	HCO_3^-	CO_3^{2-}
γ_F	0.615	0.376
γ_T	0.500	0.030

where $\gamma_T = \gamma_F m_F / m_T$ and the molal ratio are obtained from Hawley (1973).

The Stability of Triple Ions. It is of interest to compare the stability of triple ions to that of ion pairs to ascertain their relative probabilities of occurrence. This is done next by means of a very simple model.

The interaction energy when the triple ion $Na_2CO_3^0$ is formed by the reaction

$$2Na^+ + CO_3^{2-} = Na_2CO_3^0 \quad (5.364)$$

is

$$\Delta E_{Na_2CO_3} = 2\Delta_{Na^+ - CO_3^{2-}} + \Delta E_{Na^+ - Na^+} \quad (5.365)$$

or

$$\Delta E_{Na_2CO_3} = -2\frac{z_{Na}z_{CO_3}e^2}{D_e r} + \frac{z_{Na}z_{Na}e^2}{2D_e r} \quad (5.366)$$

where r is the average distance between Na^+ and CO_3^{2-} in ion pairs and triplets. The ΔE's are also electrical free energies ΔF_{el}.

The hydrated radii, or rather, the Debye distances of closest approach are, in Å,

$$r_{Na} = 2 \quad r_{Mg} = 4 \quad r_{Ca} = 3 \quad r_{CO_3} = 2.2$$

$$(5.367)$$

TABLE 5.46 Values of the stoichiometric association constants used to calculate the major chemical species in seawater of 34.8‰ salinity at 25°C (Hawley, 1973; Pytkowicz and Hawley, 1974).

	SO_4^{2-}	HCO_3^-	CO_3^{2-}	$Mg_2CO_3^{2+}$	$MgCaCO_3^{2+}$	F^-
Na^+	2.02 ± 0.03^a	0.28 ± 0.01	4.25 ± 0.03			
Mg^{2+}	10.20 ± 0.5^a	1.62 ± 0.03	112.30 ± 0.5	386.6 ± 1.7	104.2 ± 4	18.8
Ca^{2+}	10.80 ± 0.7^a	1.96 ± 0.03	279.60 ± 0.5			4.2
K^+	1.03^*					

[a] From Kester and Pytkowicz (1969).

TABLE 5.47 Ionic species in 34.8‰ salinity seawater at 25°C and pH 8.10 calculated from the stoichiometric association constants (Hawley, 1973; Pytkowicz and Hawley, 1974).

	Na^+	Mg^{2+}	Ca^{2+}	K^+
Total molality	0.4822	0.05489	0.01063	0.01062
% Free metal	97.7	89.2	88.5	98.9
% MSO_4	2.2	10.3	10.8	1.1
% $MHCO_3$	0.1	0.3	0.3	
% MCO_3	0.0	0.1	0.3	
% Mg_2CO_3		0.0		
% $MgCaCO_3$		0.0	0.1	
% MF		0.1	0.0	

	SO_4^{2-}	HCO_3^-	CO_3^{2-}	F^-
Total molality	0.02906	0.00213	0.000171	0.00008
% Free anion	39.0	81.3	8.0	51.0
% NaX	37.1	10.7	16.0	
% MgX	19.5	6.5	43.9	47.0
% CaX	4.0	1.5	21.0	2.0
% KX	0.4			
% Mg_2CO_3			7.4	
% $MgCaCO_3$			3.8	

so that

$$r_{Na-CO_3} = 4.2, \quad r_{Mg-CO_3} = 6.2, \quad \text{and so on}$$

$$(5.368)$$

Then, the interaction energies are, if r is converted to cm,

$Na_2CO_3^0$	$NaMgCO_3^+$	$NaCaCO_3^+$
-0.83×10^{-8}	-1.29	-1.44

in units of e^2/D_e. Thus, the electrical free energy of triple ions leads to larger values of K^* as can be seen from $\Delta F_{el} = -RT \ln K^*$. The calculated values of K^* for triple ions are

$Na_2CO_3^0$	$NaMgCO_3^+$	$NaCaCO_3^+$
6.76	19.6	27.5

They show that the last three are the most important. The actual values should not be taken seriously because the effects of all the other species in solution on those displayed were neglected.

The Johnson Correction. Johnson (1979) made a provisional correction for $K^{*'}$ of HCO_3^- and of the CO_3^{2-} association to include the effect of Cl^- ion pairs.

In the case of HCO_3^- the values of $K^{*'}_{MHCO_3}$ were obtained in the earlier work by fitting measured values of K_1' in several solutions to an equation linear in the free cation concentrations. Johnson (1979) corrected the values of m_F for Cl^- association, a first needed step, but could not reduce numerically the solutions to the same I_e. The results

$MgCaCO_3^{2+}$	$Mg_2CO_3^{2+}$	$Ca_2CO_3^{2+}$
-1.70	-1.54	-1.87

are presented in Table 5.48. The values of Hawley for triple ions were retained.

No single trend was found in going from the uncorrected to the corrected data. This may be due to

$MgCaCO_3^{2+}$	$Mg_2CO_3^{2+}$	$Ca_2CO_3^{2+}$
50.1	34.7	74.1

the complex changes in m_F of the cations when the chloride ion pairs are included in Equation (5.362).

The Method of Berner (1965). This method will soon be compared to that of Pytkowicz (1975) who also did not have to use ion pairs in order to calculate $(\gamma_{HCO_3})_T$ and $(\gamma_{CO_3})_T$.

Berner determined the total activity coefficients of Ca^{2+}, HCO_3^- and CO_3^{2-} in seawater and obtained

TABLE 5.48 The values of K^* corrected by Johnson (1969; also see text).

Cation	HCO_3^-	CO_3^{2-}
Na^+	0.28	1.8
K^+	0.34	2.2
Mg^{2+}	2.5	140
Ca^{2+}	3.5	230

the results shown in Table 5.49. Note that the natural seawater is a surface one diluted by rain and is not quite representative of open oceanic waters for which Cl‰ $\cong$ 19. The recipe of Lyman and Fleming (1940) has since been updated by that of Kester et al. (1967) and has been in use in our work at Oregon State University.

Berner's procedure was as follows. The thermodynamic equilibrium constants for the reactions are

$$CO_2 + H_2O = H^+ + HCO_3^- \qquad (5.369)$$

$$CO_2 + H_2O = 2H^+ + CO_3^{2-} \qquad (5.370)$$

$$K_1 H_h = \frac{a_H a_{HCO_3}}{pCO_2 a_w} = 10^{-7.81} \qquad (5.371)$$

$$K_1 K_2 K_h = \frac{a_H^2 a_{CO_3}}{pCO_2 a_w} = 10^{-18.14} \qquad (5.372)$$

where K_h is the hydration constant. pCO_2 can be used because it equals the fugacity of CO_2 at these low pressures of the gas.

The carbonate alkalinity is $CA = [HCO_3^-] + 2[CO_3^{2-}]$. From (5.371) and (5.372) plus $a_w \cong 0.98$ (Robinson, 1954),

$$CA = \frac{a_{HCO_3}}{(\gamma_{HCO_3})_T} + \frac{2a_{CO_3}}{(\gamma_{CO_3})_T}$$

$$= \frac{pCO_2}{a_H} \left[\frac{10^{-7.82}}{(\gamma_{HCO_3})_T} + \frac{10^{-17.85}}{(\gamma_{CO_3})_T a_H} \right] \qquad (5.373)$$

Berner assumed that CA is independent of the pCO_2 but this is only correct to within a few percent. We shall see that $TA = CA + BA + [OH^-]_T - [H^+]_T$ where BA is the borate alkalinity $[B(OH)_4^-]$, is the invariant quantity with changing pCO_2.

Equation (5.248) can be solved, within the scope

TABLE 5.49 Activity coefficients of Ca^{2+}, HCO_3^-, and CO_3^{2-} in seawater at 25°C and 1 atm (Berner, 1965).

Artificial Seawater[a]	19‰ Cl	$TA = 2.36 \times 10^{-3}$ (equiv/L-SW)
$(\gamma_{HCO_3})_T$ 0.550	$(\gamma_{CO_3})_T$ 0.021	$(\gamma_{Ca})_T$ 0.203
Natural Seawater	17‰ Cl	$TA = 2.09 \times 10^{-3}$
0.561	0.024	0.223

[a] Lyman and Fleming (1940).

of Berner's assumption, through the measurement of the pH in two water samples with the same CA but equilibrated with CO_2 of different partial pressures. One then has two equations such as (5.373), with the activity coefficients as the two unknowns.

$(\gamma_{Ca})_T$ was determined through the solubility of calcite. It was assumed, in contrast to results shown in Chapter 4, Volume II that pure calcite without $MgCO_3$ on its surface can exist in seawater.

The relation used for the determination of $(\gamma_{Ca})_T$ was

$$K_s = 10^{-8.34}$$

$$= (\gamma_{Ca})_T (\gamma_{CO_3})_T (m_{Ca})_T (m_{CO_3})_T \qquad (5.374)$$

$a_{CO_3} = (\gamma_{CO_3})_T (m_{CO_3})_T$ where the concentration of CO_3^{2-} was obtained from the equations of the CO_2 system (see Chapter 1, Volume II). m_{Ca} at saturation is its initial value plus $0.5 \Delta CA = 0.5(CA_e - CA_i)$. Actually ΔTA should be used in accurate work (see Chapters 1, 2, 4, and 5, Volume II). The subscripts e and i refer to equilibrium and initial values. The results of Berner are in fair agreement with those of Garrels and Thompson (1962) in spite of the radical assumption used by the latter. At Cl‰ = 19

	$(\gamma_{HCO_3})_T$	$(\gamma_{CO_3})_T$	$(\gamma_{Ca})_T$
Berner (1965)	0.550	0.021	0.203
Garrels and Thompson (1962)	0.470	0.018	0.250

The Method of Pytkowicz (1975). I used the relations

$$(\gamma_{HCO_3})_T = \frac{K_1}{K_1'} \gamma_{CO_2} a_w k_H' \qquad (5.375)$$

TABLE 5.50 Activity of water in seawater at 25°C selected from Robinson (1954) and expressed versus the salinity.

$S‰$	a_w
25	0.958
27	0.954
29	0.951
31	0.947
33	0.943
35	0.940
37	0.937

$$(\gamma_{CO_3})_T = \frac{K_2}{K_2'}(\gamma_{HCO_3})_T k_H' \qquad (5.376)$$

obtained by rearranging the relations between the apparent and the thermodynamic dissociation constants of carbonic acid (see Chapters 1 and 2, Volume II).

K_1 and K_2 were obtained from Harned and Scholer (1941) and Harned and Owen (1958), K_1' and K_2' from Hawley and Pytkowicz (1973), $(\gamma_H)_T$ from Culberson et al. (1970), a_w was based on Robinson (1954) (also see Table 5.50), and γ_{CO_2} obtained from the CO_2 solubility data of Murray and Riley (1971). I used $k_H'(\gamma_H)_T = 0.98$ instead of $k_H' = 1.134$ in my paper by mistake but the resulting γ's have been corrected in Table 5.51.

a_w in seawater is a useful quantity since it reveals the effect of salts on the water and, specifically, on the vapor pressure. The decrease in vapor pressure results from the binding of water to ions by hydration, the breaking of water bonds, and the generation of monomers in the transition zone between ions and the bulk water. Results are shown in Table 5.50.

The equation of Robinson is

$$\frac{p^0 - p}{p} = 9.206 \times 10^{-4}\ Cl‰$$

$$+\ 2.36 \times 10^{-6}\ (Cl‰)^2 \qquad (5.377)$$

The effect of T on a_w is slight since at 25°C the vapor pressure of seawater corresponds to that of a 0.5880 molal solution of NaCl, whereas at -1.974°C it is equivalent to 0.5839. The term a_w is given by p/p_0 where p is the vapor pressure of water in seawater and p_0 is that of pure water.

TABLE 5.51 Total activities, γ_T, of bicarbonate and carbonate ions in seawater at 25°C (Pytkowicz, 1975).

$S‰$	$(\gamma_{HCO_3})_T$	$(\gamma_{CO_3})_T$
25	0.621	0.0486
26	0.615	0.0465
27	0.609	0.0448
28	0.603	0.0431
29	0.598	0.0417
30	0.592	0.0403
31	0.586	0.0389
32	0.582	0.0375
33	0.577	0.0363
34	0.573	0.0352
35	0.568	0.0340
36	0.565	0.0330
37	0.561	0.0319

f_{CO_2} was calculated from

$$f_{CO_2} = \frac{B}{B_0} \qquad (5.378)$$

where B_0 and B are the Bunsen coefficients in distilled water and seawater. The Bunsen coefficient is one of the various measures for the solubilities of gases.

In addition to the values of $(\gamma_{HCO_3})_T$ and $(\gamma_{CO_3})_T$, I reported the effects of temperature on $(\gamma_{HCO_3})_T$ and $(CO_3)_T$ in Pytkowicz (1975). The low values of $(\gamma_{CO_3})_T$ in Table 5.50 are due not only to the unspecific (Debye–Hückel type) effect of the double charges but also result from the ion pairs formed by CO_3^{2-} with the major cations of seawater, which are stronger than those of HCO_3^-. Note that the entries in Table 5.50 are independent of any one ion-pairing model. The values are smaller than those of Walker et al. (1927) due to the extensive ion-pair formation in seawater.

$(\gamma_{Ca})_T$ can be obtained from the equation

$$(\gamma_{Ca})_T = \frac{K_{sO}}{K_s}\frac{1}{(\gamma_{CO_3})_T} \qquad (5.379)$$

with K_s given by

$$\log K_s = -8.400 + 0.6325\ (S‰)^{1/3} \qquad (5.380)$$

at 25°C (Ingle et al., 1973). I could just as well have used the result of Plath et al. (1980). The value of

TABLE 5.52 Activity coefficients of CO_2 in seawater.

$S\%$	0	5	10	15	20	25	30
25	1.14	1.14	1.13	1.13	1.12	1.12	1.12
27	1.15	1.15	1.14	1.14	1.13	1.13	1.13
29	1.16	1.16	1.15	1.15	1.14	1.14	1.14
31	1.17	1.17	1.17	1.17	1.15	1.15	1.15
33	1.19	1.19	1.18	1.18	1.16	1.16	1.16
35	1.20	1.20	1.19	1.19	1.18	1.17	1.17
37	1.21	1.21	1.21	1.20	1.19	1.18	1.18

$K_{sO} = 3.98 \times 10^{-9}$ was obtained from Langmuir (1968). The results were:

$S\%$	25	27	29	31	33	35	37
$(\gamma_{Ca})_T$	0.327	0.318	0.308	0.300	0.297	0.290	0.280

The values of γ_{CO_2} in seawater, obtained as by-products of this work, are shown in Table 5.52.

Free activity coefficients can be calculated from the equations of Hawley (1973) and Pytkowicz and Hawley (1974). For $[HCO_3^-]_F$,

$$(\gamma_{HCO_3})_F = \frac{K_1 \gamma_{CO_2} a_w k'_H}{K'_1}$$

$$[1 + K^{*\prime}_{NaHCO_3}[Na^+]_F + \cdots] \qquad (5.381)$$

The results are $(\gamma_{HCO_3})_F = 0.615$ and $(\gamma_{CO_3})_F = 0.376$. More exact values require correction of the $K^{*\prime}$ values and the free species present in the solutions by taking into account Cl^- ion pairs.

The next example illustrates the effects of compositional changes on free ions and ion pairs. I do not recall the details of this rough calculation but present it anyway to illustrate trends.

EXAMPLE 5.12. EFFECT OF PURE WATER COMPOSITION ON THE SPECIATION INCLUDING HCO_3^- AND CO_3^{2-} ION PAIRS. Table 5.53 is another example which does not include Cl^- ion pairing for simplicity but which illustrates trends.

Results of Butler and Huston. Butler and Huston (1970) also measured ion pairing and the behavior of activity coefficients potentiometrically. They used sodium amalgam as well as glass electrodes and the solutions contained $NaCl-NaHCO_3-H_2O$ and $NaCl-Na_2CO_3-H_2O$. The values of the logarithms of the association constants, measured at 25°C,

were $\log K^{*\prime}_{NaHCO_3} = 0.27$ and $\log K^{*\prime}_{NaCO_3} = 0.96$ at $I = 1.0$, while at $I = 0.50$ the values were 0.14 and 0.77.

The emf results from titrations in pure H_2O-NaCl and in solutions containing HCO_3^- and CO_3^{2-} were interpreted in terms of $NaHCO_3^0$ and $NaCO_3^-$ ion pairs. These titrations consisted in the addition of $NaHCO_3$ or Na_2CO_3 to $H_2O-NaCl$ at constant I to determine the changes in $(\gamma_{\pm NaCl})_T$ and the values of the association constants.

A sodium-sensitive glass electrode and a silver–silver chloride reference electrode were used to determine and then interpret the γ's. The equation for the potential is

$$E = E^0 + \frac{RT}{F'} \ln[(\gamma^2_{\pm NaCl})_T [Na^+]_T [Cl^-]_T] \quad (5.382)$$

E^0 was evaluated from the known composition and the $\gamma_{\pm NaCl}$ of a $NaCl-H_2O$ reference solution. x in (5.260) refers to HCO_3^- or CO_3^{2-}. Then, $(\gamma_{\pm NaCl})_T$ in the solutions of interest was calculated from the measured E and the known E^0, $[Na^+]_T$, and $[Cl^-]_T$.

Harned's rule for $H_2O-NaHCO_3$ was found to be of the form

$$\log(\gamma_{\pm NaCl})_T = \log(\gamma_{\pm NaCl})_T(0)$$
$$- \alpha I(1 - X_1) \qquad (5.383)$$

It is valid from $I = 0.5$ to 3.0 with $\alpha = 0.047$. The term on the left is the γ in the solutions of interest and X_1 is the mole fraction of NaCl.

The interpretation of emf's is similar to that of Pytkowicz and Kester (1969) for $H_2O-NaCl-Na_2SO_4$ in that Harned's rule behavior was related to ion pairing although the effective ionic strength was not used. Butler and Huston found that α for the carbonate case also has a value of 0.047.

This work is not exact because the effects of $NaCl^0$ ion-pair formation on the effective ionic strength I_e and on the competition of HCO_3^- and CO_3^{2-} with Cl^- for Na^+ ions were not taken into consideration in the interpretation of changes in $\gamma_{\pm NaCl}$ at constant I.

The results of Butler and Irma (1970) can be used in waters that are essentially made of NaCl plus the CO_2 species.

TABLE 5.53 Effects of compositional variations[a] at $I = 0.72$, pH $= 78$, $K_2' = 7.64 \times 10_3^{-10}$, 24°C.

	Ion	Total Molality	% Free	% MSO_1	% $MHCO_3$	% MCO_3	% NaA	% CaA	% MgA
Normal SW	Na	0.482	97.69	2.25	0.05	0.01			
	Ca	0.0106	88.71	10.92	0.31	0.06			
	Mg	0.0549	89.25	10.38	0.26	0.11			
	SO₄	0.0291	39.18				37.26	3.98	19.58
	HCO₃	0.00241[b]	81.32				10.72	1.50	6.46
	CO₃		8.96				28.86	6.11	56.07
$(Ca^{2+}) = 2x$	Na	0.438	97.66	2.28	0.05	0.01			
normal other	Ca	0.0212	88.57	11.06	0.31	0.06			
cations com-	Mg	0.0495	89.12	10.51	0.26	0.11			
pensate the	SO₄	0.0291	39.03				34.33	8.06	17.88
total ionic	HCO₃	0.00241[b]	81.43				9.75	3.00	5.82
strength	CO₃		9.16				26.79	12.47	51.59
Oxidizing	Na	0.487	97.96	1.98	0.05	0.01			
sediment	Ca	0.0109	89.87	9.73	0.33	0.07			
	Mg	0.0503	90.36	9.24	0.28	0.12			
	SO₄	0.0254	39.48				38.04	4.18	18.30
	HCO₃	0.00255[b]	81.54				10.89	1.57	6.00
	CO₃		9.28				30.28	6.59	53.85
	Na	0.501	98.61	1.37	11.23	0.00			
	Ca	0.0083	92.92	6.70	0.15	0.03			
	Mg	0.053	93.28	6.54	0.13	0.05			
	SO₄	0.0179	38.82				38.74	3.23	19.21
	HCO₃	0.0012[b]	81.17				11.23	1.23	6.38
	CO₃		8.98				30.35	5.02	55.64
	Na	0.504	99.74	0.19	0.06	0.01			
	Ca	0.0186	98.47	0.99	0.42	0.12			
	Mg	0.0156	98.57	0.93	0.35	0.21			
	SO₄	0.0022	42.19				42.85	8.35	6.61
	HCO₃	0.00239[b]	83.23				11.75	2.99	2.07
	CO₃		12.44				49.89	17.18	23.31

Source: Reprinted with permission from Bates, R. G., B. R. Staples, and R. A. Robinson, *Anal. Chem.* **42**, 867. Copyright © 1970, American Chemical Society.
[a] Represents $(HCO_3^-) + (CO_3^{2-})$.
[b] The three bottom compositions are taken from deep-sea drilling data.

EXAMPLE 5.13. Consider a solution that contains 0.40 m NaCl and 0.1 m NaHCO₃ and outline the determination of the CO₂ speciation. The equations are

$$K_{NaHCO_3}^{*'} = \frac{[NaHCO_3^0]}{[Na^+]_F[HCO_3^-]_F} = 1.38 \quad (i)$$

$$K_1 = \frac{a_H(\gamma_{HCO_3})_T[HCO_3^-]_T}{(\gamma_{H_2CO_3})[H_2CO_3^*]} = 4.40 \times 10^{-7} \quad (ii)$$

$$(Na^+)_T = [Na^+]_F + [NaHCO_3^-] = 0.500 \quad (iii)$$

$$[HCO_3^-]_T = [HCO_3^-]_F + [NaHCO_3^-] \quad (iv)$$

$$[H_2CO_3^*] + [HCO_3^-]_T = 0.100 \quad (v)$$

$$\gamma_{H_2CO_3} \cong 1.14 \quad (vi)$$

since it is a nonelectrolyte.

We have a system of six equations in the eight unknowns a_H, $(\gamma_{HCO_3})_T$, $(\gamma_{CO_3})_T$, $[H_2CO_3^*]$, $[HCO_3^-]_T$, $[Na^+]_F$, $[HCO_3^-]_F$, and $[NaHCO_3^0]$. It is obvious, therefore, that two measurements are required before the equations can be solved. One set of measurements worth considering is the determination of a_H and K_1' by a titration. K_1' removes the need to determine $(\gamma_{HCO_3})_T$ and $\gamma_{H_2CO_3}$ as well as Equation (vi), since it is expressed in terms of concentration.

Values of the association constants may be those of Pytkowicz and Hawley (1974) or Butler and Huston (1970).

5.7.11 The Hydration Model for Single Ions

Bates et al. (1970) used the equation of Stokes and Robinson (1948) in the form

$$\ln \gamma_{\pm(exp)} = |z_C z_A| \ln f_{DH} - \frac{h}{v} \ln a_w$$
$$- \ln[1 + 0.018(v - h)m] \quad (5.384)$$

This equation is equivalent to the one derived earlier in this book. f_{DH} is the rational Debye–Hückel activity coefficient, h is the number of water molecules solvated per mole of electrolyte, and v is the number of moles into which the electrolyte dissociates.

The interpretation of Bates et al. (1970) attributes to hydration other effects neglected by Debye and Hückel (1923), such as ion pairing, Gurney cosphere overlap, ion-cavity interactions, and core effects which shall be discussed later.

Equation (5.384) has two adjustable parameters, a_D in f_{DH} and h. They can be determined by fitting the equation to experimental values of $\gamma_\pm$. This means that, although h has theoretical significance, its numerical value is questionable. The reasons for this are that f_{DH} was used at concentrations well above those for which it is valid and because other effects mentioned above were not included.

Our ion-pairing model does not require a hydration correction because the hydration effect is implicit in the experimentally determined $(\gamma_\pm)_F$.

The following procedure was used to estimate the single-ion activity coefficients. Values of h for the five alkali chlorides, with the value of CsCl obtained by extrapolation, were plotted versus the crystal radii of the cations. The hydration number for Cl^- is supposed to be small according to literature data (0–0.9). Therefore, Bates et al. assumed h_{Cl} to be zero so that $h_{MCl} = h_M$. This, of course, is a nonthermodynamic assumption.

The use of the Gibbs–Duhem equation for 1-1 electrolytes,

$$-\left(\frac{55.51}{m}\right)d \ln a_w = d \ln (\gamma_M m)$$
$$+ d \ln a_{Cl} \quad (5.385)$$

TABLE 5.54 Molal activity coefficients of single ions according to the hydration theory, selected from Bates et al. (1970). The original tables extend to $m = 6.0$.

m	$(\gamma_+)_T$	$(\gamma_-)_T$	$(\gamma_+)_T$	$(\gamma_-)_T$	$(\gamma_+)_T$	$(\gamma_-)_T$
	HCl		NaCl		KCl	
0.1	0.807	0.785	0.783	0.773	0.773	0.768
0.2	0.788	0.746	0.744	0.726	0.722	0.714
0.5	0.812	0.706	0.701	0.661	0.659	0.639
1.0	0.940	0.697	0.697	0.620	0.623	0.586
2.0	1.421	0.717	0.756	0.590	0.610	0.538
	CsCl		MgCl$_2$		CaCl$_2$	
0.1	0.756	0.756	0.279	0.726	0.269	0.719
0.2	0.694	0.694	0.239	0.697	0.224	0.685
0.5	0.606	0.606	0.234	0.688	0.204	0.665
1.0	0.544	0.544	0.344	0.732	0.263	0.690
2.0	0.496	0.494	1.439	0.899	0.768	0.804

led, in conjunction with hydration relations, to

$$\ln \gamma_M' = \ln f_{DH} - \ln(1 - 0.018 \times 2m') \quad (5.386)$$

where m' is the molality in terms of unbound water. Also,

$$\log \gamma_M = \log \gamma_\pm + 0.00782hm\phi \quad (5.387)$$

and

$$\log_{Cl} = \log \gamma_\pm - 0.00782hm\phi \quad (5.388)$$

where the osmotic coefficient ϕ is $-\ln a_w/0.018vm$.

The results for single ions are shown in Table 5.54. The results in Table 5.54 pertain to an empirical h as was discussed earlier. They reveal the salting-in–salting-out transition, since $\gamma_\pm$ is the experimental value.

It is of interest to consider CsCl since one notes that $(\gamma_{Cs})_T = (\gamma_{Cl})_T$ rather than $(\gamma_K)_T = (\gamma_{Cl})_T$ which was used in the MacInnes convention. This equality for cesium should be viewed with care, however, because of the terms neglected by Stokes and Robinson (1948).

One problem with the hydration, the Brønsted–Guggenheim (Brønsted, 1922; Guggenheim, 1935) and the Pitzer (1973) equations, is that the Debye–Hückel, the Guntelberg, and similar equations are extrapolated to high concentrations. This is done in spite of the lack of theoretical meaning of the Debye–Hückel expression, from which the Guntelberg

TABLE 5.55 A comparison of $\gamma_{\pm G}$ and $\gamma_{\pm\exp}$ for NaCl at 25°C [$\gamma_{\pm G}$ at $I \leq 0.1$ from Hamer (1968)].

I	$\gamma_{\pm G}$	$\gamma_{\pm\exp}$
0.001	0.9646	
0.002	0.9509	
0.005	0.9252	
0.01	0.8985	
0.02	0.9642	
0.05	0.8062	
0.1	0.7531	0.778
0.2	0.701	0.735
0.3	0.665	0.710

TABLE 5.56 Single-ion activity coefficients for a sea-water Brønsted–Guggenheim model extrapolated to single ions and at 25°C and 1 atm, and not including HCO_3^- or CO_3^{2-} (Whitfield, 1973). The γ_i correspond to $(\gamma_i)_T$ in the ion-pair notation.

m at $I = 0.7$ Total Molality	γ_i		
	0.5 m	0.7 m	1.0 m
Na^+	0.4699	0.669	0.650	0.635
K^+	0.0101	0.643	0.657	0.592
Mg^{2+}	0.0536	0.232	0.217	0.206
Ca^{2+}	0.0104	0.221	0.203	0.189
Cl^-	0.5518	0.694	0.686	0.679
SO_4^{2-}	0.0281	0.150	0.122	0.100

one was derived, at $m > 10^{-3}$ (Frank and Thompson, 1952; Pytkowicz and Johnson, 1979). Still, these equations can serve in practice as empirical ones.

In the case of the Brønsted–Guggenheim relation, in a single-electrolyte solution, the interaction coefficient is

$$B_{CA} = \frac{\log \gamma_{\pm\exp} - \log \gamma_{\pm G}}{2I} \qquad (5.389)$$

where $\gamma_{\pm G}$ is the Guntelberg (1926) term.

We can see in Table 5.55 that $\gamma_{\pm G}$ decays very slowly. This means that $\gamma_{\pm G}$ has its original meaning and reflects roughly nonspecific interactions at very low concentration. In this case, the remaining terms of the Brønsted–Guggenheim and the Pitzer equations are empirical in practice in spite of their theoretical derivations. This arbitrariness continues to fairly high concentrations because $\gamma_{\pm G}$ decays slowly.

5.7.12 Brønsted–Guggenheim Approach

I shall examine the Brønsted–Guggenheim model (Brønsted, 1922, 1927; Guggenheim, 1935, 1936), in its original application to mean-activity coefficients later on. In this subsection I concentrate on the application of the model to single ions (Whitfield, 1973). First, however, I should reiterate that activity coefficients for mean activity coefficients may express a variety of effects under the guise of one of them.

In the Brønsted–Guggenheim model $\gamma_\pm$ is given by

$$\log \gamma_{\pm BG} = \log \gamma_{\pm G} + \nu B_{CA} \qquad (5.390)$$

$\gamma_{\pm(BG)}$ is equated to $\gamma_{\pm(\exp)}$ in the method. $\gamma_{\pm G}$ is the Guntelberg (1926) term $-A_D|z_C z_A|I^{0.5}/1 + I^{0.5}$, $2/\nu = 1/\nu_C + 1/\nu_A$, and B_{CA} is the interaction coefficient for C^{z+} and A^{z-}. This equation can be applied up to $m_{CA} = 0.1$ although it should be remembered that B_{CA} varies when m changes.

Whitfield wrote equations equivalent to (5.390) for single ions. He assumed that B_{CA} is independent of the composition of the solution at a given ionic strength. The resulting equations are

$$\log \gamma_C = |z_C z_A| \log \gamma_G + \sum_C B_{CA*}[A^{z-}]_T^* \qquad (5.391)$$

and

$$\log \gamma_A = |z_A z_C| \log \gamma_G + \sum_A B_{C*A}[C^{z+}]_T^* \qquad (5.392)$$

$|z_A/z_C|$ and $|z_A/z_C| \log \gamma_G$ is $-A_D z_A^2 I^{0.5}/(1 + I^{0.5})$. These equations apply to mixtures of electrolytes and the asterisks indicate quantities that are varied during the summation. Results for a seawater model are presented in Table 5.56.

Interaction coefficients for HCO_3^- and CO_3^{2-} were not available and γ_i's for these species are not presented in Table 5.56.

The procedure used by Whitfield (1973) was as follows: In the case of a symmetric electrolyte such as CA, (γ_{CA}) was obtained from the literature for the solution of the single electrolyte. γ_G was calculated and the interaction coefficient B_{CA} was obtained from Equations (5.391) and (5.392). It was assumed that B_{CA} was independent of composition at a given I so that these coefficients could be applied to seawater of a given $S‰$. This assumption is equivalent to replacing I_e by I in the ion-pair model, since it neglects the effect of specific interactions at different compositions on the electrical environment of ions.

The relationship between $K_{CA}^{*\prime}$ and B_{CA} in the simplest case of a single 1-1 electrolyte is obtainable from

$$(\gamma_C)_T = (\gamma_C)_F[1 + K_{CA}^{*\prime}[A^-]_F] \quad (5.393)$$

$$\log(\gamma_C)_T = \log(\gamma_C)_G + B_{CA}[A^-]_T \quad (5.394)$$

$(\gamma_C)_T$ and $(\gamma_C)_{BG}$, which describe nonspecific interactions, will be assumed to be equal. As

$$(\gamma_C)_T = (\gamma_C)_G + 2.3026 \exp B_{CA}[A^-]_T, \quad (5.395)$$

it follows that

$$K_{CA}^{*\prime} = \{2.3026 \exp[B_{CA}[A^-]_T] - 1\}[A^-]_F \quad (5.396)$$

The correspondence is almost 1-1 at a given ionic strength except that $[A^-]_F$, a function of I_e instead of only I, enters the equation.

Another way to derive the relationship between $K_C^{*\prime}$ and B_{CA} is to set Equation (5.393) as

$$\log \frac{(\gamma_C)_T}{(\gamma_C)_F} = \log(1 + K_{CA}^{*}[A^-]_F)$$

$$= \frac{K_{CA}^{*\prime}[A^-]_F}{2.3026} - \frac{(K_{CA}^{*\prime}[A^-]_F)^2}{4.606} \quad (5.397)$$

so that

$$B_{CA} \cong \frac{K_{CA}^{*\prime}}{2.3026} \frac{[A^-]_F}{[A^-]_T}$$

$$\cong \frac{K_{CA}^{*\prime}}{2.3026} \left[\frac{1}{1 + K_{CA}^{*\prime}[C^+]_F} \right] \quad (5.398)$$

EXAMPLE 5.14. THE RELATIONSHIP BETWEEN $K_{NaCl}^{*\prime}$ AND B_{NaCl}. In the case of NaCl at $I = 0.7$ we have, from the ion-pair model, $K_{NaCl}^{*\prime} = 0.29$ and $(Na)_F = 0.59$. Thus

$$B_{Na} = \frac{0.29}{2.3026} \left[\frac{1}{1 + 0.29(0.59)} \right] = 0.1075 \quad (i)$$

TABLE 5.57 The specific interaction model of Pitzer (1973) applied to seawater (Whitfield, 1975). Values of $(\gamma_\pm)_T$ predicted from simpler solutions are compared to experimental ones and values for single ions are presented.

	Predicted		
	From Pitzer	From Brønsted–Guggenheim	Experimental
NaCl	0.667	0.666	0.668[a]
KCl	0.648	0.649	0.645[b]
MgCl₂	0.473	0.467	
CaCl₂	0.463	0.457	
Na₂SO₄	0.373	0.378	0.378[c]
K₂SO₄	0.358	0.355	0.352[a]
MgSO₄	0.167	0.167	
CaSO₄	0.161	0.162	

Ion	Na⁺	K⁺	Mg²⁺	Ca²⁺	Cl⁻	SO₄²⁻
γ_i	0.65	0.62	0.22	0.21	0.68	0.12

[a] Gieskes (1966).
[b] Whitfield (unpublished result).
[c] Platford and Dafoe (1965).

Alternately, by the procedure of Whitfield (1973), that is, Equation (5.390),

$$B_{NaCl} = \left[\log \frac{(\gamma_{\pm NaCl})_{exp}}{(\gamma_{\pm NaCl})_G} \right] \frac{1}{(Cl^-)_T}$$

$$= \log\left(\frac{0.667}{0.2777}\right) \frac{1}{0.7 \times 2} \times \frac{1}{2.3026} = 0.1180 \quad (ii)$$

5.7.13 The Pitzer Model Adapted to Single Ions

As we shall see later, Pitzer (1973) elaborated on the model of Brønsted (1922) and Guggenheim (1935, 1936) and obtained coefficients that depend on the ionic strength. Whitfield (1975) applied the model to seawater with the results shown in Table 5.57. It is interesting to observe the insensitivity of $\gamma_\pm$ in seawater to the replacement of the Brønsted–Guggenheim model by that of Pitzer and the agreement with experimental data in spite of fairly severe assumptions made in both models. Further study of the behavior of the interaction coefficients with composition is of interest.

5.7.14 Dissociation of Water and the Values of $(\gamma_H)_T$ and $(\gamma_{OH})_T$

Water is the most extensive weak electrolyte in nature. It dissociates according to

$$H_2O = H^+ + OH^- \quad \text{or} \quad 2H_2O = H_3O^+ + OH^-$$

$$(5.399)$$

The hydronium ion H_3O^+ is the actual cation but H^+ is used for simplicity. Nearly pure water has a pH $= 7$ with $K_w = 10^{-14}$ at 25°C.

The dissociation of water is responsible for hydrolysis reactions which yield hydroxo complexes such as

$$Pb^{2+} + H_2O = PbOH^+ + H^+ \quad (5.400)$$

and ions such as

$$CO_3^{2-} + H_2O = HCO_3^- + OH^- \quad (5.401)$$

Ions such as OH^-, $PbOH^+$, and HCO_3^- play a role in the solution chemistry of trace metals (Chapter 6, Volume II) and of the CO_2 system (Chapter 1, Volume II).

The thermodynamic dissociation constant

$$K_w = \frac{(\gamma_H)_T(\gamma_{OH})_T(H^+)_T(OH^-)_T}{a_w} \quad (5.402)$$

was obtained by extrapolation of potentiometric measurements in cells of the type

$$Pt|H_2; NaOH(m); NaCl(m_2)|AgCl:Ag \quad (5.403)$$

(Harned and Owen, 1958; Robinson and Stokes, 1968). Some results are presented in Table 5.58 and were fit by the equation

$$\log K_w = -\frac{4471.33}{T} - 6.0846$$

$$+ 0.017053T \quad (5.404)$$

Harned and Owen also presented K_w in a number of salt solutions.

TABLE 5.58 Dissociation constants of water (Harned and Owen, 1958).

T (°C)	K_w	T (°C)	K_w
0	0.12×10^{-14}	25	1.01×10^{-14}
15	0.45	30	1.47
20	0.68	50	5.48

The effect of pressure $(K_w)_p/(K_w)_1$, selected from Owen and Brinkley (1941), is

P (bars)*	T (°C)	5	15	25	35
400		1.543	1.490	1.435	1.384
1000		2.816	2.585	2.385	2.175

Culberson and Pytkowicz (1973) determined the stoichiometric ion product of water in seawater and the NBS values of $(\gamma_H)_T$ and $(\gamma_{OH})_T$ in the molar scale by means of alkalinity titrations. The results are presented in Table 5.59 and agree well with those obtained by Hansson (1973) and Dickson and Riley (1979). The results of Dickson and Riley can be fit by the equation

$$pK_w' = \frac{3441.0}{T} + 2.256 - 0.709I^{0.5} \quad (5.405)$$

in mole/kg-SW and

$$pK_w' = \frac{3441.0}{T} + 2.241 - 0.9415\,(S‰)^{0.5} \quad (5.406)$$

in molal units. T is the absolute temperature.

5.8 THERMODYNAMICS OF ELECTROLYTE SOLUTIONS

In this section I follow in part Stokes (1979).

5.8.1 Gibbs–Duhem Equation

The important Gibbs–Duhem equation for a generic property X in a phase is

$$\sum_i n_i|d\bar{X}_i = 0 \quad (5.407)$$

*1 bar = 0.9869 atm.

TABLE 5.59 Measured values of $(\gamma_H)_T$, $(\gamma_{OH})_T$, and $K'_w \doteq [H^+]_T[OH^-]_T$ in the molar scale in seawater at 25°C. The activity coefficients are in the NBS scale and are selected from Culberson and Pytkowicz (1973).

		1.99			25.02		
$S\%_0$	T (°C)	$(\gamma_H)_T$	$(\gamma_{OH})_T$	pK'_w	$(\gamma_H)_T$	$(\gamma_{OH})_T$	pK'_w
19.90			0.346	14.277		0.304	13.329
26.68		0.773			0.699		
26.87			0.290	14.214		0.255	13.255

$S\%_0$	T (°C)	$(\gamma_H)_T$
19.77	1.99	0.754
26.68	1.99	0.773
34.58	1.99	0.803
34.58	13.02	0.771
19.77	25.02	0.698
19.77	26.68	0.701
34.58	25.02	0.710
34.58	34.97	0.669

with

$$d\bar{X}_i = d\left(\frac{\partial X}{\partial n_i}\right)_{T,P,n_j} \tag{5.408}$$

It relates with the variations in extensive properties within the phase. When $G = X$, the equation is $\sum n_i d\mu_i = 0$.

The Gibbs–Duhem equation $\mu_w\, dn_w + \mu_s dn_s = 0$ is used in the following way for a single-electrolyte solution. The practical osmotic coefficient ϕ and this equation yield

$$\ln(\gamma_{\pm s})_T = \phi - 1 + 2\int_0^m \frac{\phi - 1}{m^{0.5}}\, dm \tag{5.409}$$

s represents an electrolyte. Accurate values of $\gamma_{\pm s}$ can be obtained down to 0.1 m by this method and to 10^{-3} m potentiometrically. Thus, ϕ is primarily a tool from which one obtains f_w. It is useful because f_w and a_w are insensitive to the concentration where $m_j \to 0$ but ϕ is not.

ϕ is determined isopiestically by measuring the vapor pressure of water. Known amounts of standard NaCl and the unknown are placed in two dishes and the observer waits until the water vapor flux between dishes stops. The equilibrium molalities are obtained by weighting the dishes. Then,

$$\nu m_s = \nu_{ref} m_{ref} \phi_{ref} \tag{5.410}$$

and

$$\phi = \frac{\nu_{ref} m_{ref}}{\nu m_s} \phi_{ref} \tag{5.411}$$

m_{ref}/m is known as the isopiestic ratio, with m_{ref} and m being such that they yield the same equilibrium vapor pressure for the reference and the unknown solutions. Then, (5.409) can be used in conjunction with m obtained from Equation (5.412) to calculate $\gamma_{\pm(exp)}$ versus m. $\gamma_{\pm(exp)}$ corresponds to $(\gamma_{\pm})_T$. Some typical results for NaCl are

$$m = 0.1 \qquad (\gamma_{\pm NaCl})_T = 0.7784 \qquad \phi = 0.9324$$

$$m = 0.5 \qquad (\gamma_{\pm NaCl})_T = 0.6811 \qquad \phi = 0.9209$$

The Gibbs–Duhem equation also permits us to calculate properties such as γ and ϕ (the practical osmotic coefficient) not only for one component from another in two-component solutions but also in multicomponent solutions from the value of γ or ϕ for one of them. Then, $\phi = 1000 \ln a_w/(MW)_w \sum \nu_j m_j$. j represents the ionic components of the solution.

It may appear that, in the case of multicomponent solutions with n components, the chemical potentials of $n - 1$ components must be determined for that of the nth component to be calculated from the Gibbs–Duhem equation. Darken (1950) showed that the reverse is true. The values of μ_j for $n - 1$ components can be determined if μ_i for the ith component is known at all possible compositions of the solution (Lewis and Randall, 1961). McKay (1952, 1953) applied the method of Darken to three-component solutions in terms of molalities.

Examples of the use of the Gibbs–Duhem equation for multicomponent solutions can be found in McKay (1952, 1953), McKay and Perring (1953) and Robinson and Stokes (1968).

5.8.2 Excess Thermodynamic Functions of Solutions

Excess functions reflect deviations from ideality and help us understand the behavior of solutions, that is, the interactions that cause the departures. The functions can of course be related to another measure of nonideality, namely, the activity coefficient.

The ideal entropy of mixing of two solutions is given by

$$\Delta S_{\text{mix}} = S_{\text{soln}} - n_1 \bar{S}_1^0 - n_2 \bar{S}_2^0 \quad (5.412)$$

with $S_i = \bar{S}_i^0 - RT \ln X_i$ instead of $S_i = S_i^0 - RT \ln a_i$. Thus,

$$n_1 \bar{S}_1^0 = n_1 S_1 + n_1 R \ln X_1 \quad (5.413)$$

$$n_2 \bar{S}_2^0 = n_2 S_2 + n_2 R \ln X_2 \quad (5.414)$$

and, adding Equations (5.413) and (5.414),

$$S_{\text{mix}}^i n_1 S_1 + n_2 S_2 - n_1 \bar{S}_1^0 - n_2 \bar{S}_2^0$$
$$= R(n_1 \ln X_1 + n_2 \ln X_2) \quad (5.415)$$

I remind you that the term "ideal solution" for electrolyte solutions corresponds to $\mu_i = \mu_i^0 + RT \ln m_i$ and is used in two senses. It is truly ideal for the ficticious standard state which obeys Raoult's law but it follows Henry's law at the reference state of infinite dilution. It is the latter that counts because γ's at m are determined relative to $m \to 0$.

Athermal solutions are like ideal ones in that $\Delta H_{\text{mix}} = 0$ but ΔS_{mix} does not obey Equation (5.415). This is due to the difference in the sizes and shapes of the components so that ΔS_{mix} is not purely configurational. The condition $\Delta H_{\text{mix}} = 0$ arises because the bonds between the components i–i, j–j, and i–j have similar strengths.

Regular solutions have the ΔS_{mix} term given by the ideal one but $\Delta H_{\text{mix}} \neq 0$. This concept applies to some mixtures of liquid nonelectrolytes.

Electrolyte solutions at moderate concentrations may perhaps exhibit a regular behavior for that part of the water–electrolyte interactions which would occur for the uncharged but solvated ions (Robinson and Stokes, 1968). This approach, in which the Debye–Hückel and the regular solution effects on activity coefficients would be used simultaneously, has not yet been tested.

These functions, when determined, yield insight into the effect of mixing on the properties of electrolytes through traditional and statistical thermodynamics. One such insight is the determination of the γ_i's due to interactions on mixing. An electrolyte solution with 1 kg of solvent (n_w moles) and Σm_i moles of electrolyte solutes has

$$F_{\text{soln}} = n_w \mu_w + \Sigma m_i \mu_i \quad (5.416)$$

If one expands μ in terms of $\mu = \mu^0 + RT \ln a$ (Lewis and Randall, 1961),

$$\Delta F_{\text{soln}} = 2RT\left(- m\phi + \sum_i m_i \phi_i + \sum_i m_i \ln \frac{\gamma_i m_i}{m \gamma_{i(0)}}\right) \quad (5.417)$$

with $f_i = m_i/m$ and $m = \Sigma_i m_i$. The term ΔF of solution can also be called the ΔF of mixing. $\gamma_{i(0)}$ is the activity coefficient of i when i is the only electrolyte present.

At infinite dilution one can obtain the expression for ideal mixing

$$\Delta F_{\text{mix}}^{(i)} = 2RT \Sigma m_i \ln f_i \quad (5.418)$$

so that the excess property is (5.417) minus (5.418)

$$\Delta F_{\text{mix}}^{(\text{exc})} = 2mRT\left(-\phi + \sum_i f_i \phi_i + \sum_i f_i \ln \frac{\gamma_i}{\gamma_{i(0)}}\right) \quad (5.419)$$

Also

$$\Delta \bar{S}_{\text{soln}}^{(\text{exc})} = -\left(\frac{\partial F_{\text{soln}}^{(\text{exc})}}{\partial T}\right)_{P,m} \quad (5.420)$$

$$\Delta \bar{V}_{\text{soln}}^{(\text{exc})} = \left(\frac{\partial F_{\text{soln}}^{(\text{exc})}}{\partial P} \right)_{T,m} \qquad (5.421)$$

$$\Delta \bar{H}_{\text{soln}}^{(\text{exc})} = - RT^2 \left(\frac{\partial F_{\text{soln}}^{(\text{exc})}}{\partial T} \right)_{P,m} \qquad (5.422)$$

The determination of excess functions and specific calculations of osmotic and activity coefficients transcend the scope of this book and the interested reader, provided with the above background, may wish to consult thermodynamic texts such as Lewis and Randall (1961) and Robinson and Stokes (1968).

5.8.3 A Further Word on Harned's Rule

This rule is of interest not only in the study of $(\gamma_{\pm i})_T$'s but also in their interpretation and for the determination of $\Delta F_{\text{mix}}^{(\text{exc})}$ through $\gamma_{\pm i(0)}$ terms.

Let us consider two dissolved electrolytes 2 and 3. The empirical Harned's rule, which applies to systems of total constant molality $m = m_2 + m_3$, is $\log(\gamma_{\pm 2})_T = \log \gamma_{\pm 2(0)} - \alpha_{23} m_3$. As we saw earlier α_B is the Harned rule coefficient and $\gamma_{B(O)}$ is γ_B if only B is present. At times additional terms are required for fitting the rule to experiments as in the case of

$$\log(\gamma_{\pm 2})_T = \log \gamma_{\pm 2(0)} - \alpha_{23} m_3$$
$$- \beta_{23} m_3^2 - \delta_{23} m_3^3 - \cdots \qquad (5.423)$$

When $m = m_3 = $ total molality

$$\log \gamma_{(0)2} = \log_{2(0)} - \alpha_{23} m \qquad (5.424)$$

if the higher power terms are neglected. Then

$$\log(\gamma_{\pm 3})_T = \log \gamma_{\pm 3(0)} - \alpha_{32} m_2$$
$$= \log \gamma_{(0)3} + \alpha_{32} m_3 \qquad (5.425)$$

(0)2 or (0)3 indicates electrolyte 3 or electrolyte 2 as being the only one in the solution while 2(0) or 3(0) indicate the reverse.

An example of Harned's rule is presented below:

EXAMPLE 5.15. HARNED'S RULE FOR HCl–KCl

$m = 2$ $\log(\gamma_{\text{HCl}})_T = 0.00358 - 0.0580\, m_{\text{KCl}}$ (i)

m_{HCl}	0.1	0.5	1.0	2.0
γ_{obs}	0.7838	0.8243	0.8822	1.008
γ_{calc}	0.7823	0.8252	0.8822	1.008

This example shows how well $(\gamma_{\pm \text{HCl}})_T$ can be fit over the 0–2 m range by Harned's equation calculated from one of the preceding relations. The equation can be obtained, for example, by measuring $(\gamma_{\pm 2})_T$ and $\gamma_{2(0)}$ in the equation $\log(\gamma_{\pm 2})_T = \log_{\pm 2(0)} - \alpha_{23} m_3$ and obtaining $\alpha_{23} = 0.0580$.

5.8.4 Some Properties of Mixed Electrolyte Solutions

Two further topics on activity coefficients in mixed electrolyte solutions will be touched on here. I shall interpret briefly a plot of activity coefficients, to understand the effects of mixed electrolytes on γ's (Figure 5.22), and the cross differentiation which is the foundation of the Brønsted–Guggenheim equation (Robinson and Stokes, 1968). This equation will be studied later.

One observes that γ_{NaCl} in a NaCl–HCl mixture is increased by the substitution of HCl for NaCl at a given I. An important result is that γ_{HCl} for a trace of HCl in NaCl is equal to that of NaCl in essentially pure HCl, as is shown by the curve marked $X_{\text{HCl}} = 0$, $X_{\text{NaCl}} = 0$ in Figure 5.22 (Robinson and Stokes, 1968). The values of γ_{NaCl} and γ_{HCl} for $X_{\text{NaCl}} = 1$ and $X_{\text{HCl}} = 1$ (the bottom and the top curves) are quite different from the trace values even though I is the same at a given $m = m_{\text{HCl}} + m_{\text{NaCl}}$. This may give the impression that the extended ionic strength principle of Pytkowicz and Kester (1969) is incorrect but the reason for the apparent discrepancy is that γ_T and I_T are being used instead of γ_F and I_e (Johnson and Pytkowicz, 1979a). In other words, ion-pair effects are not being taken into account.

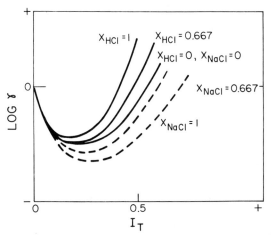

FIGURE 5.22 Values of γ_{HCl} and γ_{NaCl} versus X_i and $m^{0.5} = I_T^{0.5}$ [sketch based on Robinson and Stokes (1968)].

In the case of HCl–NaCl at $m = 1.0$

$$\gamma_{HCl(0)} = 0.809 \qquad \gamma_{(0)HCl} = 0.752$$

$$\gamma_{NaCl(0)} = 0.657 \qquad \gamma_{(0)NaCl} = 0.751$$

The trace γ's are not quite as close to each other for other pairs of electrolytes.

An important equation of use in the Brønsted–Guggenheim theory is derived next (Robinson and Stokes, 1968). First,

$$dF = \mu_2 dn_2 + \mu_3 dn_3 \qquad (5.426)$$

and, because it is a total differential,

$$\left(\frac{\partial \mu_2}{\partial m_3}\right)_{m^2} = \left(\frac{\partial \mu_3}{\partial m_2}\right)_{m^3} \qquad (5.427)$$

which leads to

$$\left(\frac{\partial \ln \gamma_2}{\partial m_3}\right)_{m^2} = \left(\frac{\partial \ln \gamma_3}{\partial m_2}\right)_{m^3} \qquad (5.428)$$

Obviously the data in Figure 5.22 obey this relation.

Examples of determinations of γ's, α's, and $\Delta F_{mix}^{(exc)}$ values in multielectrolyte solutions can be found in Robinson and Bower (1966) and Robinson and Covington (1968). These workers employed the isopiestic method and the traditional thermodynamic relations outlined earlier in this section. They worked with mixtures containing up to three electrolytes.

5.8.5 Ionic Media

Ionic media are solutions in which high concentrations of a background electrolyte maintain I_e essentially constant while the concentration of the electrolyte of interest changes during a chemical process.

EXAMPLE 5.16. SEAWATER AS AN IONIC MEDIUM. An example of such a medium is seawater considered in terms of the dissolution or removal of $CaCO_3$. Often surface oceanic waters are roughly twofold supersaturated with $CaCO_3$ and remain in a metastable state (see Chapter 5, Volume II). They can be brought to saturation in the laboratory for solubility determinations. Typical concentration changes during such events are

(Ca^{2+})	0.010–0.0098 m	2% change
(CO^{2-})	0.0002–0.0001 m	100% change

The I_e of seawater is about 0.5 m so that it changes only by 0.06%.

Stoichiometric equilibrium constants are often determined in ionic media such as seawater, $HClO_4$, $NaClO_4$, and NaCl at $I = 0.7$ or 1.0. It is important, therefore, to find out under what conditions such constants, determined in one ionic medium, can be applied to another one which can be an ionic medium or a variable one.

First, however, it is necessary to ascertain the behavior of activity coefficients in ionic media, designated by im. Free coefficients, $(\gamma_\pm)_F$, of electrolytes of interest obviously remain invariant because they depend only on I_e and not on the composition of the solution at a given (T, P) set.

Let us use the following notation. $(C_{im})_j$ and $(A_{im})_j$ are the cations and anions of im while C_1 and A_1 are the cation and anion of the minor constituent of interest. Charges are omitted for simplicity.

$(\gamma_{\pm C_1 A_1})_F$ can be set as unity at infinite dilution of $C_1 A_1$ in im because we have a freedom of choice in selecting the reference state. $(\gamma_{\pm C_1 A_1})_F$ will remain as unity with changing $(C_1 A_1)_T$ due to its constancy at a given I_e. $(\gamma_{\pm C_1 A_1})_T$ is related to the free coefficient, as shown earlier in this chapter, by

$$(\gamma_{\pm C_1 A_1})_T^2 = (\gamma_{\pm C_1 A_1})_F^2 [1 + K_{C_1 A_1}^{*'}[A_1]_F$$

$$+ \sum_j K_{C_1 A_j}^{*'}[A_1]_j] \times [1 + K_{C_1 A_1}^{*}[C_1]_F$$

$$+ \sum_j K_{C_j A_1}^{*'}[C_i]_j] \qquad (5.429)$$

This expression (Pytkowicz et al., 1974) includes interactions of the ions of $C_1 A_1$ with those of im but not those between the ions of im. At infinite dilution of $C_1 A_1$ in im we have $(\gamma_{\pm C_1 A_1})_T = (\gamma_{\pm CA})_F = 1$. $(\gamma_{\pm CA})_T$, however, varies when the concentration of $C_1 A_1$ changes, as in the usual infinite dilution convention.

EXAMPLE 5.17. VARIATION OF $(\gamma_{\pm C_1 C_1})_T$ WITH THE CONCENTRATION OF $C_1 A_1$. Let us examine a fictitious but illustrative example. Consider the following conditions: $[C_{im} A_{im}]_T = 0.5$ where C_{im} and A_{im} are the ions of the ionic medium. $K_{C_1 A_1}^{*'} = 10^{-4}$, $K_{C_1 A_{im}}^{*'} = 10^{-3}$, and $K_{C_{im} A_1}^{*'} = 10^{-2}$.

$$K_{C_1 A_1}^{*'}[A_1]_F + K_{C_1 A_{im}}^{*'}[A_{im}]_F$$
$$\uparrow$$
$$K_{C_1 A_1}^{*'}[C_1]_F + K_{im}^{*'}[C_{im}]_F \qquad (i)$$

To simplify the calculation I shall assume that the free concentrations in this equation equal the total ones. Thus, I shall be presenting the first step in what should be an iteration which would eventually the actual free concentrations.

The results are, with $(\gamma_{\pm C_1 A_1})_F = 1$,

$(C_1 A_1)_T$	0	0.1	0.2
$\gamma_{\pm T}$	1	0.997	0.997

We observe that $(\gamma_\pm)_T$ varies little under the chosen conditions. This is the case for the usual minor changes in the concentrations of the major constituents of seawater.

I shall show in Chapter 6, Volume II, when we study the speciation of some heavy metals such as Cd and Pb (Sipos et al., 1979), that the extent of complexation of metals is quite sensitive to ion pairs formed between the constituents of the ionic medium. This occurs because these interactions affect I_e and the free concentrations. This results from the changes in the amounts of the ionic medium ions C_j and A_j available for complexation after they form ion pairs.

Let us next consider the stoichiometric equilibrium constants and the extent to which they can be determined in one ionic medium and used in another solution. I must express these constants in terms of the thermodynamic ones for this purpose because these constants depend on the temperature and pressure but not on I_e and the composition of the medium.

1. The apparent dissociation constant of an acid HA is (see Chapter 1, Volume II)

$$K'_{HA} = \frac{ka_H[A^-]_T}{[HA]} = K_{HA}\frac{\gamma_{HA}}{(\gamma_A)_T} \quad (5.430)$$

where $\gamma_{HA} = (\gamma_{HA})_T = (\gamma_{HA})_F$ because a neutral entity does not form ion pairs. γ_{HA} refers to the actual species HA in solution and, therefore, is not written as $(\gamma_{\pm HA})^2$.

2. The stoichiometric solubility product, in the case of $CaCO_{3(s)}$, is

$$K_s = [Ca^{2+}]_T[CO_3^{2-}]_T = \frac{K_{sO}}{(\gamma_{\pm CaCO_3})_T^2}$$

$$= \frac{K_{sO}}{[(\gamma_{Ca})_T(\gamma_{CO_3})_T]} \quad (5.431)$$

3. The free stoichiometric solubility product is (see Chapter 4, Volume II)

$$(K_s)_F = [Ca^{2+}]_F[CO_3^{2-}]_F$$

$$= \frac{K_{sO}}{[(\gamma_{Ca})_F(\gamma_{CO_3})_F]} \quad (5.432)$$

4. The stoichiometric association constant for ion pairs, illustrated by $CaCO_3^0$, is

$$K^{*'}_{CaCO_3} = \frac{[CaCO_3^0]}{[Ca^{2+}]_F[CO_3^{2-}]_F}$$

$$= K^*_{CaCO_3}\frac{(\gamma_{Ca})_F(\gamma_{CO_3})_F}{\gamma_{CaCO_3}} \quad (5.433)$$

5. The stoichiometric stability constant for complexes, illustrated by $PbCl_2^0$, is

$$\beta'_2 = \frac{[PbCl_2^0]}{[Pb^{2+}]_F[Cl^-]_F^2} = \beta_2\frac{(\gamma_{Pb})_F(\gamma_{Cl})_F^2}{\gamma_{PbCl_2^0}}$$

$$= \frac{(\gamma_{\pm PbCl_2})_F^3}{\gamma_{PbCl_2^0}} \quad (5.434)$$

6. For redox equilibria

$$K'_{redox} = \frac{[oxid]_T}{[red]_T} = K_{redox}\frac{(\gamma_{red})_T}{(\gamma_{oxid})_T} \quad (5.435)$$

where oxid and red indicate the oxidized and the reduced forms, for example, Fe^{3+} and Fe^{2+}.

Therefore, as

K or β	Depends only on T and P
$(\gamma_\pm)_F$	Depends on T, P, and I_e
$(\gamma_\pm)_T$	Depends on T, P, I_e, and the composition

the constants

K'_{HA}, K_s, and K'_{redox}	Depend on T, P, I_e, and the composition
$(K_s)_F$, $K^{*'}$, and β'	Depend on T, P, and I_e

These results mean that

K'_{HA}, K_s, and K'_{redox}	Must be determined in the medium in which they will be used
$(K_s)_F$, $K^{*'}$, and β'	Can be determined in one medium and used in any medium with the same I_e

The use of ionic media for the determinations of equilibrium constants provides invariant values of $(\gamma_\pm)_F$ for the electrolyte of interest during chemical processes, when $(K_s)_F$ is determined as in the case of $CaCO_3$. The same is true for Pb and Cd complexes and for the dissociation constants of carbonic acid (see Chapter 1, Volume II).

It is important to realize in ionic medium work that NaCl and $NaClO_4$ form ion pairs $NaCl^0$ and $NaClO_4^0$ (Johnson and Pytkowicz, 1979). Such pairs affect I_e and the availability of Cl^- and ClO_4^- to complex trace metals. These effects must be taken into account [see Sipos et al. (1980) in Chapter 8, Volume II]. On the other hand, $HClO_4$ appears to be essentially completely dissociated so that the use of $HClO_4$ as the background (im) electrolyte simplifies the determinations of $K^{*\prime}$, β', and $(K_s)_F$.

The use of measured stoichiometric constants is recommended when time permits their determinations or they are available because then it is not necessary to obtain γ's for the ions and for the ion pairs or complexes.

Total activity coefficients are used, for example, for the calculation of apparent dissociation constants and the stoichiometric solubility products from their thermodynamic counterparts. Thus,

$$K_s = \frac{K_{sO}a_{(s)}}{(\gamma_{Ca})_T(\gamma_{CO_3})_T} \tag{5.436}$$

$a_{(s)}$ is the activity of the solid $CaCO_3$ and is unity for a pure carbonate and $(\gamma_{Ca})_T(\gamma_{CO_3})_T = (\gamma_{\pm CaCO_3})_T^2$.

I do not recommend this procedure for ionic media but do advise the direct measurement of K_s in the ionic medium of interest.

Let us see how sensitive $(\gamma_{CO_3})_T$ is in contrast to $(\gamma_{CO_3})_F$ which is fixed to changes in composition. The stoichiometric association constants involving CO_3^{2-} are of the type

$$K' = \frac{[MCO_3]}{[M]_F[CO_3^{2-}]_F} \tag{5.437}$$

Thus,

$$\frac{[CO_3^{2-}]_T}{[CO_3^{2-}]_F} = \frac{(\gamma_{CO_3})_F}{(\gamma_{CO_3})_T}$$
$$= 1 + \sum_M \frac{[MCO_3]}{[CO_3^{2-}]_T}$$
$$= 1 + \sum_M K^*_{MCO_3}[M]_F \tag{5.438}$$

Therefore, $(\gamma_{CO_3})_T$ only varies of $[M]_F$ changes appreciably. M represents Na^+, K^+, Ca^{2+}, and Mg^{2+} and their concentrations seldom vary appreciably in natural processes.

5.9 MEAN ACTIVITY COEFFICIENTS OF ELECTROLYTES

5.9.1 Ion Pairs

We examined earlier the single-ion approach to ion pairs. That approach has the drawback of requiring a nonthermodynamic assumption. In this section I consider methods that are directly related to $\gamma_\pm$. Emphasis is placed on the work of Johnson (1979) and Johnson and Pytkowicz (1978, 1979a, 1979b) because their model includes Cl^- association with H^+, Na^+, K^+, Mg^{2+}, and Ca^{2+}.

Two key techniques used in activity coefficient work, namely, isopiestic and potentiometric measurements, were examined earlier. In this chapter I present other techniques, such as conductance, and further material on potentiometric approach. This will be done specifically from the standpoint of ion pairs.

First, however, I shall mention some evidence that supports the ion-pair concept. The theoretical results of Bjerrum (1926), Fuoss (1958), and Kester and Pytkowicz (1975) lead one to expect such pairs. This is also a matter of common sense since particles with thermal motion and coulombic attraction are bound to spend a fraction of the time together. Thus, at any one time a number of flickering pairs are present. Other approaches, such as sound attenuation (Fisher, 1978) and Raman spectroscopy confirm experimentally the presence of pairs. Self-consistent potentiometric results were obtained by Pytkowicz and Kester (1969) and by Johnson and Pytkowicz as $(\gamma_\pm)_F$ and $K^{*\prime}$ obtained in simple solutions led to the proper prediction of $(\gamma_\pm)_T$ in complex ones. $(\gamma_\pm)_T$ includes all properties: nonspecific interactions, hard-core repulsion, cosphere overlap, changes in dielectric constant, and hydration effects. $(\gamma_\pm)_F$, on the other hand, reflects all of these effects minus the ion-pairing one and is a formally valid construct which is obtained operationally.

Electrical Conductance. The resistance of a conductor is given by

$$R_{el} = \rho_{el}l/A \text{ ohms} \tag{5.439}$$

ρ_{el} is the resistivity (specific resistance), l the length in cm, and A the cross section in cm^2. The specific conductance K_{el} is $1/\rho_{el}$ and the conductance is

$$C_{el} = \frac{1}{R_{el}} = \frac{K_{el}A}{l} \quad ohms^{-1} \quad (5.440)$$

The equivalent conductance Λ is that of a solution of volume V containing 1 g-equiv of the electrolyte and placed between two electrodes 1 cm apart.

$$\Lambda = K_{el}V = \frac{1000 K_{el}}{cn_e} \quad ohms/cm^2 \quad (5.441)$$

c is the molarity of the solution and n_e the number of equivalents per mole. The factor 1000 appears because c is in moles/L while V is in mL. The term $1000/cn_e$ equals V because $1000/cn_e$ is the number of mL/g-equiv.

NaCl^0 was found to be 0.92, with $a_D = 6.11$ Å, and K^* for KCl^0 was 149.90 with $a_D = 5.655$ Å.

The extent of ion pairing was obtained from the difference between the calculated and the measured conductance and is subject to errors. The reason for this is that the curve-fitting parameter a_D is not the true distance of closest approach. Furthermore, the Debye–Hückel equation is curve fitted to measured $\gamma_{\pm}$'s and yields, therefore, $(\gamma_{\pm})_T$'s that already include the effect of ion pairs.

Other applications of the conductance approach can be found in Fisher (1978) and Fisher and Fox (1977), who examined the effect of pressure on the conductance and ion pairing and in D'Aprano (1971) who calculated the association of alkali perchlorates in water at 25°C. He used a modified Fuoss and Hsia (1967) equation in which the Debye–Hückel approach was dropped. The values of K^* were found to be

LiClO$_4$	NaClO$_4$	KClO$_4$	RbClO$_4$	CsClO$_4$
-0.07 ± 0.10	0.19 ± 0.10	0.99 ± 0.03	1.35 ± 0.03	1.69 ± 0.09

The ion-pair approach makes use of the fact that the degree of dissociation is given by

$$\alpha_{el} = \frac{c_F}{c_T} = \frac{\Lambda_T}{\Lambda_F} \quad (5.442)$$

when only ion pairs with zero net charge are present. These are assumed not to contribute to the conductance. Λ_F represents the equivalent conductance of free ions and Λ_T is the value that is measured experimentally. The ratio permits the elimination of terms that are not needed.

Λ_F is usually calculated theoretically as, for example (Fuoss and Onsager, 1957), by

$$\Lambda_F = \Lambda^0 - A_F c_F^{0.5} + B_F c_F \log c_F + D_F c_F \quad (5.443)$$

where the parameters A_F, B_F, and so on are calculated theoretically. The main problem with this method is that the Fuoss–Onsager equation is related to the extended Debye–Hückel approach. Therefore, it is strictly a semiempirical equation which breaks down at $c > 0.1\ M$.

Chin and Fuoss (1968) measured the conductance of NaCl and KCl in water at 25°C and at concentrations up to 1 N. They concluded that their results showed the formation of solvent-separated ion pairs, that is, of outer sphere contact. K^* for

In general terms the values of K^* are smaller than is to be expected from the small charges and large crystal radii of the ions. The crystal radius of Li$^+$ is smaller than that of Cs$^+$ so that the former is the more hydrated one. This is due to its larger electric field which results from its greater charge density. Therefore, the hydrated Li$^+$ is larger than the hydrated Cs$^+$, as is shown by the trend in K^*.

Solubilities. The determination of the extent of MgSO$_4^0$ formation by the use of the solubility of Mg(OH)$_2$ was described earlier. Another example is that of Elgquist and Wedborg (1974). Their method, explained in terms of stoichiometric solubilities, was as follows. For a symmetric solid which dissociates according to $CA_{(s)} = C + A$, K_{sO} is given by

$$a_{(s)}K_{sO} = (\gamma_C)_T(\gamma_A)_T[C^+]_T[A^-]_T$$

$$= (\gamma_C)_F(\gamma_A)_F[C^+]_F[A^-]_F \quad (5.444)$$

K_s can be defined in two valid ways:

$$K_s = [Ca^{2+}]_T[SO_4^{2-}]_T\, a_w^2$$

$$\cong [Ca^{2+}]_T[SO_4^{2-}]_T \quad (5.445)$$

as $a_w \cong 1$, and

$$K_s = [Ca^{2+}]_T[SO_4^{2-}]^2[H_2O]^2 \qquad (5.446)$$

I shall use the upper definition.

Elgquist and Wedborg (1975) examined the variation of K_s when the compositions of solutions roughly similar to seawater were changed at a constant total ionic strength. They did not take into account the effective ionic strength. To a first approximation the constant I means that the γ_F's for given ions are the same in all solutions. They further assumed that the only ion pairs that would form would be $MgCl^+$, $CaCl^+$, $NaSO_4^-$, $MgSO_4^0$, and $CaSO_4^0$.

The speciation was obtained by solving a system of mass balance and equilibrium equations which, for SO_4^{2-} as an example, would be

$$[SO_4^{2-}]_T = [SO_4^{2-}]_F + [NaSO_4^-]$$
$$+ [MgSO_4^0] + [CaSO_4^0] \qquad (5.447)$$

being nearly unity. In reality, it is γ_F that is the invariant quantity at a given I_e while γ_T does vary somewhat. This small variation can be seen from the simplified equation for only one electrolyte

$$\frac{(\gamma_A)_T}{(\gamma_A)_F} = 1 + K_{CA}^{*'}[C]_F \qquad (5.451)$$

when $[C]_F$ is large, as in the case of $[Na^+]$ or of another major cation of seawater. Therefore, the stoichiometric solubility product (K_s) in the infinite dilution scale and the thermodynamic one in the ionic medium scale are essentially the same, that is,

$$K_s = [Ca^{2+}]_T[CO_3^{2-}]_T[H_2O]^2 = K_{sO}^{(im)} \qquad (5.452)$$

The association constants obtained from Elgquist and Wedborg (1975) are shown in the upper line while those of Johnson and Pytkowicz (1979) are presented in the bottom line.

The values of K_{SP}^{im}

	$NaSO_4^-$	KSO_4^-	$MgSO_4^0$	$CaSO_4^0$	$MgCl^+$	$CaCl^+$	$NaCl^0$
$K^{*(im)}$	1.22	1.84	12.3	30.6	0.48	1.20	—
$K^{*'}$	9.78	—	19.4	41.7	1.91	2.24	0.32

$$K^{*'} = \frac{[CaSO_4^0]}{[Ca^{2+}]_F[SO_4^{2-}]_F} \qquad (5.448)$$

$$(K_s)_F = [Ca^{2+}]_F[SO_4^{2-}]_F \qquad (5.449)$$

The last equation can be written in terms of free concentrations. Then,

$$(K_s)_F = K_s \frac{(\gamma_{Ca})_T(\gamma_{SO_4})_T}{(\gamma_{Ca})_F(\gamma_{SO_4})_F}$$
$$= K_s \frac{[Ca^{2+}]_F[SO_4^{2-}]_F}{[Ca^{2+}]_T[SO_4^{2-}]_T} \qquad (5.450)$$

where K_s is the measured quantity. The set of equations such as (5.446) and (5.447) plus (5.449) is solved simultaneously for all the ions present and the speciation is obtained.

Actually, Elgquist and Wedborg (1975) used the Swedish ionic medium scale in which the ionic medium, such as seawater, is the reference state. The total activity coefficients of the ions of interest change little in the im and can, therefore, be set as

The difference between our results and those of Elgquist and Wedborg is not due to the difference in scales because $K^{*'} = K^{*(im)}$ by definition. It is due in part to the fact that these workers neglected $NaCl^0$ pairs, which play an important role in seawater due to the high $[Na^+]_T$, and did not use I_e.

Raman Spectroscopy. Raman spectra do not show coulombic entities such as ion pairs directly but only covalent species. Daly et al. (1972) used this property to demonstrate indirectly the effect of the competition of H^+, Ca^{2+}, and Mg^{2+} for SO_4^{2-} on the covalent HSO_4^- peak.

First, Raman spectra of Na_2SO_4 and $MgSO_4$ solutions failed to show $NaSO_4^-$ and $MgSO_4^0$ peaks. This indicated the coulombic nature of these ion pairs.

Their existence was proven in the following way. HCl was added to the sulfate solutions to generate HSO_4^- which, being covalent, yielded a strong Raman band. Further addition of Na_2SO_4 and $MgSO_4$ weakened this band with an increase in the SO_4^{2-} band. This means that Na^+ and Mg^{2+} compete with

H^+ for SO_4^{2-} through the formation of $NaSO_4^-$ and $MgSO_4^0$.

The competing reactions are

$$H^+ + SO_4^{2-} = HSO_4^- \qquad (5.453)$$

$$Na^+ + SO_4^{2-} = NaSO_4^- \qquad (5.454)$$

$$Mg^{2+} + SO_4^{2-} = MgSO_4^0 \qquad (5.455)$$

or

$$HSO_4^- + Na^+ = NaSO_4^- + H^+ (5.456)$$

$$HSO_4^- + Mg^{2+} = MgSO_4^0 + H^+ (5.457)$$

The concentrations of the ion pairs were calculated from the data of Kester and Pytkowicz (1969). The resulting estimated shifts in the integrated bands of HSO_4^- and SO_4^{2-} were in good agreement with the measured ones.

Again I must remind the reader that the model of Kester and Pytkowicz (1969) works because Cl^- pairs, which were not considered, lead to higher values of $K^*_{CSO_4}$ but to a smaller availability of the cations, so that there is a cancellation of errors and $[C]_F$ is about right.

Davies and Oliver (1973) reached the opposite conclusion from Raman data, namely, that ion pairs are inner sphere entities without water of hydration. This is difficult to accept because the energy of hydration is higher than the interionic coulombic energy.

An Application of the Fuoss Theory.

Fuoss (1958) presented a theory of ion-pair equilibrium, based on the work of Boltzmann (1868). He felt that this was necessary due to a drawback in the Bjerrum (1926) approach. This problem is the arbitrary cutoff of ion pairs at the interionic distance of minimum energy. It leads to the acceptance of ion pairs made of ions that are not in contact.

Fuoss, to make certain that only ions in contact were considered as pairs, visualized the cations as changed spheres of radius a_D. The anions are visualized as point charges on or within the surface of the cation.

The concentration of ion pairs is proportional to the volume occupied by free cations since a high number of cations increases the probability of ion-pair formation, increases with the coulombic energy of interaction E between cations and anions, and is proportional to the Boltzmann distribution function $\exp(-E/k_BT)$. This leads to

$$K^* = \frac{4\pi N_A}{3000} a_D^3 \, e^{-E/k_BT} \qquad (5.458)$$

a_D in theory is the Debye–Hückel distance of closest approach for hydrated ions.

Kester and Pytkowicz (1969) calculated values of $K^{*\prime}$ from the Fuoss equation and estimated γ's and compared the results to those determined experimentally (Table 5.60). It can be seen that the agreement is fair for most but not all species. This is to be expected because of the simplified model of Fuoss and because Kester and Pytkowicz (1969) and Pytkowicz and Hawley (1974) were not aware of Cl^- ion pairs when they calculated $K^{*\prime}$ for SO_4^{2-}, HCO_3^-, and CO_3^{2-} ion pairs.

The most important results were the magnitudes of $K^{*\prime}$ for Cl^- ion pairs, as they led to the work of Johnson (1979) and Johnson and Pytkowicz (1978, 1979a, 1979b). I have no explanation for the large discrepancies for $MgCO_3^0$ and $CaCO_3^0$.

TABLE 5.60 A comparison of values of $K^{*\prime}$ obtained experimentally at $I = 0.7$ and by means of the Fuoss (1958) theory, according to Kester and Pytkowicz (1975).

Species	$K^{*\prime}_{(Fuoss)}$	$K^{*\prime}_{(exp)}$
$NaSO_4^-$	2.1	2.02[a]
$MgSO_4^0$	4.6	10.2[b]
$CaSO_4^0$	6.5	10.8[c]
$NaHCO_3^0$	0.46	0.28[d]
$MgHCO_3^+$	2.8	1.62[d]
$CaHCO_3^+$	2.4	1.96[d]
$NaCO_3^-$	2.0	4.25[d]
$MgCO_3^0$	3.9	112.3[d]
$CaCO_3^0$	4.9	279.6[d]
NaF^0	0.40	0.046[e]
MgF^+	2.7	19.4[e]
CaF^+	2.4	4.5[e]
$NaB(OH)_4^0$	0.46	0.53[f]
$MgB(OH)_4^0$	2.8	8.03[f]
$CaB(OH)_4^0$	2.4	13.0[f]
$NaCl^0$	0.43	—
$MgCl^+$	2.5	—
$CaCl^+$	2.3	—

[a] Pytkowicz and Kester (1969).
[b] Kester and Pytkowicz (1968).
[c] Kester and Pytkowicz (1969).
[d] Pytkowicz and Hawley (1974).
[e] Miller and Kester (1975).
[f] Byrne and Kester (1974).

TABLE 5.61 Comparisons between values of K^* obtained from the Fuoss theory for various diameters a_{KJ} with those measured by Johnson (1979). Ranges of values for the thermodynamic association constants were obtained by extrapolation of stoichiometric constants by two methods; the direct method and the use of $\gamma_{\pm F}$ and γ_{ip} in conjunction with $K^{*'}$.

	$a_{KJ} = r_c + r_A$	$a_{KJ} = r_c + r_A + 1.38$	$a_{KJ} = r_c + r_A + 2.76$	Measured
HCl^0	0.77	1.15	1.77	0.31–0.52
$NaCl^0$	0.71	1.01	1.55	0.58–0.93
KCl^0	0.76	1.13	1.74	0.61–1.01

Johnson (1979) compared his experimental results for chlorides, converted into thermodynamic results by extrapolation, with those calculated from the Fuoss theory. He assumed one case in which the ions were separated by a water molecule with one in which cosphere overlap corresponded to the separation of the ions by half a water molecule, that is, $a_{KJ} = r_c + 0.5 \times 2.76$ Å where a_{KJ} is the Johnson diameter and r_c is the crystal radius. The results are shown in Table 5.61. There is fair agreement considering the assumptions involved in the Fuoss theory.

The Johnson–Pytkowicz Model. We have already seen a number of examples of the potentiometric method and will now examine the one due to Johnson (1979) and Johnson and Pytkowicz (1978, 1979a). Our method does not require a separation of ions since we deal with mean quantities throughout.

This method is based on the fact that $HClO_4$ can be considered completely dissociated since its K^* is 10^{-7} (Kossiakov and Harker, 1938). Our results are insensitive to $K^*_{HClO_4}$ even if we vary it from 10^{-7} to 10^{-2}.

HCl–HClO₄. The following equations were used for the determination of the activity coefficients and the speciation in HCl solutions, based on data in HCl–HClO₄ solutions.

$$\frac{\{HCl\}}{\{HCl\}^{(m)}} = \frac{[H^+]_F[Cl^-]_F}{[H^+]_F^{(m)}[Cl^-]_F^{(m)}} \quad (5.459)$$

The free activity coefficients are canceled because the single-electrolyte (HCl) solution and the mixed one (HCl + HClO₄), which is indicated by the superscript (m), are at the same effective ionic strength.

$[HCl]_T$ is not known at this point because I_e has not yet been determined, and is one of the variables. The equality of I_e in HCl and in HClO₄ is just a condition imposed on the equations at this point.

In the HCl solution $[H^+]_T = [Cl^-]_T$ so that

$$\begin{aligned} a_{HCl} &= (\gamma_{\pm HCl})_T^2 [Cl^-]_T \\ &= [(\gamma_{\pm HCl})_T (H^+)_T]^2 \end{aligned} \quad (5.460)$$

As I_e is the same in both solutions,

$$I_e = 0.5([H^+]_F + [Cl^-]_F]$$

$$= 0.5[[H^+]_F^{(m)} + [Cl^-]_F^{(m)} + [ClO_4^-]^{(m)}] = I_e^{(m)}$$

$$(5.461)$$

Therefore,

$$I_e = I_e^{(m)} = [H^+]_F = [H^+]_F^{(m)} \quad (5.462)$$

The mass balance equations for the free and the ion-pair concentrations are

$$[H^+]_T^{(m)} = [H^+]_F^{(m)} + [HCl^0]^{(m)} + [HCl_4^0]^{(m)} \quad (5.463)$$

$$[Cl^-]_T^{(m)} = [Cl^-]_F^{(m)} + [HCl^0]^{(m)} \quad (5.464)$$

$$[ClO_4]_T^{(m)} = [ClO_4^-]^{(m)} + [HClO_4^0]^{(m)} \quad (5.465)$$

$$[H_2]_T = [H^+]_F + [HCl_0] \quad (5.466)$$

$$[Cl^-]_F = [H^+]_F \quad (5.467)$$

Furthermore,

$$K^{*'}_{HClO_4} = \frac{[HClO_4^0]}{[H^+]_F^{(m)}[ClO_4^-]_F^{(m)}} = 10^{-7} \quad (5.468)$$

$$K^{*'}_{HCl} = \frac{[HCl^0]}{[H^+]_F[Cl^-]_F} = K^* \frac{(\gamma_H)_F(\gamma_{Cl})_F}{\gamma_{HCl}} \quad (5.469)$$

As K^*, γ_F, and γ_{ip} depend on P, T, and I_e but not on the composition of the medium, $K_{HCl}^{*\prime}$ is the same in the two solutions. Therefore,

$$\frac{[HCl^0]}{[H^+]_F^{(m)}[Cl^-]_F^{(m)}} = \frac{[HCl^0]}{[H^+]_F[Cl^-]_F} = K_{HCl}^{*\prime}$$

(5.470)

Thus, there is a system of 10 independent nonlinear equations which can be solved for the 10 unknowns $[H^+]_T$, $[Cl^-]_F$, $\{HCl^0\}$, $[H^+]_F$, $[HCl]$, $[H^+]_F^{(m)}$, $[Cl^-]_F^{(m)}$, $[ClO_4^-]_F^{(m)}$, $[HCl^0]^{(m)}$, and $[HClO_4^0]^{(m)}$. The system of equations was solved by the use of published data on $\{HCl\}^{(m)}$. The solution also required an empirical relation between $(\gamma_{\pm HCl})_T$ and m_{HCl} for equation (5.470).

The resulting values of $K_{HCl}^{*\prime}$ obtained from various sets of activity coefficients $(\gamma_{\pm HCl})_T$ are presented in Table 5.62.

Further Association Constants. Further association constants were then determined by a similar method to that used above. As an example, activity coefficients in HCl–NaCl were used to determine $K_{NaCl}^{*\prime}$.

Thermodynamic association constants for chlorides, obtained by extrapolation to infinite dilution, are shown in Table 5.63. Our results were expressed in molarities and so were the literature values for $(\gamma_\pm)_T$ and concentrations, because this conversion from molarities to molalities required the densities of pure solutions which we calculated from

$$\rho = a_\rho + b_\rho I_T$$

(5.471)

and the Culberson equation was used to calculate $(\gamma_\pm)_T$. Constants for Equation (5.471) and for the Culberson equation are shown in Table 5.64.

The results for K^* in Figure 5.22 agree with the qualitative prediction of Kay (1962) and Prue (1966). These authors postulated that the association constants of the alkali and alkaline earth chloride ion pairs would be in the increasing order Li < Na < K < Cs, and Mg < Ca < Sr < Ba, due to the effects of hydration. Prue reasoned that a small cation can form a stronger bond with a dipolar solvent molecule than with an anion, unless the anion is very small. However, as the size of the cation increases, the ion-solvent bond strength would decrease and this allows the anion to come closer.

Therefore, the distance of closest approach decreases when the cation radius increases, causing a larger association constant.

The values of $K^{*\prime}$, the stoichiometric association constants, were fitted to

$$\ln K^* = A_J + B_J I_e + C_J I_e^2$$

(5.472)

The coefficients are shown in Table 5.65.

Values of $K^{*\prime}$ are shown in Tables 5.66 through 5.74. The association constant of $HClO_4^0$ is less than that of HCl^0 despite the larger crystal radius of ClO_4^-. This fact must be interpreted as showing that Cl^- and ClO_4^- are less strongly hydrated than cations. Therefore, the association constant of H^+ will be inversely proportional to the crystal radius of the anion. While this is true for association with very strongly hydrated cations such as H^+ and Li^+, caution should be used when extending this argument to larger cations. Prue argued that the association constant of a large cation and anion, both of which are weakly hydrated, may be reinforced by ion-induced dipole and dispersion forces. The thermodynamic association constants of K^+ and Cs^+ with ClO_4^-, which have been determined from conductivity measurements (D'Aprano, 1971) are somewhat larger than the corresponding Cl^- association constants, a result that lends support to this argument. The low value of the association constant of HCl^0 is probably due to the strong hydration of the proton (Kortum, 1965). Further work is required to develop a quantitative model of ion association that incorporates the effects of the solvent in the manner suggested by Prue.

The stoichiometric association constants that we reported in this work may be used to calculate the speciation in electrolyte solutions, including mixtures, without estimating any activity coefficients (Table 5.75). The procedure for calculating the speciation in solutions has been discussed earlier (Johnson and Pytkowicz, 1979a). The free concentrations of the cations in pure solutions of all the electrolytes that were examined are shown in Table 5.76 for rounded values of the total ionic strength. Although all calculations were performed using molar concentrations, the results have been converted to the molal scale for convenience. Up to 55% of the cations in solutions of 2-1 salts and 35% of the cations in solutions of 1-1 salts may be present in ion pairs with chloride.

Thermodynamic values are shown in Table 5.77.

TABLE 5.62 The results obtained for the stoichiometric association constant of HCl^0. $[HClO_4^0]$ is less than 10^{-7} M (Johnson, 1979).

$(\gamma_{\pm HCl})_T$	$(H)_T^{(m)}$	$[Cl]_T^{(m)}$	$[HCl^0]^{(m)}$	I_e	K_{HCl}^*	Source of $(\gamma_{HCl})_T$
0.880	0.9503	0.0950	0.00998	0.9398	0.125	Murdoch and Barton (1933)
0.871	0.9487	0.2372	0.02551	0.9236	0.131	Murdoch and Barton (1933)
0.853	0.9554	0.4777	0.04907	0.9068	0.126	Murdoch and Barton (1933)
0.823	0.6829	0.06829	0.00624	0.6763	0.149	Murdoch and Barton (1933)
0.825	0.7262	0.1815	0.01773	0.7081	0.153	Murdoch and Barton (1933)
0.790	0.3916	0.03917	0.00321	0.3884	0.230	Murdoch and Barton (1933)
0.783	0.3904	0.09687	0.00738	0.3831	0.215	Murdoch and Barton (1933)
0.774	0.5033	0.2516	0.01148	0.4916	0.0972	Murdoch and Barton (1933)
0.770	0.2949	0.1473	0.00774	0.2868	0.193	Murdoch and Barton (1933)
0.814	0.09900	0.00981	0.00052	0.09847	0.574	Murdoch and Barton (1933)
0.809	0.09716	0.02429	0.00105	0.09611	0.469	Murdoch and Barton (1933)
0.808	0.09585	0.04793	0.00264	0.09319	0.624	Murdoch and Barton (1933)
0.873	0.9537	0.1907	0.01968	0.9337	0.123	Storonkin, Lagunov, and Belokoskov (1956)
0.859	0.8599	0.09554	0.00983	0.8504	0.135	Storonkin, Lagunov, and Belokoskov (1956)
0.856	0.8317	0.06692	0.00686	0.8251	0.139	Storonkin, Lagunov, and Belokoskov (1956)
0.853	0.8128	0.04781	0.00487	0.8076	0.140	Storonkin, Lagunov, and Belokoskov (1956)
0.848	0.7940	0.02870	0.00282	0.7916	0.137	Storonkin, Lagunov, and Belokoskov (1956)
0.844	0.7846	0.01914	0.00179	0.7831	0.132	Storonkin, Lagunov, and Belokoskov (1956)
0.850	0.7751	0.00957	0.00100	0.7738	0.151	Storonkin, Lagunov, and Belokoskov (1956)
0.778	0.5896	0.4913	0.04217	0.5474	0.172	Storonkin, Lagunov, and Belokoskov (1956)
0.768	0.3946	0.2959	0.02365	0.3708	0.234	Storonkin, Lagunov, and Belokoskov (1956)
0.768	0.2965	0.1977	0.01232	0.2842	0.234	Storonkin, Lagunov, and Belokoskov (1956)
0.773	0.1981	0.09903	0.00265	0.1954	0.141	Storonkin, Lagunov, and Belokoskov (1956)
0.782	0.1685	0.06936	0.00261	0.1659	0.236	Storonkin, Lagunov, and Belokoskov (1956)
0.789	0.1487	0.04957	0.00210	0.1466	0.301	Storonkin, Lagunov, and Belokoskov (1956)
0.794	0.1289	0.02975	0.00103	0.1278	0.281	Storonkin, Lagunov, and Belokoskov (1956)
0.796	0.1091	0.00992	0.00012	0.1090	0.110	Storonkin, Lagunov, and Belokoskov (1956)
0.902	1.0110	0.08459	0.00958	1.0010	0.128	Vasil'ev and Glavina (1976)
0.906	1.0027	0.08450	0.01023	0.9929	0.139	Vasil'ev and Glavina (1976)
0.883	0.9111	0.08562	0.01036	0.9003	0.153	Vasil'ev and Glavina (1976)
0.895	0.9886	0.1702	0.02074	0.9683	0.143	Vasil'ev and Glavina (1976)

TABLE 5.62   (*Continued*)

$(\gamma_{\pm HCl})_T$	$(H)_T^{(m)}$	$[Cl]_T^{(m)}$	$[HCl^0]^{(m)}$	I_e	K_{HCl}^*	Source of $(\gamma_{HCl})_T$
0.819	0.5946	0.08540	0.00938	0.5851	0.211	Vasil'ev and Glavina (1976)
0.829	0.6015	0.08543	0.01095	0.5903	0.249	Vasil'ev and Glavina (1976)
0.814	0.6007	0.1701	0.01961	0.5814	0.224	Vasil'ev and Glavina (1976)
0.811	0.5956	0.1708	0.01851	0.5772	0.210	Vasil'ev and Glavina (1976)
0.807	0.5942	0.3472	0.05578	0.5381	0.355	Vasil'ev and Glavina (1976)
0.804	0.5935	0.4367	0.09408	0.4993	0.550	Vasil'ev and Glavina (1976)
0.878	0.9482	0.00948	0.00089	0.9469	0.110	Macaskill and Pethybridge (1977)
0.868	0.9537	0.1907	0.01792	0.9362	0.111	Macaskill and Pethybridge (1977)
0.858	0.9597	0.3839	0.03665	0.9227	0.114	Macaskill and Pethybridge (1977)
0.848	0.9658	0.5795	0.05710	0.9083	0.120	Macaskill and Pethybridge (1977)
0.837	0.9720	0.7776	0.08092	0.8907	0.130	Macaskill and Pethybridge (1977)
0.802	0.09927	0.00993	0.00016	0.09911	0.170	Macaskill and Pethybridge (1977)
0.802	0.09930	0.01986	0.00032	0.09902	0.167	Macaskill and Pethybridge (1977)
0.800	0.09937	0.03975	0.00041	0.09897	0.106	Macaskill and Pethybridge (1977)
0.799	0.09943	0.05966	0.00085	0.09858	0.147	Macaskill and Pethybridge (1977)
0.798	0.09949	0.07960	0.00215	0.09730	0.285	Macaskill and Pethybridge (1977)

The work was also extended to sulfates. All $K^{*\prime}$ values for sulfates could be fit by expressions of the type

$$\ln K_{MX}^{*\prime} = a_{MX} + b_{MX}I_e \quad (5.473)$$

The fitting constants are presented in Table 5.78.

Values of $(\gamma_\pm)_F$ for this work were calculated for pure NaCl solutions by Cole and Pytkowicz. The method used was as follows:

1. $(\gamma_\pm)_T$ versus I was obtained from the literature.
2. $K^{*\prime}$ versus I_e was first obtained from Equation (5.473) by setting $I_e = I$.
3. $[NaCl^0]$ was calculated from $K_{NaCl}^{*\prime} = x/([NaCl] - x)^2$ with $x = [NaCl^0]$.
4. x was used to recalculate I and x by iteration until convergence was obtained.

5. $(\gamma_{\pm NaCl})_F$ was calculated from $(\gamma_{\pm NaCl})_F^2 = (\gamma_{\pm NaCl})_T^2[NaCl]_T/[NaCl]_F$.

Activity coefficients calculated for NaCl and for KCl solutions by the Johnson (1979) model are shown in Table 5.79.

The results are plotted in Figures 5.23 and 5.24. They reveal the depression of $(\gamma_\pm)_T$ due to ion association. Note that $(\gamma_\pm)_F$ is the quantity which is independent of composition at a given ionic strength and is used for the determination of $K^{*\prime}$ and β'.

TABLE 5.63   Thermodynamic association constants for chlorides.

Ion Pair	K_{MCl}	Ion Pair	K_{MCl}	Ion Pair	K_{MCl}
LiCl^0	0.62	NH$_4$Cl^0	1.23	CaCl$^+$	5.2
NaCl^0	0.89	Me$_4$NCl^0	2.1	SrCl$^+$	5.5
KCl^0	1.16	Et$_4$NCl^0	1.9	BaCl$^+$	5.7
CsCl^0	1.6	MgCl$^+$	4.1	MnCl$^+$	5.7

TABLE 5.64 Constants for Equation 5.471 (Johnson, 1979) and for the Culberson equation.

	Equation (5.471)			Equation (5.000)				
Electrolyte	a_ρ	b_ρ	Range of I	B_C	C_C	D_C	E_E	Range of I
HCl	0.9976	0.0166	0–1.8	1.212	0.2097	−0.1292	0.0585	0–1.0
LiCl	0.9976	0.0224	0–2.0	1.103	0.2019	−0.1095	0.0389	0–1.4
NaCl	0.9978	0.0381	0–1.5	1.350	0.0437	−0.0094	0	0–1.0
KCl	0.9981	0.0430	0–1.5	1.307	0	0	0.0021	0–1.0
CsCl	0.9987	0.1194	0–1.5	1.541	−0.1753	0.1489	−0.0372	0–1.4
NH$_4$Cl	0.9976	0.0144	0–2.0	1.449	−0.0452	0.0460	−0.0122	0–1.4
(Me)$_4$NCl	0.9973	0	0.1					
(Et)$_4$NCl	0.9970	0	0.1					
MgCl$_2$	0.9974	0.0245	0–1.8	1.800	−0.0337	0.1156	−0.04101	0–1.2
CaCl$_2$	0.9973	0.0290	0–1.8	1.501	0.0790	−0.0155	0	0–1.2
SrCl$_2$	0.9976	0.0446	0–2.1	1.134	0.3142	−0.2812	0.0910	0–1.8
BaCl$_2$	0.9975	0.0583	0–1.5	1.252	0.2095	−0.2000	0.0667	0–1.8
MnCl$_2$	0.9973	0.0334	0–1.5	1.103	0.3513	−0.3166	0.1030	0–1.8

Speciation and Activity Coefficients in Seawater. The speciation of the major constituents of seawater, excluding HCO_3^- and CO_3^{2-}, was obtained from values of $K^{*\prime}$ equilibrium and mass balance equations and is presented in Table 5.80. The absence of the two anions has only a slight effect on the speciation of Na^+, K^+, Mg^{2+}, Ca^{2+}, Cl^-, and SO_4^{2-} because these ions are present in much larger amounts than HCO_3^- and CO_3^{2-} (Johnson and Pytkowicz, 1979a).

Values of $(\gamma_\pm)_T$ in seawater, calculated from values in single-electrolyte solutions and associa-

tion constants, are compared to measured values in Table 5.81. The agreement is quite good except for Na$_2$SO$_4$.

Cl^- ions do not associate as strongly as SO_4^{2-} with the major cations of seawater because of their single charge. They do, however, associate more extensively due to their larger concentration.

Calculations for the Ion-Pair Model of Johnson and Pytkowicz (1978, 1979a, 1979b). One may wish to calculate one or more of the following quantities for an electrolyte in multielectrolyte solutions:

TABLE 5.65 Constants of Equation (5.472). $S_{y/x}$ gives the standard error of the estimate, expressed as a percentage of K^*. This is approximately the uncertainty in the predicted values of $K^{*\prime}$. Results for HCl^0 are from Khoo, Chan, and Lini (1977b).

Ion Pair	A_J	B_J	C_J	Range of Fit (I_e)	$S_{y/x}$
HCl^0	−1.179	−0.982	0	0.09–1.0	7.9
LiCl^0	−0.979	−1.205	0	0.09–1.0	5.5
NaCl^0	−0.570	−0.970	0	0.09–0.88	4.7
KCl^0	−0.280	−0.718	0	0.09–0.80	3.7
CsCl^0	−0.104	−0.127	0	0.3 −0.64	0.7
MgCl$^+$	0.736	0.028	−0.467	0.09–0.75	2.1
CaCl$^+$	0.961	−0.004	−0.426	0.09–0.68	2.0
SrCl$^+$	0.990	0.25	−0.638	0.09–0.74	1.4
BaCl$^+$	1.063	0.13	−0.338	0.09–0.73	2.2
NH$_4$Cl^0	−0.277	−0.617	0	0.09–0.80	2.6
Me$_4$NCl^0	0.157	0	0	0.095	3.4
Et$_4$NCl^0	0.064	0	0	0.092	2.0
MnCl$^+$	1.040	−0.037	−0.502	0.09–0.73	1.7

TABLE 5.66 Stoichiometric association constants obtained for $LiCl^0$ at 25°C.

I_e	$K_{LiCl}^{*'}$	$[Li]_T^{(m)}$	$[Cl]_T^{(m)}$	$(\gamma_{\pm HCl})_T^{(m)}$	Reference[a]
0.0965	0.334	0.08961	0.09956	0.7952	(1)
0.0967	0.342	0.04978	0.09956	0.7958	
0.8776	0.131	0.8807	0.9785	0.822	(2)
0.8791	0.129	0.9687	0.9785	0.823	
0.8837	0.121	0.9736	0.9785	0.826	
0.8851	0.119	0.9776	0.9785	0.827	
0.2778	0.274	0.1987	0.2979	0.758	(3)
0.3744	0.241	0.3982	0.4081	0.757	
0.4269	0.203	0.3650	0.4639	0.763	
0.6253	0.185	0.6878	0.6976	0.778	
0.7079	0.164	0.6914	0.7896	0.792	
0.8821	0.141	0.9822	0.9920	0.821	
0.9675	0.122	0.9834	1.0811	0.842	
0.8733	0.132	0.9785	0.9804	0.822	(4)

[a]The references are for the activity coefficients in binary solutions and are listed at the end of Table 5.74. (m) indicates a mixture of two electrolytes, one of which is HCl, used in the method presented in the text.

TABLE 5.67 Stoichiometric association constants obtained for $NaCl^0$ at 25°C.

I_e	$K_{NaCl}^{*'}$	$[Na]_T^{(m)}$	$[Cl]_T^{(m)}$	$(\gamma_{\pm HCl})_T^{(m)}$	Reference
0.0952	0.509	0.08962	0.09958	0.7902	(5)
0.0956	0.493	0.06970	0.09957	0.7920	
0.0960	0.485	0.04978	0.09957	0.7935	
0.0964	0.477	0.02987	0.09957	0.7949	
0.09684	0.378	0.00996	0.09956	0.7965	
0.3331	0.418	0.3395	0.3772	0.7394	
0.3376	0.405	0.2640	0.1132	0.7446	
0.3416	0.395	0.1886	0.3772	0.7492	
0.3453	0.386	0.1131	0.3772	0.7535	
0.3493	0.322	0.0377	0.3771	0.7581	
0.5610	0.340	0.5963	0.6626	0.7463	
0.5711	0.329	0.4638	0.6626	0.7543	
0.5806	0.318	0.3313	0.6626	0.7619	
0.5905	0.292	0.1988	0.6625	0.7698	
0.5993	0.225	0.0663	0.6625	0.7768	
0.7106	0.303	0.7699	0.8554	0.7614	
0.7250	0.294	0.5988	0.8554	0.7714	
0.7390	0.282	0.4292	0.8554	0.7813	
0.7530	0.265	0.2566	0.8553	0.7911	

TABLE 5.68 Stoichiometric association constants obtained for KCl^0 at 25°C.

I_e	$K_{KCl}^{*'}$	$[K]_T^{(m)}$	$[Cl]_T^{(m)}$	$(y_{\pm HCl})_T^{(m)}$	Reference
0.0936	0.722	0.08955	0.09950	0.7845	(6)
0.0943	0.721	0.06966	0.09952	0.7872	
0.0952	0.687	0.04977	0.09953	0.7904	
0.0958	0.698	0.02986	0.09954	0.7929	
0.0966	0.670	0.00996	0.09956	0.7956	
0.3186	0.606	0.3383	0.3759	0.7240	
0.3257	0.597	0.2633	0.3761	0.7320	
0.3333	0.572	0.1882	0.3764	0.7407	
0.3402	0.564	0.1130	0.3767	0.7484	
0.3477	0.477	0.0377	0.3770	0.7568	
0.5230	0.536	0.5927	0.6585	0.7188	
0.5401	0.523	0.4616	0.6594	0.7321	
0.5569	0.515	0.3301	0.6603	0.7452	
0.5748	0.497	0.1984	0.6612	0.7590	
0.5932	0.441	0.0662	0.6621	0.7732	
0.6529	0.498	0.7638	0.8486	0.7251	
0.6779	0.486	0.5951	0.8501	0.7420	
0.7036	0.483	0.4258	0.8516	0.7595	
0.7299	0.457	0.2559	0.8531	0.7774	
0.7562	0.453	0.0855	0.8546	0.7951	

TABLE 5.69 Stoichiometric association constants obtained for $CsCl^0$ at 25°C.

I_e	$K_{CsCl}^{*'}$	$[Cs]_T^{(m)}$	$[Cl]_T^{(m)}$	$(y_{'HCl})_T^{(m)}$	Reference
0.1009	0.874	0.09936	0.10929	0.775	(7)
0.1804	0.872	0.1978	0.2077	0.735	
0.3147	0.899	0.3921	0.4019	0.696	
0.4866	0.859	0.6776	0.6873	0.675	
0.6387	0.8119	0.9568	0.9664	0.670	

TABLE 5.70 Stoichiometric association constants obtained for $MgCl^+$ at 25°C.

I_e	$K_{MgCl}^{*'}$	$[Mg]_T^{(m)}$	$[Cl]_T^{(m)}$	$(y_{\pm HCl})_T^{(m)}$	Reference
0.0922	2.08	0.03027	0.06938	0.7900	(8)
0.0926	2.07	0.02685	0.07280	0.7910	
0.0932	2.12	0.02015	0.07947	0.7924	
0.0943	2.10	0.01345	0.08615	0.7941	
0.0953	2.37	0.00652	0.09306	0.7952	
0.3841	2.00	0.1492	0.3474	0.7290	
0.3883	2.04	0.1326	0.3637	0.7324	
0.4018	1.98	0.1004	0.3954	0.7412	
0.4180	1.91	0.06654	0.4286	0.7498	
0.4367	1.78	0.03136	0.4631	0.7581	
0.6954	1.71	0.2982	0.6910	0.7457	
0.7124	1.67	0.2652	0.7229	0.7551	
0.7484	1.62	0.1970	0.7886	0.7726	
0.7892	1.61	0.1272	0.8561	0.7900	
0.8376	1.33	0.06316	0.9177	0.8096	

TABLE 5.71 Stoichiometric association constants obtained for $CaCl^+$ at 25°C.

I_e	$K^{*\prime}_{CaCl}$	$[Ca]^{(m)}_T$	$[Cl]^{(m)}_T$	$(y_{\pm HCl})^{(m)}_T$	Reference
0.0908	2.60	0.02989	0.06975	0.7863	(9)
0.0911	2.64	0.02657	0.07306	0.7874	
0.0921	2.68	0.01992	0.07969	0.7899	
0.0935	2.64	0.01328	0.08632	0.7925	
0.0949	2.91	0.00664	0.09294	0.7944	
0.3727	2.40	0.1489	0.3475	0.7207	
0.3785	2.42	0.1323	0.3638	0.7256	
0.3924	2.46	0.09912	0.3965	0.7351	
0.4133	2.23	0.06600	0.4290	0.7469	
0.4332	2.11	0.03296	0.4615	0.7562	
0.6658	2.15	0.2966	0.6920	0.7305	
0.6860	2.09	0.2633	0.7241	0.7418	
0.7278	2.02	0.1970	0.7881	0.7626	
0.8857	1.91	0.1310	0.8518	0.7837	
0.8297	1.61	0.06554	0.9152	0.8060	

$(\gamma_\pm)_F$, $(\gamma_\pm)_T$, $K^{*\prime}$, and m_F. The values of m_T are given by the composition of the solution. Our data for these calculations are limited at present to 25°C and 1 atm.

EXAMPLE 5.18. ALGEBRAIC STATEMENT OF THE PROBLEM. Let us assume that we have a NaCl–$CaSO_4$ solution with $I = 0.5$ and that we wish to calculate $(\gamma_{\pm NaCl})_F$, $K^{*\prime}$ for $NaCl^0$, $CaCl^+$, $NaSO_4^-$, and $CaSO_4^0$, and the speciation, that is, the concentrations of the free ions Na^+, Ca^{2+}, Cl^-, SO_4^{2-} and the concentrations of the ion pairs.

1. First we obtain

$$K^{*\prime}_{NaCl} = \frac{[NaCl^0]}{[Na^+]_F[Cl^-]_F} \tag{i}$$

$$K^{*\prime}_{NaSO_4} = \frac{[NaSO_4^-]}{[Na^+]_F[SO_4^{2-}]_F} \tag{ii}$$

$$K^{*\prime}_{CaCl} = \frac{[CaCl^+]}{[Ca^{2+}]_F[Cl^-]_F} \tag{iii}$$

TABLE 5.72 Stoichiometric association constants obtained for $SrCl^+$ at 25°C.

I_e	$K^{*\prime}_{SrCl}$	$[Sr]^{(m)}_T$	$[Cl]^{(m)}_T$	$(y_{\pm HCl})^{(m)}_T$	Reference
0.0289	3.46	0.00665	0.02327	0.8518	(10)
0.0471	3.09	0.01329	0.03656	0.8240	
0.0648	2.91	0.01994	0.04985	0.8051	
0.0904	2.74	0.02990	0.06977	0.7853	
0.1676	2.75	0.06307	0.1361	0.7469	
0.2364	2.76	0.09619	0.2023	0.7274	
0.3606	2.70	0.1623	0.3345	0.7105	
0.4726	2.65	0.2282	0.4662	0.7056	
0.6378	2.39	0.3266	0.6632	0.7142	
0.6496	2.44	0.2966	0.6921	0.7221	(11)
0.6690	2.42	0.2634	0.7242	0.7332	
0.6939	2.32	0.2302	0.7563	0.7461	
0.7153	2.33	0.1971	0.7883	0.7564	
0.7395	2.30	0.1640	0.8202	0.7675	
0.0900	2.91	0.02990	0.06977	0.7841	(12)
0.09056	3.12	0.02325	0.07639	0.7863	
0.0955	3.97	0.00332	0.09626	0.7950	

TABLE 5.73 Stoichiometric association constants obtained for $BaCl^+$ at 25°C.

I_e	$K^{*'}_{BaCl}$	$[Ba]_T^{(m)}$	$[Cl]_T^{(m)}$	$(y_{\pm HCl})_T^{(m)}$	Reference
0.6179	2.73	0.3258	0.6614	0.7042	(13)
0.6318	2.78	0.2959	0.6905	0.7135	
0.6549	2.71	0.2628	0.7227	0.7266	
0.6756	2.75	0.2298	0.7549	0.7375	
0.6999	2.75	0.1968	0.7870	0.7493	
0.7268	2.71	0.1638	0.8191	0.7617	
0.0894	3.13	0.02989	0.06974	0.7826	(12)
0.0906	3.11	0.02324	0.07637	0.7863	
0.0919	3.19	0.01660	0.08300	0.7893	
0.0938	3.09	0.009958	0.08963	0.7927	
0.0961	2.22	0.003319	0.09625	0.7961	
0.0898	2.95	0.03033	0.06930	0.7836	(14)
0.0901	2.98	0.02750	0.07213	0.7849	
0.0916	2.82	0.02064	0.07897	0.7889	
0.0931	2.79	0.01414	0.0854	0.7918	
0.0950	3.05	0.00601	0.09356	0.7945	
0.3611	2.86	0.1491	0.3468	0.7124	
0.3658	2.92	0.1351	0.3606	0.7167	
0.3843	2.95	0.09844	0.3968	0.7299	
0.4104	2.60	0.06272	0.4320	0.7450	
0.4388	2.37	0.02136	0.4729	0.7584	
0.6354	2.70	0.2959	0.6904	0.7154	
0.6627	2.62	0.2569	0.7284	0.7306	
0.6979	2.61	0.2057	0.7783	0.7484	
0.7620	2.52	0.1267	0.8554	0.7772	
0.8430	2.07	0.03998	0.9397	0.8108	

TABLE 5.74 Stoichiometric association constants obtained for NH_4Cl^0 at 25°C.

I_e	$K^{*'}_{NH_4Cl}$	$[NH_4]_T^{(m)}$	$[Cl]_T^{(m)}$	$(y_{\pm HCl})_T^{(m)}$	Reference
0.0244	0.976	0.02235	0.02492	0.8595	(15)
0.0245	0.952	0.01690	0.02492	0.8610	
0.0245	0.943	0.01207	0.02492	0.8622	
0.0246	0.961	0.00766	0.02492	0.8632	
0.0247	0.674	0.00332	0.02493	0.8646	
0.0481	0.779	0.04403	0.04979	0.8237	
0.0483	0.819	0.03278	0.04980	0.8251	
0.0486	0.760	0.02358	0.04981	0.8272	
0.0487	0.843	0.01529	0.04981	0.8281	
0.0489	1.47	0.00504	0.04982	0.8290	
0.0483	0.750	0.03833	0.04979	0.8250	
0.0485	0.761	0.02924	0.04980	0.8263	
0.0490	0.418	0.01791	0.04981	0.8302	
0.0489	0.899	0.00903	0.04982	0.8291	
0.0490	0.934	0.00537	0.04982	0.8298	
0.0934	0.737	0.08908	0.09939	0.7842	
0.0943	0.715	0.06941	0.09943	0.7874	
0.0950	0.703	0.05054	0.09946	0.7901	
0.0956	0.691	0.03558	0.09949	0.7922	
0.0967	0.548	0.01073	0.09954	0.7959	

TABLE 5.74 (*Continued*)

I_e	$K^{*'}_{\mathrm{NH_4Cl}}$	$[\mathrm{NH_4}]^{(m)}_T$	$[\mathrm{Cl}]^{(m)}_T$	$(y_{\pm\mathrm{HCl}})^{(m)}_T$	Reference
0.0937	0.739	0.08010	0.09941	0.7854	
0.0947	0.716	0.05796	0.09945	0.7889	
0.0953	0.725	0.04093	0.09948	0.7911	
0.0962	0.708	0.01861	0.09952	0.7943	
0.0965	0.614	0.01334	0.09954	0.7954	
0.2174	0.673	0.2228	0.2471	0.7369	
0.2220	0.685	0.1564	0.2474	0.7440	
0.2250	0.684	0.1184	0.2476	0.7486	
0.2287	0.653	0.07771	0.2478	0.7542	
0.2324	0.620	0.03140	0.2480	0.7599	
0.4056	0.605	0.3907	0.4903	0.7206	
0.4183	0.593	0.2856	0.4912	0.7322	
0.4292	0.572	0.2013	0.4920	0.7421	
0.4418	0.541	0.1002	0.4930	0.7535	
0.4488	0.607	0.03440	0.4936	0.7596	
0.7408	0.490	0.7652	0.9643	0.7370	
0.7778	0.464	0.5642	0.9680	0.7600	
0.8101	0.446	0.3832	0.9714	0.7801	
0.8399	0.434	0.2142	0.9744	0.7986	
0.8578	0.423	0.1150	0.9763	0.8098	
0.7250	0.500	0.8517	0.9627	0.7272	
0.7628	0.474	0.6470	0.9665	0.7507	
0.7919	0.463	0.4785	0.9696	0.7687	
0.8243	0.462	0.2872	0.9731	0.7888	
0.8626	0.359	0.1055	0.9765	0.8129	
0.0937	0.693	0.08945	0.09939	0.7854	(13)
0.0945	0.680	0.06960	0.09943	0.7881	
0.0951	0.694	0.04973	0.09947	0.7904	
0.0958	0.710	0.02985	0.09951	0.7928	
0.0968	0.383	0.00995	0.09954	0.7965	
0.4027	0.579	0.4408	0.4898	0.7186	
0.4134	0.573	0.3435	0.4907	0.7284	
0.4248	0.562	0.2458	0.4916	0.7387	
0.4363	0.548	0.1478	0.4925	0.7491	
0.4490	0.444	0.04935	0.4935	0.7605	
0.7267	0.486	0.8661	0.9624	0.7297	
0.7590	0.471	0.6761	0.9659	0.7498	
0.7924	0.454	0.4847	0.9695	0.7707	
0.8247	0.452	0.2919	0.9731	0.7908	
0.8619	0.400	0.09767	0.9767	0.8142	

$$K^{*'}_{\mathrm{CaSO_4}} = \frac{[\mathrm{CaSO_4^0}]}{[\mathrm{Ca^{2+}}]_F[\mathrm{SO_4^{2-}}]_F} \qquad \text{(iv)}$$

from the previous tables or from Equation (5.472). The values of $K^{*'}$ are obtained at $I = 0.5$ for the first pass.

2. Next, we have the relations

$$[\mathrm{Na^+}]_T = [\mathrm{Na^+}]_F + [\mathrm{NaCl^0}] + [\mathrm{NaSO_4^-}] \qquad \text{(v)}$$

$$[\mathrm{Ca^{2+}}]_T = [\mathrm{Ca^{2+}}]_F + [\mathrm{CaCl^+}] + [\mathrm{CaSO_4^0}] \qquad \text{(vi)}$$

$$[\mathrm{Cl^-}]_T = [\mathrm{Cl^-}]_F + [\mathrm{NaCl^0}] + [\mathrm{CaCl^+}] \qquad \text{(vii)}$$

$$[\mathrm{SO_4^{2-}}]_T = [\mathrm{SO_4^{2-}}]_F + [\mathrm{NaSO_4^-}] + [\mathrm{CaSO_4^0}] \qquad \text{(viii)}$$

The unknowns are $[\mathrm{NaCl^0}]$, $[\mathrm{NaSO_4^-}]$, $[\mathrm{CaCl^+}]$, $[\mathrm{CaSO_4^0}]$, $[\mathrm{Na^+}]_F$, $[\mathrm{Ca^{2+}}]_F$, $[\mathrm{Cl^-}]_F$, and $[\mathrm{SO_4^{2-}}]_F$ so that we have eight equations in eight unknowns.

TABLE 5.75 Stoichiometric association constants obtained for Mn_4Cl^+ at 25°C.

I_e	$K_{MnCl}^{*\prime}$	$[Mn]_T^{(m)}$	$[Cl]_T^{(m)}$	$(y_{\pm HCl})_T^{(m)}$	Reference
0.0906	2.86	0.02630	0.07334	0.7862	(12)
0.0912	2.83	0.02298	0.07664	0.7878	
0.0923	2.95	0.01660	0.08300	0.7902	
0.0940	2.79	0.01027	0.08932	0.7932	
0.0950	2.71	0.00689	0.09269	0.7946	
0.0902	2.83	0.02983	0.06981	0.7847	
0.0915	2.70	0.02323	0.07639	0.7884	
0.0928	2.65	0.01659	0.08302	0.7913	
0.0946	2.33	0.00996	0.08963	0.7943	
0.0960	2.50	0.00331	0.09626	0.7959	
0.3714	2.61	0.2136	0.3574	0.7203	
0.3826	2.58	0.1144	0.3814	0.7287	
0.3992	2.57	0.08258	0.4127	0.7390	
0.4187	2.49	0.05141	0.4434	0.7492	
0.4264	2.52	0.03886	0.4557	0.7527	
0.6718	2.22	0.2761	0.7115	0.7343	
0.7025	2.16	0.2275	0.7585	0.7504	
0.7477	2.06	0.1641	0.8199	0.7715	
0.7943	2.00	0.1020	0.8798	0.7913	
0.8144	1.96	0.07335	0.8966	0.7995	

(1) Guntelberg (1938).
(2) Harned (1926).
(3) Harned and Swindells (1926).
(4) Struck and Schneider (1972).
(5) Macaskill, Robinson, and Bates (1977).
(6) Macaskill and Bates (1978).
(7) Harned and Schupp (1926).
(8) Khoo, Chan, and Lim (1977a).
(9) Khoo, Chan, and Lim (1977b).
(10) Harned and Paxton (1953).
(11) Harned and Gary (1955).
(12) Downes (1970).
(13) Harned and Gary (1954).
(14) Khoo, Chan, and Lim (1978).
(15) Robinson, Roy, and Bates (1974).

TABLE 5.76 Free concentrations of the cation in pure solutions of each electrolyte. All concentrations and the total ionic strengths are on the molal scale.

I	Free Cation Concentrations										
	HCl	LiCl	NaCl	KCl	CsCl	NH_4Cl	$MgCl_2$	$CaCl_2$	$SrCl_2$	$BaCl_2$	$MnCl_2$
0.1	0.0974	0.0965	0.0953	0.0933	0.0924	0.0937	0.0295	0.0287	0.0285	0.0283	0.0284
0.2	0.1908	0.1893	0.1841	0.1789	0.1737	0.1783	0.0534	0.0512	0.0505	0.0498	0.0504
0.3	0.2816	0.2792	0.2687	0.2584	0.2474	0.2571	0.0739	0.0699	0.0684	0.0672	0.0686
0.4	0.371	0.368	0.351	0.334	0.316	0.332	0.0919	0.0861	0.0838	0.0821	0.0844
0.5	0.459	0.455	0.431	0.407	0.379	0.404	0.1083	0.1007	0.0974	0.0950	0.0987
0.6	0.547	0.543	0.511	0.479	0.440	0.474	0.1235	0.1140	0.1098	0.1066	0.1118
0.7	0.634	0.630	0.590	0.549	0.497	0.542	0.1379	0.1265	0.1213	0.1171	0.1242
0.8	0.722	0.719	0.669	0.618	0.552	0.608	0.1517	0.1385	0.1322	0.1269	0.1361
0.9	0.809	0.808	0.748	0.686	0.605	0.674	0.1652	0.1500	0.1428	0.1360	0.1477
1.0	0.898	0.897	0.827	0.754	0.657	0.740	0.1785	0.1612	0.1532	0.1448	0.1593

TABLE 5.77 Thermodynamic association constants of ion pairs at 25°C. The values listed under other results were obtained as the best fit to conductivity data.

Ion Pair	K^* This Paper	K^* Other Results	Ion Pair	K^* This Paper	K^* Other Results
$HClO_4^0$	0.01	0.1			
HCl^0	0.52	0.3, 0.0			
$LiCl^0$	0.62	0.0, 0.2	$MgCl^+$	4.1	4.5
$NaCl^0$	0.89	0.9, 0.3, 0.2	$CaCl^+$	5.2	
KCl^0	1.16	0.8, 0.4, 0.2	$SrCl^+$	5.5	
$CsCl^0$	1.6	0.4	$BaCl^+$	5.7	
NH_4Cl^0	1.23		$MnCl^+$	5.7	

3. The value of $(\gamma_{\pm NaCl})_T$ in mixtures of NaCl and $CaSO_4$ is assumed to be known from the literature. Then, $(\gamma_{\pm NaCl})_F$ can be calculated from $(\gamma_{\pm NaCl})_F^2 = (\gamma_{\pm NaCl})_T^2 (Na^+)_T (Cl^-)_T / (Na^+)_F (Cl^-)_F$.

4. Step 1 was calculated at the total ionic strength of the solution. Now we can use the results of step 2 to obtain a first pass for values at I_e. Then 1 through 3 are repeated and the iteration is carried on until the concentrations and $(\gamma_\pm)_F$ remain invariant within the desired limits.

Note then when $(\gamma_\pm)_F$ is available then one can use the above procedure to obtain $(\gamma_\pm)_T$.

Some Further Results of the Johnson and Pytkowicz (1978, 1979a, 1979b) Model and Intercomparisons. First, I present values of the trace activity coefficients of Johnson and Pytkowicz (1979a, 1979b) and Pitzer and Kim (1974), as well as of experimental ones in Table 5.82. The trace activity coefficient in the system denoted by A/B is that the

TABLE 5.78 The constants of Equation (5.473) which may be used to determine the stoichiometric association constant at any effective ionic strength. The units of the stoichiometric association constants are per mole/L. The scatter of the association constants about the line is ±5%.

Ion Pair	a_{MX}	b_{MX}	Range of Fit (I_e)
$CaCl^+$	1.073	−0.442	0.37–0.70
$NaSO_4^-$	2.28	0	0.46–0.54
KSO_4^-	2.48	0	0.46–0.52
$MgSO_4^0$	3.73	0	0.51–0.54

TABLE 5.79 Calculated values of $(\gamma_{\pm NaCl})_F$ and $(\gamma_{\pm KCl})_F$.

I_T	I_e	$(\gamma_{\pm NaCl})_F$	$(\gamma_{\pm NaCl})_T$
0.1	0.09532	0.7967	0.7778
0.2	0.18399	0.7648	0.7336
0.3	0.26857	0.7495	0.7091
0.4	0.35054	0.7404	0.6931
0.5	0.43087	0.7346	0.6819
0.6	0.51025	0.7305	0.6736
0.7	0.58916	0.7275	0.6674
0.8	0.66799	0.7251	0.6626
0.9	0.74708	0.7232	0.6589
1.0	0.82667	0.7215	0.6560
1.1	0.90700	0.7200	0.6538
1.2	0.98823	0.7186	0.6521
1.3	1.07056	0.7173	0.6508

I_e	$(\gamma_{\pm NaCl})_T$	$(\gamma_{\pm NaCl})_F$	I_e	$(\gamma_{\pm NaCl})_T$	$(\gamma_{\pm NaCl})_F$
0.1	0.7752	0.7937	0.6	0.6667	0.7271
0.2	0.7287	0.7615	0.7	0.6610	0.7242
0.3	0.7025	0.7457	0.8	0.6571	0.7220
0.4	0.6858	0.7365	0.9	0.6542	0.7201
0.5	0.6746	0.7308	1.0	0.6520	0.7184

I_T	I_e	$(\gamma_{\pm KCl})_F$	$(\gamma_{\pm KCl})_T$
0.1	0.09379	0.7937	0.7687
0.2	0.1788	0.7590	0.7177
0.3	0.2584	0.7405	0.6873
0.4	0.3338	0.7291	0.6660
0.5	0.4067	0.7205	0.6498
0.6	0.4776	0.7140	0.6370
0.7	0.5472	0.7085	0.6264
0.8	0.6158	0.7039	0.6176
0.9	0.6837	0.6999	0.6100
1.0	0.7513	0.6962	0.6035
1.1	0.8186	0.6929	0.5978
1.2	0.8860	0.6899	0.5928
1.3	0.9535	0.6870	0.5883
1.4	1.0213	0.6842	0.5844

I_e	$(\gamma_{\pm KCl})_T$	$(\gamma_{\pm KCl})_F$
0.1	0.791	0.763
0.2	0.753	0.708
0.3	0.734	0.675
0.4	0.721	0.651
0.5	0.712	0.634
0.6	0.705	0.619
0.7	0.699	0.609
0.8	0.694	0.600
0.9	0.689	0.592
1.0	0.685	0.586

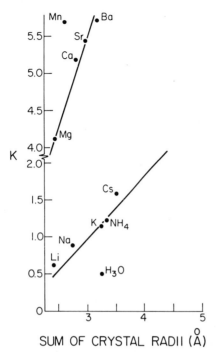

FIGURE 5.23 Thermodynamic association constants of chlorides versus the sum of the crystal radii.

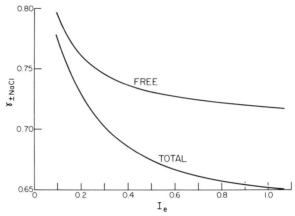

FIGURE 5.24 Free and total activity coefficients of NaCl in pure NaCl solutions. The term $(\gamma_\pm)_F$ does not depend on the composition but only on I_e.

mean total coefficient of A at infinite dilution in a solution of $I = 1.0$. The agreement of all the data is fair when one considers the extrapolations required to $I < 10^{-4}$ m.

In Table 5.83 are shown values of $(\gamma_\pm)_F$ obtained from $(\gamma_\pm)_T$ and the ion-pairing model of Johnson and Pytkowicz (1979a). The results are at $I_e = 0.6$ so that they are of use in the determination of stoichiometric equilibrium constants for the oceans.

Values of $(\gamma_\pm)_T$ calculated for electrolytes in seawater by the methods of Johnson and Pytkowicz (1979a), Whitfield (1975), and Robinson and Wood (1972) are compared to experimental values. The experimental results are from Gieskes (1966) and Platford (1962) for NaCl, Platford and DaFoe (1965) for KCl, and Whitfield (1975) for K_2SO_4 in Table 5.84. It can be seen that all three models yield values in fair agreement with the measured ones. The specific interaction and the cluster methods will be described later in this chapter. Our results were obtained from the expression

$$(\gamma_\pm)_T^2 = \frac{(\gamma_\pm)_F^2 m_F}{m_T} \qquad (5.474)$$

where $(\gamma_\pm)_F$, m_F, and m_T were calculated from our system of equations applied to seawater ($S\%_0 = 34.8\%_0$).

Experimental values of $(\gamma_{\pm NaCl})_T$ in seawater were obtained potentiometrically by Johnson (1979). They were calculated from the data by means of the expression

$$2 \log(\gamma_{\pm NaCl}^{sw})_T = \frac{\Delta E}{2.306RT/F_e}$$
$$+ \log \frac{[Na^+]_T[Cl]_T(\gamma_{\pm NaCl})_T^2}{[Na^{+sw}]_T[Cl^{-sw}]_T} \qquad (5.475)$$

where the superscript sw indicates NaCl in seawater, the absence of sw means measurements in pure NaCl, and E is the emf. A comparison of $(\gamma_{\pm NaCl})_T$ experimental, calculated from Pitzer's approach (1973) and obtained by our ion-pair model also revealed good agreement (Johnson, 1979). Ion pairing with SO_4^{2-} does not have much influence upon the results because $[SO_4^{2-}]_T$ is small relative to $[NA^+]_T$ and $[Cl^-]_T$.

The calculations of equilibrium properties of chemical systems (solubility products, vapor pressure, etc.) by the ion-pair model and their agreement with measured ones only show that the model is a convenient formal representation for what happens in solution. Transport properties (diffusion, viscosity, electrical conduction) as well as Raman spectroscopy and sound attenuation, which have already been discussed, depend directly on the solute particles, that is, on ions and ion pairs.

TABLE 5.80 Results of the calculation of the speciation of seawater. Although calculations were performed using molar concentrations, the results have been converted to the molal scale for convenience. The error in the concentrations of the ion pairs is approximately 0.6% for a corresponding error of 5% in the stoichiometric association constants.

Cation	Total Molality	% Free	% M–Cl	% M–SO_4		
Na^+	0.4822	82.97	13.31	3.72		
K^+	0.01062	78.28	17.45	4.27		
Mg^{2+}	0.05489	48.14	42.71	9.16		
Ca^{2+}	0.01063	43.54	46.93	9.53		
Anion	Total Molality	% Free	% Na–X	% K–X	% Mg–X	% Ca–X
Cl^+	0.5657	83.31	11.34	0.33	4.14	0.88
SO_4^{2-}	0.02906	15.91	61.75	1.56	17.30	3.49

TABLE 5.81 Results of the calculation of the total mean activity coefficients of salts in seawater compared to experimental values. The activity coefficients are in the molal scale. The uncertainty in the calculated activity coefficients is approximately 0.3% for an error of 5% in the stoichiometric association constant.

Electrolyte	$\gamma_\pm$ Calculated	$\gamma_\pm$ Experimental	Source
NaCl	0.661	0.668 ± 0.004	Harned (1926)
KCl	0.642	0.645 ± 0.008	Harned and Swindells (1926)
$MgCl_2$	0.467	—	
$CaCl_2$	0.458	—	
Na_2SO_4	0.354	0.378 ± 0.016	Harned and Copson (1933)
K_2SO_4	0.351	0.352 ± 0.018	Harned and Swindells (1926)
$MgSO_4$	0.129		
$CaSO_4$	0.125		

TABLE 5.82 An intercomparison of trace activity coefficients

A/B	Experimental	Johnson and Pytkowicz (1979a,b)	Pitzer and Kim (1974)
KCl/NaCl	0.620	0.636	0.626
KCl/$CaCl_2$	0.639	0.628	0.662
KCl/$MgCl_2$	0.649	0.636	0.674
NaCl/$CaCl_2$	0.671	0.643	0.682
NaCl/$MgCl_2$	0.682	0.650	0.694
$CaCl_2$/$MgCl_2$	0.461	0.460	0.461

TABLE 5.83 Free mean molal activity coefficients for a number of electrolytes at an effective ionic strength of 0.6. Total mean molar activity coefficients of the same electrolytes at a total ionic strength of 0.6 are included for comparison (Robinson and Stokes, 1968).

Electrolyte	$(\gamma_\pm)_F$ ($I_e = 0.6$)	$\gamma_{\pm T}$ ($I_T = 0.6$)
$HClO_4$	0.802	0.802
HCl	0.859	0.772
NaCl	0.805	0.681
KCl	0.808	0.648
$MgCl_2$	0.699	0.490
$CaCl_2$	0.700	0.474
Na_2SO_4	0.746	0.372
K_2SO_4	0.760	0.356

TABLE 5.84 Comparison of total mean activity coefficients (molal scale) obtained from our ion association model (Johnson and Pytkowicz, 1979a) (column a), from a specific interaction model (with permission from Whitfield, M., *Chemical Oceanography,* Vol. 1. Copyright © 1975 by Academic Press Ltd., London) (column b), and from the cluster integral model (Robinson and Wood, 1972) (column c). Experimental values are also included.

| Salt | $(\gamma_{\pm})_T$ | | | | |
	a	b	c	Experimental	Sources
NaCl	0.661	0.667	0.669	0.668 ± 0.003	Gieskes (1966)
				0.672 ± 0.007	Platford (1962)
KCl	0.642	0.648	0.639	0.645 ± 0.007	Platford and Dafoe (1965)
$MgCl_2$	0.467	0.473	0.467	—	
$CaCl_2$	0.458	0.463	—	—	
Na_2SO_4	0.354	0.373	0.363	0.378 ± 0.016	Whitfield (1975)
K_2SO_4	0.351	0.358	0.345	0.352 ± 0.018	Whitfield (1975)
$MgSO_4$	0.129	0.167	0.158	—	
$CaSO_4$	0.125	0.161	—	—	

Electrical Conductance. The ion-pair model was tested by Johnson (1979) by the use of the electrical conductance. The equivalent conductance of a solution is given by

$$\Lambda = \frac{1000\,K}{(c_{equiv})_T} \qquad (5.476)$$

as we saw earlier. K is the conductivity and $(c_{equiv})_T$ is the total equivalent concentration in terms of equiv/L. The degree of ionization of strong electrolytes in which ion pairs are present and have zero net charge is given traditionally by

$$\alpha = \frac{(c_{equiv})_F}{(c_{equiv})_T} = \frac{\Lambda}{\Lambda^*} \qquad (5.477)$$

Λ^* is the equivalent conductance expressed in terms of free ions, Λ is the measured value and, Λ^* is calculated from theoretical expressions (e.g., Fuoss and Onsager, 1957).

Such calculations cannot be done accurately. Instead, Johnson (1979) calculated the conductance of mixtures of electrolytes by assuming that the conductance is equal to the sum of the conductances of the free ions present in the solutions. He assumed, furthermore, that conductances depend only on I_e. This led to

$$\Lambda = \Lambda^* \frac{\sum_i (i)_F}{\sum_i (i)_T} \qquad (5.478)$$

where (i) represents the concentration of species i and with

$$\Lambda^* = \frac{\sum_i (i)_F \Lambda_i^*}{\sum_i (i)_F} \qquad (5.479)$$

Λ_i^* is the conductance due to i in a pure solution of i, obtainable from the literature and the speciation model, whereas the other quantities in the last two equations refer to mixtures of electrolytes. The results are shown in Table 5.85.

Thermodynamic values of the association constants, determined by the extrapolation of $K^{*'}$ to infinite dilution, are

	$HClO_4^0$	HCl^0	$NaCl^0$	KCl^0
K^*	10^{-7}	0.31–0.52	0.58–0.93	0.61–1.01

Values for γ_{ip} versus I_e were obtained from $K^*/(\gamma_{\pm})_F^2$ and are shown in Figure 5.25 while the speciation in seawater is presented in Table 5.86. The good agreement between theoretical experimental activity coefficients is due to the predominance of Cl^- over the other anions.

The Johnson–Pytkowicz Model Tested with Two Weak Electrolytes. Cole et al. (in press) tested the methods of Johnson and Pytkowicz (1978, 1979a) by applying the ion-pairing procedures to two weak acids, H_3PO_4 and HSO_4^-, for which the dissociation constants are known. It does not matter that Cole,

TABLE 5.85 Results of the calculation of the equivalent conductivity of mixtures of univalent electrolytes. The error in the calculated conductivity corresponding to an uncertainty of $\pm 10\%$ in either association constant is about 0.1%. The experimental results were obtained from Ruby and Kawai (1926).

Electrolyte A	$m_{A,T}$	Electrolyte B	$m_{B,T}$	$\Lambda_{\text{calculated}}$	$\Lambda_{\text{measured}}$
NaCl	0.800	KCl	0.200	91.3	91.14
NaCl	0.600	KCl	0.400	96.6	96.24
NaCl	0.400	KCl	0.600	101.8	101.54
NaCl	0.200	KCl	0.800	107.0	106.88
HCl	0.800	KCl	0.200	290.5	291.07
HCl	0.600	KCl	0.400	246.9	247.82
HCl	0.400	KCl	0.600	202.6	202.38
HCl	0.200	KCl	0.800	157.7	157.23
HCl	0.800	NaCl	0.200	284.5	283.19
HCl	0.600	NaCl	0.400	235.3	232.12
HCl	0.400	NaCl	0.600	185.8	182.07
HCl	0.200	NaCl	0.800	136.0	132.98

Johnson and Pytkowicz compared calculations for coulombic ion pairs of strong electrolytes with those for covalent bonds of weak acids. Formally, it is only necessary that pairs of ions be formed and that correct results for the dissociation constants of weak acids, based on the procedures for ion pairs, be obtained.

The results are presented in Table 5.87. The extrapolation was based on the equations of the type (Reardon and Langmuir, 1976)

$$\log(K^{*\prime}_{\text{HSO}_4}) - \log(\gamma_{\pm \text{H}_2\text{SO}_4})^2_F = \log K^{*\prime}_{\text{HSO}_4}$$
$$+ \log \frac{[(\gamma_{\text{H}})_F}{(\gamma_{\text{HSO}_4})_F]} \qquad (5.480)$$

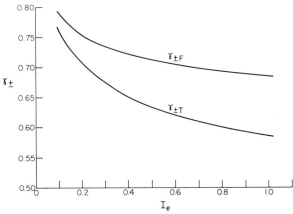

FIGURE 5.25 Free and total activity coefficients of KCl in pure KCl solutions. The term $(\gamma_{\pm \text{KCl}})_F$ does not depend on the composition but only on I_e.

The use of this equation avoids the need to use $\log[K^{*\prime}/(\gamma_{\pm})_F]$ and the large change in $(\gamma_{\pm})_F$ at low values of I_e.

5.9.2 Other Methods and Results

In this section results are surveyed, which, in addition to those described earlier, do not include full consideration of Cl^- ion pairing and/or still use the mean-salt method (MSM).

Let us first examine the MSM in terms of free quantities. Consider, for example, NaCl and KCl. In the original MSM (Garrels and Thompson, 1962) one obtained $\gamma_{\text{Na}} = (\gamma_{\pm \text{NaCl}})^2/\gamma_{\text{KCl}}$. No distinction was made between free and total quantities because Cl^- was not known to associate. Thus, experimental, that is, total quantities were used. This is not correct because $(\gamma_{\text{Cl}})_T$ in the numerator and in the denominator depend on the degrees of association of NaCl and KCl and are, therefore, different as we saw earlier.

$(\gamma)_F$ has the advantage of only depending on I_e. In terms of free quantities,

$$(\gamma_{\text{Na}})_F = \frac{(\gamma_{\pm \text{NaCl}})^2_F}{(\gamma_{\pm \text{KCl}})_F} = \frac{(\gamma_{\text{Na}})_F(\gamma_{\text{Cl}})_F}{(\gamma_{\text{Cl}})_F} \qquad (5.481)$$

If I_e is the same for the two solutions, then $(\gamma_{\text{Cl}})_F$ cancels out provided that $(\gamma_{\text{K}})_F = (\gamma_{\text{Cl}})_F$. This last equality is a new form of the MacInnes assumption (1919). It is compatible with the old one because the

TABLE 5.86 The distribution of species in an artificial seawater containing only Na^+, K^+, Mg^{2+}, Ca^{2+}, and Cl^- and calculated using our model. The column headed (*JP*) yields the values of the total mean activity coefficient for the chloride salt of each cation. The column headed (*P*) presents the total mean activity coefficients calculated from the Pitzer theory (Whitfield, 1975). The last column corresponds to values for total activity coefficients measured in natural seawater.

Ion	Total Molality	% Free	$\dfrac{(\gamma_\pm)_T}{\gamma_\pm}$ (*JP*)	$\dfrac{(\gamma_\pm)_T}{\gamma_\pm}$ (*P*)	(Experimental)
Na^+	0.4944	85.1	0.668	0.675	0.668 ± 0.003^a
K^+	0.0101	80.2	0.650	0.655	0.645 ± 0.018^b
Mg^{2+}	0.0547	50.3	0.476	0.486	
Ca^{2+}	0.0105	45.7	0.466	0.470	
Cl^-	0.6349	82.9			

^{a}Gieskes (1966).
^{b}Whitfield (1975).

ion pairs KCl^0 do not contribute toward a difference in transference numbers.

We must remember several points. First, the MacInnes assumption, if valid at all, is limited to $m \cong 0.1$. Second, $(\gamma_i)_F$ for the above species does not depend on Cl^- ion pairs as $(\gamma_i)_T$ does, except for the effect of the ion pairs on I_e. Researchers may then conclude erroneously that Cl^- ion pairing can be disregarded in equilibrium studies in, for example, seawater. This is not correct because the effect of chloride association appears in the mass balance equations of the type $[M^{z+}]_T = [M^{z+}]_F + [MCl^{z-1}] + \cdots$, which have to be solved simultaneously with association constants for determining speciation models.

Atlas (1976) and Atlas and Pytkowicz (1977) determined the speciation of dissolved inorganic phosphate, TPO_4, in seawater. The apparent dissociation constants of H_3PO_4, that is, K_1', K_2', and K_3', were measured directly in seawater. I remind you that values of K' are invariant in ionic media as was shown theoretically earlier and experimentally by Pytkowicz et al. (1974).

As $K_1' = K_1 \gamma_{H_3PO_4}/k_H'(\gamma_{H_2PO_4})_T$ and $(\gamma_{H_2PO_4})_T$ reflects the effect of ion pairs, K_1' also includes this effect. The same is true of K_2' and K_3'. Therefore, the constants can be used for the determination of the association of phosphate species.

The method used by Atlas and Pytkowicz (1977) was similar to that of Hawley (1973) and Pytkowicz and Hawley (1974) for bicarbonate and carbonate ion pairs in that the variations in apparent dissociation constants with composition were used to determine the stoichiometric association constants. A provisional model in seawater, without Cl^- association, is presented in Table 5.88.

TABLE 5.87 Thermodynamic association constants of H_3PO_4 and H_2SO_4 at 25°C determined from extrapolation by Cole et al. (in press) and by others. The error expressed in the results of Cole et al. refers to the first standard deviation of K^* values from their linear fit.

Species	log K_{CA} Cole et al. (in press)	Others	Reference
H_3PO_4	2.167 ± 0.019	2.148 ± 0.005	Pitzer and Kim (1974)
		2.161	Bjerrum and Unmack (1929)
		2.124	Nims (1934)
HSO_4^-	1.999 ± 0.006	1.994 ± 0.004	Young et al.(1978)
		1.975 ± 0.035	Covington et al. (1978)
		1.987 ± 0.004	Dunsmore and Nancollas (1958)
		1.979	Pitzer et al. (1977)

TABLE 5.88 Distribution of species for phosphate in seawater at $S\%_0 = 34.8$, $T = 25°C$, and pH = 8.

Species	Percent	Species	Percent
HPO_4^{2-}	28.7	$CaH_2PO_4^+$	0.01
$MgHPO_4^0$	41.4	$NaH_2PO_4^0$	0.1
$CaHPO_4^0$	4.7	$(PO_4^{3-})_F$	0.01
$NaHPO_4^-$	15.0	$MgPO_4^-$	1.5
$(H_2PO_4^-)_F$	0.9	$CaPO_4^-$	7.6
$MgH_2PO_4^+$	0.1	$NaPO_4^{2-}$	0.01

TABLE 5.89 Values of total activity coefficients of single ions in seawater.

Ion	$(\gamma_i)_{exp}$	Millero and Schreiber (manuscript)	Whitfield (1975)
H^+	0.626–0.699	0.688	0.782
Na^+	0.708	0.690	0.708
K^+	0.662	0.615	0.668
Mg^{2+}	0.257	0.255	0.269
Ca^{2+}	0.185–0.261	0.228	0.241
Sr^{2+}	0.188	0.231	0.239
Cl^-	0.628	0.628	0.628
F^-	0.296	0.333	0.519
OH^-	0.218	0.236	0.562
HCO_3^-	0.532	0.536	—
$B(OH)_4^-$	0.360	0.351	—
$H_2PO_4^-$	0.416	0.395	0.456
CO_3^{2-}	0.032	0.029	0.095
SO_4^{2-}	0.104	0.085	0.103

Millero and Schreiber (manuscript) estimated values of $(\gamma_i)_T$ for ions in natural waters containing a number of ions, by combining the MSM and the Pitzer (1973) association model.

These authors used the MSM in terms of the original Garrels and Thompson (1962) model and $(\gamma_i)_T$, and neglected Cl^- ion pairing. Values of K^* and $K^{*\prime}$ were obtained from the literature. When $K^{*\prime}$ was not available, K^* was extrapolated to $I = 0.7$ by means of the values $(\gamma_i)_F$ and rough estimates of γ_{ip}. The effective ionic strength was not taken into consideration. Some results are shown in Table 5.89. It is interesting that the MSM, as was used by Millero and Schreiber (manuscript) in conjunction with the Pitzer (1973) approach, yields closer agreement to $(\gamma_i)_{exp}$ than the adaptation of the Pitzer method to single ions by Whitfield (1975).

Dickson and Whitfield (manuscript) extended the model of Garrels and Thompson (1962). They used the MSM, based on $(\gamma_i)_F$ and K^*, to obtain estimates of the apparent dissociation constants of weak acids in solution from thermodynamic values.

Cl^- ion pairs were neglected but this did not affect the results unduly. The reason for this is that the primary effect of Cl^- is on $(\gamma_i)_T$ for cations and not for the anions of acids. The authors compared the dissociation constants obtained by us (Mehrbach et al., 1973), by Hansson (1973), and by them.

$k'_H a_H$ is used with K'_{HA}, a generalized dissociation constant, in the NBS scale and, therefore, the effect of HCl^0 on $(\gamma_H)_T$ does not need to be considered explicitly.

Hansson uses $[H^+]_T$ instead of $a_{H(NBS)}$, where (NBS) refers to the National Bureau of Standards pH Scale. Dickson and Whitfield, therefore, calculated $[H^+]_T$ to compare the three sets of results for the dissociation constants of H_2CO_3. This was possible because $(\gamma_H)_{T(NBS)}$, $[H^+]_T$, and $a_{H(NBS)}$ are known. k'_H was neglected. They calculated $[H^+]_T = [H^+]_F + [HSO_4^-]$ instead of $[H^+]_T = [H^+]_F +$

$[HF^0] + [HSO_4^-] + [HCl^0]$ from their speciation model. This may work as a first approximation. The results are shown in Figure 5.26 and reveal a fair to good agreement when one considers that a logarithmic plot was used.

Dyrssen and Wedborg (1973) estimated the speciation of seawater using the seawater as the ionic medium. Cl^- ion pairs were not introduced. A mixture of measured and literature values for association constants were applied to the problem. Speciation calculations were made in steps, based on the SO_4^{2-} system, the CO_2 system, and so on, instead of a simultaneous calculation with all the required equations.

0.7 m NaCl was used as a reference medium [see Pytkowicz and Kester (1969)] to calculate association in seawater. This was accomplished by observing the difference in the potentiometric behavior of the various systems in seawater and in the NaCl solution. Such a procedure enhances the error due to disregarding the effect of Cl^- ion pairs since it is based on the differences in emf's in two solutions containing different amounts of Cl^- ion pairs. I_e was not taken into consideration. Further work by the Swedish school can be found in Elgquist and Wedborg (1975). The equilibrium constants for water, CO_2, HSO_4^-, and HF are shown in Figure 5.27.

Hydration Model. We followed earlier the derivation of the hydration equation of Robinson and Stokes (1968). I should remind you at this point that

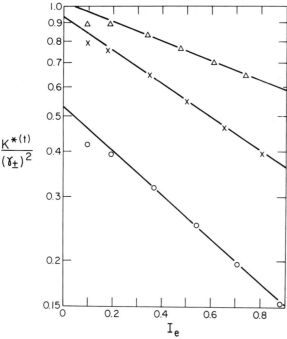

FIGURE 5.26 Values of $\gamma_{ip} = K^{*(t)}/(\gamma_{\pm})^2_F$, the activity coefficients of ion pairs, versus I_e.

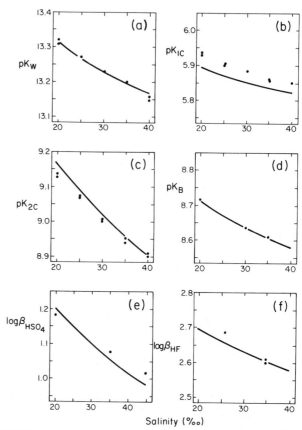

FIGURE 5.27 A comparison of the computations of Dickson and Riley (1979), solid lines, with the data of Hansson (1973).

the hydration correction is required by all equations that start from Debye–Hückel type relations. Such equations are meant to explain the deviation of $\gamma_{\pm(th)}$ from $\gamma_{\pm(exp)}$ by means of assumed nonhydration terms assumed. Although such terms have a theoretical significance, their numerical values are empirical because effects such as hydration are neglected (Brønsted, 1923; Guggenheim, 1935; Pitzer, 1973). Our ion-pairing approach, however, compares experimental data in the solution under study and in a reference solution so that the hydration effect is canceled out.

On the other hand, the hydration theory is simplistic in that it contains no terms for core overlap, ion-cavity interactions, and ion pairing.

The hydration theory is based on the fact that the experimental coulombic free energy includes the effects of the changes in the concentrations of solvent and solute at a nominal m while the Debye–Hückel (1923) free energy does not.

Stokes and Robinson (1948) extended the calculations of Robinson and Stokes (1968) to a large number of electrolytes but the fitting parameters should be considered empirical for the reasons given earlier.

Bates et al. (1970) examined hydration in "unassociated" chlorides at high ionic strengths and their results are shown in Table 5.90.

There is a problem with the above data because the authors find that

Salt	$MgCl_2$	$MgBr_2$	MgI_2
h	13.7	17.0	19.0

and the larger anions are more hydrated than the smaller ones.

Robinson and Bates (1978) later extended the work to mixtures of NaCl and $MgCl_2$.

Brønsted–Guggenheim Equation. This is a specific interaction model in which no given microscopic model, such as the ion-pair model, is ascribed to the energies of interaction.

5.9.3 The Postulate of Brønsted

Brønsted (1923) postulated from observation and theory that

$$\ln \gamma_{\pm MX} = -3\alpha m_{MX}^{0.5} - 2\beta_{MX} m_{MX} \quad (5.482)$$

TABLE 5.90 Values of a_D and h according to Bates et al.

Electrolyte	a_D (Å)	h
HCl	4.47	8.0
LiCl	4.32	7.1
NaCl	3.97	3.5
KCl	3.63	1.9
RbCl	3.49	1.2
NH$_4$Cl	3.75	1.6
MgCl$_2$	5.02	13.7
CaCl$_2$	4.73	12.0
SrCl$_2$	4.61	10.7
BaCl$_2$	4.45	7.7

Source: Reprinted with permission from Bates, R. G., B. R. Staples, and R. A. Robinson, *Anal. Chem.* **42**, 867. Copyright © 1970 American Chemical Society.

TABLE 5.91 Mean total activity coefficients $(\gamma_\pm)_T$ calculated from the model of Brønsted–Guggenheim (Guggenheim, 1935) for quasiseawater at 25°C (Whitfield, 1973).

I	$(\gamma_\pm)_T$			
	0.5 M	0.7 M	1.0 M	Experimental at $I = 0.7$
NaCl	0.681	0.668	0.657	0.668[a]
KCl	0.668	0.650	0.634	—
MgCl$_2$	0.481	0.467	0.456	—
CaCl$_2$	0.473	0.457	0.443	—
Na$_2$SO$_4$	0.406	0.376	0.343	0.378[b]
K$_2$SO$_4$	0.395	0.363	0.328	—
MgSO$_4$	0.186	0.165	0.144	—
CaSO$_4$	0.181	0.160	0.138	—

[a]Gieskes (1966).
[b]Platford and Dafoe (1965).

α turned out later, with the development of the Debye–Hückel theory, to be $|z_M z_X| A_D$, and m_{MX}, used by Brønsted for 1-1 electrolytes, became I, the ionic strength. β_{MX} is a coefficient which depends on MX. Brønsted postulated, furthermore, that only interactions between ions with unlike charges need to be taken into consideration due to the greater distance between ions with charges of the same sign.

5.9.4 The Extension by Guggenheim

Guggenheim (1935, 1936), who based his work on the postulate of Brønsted and the results of Debye and Hückel (1923), arrived at the following equation:

$$\ln \gamma_{\pm MX} = -\frac{A_D |z_M z_X| I^{0.5}}{1 + I^{0.5}} + \nu B_{MX} m_{MX}$$

(5.483)

This equation applies to a single electrolyte. The first term, discussed earlier and shown to provide poor fits even at very low concentrations, corresponds roughly to $a_D = 3$ Å. For multielectrolyte solutions

$$\ln \gamma_{\pm MX} = -\frac{A_D |z_M z_A| I^{0.5}}{1 + I^{0.5}}$$
$$+ \frac{\nu_M}{\nu_M + \nu_A} \sum_A B_{MA} m_A + \frac{\nu_M}{\nu_M + \nu_A} \sum_M B_{MA} m_M$$

(5.484)

The summations extend over all the anions A, including X, and all the cations C in solution. The B_{MA} and B_{CX} terms are obtainable in single-salt solutions and are specific interaction parameters to the extent to which the Guntelberg term represents nonspecific interactions. We saw earlier in this book that this is not entirely true. Therefore, the errors in the first term are incorporated into the B coefficients. Still, the method works well up to $m = 0.1$ (Robinson and Stokes, 1968).

For two electrolytes ($B = MX$, $C = NY$) the equation has the form (Robinson and Stokes, 1968)

$$\ln_{\pm B} = -\frac{A_D |z_M z_X| I^{0.5}}{1 + I^{0.5}}$$
$$[2X_{MA} + (B_{M'A'} + B_{MA'})(1 - X_{MA})m] \quad (5.485)$$

$X_{MA} m$ is the molality of MA and $(1 - X_{MA})m$ is that of $M'A'$.

Guggenheim (1935, 1936) developed his equation by adding the contributions of the nonspecific interaction and the specific interaction to yield the total ionic interaction free energy as $G_{\text{interact}} = G_{\text{nonspec}} + G_{\text{spec}}$.

Guggenheim concluded that his equation only applied up to $I = 0.1$ because only one term was used for each electrolyte. This means that, at higher concentrations, terms beyond B_{MX} are required even for a single solution of MX.

The values of B are valid at a given ionic strength and Equation (5.351) can be used to derive Harned's rule. Whitfield (1973) used this early model to

TABLE 5.92 Specific interaction coefficients β for the Brønsted–Guggenheim equation (Lewis and Randall, 1961) at 25°C.

HCl	0.27	$NaClO_4$	0.13
HBr	0.33	$NaBrO_3$	0.01
HI	0.36	$NaIO_3$	
$HClO_4$	0.30	$NaNO_3$	0.04
HNO_3		RbCl	0.06
LiCl	0.22	CsCl	0.0
NaF	0.07	$TlNO_3$	−0.36
NaCl	0.15		
$NaClO_3$	0.10		

calculate mean and single-ion activity coefficients in seawater. The results are shown in Table 5.91. Note that β_{MX} is related to B_{MX} by

$$\beta_{MX} = \frac{2.303 B_{MX}}{2} = 1.151 B_{MX} \quad (5.486)$$

in specific interaction studies.

Specific interaction coefficients of the Brønsted–Guggenheim equation are shown in Table 5.92.

B and $K^{*\prime}$ can be shown to be related for a single ion, such as SO_4^{2-}, by

$$B = \frac{K^{*\prime}[SO_4^{2-}]_F}{2.303[SO_4^{2-}]_T} \quad (5.487)$$

where B and $K^{*\prime}$ pertain to an interaction Na^+–SO_4^{2-}.

5.9.5 The Pitzer Theory

The method proposed by Pitzer (1973) is essentially an extension of that of Brønsted and Guggenheim and is also related to the statistical thermodynamic approach.

The Debye–Hückel theory recognizes a distance of closest approach a_D between ions but does not take into account the effects of the interactions between the hard cores of the ions on the osmotic pressure and on other properties of ionic solutions. Note that there can be cosphere overlap between ions so that the hard core may be smaller than the sum of the hydrated ionic radii.

Consider an ideal gas for which $PV = nRT$ and a real one for which a series expansion (the virial

theorem) is used, in order to better understand the meaning of the kinetic effect. For the latter

$$\frac{PV}{RT} = 1 + \frac{B_V}{V} + \frac{C_V}{V^2} + \cdots \quad (5.488)$$

is used.

The first term is the ideal one, per mole of gas. The next one, the second virial coefficient, can be shown to result from specific forces between particles. Pitzer's model (1973) is based on the virial expansion of the osmotic pressure in ionic solutions and the introduction of a term for the second virial coefficient which was not considered by Brønsted and Guggenheim. Then, equations for the osmotic coefficient and the activity coefficient versus the ionic strength were derived by him.

Thus, the Pitzer theory (1973) replaces the B_{ij} terms of the Brønsted–Guggenheim equation by expressions based on the second virial coefficient. This leads to values of B_{ij} which are sensitive to I and which permit the use of the theory for molalities higher than $m = 1.0$ instead of only to $m = 0.1$. The ionic strength dependence of activity coefficients appears to be stronger at low rather than large values of I.

The full equation for activity coefficients and the values of the required parameters for a large number of electrolytes, obtained by Pitzer and coworkers, can be found in Pitzer (1979). Whitfield (1975) used a slightly simplified equation in that he deleted, according to the Brønsted postulate, terms for interactions between ions with the same charge type (positive–positive and negative–negative). The Whitfield expression is shown below:

$$\ln \gamma_{\pm MX} = \ln \gamma'_{\pm el}$$

$$+ \frac{2\nu_M}{\nu} \sum_A m_A \left[B_{Ma} + \left(\sum_C m_C z_C \right) C_{MA} \right]$$

$$+ \frac{2\nu_X}{\nu} \sum_C m_C \left[B_{CX} + \left(\sum_C m_C z_C \right) C_{CX} \right] \quad (5.489)$$

$$+ \sum_C \sum_A m_C m_A \left[|z_M z_X| B_{CA} + \left(\frac{2\nu_M z_M}{\nu} \right) C_{CA} \right]$$

This equation includes the effect of the third virial coefficient in the last term. Its form is that of the Brønsted–Guggenheim one, except for the variation of the B_{ij} term with I and for the presence of the C_{ij} term. MX is the electrolyte of interest while C

TABLE 5.93 Parameters for the Pitzer equations (Pitzer and Mayorga, 1973, 1974). See Equation (5.486) for the relationship between B and β.

	$\beta^{(0)}$	$\beta^{(1)}$	$\beta^{(2)}$	$C^{(\phi)}$	Valid up to
NaCl	0.0765	0.2664	0	0.00127	6 m
KCl	0.04835	0.2122	0	−0.00084	4.8
MgCl$_2$	0.3524	1.6815	0	0.00519	4.5
CaCl$_2$	0.3159	1.614	0	−0.00339	2.5
Na$_2$SO$_4$	0.01958	1.113	0	0.00498	3.5
K$_2$SO$_4$	0.04995	0.7793	0	0	0.7
MgSO$_4$	0.221	3.343	−37.23	0.025	3.0
CaSO$_4$	0.20	2.65	−55.7	0	0.011

Source: Reprinted with permission from Pitzer, K. S. and G. Mayorga, *J. Phys. Chem.* **77**, 2300. Copyright © 1973 American Chemical Society.

TABLE 5.94 Values of $(\gamma_{\pm})_T$ in solutions that simulate seawater, at 25°C and $I = 0.7$.

	(1)	(2)	(3)	(4)
NaCl	0.666	0.667	0.661	0.670
KCl	0.649	0.648	0.642	0.645
MgCl$_2$	0.467	0.473	0.467	
CaCl$_2$	0.457	0.463	0.458	
Na$_2$SO$_4$	0.376	0.373	0.354	0.378
K$_2$SO$_4$	0.365	0.358	0.351	0.352
MgSO$_4$	0.167	0.167	0.129	
CaSO$_4$	0.162	0.161	0.125	

(1) Whitfield (1973) using the Brønsted–Guggenheim model.
(2) Whitfield (1975) using the Pitzer model.
(3) Johnson and Pytkowicz (1979a) ion-pair model.
(4) Experimental.

and A are the running indices for all the other cations and anions present. For a single-electrolyte solution

$$\ln \gamma_{\pm MX} = \ln \gamma'_{\pm el} + B_{MX} m_{MX}$$
$$+ v(v_M v_X)^{0.5} C_{\pm MX} m_{MX} \tag{5.490}$$

Whitfield (1975) uses the values of the parameters for the above equations from Pitzer and Mayorga (1973, 1974) which are shown in Table 5.93. Extensive tabulations of the parameters can be found in Pitzer (1979).

Some values of calculated and experimental activity coefficients are shown in Tables 5.94 and 5.95. The values of $\gamma_{\pm T}$ and $(\gamma_i)_T$ are calculated from parameters obtained in simple solutions. The absence of HCO_3^- and CO_3^{2-} from the artificial seawaters has only a slight effect upon the γ's due to the low concentrations of these two ions.

5.10 METHOD OF THE CANONICAL ENSEMBLE

5.10.1 The Microcanonical Ensemble

The material that follows will only be of use to those readers who wish to obtain a foundation for a deep understanding of the application of statistical mechanics to solutions.

The introductory part to this section is a continuation of the statistical thermodynamics treated in Chapter 4. It is used for interpretations of $(\gamma_{\pm})_T$.

In the microcanonical ensemble N, the number of particles, V, the volume of the system, and E, the internal energy are fixed. The physical system is isolated and so are the members of the ensemble that represent the possible states of the system. I use the term "isolated" to mean that the system is surrounded by a rigid, impermeable, and adiabatic wall.

For this ensemble

$$S = k_B \ln \Omega(N, V, E) \tag{5.491}$$

where Ω is the number of states available in the ensemble. Each state of the system and member of the ensemble corresponds to a distribution of the particles of the physical system among energy levels in such a way as to have the total energy E.

The values of μ_i and P for this ensemble are

$$-\frac{\mu_i}{k_B T} = \left(\frac{\partial \ln \Omega}{\partial n_i}\right) E, V, N_j \tag{5.492}$$

and

$$\frac{P}{k_B T} = \left(\frac{\partial \ln \Omega}{\partial V}\right) E, N \tag{5.493}$$

where i represents one kind of particle.

5.10.2 The Canonical Ensemble

The walls of the ensemble are rigid adiabatic and those of the members are rigid diathermic. Thus, heat can be transferred among all members. Weak interactions between the particles of the system are permitted. The canonical ensemble can be inter-

TABLE 5.95 Estimated single-ion activity coefficients $(\gamma_i)_T$ at 25°C in artificial seawaters of $I = 0.7$.

	Whitfield (1973)	Whitfield (1975)	Millero and Schreiber (manuscript)	Based on Experiments
Na^+	0.65	0.64	0.708	0.708
K^+	0.62	0.61	0.668	0.662
Mg^{2+}	0.22	0.22	0.269	0.257
Ca^{2+}	0.20	0.21	0.241	0.22
Cl^-	0.69	0.69	0.628	0.628
SO_4^{2-}	0.12	0.13	0.103	0.104

preted as corresponding to each particle of the physical system or as the N particles being a member of the ensemble.

The ensemble has the Boltzmann distribution

$$\frac{N_i}{N} = \frac{e^{-\varepsilon_i/k_B T}}{\sum_i e^{-\varepsilon_i/k_B T}} \qquad (5.494)$$

when particles are considered to be members. ε_i is the energy per particle of type i. The average ε is

$$\bar{\varepsilon} = \frac{\sum_i \varepsilon_i N_i}{\sum_i N_i} \qquad (5.495)$$

In the continuum notation when N particles constitute a member,

$$\bar{E} = \frac{\int_V E e^{-E/k_B T} dV}{\int_V e^{-E/k_B T} dV} \qquad (5.496)$$

where the denominator is the partition function. V is the volume in the phase space of ordinates and momenta. The configuration integral, a useful function, is

$$\int e^{-E_p/k_B T} dq_1 \cdots dq_N \qquad (5.497)$$

The grand canonical ensemble has (V, T, μ) constant but will not be discussed here. Next, let us apply the canonical ensemble to electrolyte solutions.

5.10.3 Applications of the Canonical Ensemble

Early efforts to apply the method of the canonical ensemble were reviewed by Prigogine (1939) who

concluded that the canonical ensemble cannot be applied rigorously to concentrated solutions. I shall present an additional and serious inadequacy of the canonical method in this section. Prigogine, furthermore, concluded that concentrated solutions could only be modeled rigorously in terms of ionic lattices and first proposed the concept of a degree of order that depends on the concentration. His lattice model, as we shall see later, is conceptually unsatisfactory because he felt compelled as did Kirkwood (1936), to obtain an equation that could be reduced to that of Debye and Hückel at low concentrations.

Let us examine the main features of the method of the canonical ensemble as used by Fowler (1925), Kirkwood (1936), Fowler and Guggenheim (1939), Kirkwood and Poirier (1954), and Frank and Thompson (1959b). I will not attempt to isolate their individual procedures because the overlap among them is large and would lead to excessive repetition, except in the ways in which these authors, and later Glueckauf (1969), used their results to infer the possibility of quasilattices in solution.

In essence the method consists in fixing two ions or larger clusters at a time and calculating the total energy of interaction among all the ions in solution when the ions outside of the cluster of two or more are allowed to assume all possible configurations. One then calculates a Boltzmann-type probability for finding the fixed ions at a certain distance from each other, a generalization of the procedure of Debye and Hückel. This leads to the distribution of net charge density, a generalized Poisson equation and, therefore, to the average potential of interaction and to activity coefficients. The main weakness of the method is that limited numbers of ions within the clusters (the central ion in the simple Debye–Hückel case) are assumed to control the ion cloud while in reality each and every ion affects each and every other ion, a feature of the lattice model.

Consider N ions of various types in a solution of

fixed volume V at a given temperature T, the conditions for a canonical ensemble. The total coulombic energy of interaction for any one configuration of all the ions, which is a member of the ensemble, is

$$E^{(c)} = G' = \sum_{j+i}^{N} \sum_{i=1}^{N} \frac{z_i z_j}{2 D_e r_{ij}} r_{ij} \quad (5.498)$$

with $r_{ij} \geqslant b$. The factor 2 appears so that the i–j bond is not counted twice in the summation. In strongly interacting systems, it is not exactly true that $E^{(c)}$ can be built only from binary interaction terms.

Next, consider two specific ions α and β. The argument that follows is only rigorous if there are no long-range forces in the system (Fowler, 1925), and is, therefore, only approximate for the coulombic interaction of dissolved ions. The probability of finding β in a volume element dV anywhere in the solution is dV/V. If, however, we fix α in a volume element of dV_α, then the probability of finding β in a volume element dV_β situated at a distance $r_{\alpha\beta}$ from α is

$$\alpha\beta_{\bar{p}(r_{\alpha\beta})} = e^{-\bar{E}_{\alpha\beta}^{(c)}/k_B T} \frac{dV_\beta}{V} \quad (5.499)$$

$E_{\alpha\beta}^{(c)}$ represents the average energy due to all ions when α and β are fixed and the remaining $N - 2$ ions are allowed to occupy all possible configurations within V minus that part of the solution occupied by the hard cores of the ions. In other words, $E_{\alpha\beta}^{(c)}$ is the work required to bring β from infinity to a given site distant $r_{\alpha\beta}$ from α, averaged over all possible configurations of the other ions. The term hard core is used in the sense that the potential is infinite for an interionic distance smaller than b and finite for $r_{ij} \geqslant b$.

Then,

$$e^{-E_{\alpha\beta}^{(c)}/k_B T} = \frac{V^2 \int \cdots \int e^{-E^{(c)}/k_B T} dV^{(N-2)}}{\int \cdots \int e^{-E^{(c)}/k_B T} dV^{(N)}} \quad (5.500)$$

V^2 is a normalizing factor because the integral in the numerator is extended only over the volumes occupied by the $N - 2$ ions and excludes dV_α and dV_β. The denominator is the configurational integral

$$Z(T, V, N) = \int \cdots \int e^{-E^{(c)}/k_B T} dV^{(N)} \quad (5.501)$$

with

$$dV^{(N)} = \Pi_A \, dV_A \, \Pi_B \, dV_B \cdots \quad (5.502)$$

for N_A ions of type A, N_B ions of type B, and so on, with $N = N_A + N_B + \cdots$. The symbol Π_i indicates a product over all ions of type i.

If we take the logarithms of the terms in Equation (5.500) and differentiate the result relative to x_β, one of the coordinates of the position occupied by β, then

$$\frac{\partial \bar{E}_{\alpha\beta}^{(c)}}{\partial x_\beta} = \frac{\int \cdots \int \frac{\partial E^{(c)}}{\partial x_\beta} e^{-E^{(c)}/k_B T} dV^{(N-2)}}{\int \cdots \int e^{-E^{(c)}/k_B T} dV^{(N-2)}} \quad (5.503)$$

as $Z(T, V, N)$ extends over all possible configurations and is, therefore, independent of x_β.

If $\alpha\beta_{\bar{Y}}$ represents the average value of a generic property which depends on the positions of the ions α and β, and on all possible configurations of the other ions, then

$$\alpha\beta_{\bar{Y}} = \frac{\int \cdots \int Y e^{-E^{(c)}/k_B T} dV^{(N-2)}}{\int \cdots \int e^{-E^{(c)}/k_B T} dV^{(N-2)}} \quad (5.504)$$

as is shown for example, by Fowler and Guggenheim (1939). By comparing Equations (5.503) and (5.504) we see, therefore, that $\bar{E}_{\alpha\beta}^{(c)}$ is the coulombic potential energy of the average force acting on the ion β, situated at a distance $r_{\alpha\beta}$ from α, when α and β are held fixed while all the remaining $N - 2$ ions are allowed to occupy all possible configurations.

I will make an important critique at this time before proceeding with the development of the canonical method. The distribution function $\exp(E_{\alpha\beta}^{(c)}/k_B T)$ which enters into the probability of finding ion β at a distance $r_{\alpha\beta}$ from α is a form of the Boltzmann law in which the effects of pairs of ions on all the other ions are taken into consideration, whereas in the Debye–Hückel theory only the effects of single ions (the central ions J) were included. The Debye–Hückel theory was limited in that the particular form of the Boltzmann law used with it was rigorously valid only for ions that did not interact except by means of elastic collisions. The pairwise form of the Boltzmann law used in the more general canonical approach is limited in that no long-range forces, such as those encountered in ionic solutions, are rigorously acceptable (Fowler and Guggenheim, 1939).

An even more serious limitation is that Equation (5.500) implies that only the ions α and β shape the ion cloud, an extension of the Debye–Hückel case in which J (or α in the nomenclature being used now by me) shaped it. The treatment is still not entirely rigorous even if clusters containing more than two ions are used because each and every ion in the solution affects each and every other ion, due to the presence of long-range forces. Hence, all ions simultaneously affect and are part of the ionic cloud and the only rigorous solution, which takes all ions into consideration, can only be reached by a lattice-type model. Although I reached this conclusion independently, the credit for first obtaining it must be given to Professor I. Prigogine.

Let us now continue with the canonical method because, although it does not yield an exact solution, it does point out the existence of lattices at high concentrations and is, therefore, an integral part of this chapter.

The fundamental approximation of Debye and Hückel corresponds in the canonical notation to

$$\bar{E}^{(c)}_{\alpha\beta} = z_\beta e^{\alpha\bar{V}_{(r_{\alpha\beta})}} \tag{5.505}$$

where $\alpha\bar{V}_{(r_{\alpha\beta})}$ is the average potential at a point distant $r_{\alpha\beta}$ from α due to all possible configurations of $N-1$ ions when only α, the central ion, is fixed as the center of the system of coordinates and is the only ion to affect the continuous ion cloud. The canonical assumption

$$\bar{E}^{(c)}_{\alpha\beta} = z_\beta e^{\alpha\beta\bar{V}_{(r_{\alpha\beta})}} \tag{5.506}$$

does reduce to Equation (5.505) at very high dilutions (Kirkwood, 1936).

The average charge density at dV_β in the canonical ensemble is

$$\alpha\beta\bar{\rho}_{(r_{\alpha\beta})} = \frac{\sum\limits_{\beta} z_\beta e^{\alpha\beta\bar{\rho}_{(r_{\alpha\beta})}}}{dV_\beta} \tag{5.507}$$

which, from Equation (5.499), is

$$\alpha\beta\bar{\rho}_{(r_{\alpha\beta})} = \frac{\sum\limits_{\beta} z_\beta e\, e^{-\bar{E}^{(c)}_{\alpha\beta}/k_BT}}{V} \tag{5.508}$$

The summation extends over all ions of the type β.

The Poisson equation then becomes

$$\nabla^2_\beta(E^{(c)}) = -\frac{4\Pi z_\beta e^{\alpha\beta\bar{\rho}_{(r_{\alpha\beta})}}}{D_e} \tag{5.509}$$

$\nabla^2_\beta(E^{(c)})$ represents the Laplacian of $E^{(c)}$ with reference to the coordinates of β. Furthermore, as was shown by Kirkwood (1936) and by Frank and Thompson (1959b),

$$\nabla^2[\alpha\beta\bar{V}_{(r_{\alpha\beta})}] = \nabla^2_\beta(E^{(c)})$$
$$-\frac{z^2_\beta e^2}{k_B\bar{T}}\sum[\alpha\beta\bar{E}^2_{x_\beta} - (\alpha\beta\bar{E}_{x_\beta})^2] \tag{5.510}$$

$\alpha\beta E_{x_\beta}$ represents the average electric field strength (not to be confused with the energy $E^{(c)}_{\alpha\beta}$ used earlier) along the coordinate x_β when α and β are fixed and the summation extends over all the coordinates x, y, and z. The term in the brackets represents the effect of fluctuations, that is, the root mean-square fluctuations of the components of E. This quantity cannot be evaluated precisely. Note that $E^{(c)}$ represents an electrical energy while E refers to the electric field.

Let us examine next how various authors interpreted the solutions of the Poisson equation in terms of the existence of lattices at high concentrations.

5.10.4 The Kirkwood Interpretation

Kirkwood (1936) and Kirkwood and Poirier (1954) found solutions for the Poisson equation which have real roots for $\kappa a_D < 1.03$. When $\kappa a_D \geqslant 1.03$, that is, at high concentrations, the roots move into the complex plane as complex conjugates. Then, the potential oscillates in a manner characteristic of the potential of mean forces in the liquid state. Kirkwood had already reached this conclusion in his earlier work and thought that the distribution function for the ions approached the form for the liquid state. He decided, therefore, that an ionic lattice of the crystalline type was unlikely. No proof was offered for this opinion and, furthermore, we have seen that the canonical method that was used by Kirkwood is not rigorous. To this lack of rigor must be added the mathematical difficulties that force workers who use the canonical approach to seek approximate solutions at high concentrations. The approximate nature of the solutions is accentuated because the form of the repulsion forces and the role of fluctuations are not known quantitatively.

5.10.5 The Method of Frank and Thompson

Frank and Thompson (1959a) reasoned that if the ion β in dV_β was subdivided into many smaller ions in a thought process then in the limit a situation would arise in which each fragment of β would make such a small contribution to the total charge, the potential, and the field strength, that $\alpha\beta_{\bar{p}(r_{\alpha\beta})}$ could be replaced by $\beta_{\bar{p}(r_{\alpha\beta})}$. In other words, only α would have a sizable effect on the ion cloud. Furthermore, the fluctuation terms in Equation (5.510) would result from so many contributions by positive and negative microparticles that the effect of fluctuations would be negligible. Note that the quasicancellation of fluctuations may perhaps be understandable from the standpoint of their effect on a central ion α, but that the same is not true in a lattice in which all ions are considered simultaneously. Under the conditions envisioned by Frank and Thompson, Equation (5.509) would be reduced to the Debye–Hückel form.

Their reasoning was then translated to actual whole ions, to provide a test of the limits of validity of the Debye–Hückel theory. Frank and Thompson examined this problem by calculating the fraction of the total cloud effect on the central ion which is produced by that part of the cloud which lies within a given distance from the central ion. As examples, at $c = 10^{-2}$ moles–L an average of 0.8 of an ion lies within a shell 25.1 Å thick around the central ion and produces 50% of the cloud effect on the potential, whereas at $c = 10^{-8}$ moles/L only 0.194 of an ion is present in a shell 1565 Å thick and it produces 5% of the total ion-cloud effect. The Debye–Hückel theory obviously cannot be valid at $c = 10^{-2}$ moles/L because the single ion (actually 0.8 of an ion) that is present within a shell 25.1 Å thick contributes such a large fraction of the ion-cloud potential on the central ion that it will necessarily introduce a contribution due to its fluctuations. Furthermore, such an ion cannot produce the time-average spherical charge distribution required by the Debye–Hückel theory.

Frank and Thompson then introduced the measure of how fine grained a solution is by comparing r_0, the average nearest neighbor distance, and $1/\kappa$, the thickness of the ion atmosphere. A large value of $(1/\kappa)/r_0$ implies a fine-grained solution and, as this ratio is unity at $c = 0.003$, this concentration was chosen as being definitely the upper limit of validity of the Debye–Hückel theory. Note that the examples quoted in the previous paragraph roughly parallel the increase in $1/\kappa$ with decreasing concentration and the concurrent increase in $(1/\kappa)r_0$. The concept of a fine-grained cloud should not be confused with the mathematical concepts of dense and continuous distributions. Therefore, differentiations and integrations of discretely varying quantities, such as the actual charge density in a fine-grained cloud, which is modeled as continuous, are not mathematically rigorous. The lattice model used by us is more nearly correct in that ions are considered to be isolated entities although infinite series are used for the calculation of the Madelung constant. This should be a good approximation since the number of ions in a kg of water is usually very large.

Frank and Thompson then pursued the question of an acceptable model at $c > 0.003$ and, as was mentioned earlier, concluded that a law cubic in the molarity c was correct. This was interpreted by them and others as an indication of local long-range order at concentrations as low as $c = 0.001$ with the replacement of long-range order of a Debye–Hückel type by liquidlike nearest neighbor interactions. However, such a relationship lacks physical significance in the range of concentrations examined by Frank and Thompson because in this range a' and b' are not expected to be theoretically constant and an inflexion occurs which only simulates a straight line (see Pytkowicz and Johnson, 1979).

Glueckauf observed that, roughly speaking, the Kirkwood solution of the Poisson equation for 1-1 electrolytes can be represented by the approximate equation

$$\ln \gamma_{\pm s} = -A_D \kappa a_D \left(\frac{1 + 0.5\kappa a_D}{1 + \kappa a_D}\right)^{0.5} \quad (5.511)$$

from $\kappa a_D = 0.13$ (i.e., $m \cong 0.01$) to $\kappa a_D = 1.0$ (i.e., $m \cong 0.6$). Furthermore, he concluded that Equation (5.511) begins to deviate slightly from that of Kirkwood at about $\kappa a_D = 0.8$ ($m \cong 0.4$) and, as m increases further, yields results that correspond to an equation of the type

$$-\ln \gamma_{\pm s} = kc^{1/3} \quad (5.512)$$

where k is a proportionality constant. Glueckauf expected such an equation because, on the basis of the conclusions of Frank and Thompson and from his interpretation of the suggestions of Kirkwood (1936), he anticipated a quasilattice at $\kappa a_D \geqslant 0.8$ (i.e., $m \geqslant 0.4$). We have seen earlier that equations such as (5.512) have no theoretical meaning unless they become representative at values of m well above the transition from salting-in to salting-out. We shall see shortly that even then such equations do not describe the behavior of activity coefficients.

5.10.6 Cluster Integrals

Next, I shall briefly mention the results of a theoretical statistical thermodynamic school founded by McMillan and Meyer (1945). These authors used the correlation function g_{ij} which gives the probability of finding an ion j at a distance r from i. This function is of course related to the energy of interaction of i and j and is zero for ideal gases. It is related to the Boltzmann factor, for very weak interactions, by

$$g_{ij} = \exp \frac{-E_{ij}}{RT} \qquad (5.513)$$

In dense fluids and in electrolyte solutions there are not only pairwise interactions but also higher order ones. Still, the above equation can be solved if E_{ij} is replaced by an effective energy $E_{ij}^{(e)}$. The central problem of this approach is the evaluation of $E_{ij}^{(e)}$. Once this term is known as a function of the interionic distance, the partition function and, hence, the thermodynamic properties, can be calculated.

Several methods have been employed for the determination of $E_{ij}^{(e)}$. In the cluster integral expansion the numbers of pairwise interactions for larger and larger clusters of ions are counted and, from the interaction energy between pairs weighted by their frequency, one arrives at an approximate solution. As an example, for ion triplets there are the following configurations.

The open terms diverge while the closed ones converge. The overall value of a thermodynamic function converges when the Boltzmann term, with the overall energy, is integrated.

In another technique, the Monte Carlo (Card and Valleau, 1970), a computer determines an enormous number of possible configurations and energies but is limited to a small number of ions. The computer then calculates $E_{ij}^{(e)}$.

The form of the pairwise energy is a challenging one. One may use as a model

$$E_{ij}^{(e)} = \frac{z_i z_j \varepsilon^2}{D_e r} + \text{CORE} + \text{CAVITY}$$
$$+ \text{GURNEY} \qquad (5.514)$$

as was done by Ramanathan and Friedman (1971). The CORE term represents the repulsion between ions and represents a hard core if the ion, like a billiard ball, cannot be penetrated by its counterion. The model is called the primitive one if only the first two terms of the above equation are taken into consideration. Ramanathan and Friedman used the Gurney cosphere overlap as a curve-fitting parameter to adjust the theoretical to the experimental activity coefficient.

5.10.7 The Method of Harned and Robinson

An extensive examination of activity and osmotic coefficients in multielectrolytes was presented by Harned and Robinson (1940). The core of the book is the use of series expansions and multiterm expressions to make the transition from binary (water plus electrolyte) to multicomponent solutions. The approach lacks the theoretical cohesion given to multielectrolyte solutions by the models of Pitzer (1973) and Johnson and Pytkowicz (1978, 1979a, 1979b).

The uses of the cluster integral expansion and the radial distribution function can be found in Mazo and Mov (1979) and more specific derivations and calculations are present in Ramanathan and Friedman (1971), Reilly et al. (1971), and Robinson and Wood (1972). Robinson and Wood obtained values for seawater in fair agreement with those calculated from the other models.

5.11 A WORD ON PARTIAL MOLAL VOLUMES IN SEAWATER

I present in this section estimates of partial molal volumes in seawater because they are relevant to the study of equilibria in seawater at pressure, to be studied in Chapters 2 and 5, Volume II.

These approaches are quite different from the simple thermodynamic method for dilute solutions (Owen and Brinkley, 1941). Duedall (1966) determined conventional partial equivalent volumes of the major ions of seawater relative to $\bar{v}_{Na}$. Thus,

$$\bar{v}_{K-Na} = \bar{v}_{KCl} - \bar{v}_{NaCl} \qquad (5.515)$$

and

$$\bar{v}_{SO_4-Cl} = \bar{v}_{Na_2SO_4} - \bar{v}_{NaCl} \qquad (5.516)$$

This method does not lend itself to the study of the effect of pressure on equilibria. Thus, for a reaction $CA_{(s)} = C + A$, the value of $\Delta\bar{v}$ is $\bar{v}_C + \bar{v}_A - v_{CA}$, where the last term is the specific volume. $\partial \ln K_s/\partial P$ is $-\Delta\bar{v}/RT$. The Duedall terms yield $\Delta\bar{v}_D = \bar{v}_C + \bar{v}_A + 2\bar{v}_{Cl} - v_{CA}$.

Duedall (1968) measured values for the $\bar{v}$ of 16 salts in seawater. These results are useful for calculations of the effects of pressure on equilibria in homogeneous solutions and in the case of the solubility of pure solids without surface coatings.

Millero (1969) introduced a model in which $\bar{v}^0$ (ion) = $\bar{v}^0$ (int) + $\bar{v}^0$ (electr). The term $\bar{v}^0$ (int) is the intrinsic volume of the ion, related to the crystal radius plus the change in volume due to the local breakdown in the structure of water. This breakdown occurs between the hydrated and the bulk water. The symbol $\bar{v}^0$ (electr) corresponds to the electrostriction during the hydration of the ion. Millero then calculated the values of $\bar{v}^0$ (ion)^{sw} for ions at the seawater reference state from $\bar{v}^0$ (ion). This state is one of infinite dilution of the ion in the seawater. A convention was required to obtain $\bar{v}^0$

(ion)^{sw} for single ions. $\bar{v}^0$ (H$^+$)^{sw} = -3.7 mL/mole was selected in order to obtain a smooth curve for $\bar{v}^0$ (ion)^{sw} versus r (ion)^3. Thus, the approach is empirical in practice like single-ion γ's.

Millero (1969) then used departures from an expected correlation to calculate $\bar{v}$'s for free ions and, later on, to calculate the effect of pressure on various properties such as the extent of ionic association (e.g., Millero, 1971).

5.12 INTERCOMPARISONS

Intercomparisons between values of $(\gamma_\pm)_T$, $(\gamma_i)_T$, and of the speciation of seawater are shown in Tables 5.96, 5.97, and 5.98.

The fair agreement between the values of $(\gamma_\pm)_T$ occurs because the methods yield $(\gamma_\pm)_F$ or specific interactive coefficients which absorb inaccuracies in the theories. In the case of ion pairs the methods are such that, if $(m_F/m_T)^{0.5} = x\%$ of the true value for a symmetric electrolyte, then $(\gamma_\pm)_T/(\gamma_\pm)_F$ is also $x\%$ of its correct value. Therefore, $(\gamma_\pm)_T = (\gamma_\pm)_F(m_F/m_T)^{0.5}$ is insensitive to errors.

The large difference in Table 5.97 is due to the formation of Cl$^-$ ion pairs found in the work of Johnson (1979). Different theories yield different speciations which are observable primarily in the low-concentration anions. The speciations are important in the explanations of the physicochemical properties of solutions.

TABLE 5.96 Intercomparison of $(\gamma_\pm)_T$ results for seawater of $S‰ \cong 35$ and $T = 25°C$.

Salt	(1)	(2)	(3)	(4)	(5)	(6)	(7)
KCl	0.648	0.627	0.649	0.648	0.648	0.621	0.645
NaCl	0.661	0.666	0.666	0.661	0.661	0.659	0.667
MgCl$_2$	0.467	0.454	0.467		0.467	0.463	
CaCl$_2$	0.458	0.455	0.457			0.445	
Na$_2$SO$_4$	0.354	0.323	0.378	0.354	0.366	0.350	0.378
K$_2$SO$_4$	0.357	0.299	0.365	0.347	0.345	0.324	0.352
MgSO$_4$	0.129	0.131	0.167		0.158	0.150	
CaSO$_4$	0.125	0.127	0.162	0.151		0.142	0.136

(1) Pytkowicz, Atlas, and Culberson (1977)—ion pairing without chloride pairs.
(2) Berner (1971)—mean-salt method.
(3) Whitfield (1975)—specific interaction model.
(4) Johnson and Pytkowicz (1979a)—ion pairing with chloride pairs.
(5) Robinson and Bates (1978)—hydration theory.
(6) Kester and Pytkowicz (1969)—ion pairing.
(7) Experimental.

TABLE 5.97 Total activity coefficients of single ions at about 35‰ salinity and 25°C.

Reference	K^+	Na^+	Ca^{2+}	Mg^{2+}	Cl^-	SO_4^{2-}	HCO_3^-	CO_3^{2-}
Pytkowicz, Atlas, and Culberson (1977)	0.618	0.695	0.225	0.254	0.625	0.084	0.501	0.030
Berner (1971)	0.624	0.703	0.237	0.252	0.630	0.068		
van Breemen (1973)	0.620	0.695	0.228	0.254	0.630	0.090		
Whitfield (1973)	0.617	0.650	0.203	0.217	0.686	0.122		
Leyendekkers (1973)	0.630	0.680	0.214	0.234	0.658	0.108		

5.13 CONCLUSIONS AND RECOMMENDATIONS

5.13.1 Standard and Reference States

Standard State

For a solvent—the pure substance.
For a solute—ideal behavior with $\gamma = 1$ and $m = 1$.

This is a fictitious state. γ is the activity coefficient and m is the molality.

Reference State

For a solute—infinite dilution for which $\gamma = 1$.

Determinations of γ at a molality m are usually calculated from the reference state to m.

5.13.2 Activities, Activity Coefficients, and the Chemical Potential

The chemical potential $\mu = (\partial F/\partial n_i)_{T,P,n_j}$ is related to the activity a_i by

$$\mu_i = \mu_i^0 + RT \ln a_i \qquad a_i = \gamma_i m_i$$

where μ_i^0 is the value of μ at the standard state.

5.13.3 Mean Quantities for an Electrolyte $C_c A_a$

Activity

$$a_\pm = [(a_C)^{v^+}(a_A)^{v^-}]^{1/v} \qquad v = v_+ + v_-$$

Activity Coefficient of an Electrolyte

$$\gamma_\pm = \frac{a_\pm}{m_\pm} = [(\gamma_C)^{v^+}(\gamma_A)^{v^-}]^{1/v}$$

5.13.4 Free and Total Activity Coefficients

The coefficients are related by

$$a = \gamma_F m_F = \gamma_T m_T$$

F indicates free ions and T applies to free ions plus ion pairs. This equation can be used for γ_i, the coefficient for a single ion, or for $\gamma_\pm$.

5.13.5 Equations for Activity Coefficients

In general, theories of activity coefficients do not take into account all the processes that occur in solution. These processes are the nonspecific interactions studied by Debye and Hückel, changes in dielectric constants, hydration, ion pairing or specific interactions, ion-cavity interactions, cosphere overlap, and so on.

Most but not all the theories of activity coefficients are developed for mean activity coefficients, as is the case for the Debye–Hückel one, and can be adapted, through a nonthermodynamic assumption, to the activity coefficients of single ions. A few approaches are aimed directly at single ions. I recommend that the use of single ions be avoided when possible as only values of $\gamma_\pm$ can be verified directly by experiments.

Note that the Debye–Hückel limiting law has no fitting constants such as a_D in the extended version. It is, therefore, truly predictive and is recommended for $I < 10^{-3}$.

The Debye–Hückel Equations. The equations were derived theoretically for nonspecific interactions among ions due to coulombic forces. There are forms for mean and for individual ion coefficients, the former being

$$\ln \gamma_{\pm(DH)} = -\frac{|z_C z_A| A_D I^{0.5}}{1 + B_D a_D I^{0.5}}$$

TABLE 5.98 Speciation of seawater (percentages) obtained, in vertical order, by Garrels and Thompson (1962), Kester and Pytkowicz (1969), Hawley (1973), and Johnson (1979). Triple ions and chloride ion pairs are not shown but are counted in their effects on other ions.

| | Cations | | | |
	Na	Mg	Ca	K
Free metal	99	87	91	99
	97.7	89.0	88.5	98.8
	97.7	89.2	88.5	98.9
	84.0	50.9	45.8	78.5
MSO_4	1.2	11	8	1
	2.2	10.3	10.8	1.2
	2.2	10.3	10.8	1.1
	3.8	9.7	10.0	4.3
$MHCO_3$	0.01	0.1	1	
	0.03	0.6	0.6	
	0.1	0.3	0.3	
	0.0	0.2	0.3	
MCO_3		0.3	0.2	
		0.13	0.07	
		0.1	0.3	
		0.0	0.1	
MF		0.1		

| | Anions | | | |
	SO_4	HCO_3	CO_3	F
Free anion	54	69	9	
	39.0	70.0	9.1	
	39.0	81.3	8.0	51.0
NaX	21	8	17	
	37.2	8.6	17.3	
	37.1	10.7	16.0	
MgX	21.5	19	67	
	19.4	17.8	67.3	
	19.5	6.5	43.9	47.0
CaX	3	4	7	
	4.0	3.3	6.4	
	4.0	1.5	21.0	2.0
KX	0.4			
	0.4			
Mg_2CO_3			7.4	
$MgCaCO_3$			3.8	

The limiting case, $\ln \gamma_{\pm(DH)} = -|z_C z_A| A_D I^{0.5}$ is valid to $I = 10^{-3}$ while the extended one, which is empirical in practice, works to about $I = 0.1$. The term $I = 0.5\Sigma m_i z_i^2$ is the ionic strength.

The equations of Debye–Hückel are only valid for single-electrolyte solutions and yield γ_T.

The Hückel Equation. The Hückel (1925) equation yields little improvement over the Debye–Hückel one in practice. In theory it adds a salting-out term to the earlier equations. It only works well, however, up to $I = 0.1$ which is well within the salting-in region. It is only valid for single-electrolyte solutions.

The Guntelberg Equation. The Guntelberg equation (1926) is a simplified version of the Debye–Hückel one, adapted for use in multielectrolyte solutions (Guntelberg, 1926). It is a poor predictor of values of γ even in single-electrolyte solutions at $m < 0.05$.

The Guggenheim Relation. A better equation than that of Hückel is the one proposed by Guggenheim (1935). It also has a second term linear in I but can be applied to $m = 0.1$ in mixed electrolyte solutions.

The Culberson Equation. This equation (Pytkowicz et al., 1978) provides an empirical fit which, for single-electrolyte solutions, extends from $I = 0.1$ to 1.0. It is a useful equation for interpolations and for computer calculations.

Hydration Equation. This equation was derived by Robinson and Stokes (1968). It adds the effects of hydration on the concentrations of water and ions but does not account for other processes. In practice it is an empirical equation with two adjustable parameters, a_D and h, which works well over the full concentration range ($I < 10^{-3}$ to $I = 6.0$). It loses theoretical significance in mixed electrolyte solutions due to the use of a_D.

Davies Equation. The Davies (1962) equation is recommended in mixed electrolyte solutions for $I < 0.1$ and sometimes for larger values. It includes ionic association and becomes predictive when the degree of association is known. Remember that $\gamma_{\pm(D)}/\alpha$ should be used, where α is the degree of dissociation, if $(\gamma_\pm)_T$ is required.

Other Methods. A number of other methods are presented in the text. The ion-pair approach is described in the next section.

The specific interaction methods of Brønsted–Guggenheim (Brønsted, 1923, 1927; Guggenheim, 1935, 1936) and of Pitzer (1973) work quite well in practice (Whitfield, 1973, 1975) even though the

Guntelberg-type term is used in conjunction with the kinetic core term to describe nonspecific interactions even at high concentrations. The results of Pitzer and the ion-pair ones of Johnson and Pytkowicz (1978, 1979a, 1979b) are recommended for mixed electrolyte solutions.

Berner (1965) and Pytkowicz (1975) obtained the total activity coefficients of HCO_3^- and CO_3^{2-} with a minimum of assumptions.

Speciation Models

The Model of Garrels and Thompson. Garrels and Thompson (1962) first applied the ion-pair concept to seawater and a considerable amount of work using their assumptions was done later. The method uses single-ion activity coefficients.

In essence, thermodynamic association constants are obtained from the literature and free activity coefficients of single ions and ion pairs are estimated roughly. Some of the activity coefficients are calculated by the mean-salt method, based on the MacInnes (1919) assumption to the effect that $\gamma_K = \gamma_{Cl}$. These results, in conjunction with mass balance equations, yield speciation models of solutions as well as total activity coefficients. This method is fast and does not require many measurements needed in more careful work. It is only recommended for rough estimates of the speciation in natural waters.

The OSU Approach. Pytkowicz and Kester (1969), Kester and Pytkowicz (1969), Hawley (1973), and Hawley and Pytkowicz (1973) obtained the speciation of seawater experimentally and with fewer assumptions than those of Garrels and Thompson (1962). All these authors, however, neglected Cl^- ion pairs which were introduced by Johnson (1979) and Johnson and Pytkowicz (1978, 1979a, 1979b).

The specific interaction method of Pitzer (1973) and the ion-pair method of Johnson and Pytkowicz (1978, 1979a) yield activity coefficients in and properties of complex solutions from data in simpler methods.

Our method does not require earlier assumptions because it is based on potentiometric data referred to $HClO_4$ as a completely dissociated electrolyte.

$(\gamma_\pm)_F$ in the ion-pair model includes all processes, such as nonspecific interactions, hydration, changes in the dielectric constant, ion-cavity interactions, cosphere overlap, and so on, with the exception of ion pairing which is introduced in $(\gamma_\pm)_T$.

The existence of several ion pairs has been demonstrated theoretically (Kester and Pytkowicz, 1975), by sound attenuation (Fisher, 1967), Raman spectroscopy (Daley et al., 1972), and so on. That of Cl^- ion pairs is to be expected because thermal motion and coulombic attraction lead a fraction of the ions to be in contact at any one time. The association constants are small but the concentration of Cl^- is large in media such as seawater.

Usefulness of Activity Coefficients for Equilibria in Solutions. Equilibria in solutions may be approached from thermodynamic, apparent, and stoichiometric equilibrium constants.

In the case of an acid HA,

$$K = \frac{a_H a_A}{a_{HA}} = \frac{(\gamma_H)_T (\gamma_A)_T}{\gamma_{HA}} \frac{(m_H)_T (m_A)_T}{m_{HA}}$$

$$= \frac{(\gamma_H)_T (\gamma_A)_T}{\gamma_{HA}} K'' = \frac{a_H (\gamma_A)_T}{\gamma_{HA}} K'$$

K is the thermodynamic, K'' is the stoichiometric, and K' is the so-called "apparent" dissociation constant.

The recommended procedures in ionic media are the direct determinations or the use of tabulated values of K' and K'', to avoid estimates of single-activity coefficients such as $(\gamma_A)_T$. If K' and K'' are not available and there is no time to determine them, then K can be used in conjunction with $(\gamma_A)_T$ and γ_{HA}.

Similar considerations apply to solubility products for which mean total activity coefficients are used, and for other types of equilibria.

SUMMARY

The summary is contained in the section on conclusions and recommendations (Section 5.13).

REFERENCES

Åkerlöf, G. (1937). *J. Phys. Chem.* **41,** 1053.

Arrhenius, S. A. (1887). *Z. Phys. Chem.* **1,** 481.

Atlas, E. L. (1976). Ph.D. Thesis, Oregon State University, Corvallis.

Atlas, E., and R. M. Pytkowicz (1977). *Limnol. Oceanogr.* **22,** 290.

Bates, R. G., B. R. Staples, and R. A. Robinson (1970). *Anal. Chem.* **42**, 867.

Berner, R. A. (1965). *Geochim. Cosmochim. Acta* **29**, 947.

Berner, R. A. (1971). *Principles of Chemical Sedimentology,* McGraw-Hill, New York.

Bjerrum, M. (1926). *Z. Physik. Chem.* **119.**

Bjerrum. M., and A. Unmack (1929). *Danske Vidensk. Selsk. Math-Fys.* **9,** 1.

Bockris, J. O'M., and A. K. N. Reddy (1970). *Modern Electrochemistry,* Vol. 1, Plenum Press, New York.

Boltzmann, L. (1868). *Wiener Berichte* **58**, 517.

Brønsted, J. N. (1922). *J. Am. Chem. Soc.* **44**, 877.

Brown, A. S., and D. A. MacInnes (1935). *J. Am. Chem. Soc.* **57**, 1356.

Butler, J. N., and Huston, R. (1970). *J. Phys. Chem.* **74**, 2976.

Card, D. N., and J. P. Valleau (1970). *J. Chem. Phys.* **52**, 6232.

Chin, Y.-C., and R. M. Fuoss (1968). *J. Phys. Chem.* **72**, 4123.

Covington, A. K., M. A. Hakeem, and W. F. K. Wynne-Jones (1963). *J. Chem. Soc.,* 4394.

Culberson, C., R. M. Pytkowicz, and J. E. Hawley (1970). *J. Mar. Res.* **28,** 15.

Culberson, C., and R. M. Pytkowicz (1973). *Mar. Chem.* **1**, 309.

Culberson, C., G. Latham, and R. G. Bates (1978). *J. Phys. Chem.* **82**, 2693.

Daly, F. P., C. W. Brown, and D. R. Kester (1972). *J. Phys. Chem.* **76**, 3664.

D'Aprano, A. (1971). *J. Phys. Chem.* **75**, 3290.

Darken, L. S. (1950). *J. Am. Chem. Soc.* **72**, 2909.

Davies, C. W. (1938). *J. Chem. Soc.,* 2093.

Davies, C. W. (1962). *Ion Association,* Butterworths, London.

Davies, C. W., and A. L. Jones (1955). *Trans. Faraday Soc.* **51**, 812.

Davies, A. R., and B. G. Oliver (1973). *J. Phys. Chem.* **77**, 1315.

Debye, P., and E. Hückel (1923). *Physik. Z.* **9**, 185.

Debye, P., and J. McCauley (1925). *Physik. Z.* **26**, 22.

Dickson, A. G., and J. P. Riley (1979). *Mar. Chem.* **7**, 89.

Downes, C. J. (1970). *J. Phys. Chem.* **74**, 2153.

Duedall, I. (1966). M.S. Thesis, Oregon State University, Corvallis.

Duedall, I. (1968). Environ. Sci. Tech. **2**, 706.

Dunsmore, H. S., and G. H. Nancollas (1958). *J. Chem. Soc.,* 4144.

Dyrssen, D., and M. Wedborg (1973). In *The Sea,* Vol. 5, E. D. Goldberg, Ed., Interscience, New York, p. 181.

Elgquist, B., and M. Wedborg (1975). *Mar. Chem.* **3**, 215.

Fisher, F. H. (1967). *Science* **157**, 823.

Fisher, F. H. (1978). *J. Soln. Chem.* **7**, 897.

Fisher, F. H., and A. P. Fox (1977). *J. Soln. Chem.* **6**, 641.

Fowler, R. H. (1925). *Proc. Cambridge Phil. Soc.* **22**, 861.

Fowler, R. H., and E. A. Guggenheim (1939). *Statistical Thermodynamics,* Cambridge University Press, Cambridge.

Frank, H. S., and M. W. Wen (1957). *Discuss. Faraday Soc.* **24**, 133.

Frank, H. S., and P. T. Thompson (1959a). *J. Chem. Phys.* **31**, 1086.

Frank, H. S., and P. T. Thompson (1959b). In *The Structure of Electrolyte Solutions,* W. J. Hamer, Ed., Wiley, New York.

Fuoss, R. M. (1958). *J. Am. Chem. Soc.* **80**, 5059.

Fuoss, R. M., and L. Onsager (1957). *J. Phys. Chem.* **61**, 668.

Fuoss, R. M., and K.-L. Hsia (1967). *Proceed. Nat'l Acad. Sci.* **57**, 1550.

Garrels, R. M., M. E. Thompson, and R. Siever (1960). *Am. J. Sci.* **259**, 24.

Garrels, R. M., and M. E. Thompson (1962). *Amer. J. Sci.* **260**, 57.

Gieskes, J. M. T. M. (1966). *Zeitscher. Physik. Chemie Neve Folge* **50**, 78.

Glandsdorff, P., and I. Prigogine (1971). *Thermodynamic Theory of Structure, Stability and Fluctuations,* Interscience, New York.

Glueckauf, E. (1969). *Proc. Royal Soc. Ser. A.* **310**, 449.

Greenwald, I. (1941). *J. Biol. Chem.* **141**, 789.

Guggenheim, E. A. (1935). *Philos. Mag.* **19**, 588.

Guggenheim, E. A. (1936). *Applications of Statistical Mechanics,* Clarendon Press, Oxford.

Guntelberg, E. (1926). *Z. Physik. Chem.* **123**, 199.

Gurney, R. W. (1962). *Ionic Processes in Solutions,* Dover, New York.

Hamer, W. J. (1959). *The Structure of Electrolyte Solutions,* Wiley, New York.

Hamer, W. J. (1968). *Theoretical Mean Activity Coefficients of Strong Electrolytes in Aqueous Solutions from 0 to 100°C.,* National Standard Reference Data Series, National Bureau of Standards, Publ. No. 24, Washington.

Hansson, I. (1973). *Deep-Sea Res.* **20**, 461.

Harned, H. S. (1926). *J. Am. Chem. Soc.* **48**, 326.

Harned, H. S., and F. E. Swindells (1926). *J. Am. Chem. Soc.* **48**, 126.

Harned, H. S., and O. E. Schupp (1926). *J. Am. Chem. Soc.* **52**, 3892.

Harned, H. S., and H. R. Copson (1933). *J. Am. Chem. Soc.* **55**, 2206.

Harned, H. S., and R. A. Robinson (1940). *Trans. Faraday Soc.* **36**, 973.

Harned, H. S., and T. R. Paxton (1953). *J. Phys. Chem.* **57**, 531.

Harned, H. S., and R. Gary (1954). *J. Am. Chem. Soc.* **76**, 5924.

Harned, H. S., and R. Gary (1955). *J. Am. Chem. Soc.* **77**, 1994.

Harned, H. S., and B. B. Owen (1958). *The Physical Chemistry of Electrolyte Solutions,* Reinhold, New York.

Hasted, J. B., R. M. Ritson, and C. N. Collins (1948). *J. Chem. Phys.* **16**, 1.

Hawley, J. E. (1973). Ph.D. Thesis, Oregon State University, Corvallis.

Hawley, J. E., and R. M. Pytkowicz (1973). *Mar. Chem.* **1**, 245.

Horne, R. A. (1969). *Marine Chemistry: The Structure of Water and the Chemistry of the Hydrosphere,* Interscience, New York.

Hostetler (1963). *J. Phys. Chem.* **67**, 720.

Hückel, E. (1925). *Phys. Z.* **26**, 93.

Ingle, S. E., C. H. Culberson, J. E. Hawley, and R. M. Pytkowicz (1973). *Mar. Chem.* **1**, 295.

Johnson, K. S. (1979). Ph.D. Thesis, Oregon State University, Corvallis.

Johnson, K. S., and R. M. Pytkowicz (1978). *Am. J. Sci.* **278,** 1428.

Johnson, K. S., and R. M. Pytkowicz (1979a). *Mar. Chem.* **8,** 87.

Johnson, K. S., and R. M. Pytkowicz (1979b). In *Activity Coefficients in Electrolyte Solutions,* Vol. 2, R. M. Pytkowicz, Ed., CRC Press, Boca Raton, p. 1.

Johnson, K. S., R. M. Pytkowicz and C. S. Wong (1979). *Limnol. Oceanogr.* **24,** 474.

Joule, J. P., and W. Thomson (1853). *Proc. Royal Soc. (London)* **143,** 357.

Kay, R. L. (1962). In *Electrolytes,* B. Pesce, Ed., Pergamon Press, New York, p. 119.

Kester, D. R. (1970). Ph.D. Thesis, Oregon State University, Corvallis.

Kester, D. R. (1975). In *Chemical Oceanography,* Vol. 1, J. P. Riley and G. Skirrow, Eds., Academic Press, New York, p. 497.

Kester, D. R., I. W. Duedall, D. N. Connors, and R. M. Pytkowicz (1967). *Limnol. Oceanogr.* **12,** 176.

Kester, D. R., and R. M. Pytkowicz (1968). *J. Geophys. Res.* **73,** 5421.

Kester, D. R., and R. M. Pytkowicz (1969). *Limnol. Oceanogr.* **14,** 686.

Kester, D. R., and R. M. Pytkowicz (1975). *Mar. Chem.* **3,** 365.

Khoo, K. H., C.-Y. Chan, and T.-K. Lim (1977a). *J. Solut. Chem.* **6,** 855.

Khoo, K. H., C.-Y. Chan, and T.-K. Lim (1977b). *J. Solut. Chem.* **6,** 651.

Khoo, K. H., C.-Y. Chan, and T.-K. Lim (1978). *Faraday Trans.* **174,** 837.

Kirkwood, J. (1936). *Chem. Rev.* **19,** 275.

Kirkwood, J. G., and J. C. Poirier (1954). *J. Phys. Chem.* **58,** 591.

Kittel, C. (1959). *Introduction to Solid State Physics,* Wiley, New York.

Kortum, G. (1965). *Treatise on Electrochemistry.* Elsevier, Amsterdam.

Kossiakov, A., and D. Harker (1938). *J. Am. Chem. Soc.* **60,** 2047.

Lafon, G. M. (1969). Ph.D. Thesis, Northwestern University, Evanston.

Langmuir, D. (1968). *Geochim. Cosmochim. Acta* **32,** 835.

Latimer, W. M. (1952). *Oxidation Potentials,* Prentice-Hall, New York.

LeChatelier, H. (1885). *Compt. Rend.* **100,** 441.

Lewis, G. N. (1923). *Valence and Structure of Atoms and Molecules,* Chemical Catalog Co., New York.

Lewis, G. N., T. B. Brighton, and R. L. Sebastian (1917). *J. Am. Chem. Soc.* **39,** 2245.

Lewis, G. N., and M. Randall (1921). *J. Am. Chem. Soc.* **43,** 1112.

Lewis, G. N., and M. Randall (1961). *Thermodynamics,* revised by K. S. Pitzer and L. Brewer, 2nd Ed., McGraw-Hill, New York.

Leyendekkers, J. V. (1973). *Mar. Chem.* **1,** 75.

Lyman, J., and R. H. Fleming (1940). *J. Mar. Res.* **3,** 134.

Macaskill, J. B., and A. D. Pethybridge (1977). *J. Chem. Thermodyn.* **9,** 239.

Macaskill, J. B., R. A. Robinson, and R. G. Bates (1977). *J. Soln. Chem.* **6,** 385.

Macaskill, J. B., and R. G. Bates (1978). *J. Soln. Chem.* **7,** 433.

MacInnes, D. A. (1919). *J. Am. Chem. Soc.* **41,** 1086.

Malmberg, C. G., and A. A. Margott (1956). *J. Res. NBS* **56,** 1.

Markham, A. M., and K. A. Kobe (1941). *Am. Chem. Soc. J.* **63,** 449.

Mazo, R. M., and C. Y. Mov (1979). In *Activity Coefficients of Electrolyte Solutions,* Vol. 1, R. M. Pytkowicz, Ed., CRC Press, Boca Raton, p. 29.

McKay, H. A. C. (1952). *Nature* **169,** 464.

McKay, H. A. C. (1953). *Trans. Faraday Soc.* **49,** 237.

McKay, H. A. C., and J. K. Perring (1953). *Trans. Faraday Soc.* **49,** 163.

McMillan, W. G., and J. E. Meyer (1945). *J. Chem. Phys.* **13,** 276.

Mehrbach, C., C. H. Culberson, J. E. Hawley, and R. M. Pytkowicz (1973). *Limnol. Oceanogr.* **18,** 897.

Millero, F. J. (1969). *Limnol. Oceanogr.* **14,** 376.

Murdoch, P. G., and R. C. Barton (1933). *J. Am. Chem. Soc.* **55,** 4074.

Murkejee, P. (1966). *J. Phys. Chem.* **70,** 2708.

Murray, C. N., and J. P. Riley (1971). *Deep-Sea Res.* **18,** 533.

Nancollas, G. H., and N. Purdie (1964). *Quart. Rev. (London)* **18,** 1.

Nims, F. L. (1934). *J. Am. Chem. Soc.* **56,** 1110.

Owen, B. B., and S. R. Brinkley, Jr. (1941). *Chem. Revs.* **29,** 461.

Pauling, L. (1948). *The Nature of the Chemical Bond,* Cornell University Press, Ithaca.

Pitzer, K. S. (1973). *J. Phys. Chem.* **77,** 268.

Pitzer, K. S. (1979). In *Activity Coefficients in Electrolyte Solutions,* Vol. 1, R. M. Pytkowicz, Ed., CRC Press, Boca Raton, p. 157.

Pitzer, K. S., and G. Mayorga (1973). *J. Phys. Chem.* **77,** 2300.

Pitzer, K. S., and G. Mayorga (1974). *J. Soln. Chem.* **3,** 539.

Pitzer, K. S., and J. J. Kim (1974). *J. Am. Chem. Soc.* **96,** 5701.

Pitzer, K. S., R. N. Roy, and L. F. Silvester (1977). *J. Am. Chem. Soc.* **99,** 4930.

Platford, R. F. (1962). *J. Mar. Res.* **20,** 55.

Platford, R. F. (1965). *J. Mar. Res.* **23,** 55.

Platford, R. F., and T. Dafoe (1965). *J. Mar. Res.* **23,** 63.

Plath, D. C., K. S. Johnson, and R. M. Pytkowicz (1980). *Mar. Chem.* **10,** 9.

Prigogine, I. (1939). *Contributions à la Theorie des Electrolytes Forts,* Gauthier-Villars, Paris.

Prue, J. E. (1966). In *Chemical Physics of Ionic Solutions,* B. E. Conway and R. G. Barradar, Eds., Wiley, New York, p. 163.

Pytkowicz, R. M. (1975). *Limnol. Oceanogr.* **20,** 971.

Pytkowicz, R. M. (1979). In *Activity Coefficients in Electrolyte*

Solutions, R. M. Pytkowicz, Ed., CRC Press, Boca Raton, p. 301.

Pytkowicz, R. M., I. W. Duedall, and D. N. Connors (1966). *Science* **152**, 640.

Pytkowicz, R. M., and R. Gates (1968). *Science* **156**, 690.

Pytkowicz, R. M., and D. R. Kester (1969). *Am. J. Sci.* **267**, 217.

Pytkowicz, R. M., and J. E. Hawley (1974). *Limnol. Oceanogr.* **19**, 223.

Pytkowicz, R. M., S. E. Ingle, and C. Mehrbach (1974). *Limnol. Oceanogr.* **19**, 665.

Pytkowicz, R. M., E. Atlas, and C. H. Culberson (1975). In *Marine Chemistry in the Coastal Environment*, T. M. Church, Ed., *ACS Symposium Series* **18**, American Chemical Society, Washington, p. 1.

Pytkowicz, R. M., K. Johnson, and C. Curtis (1976). *Oregon State University School of Oceanography Report* **76**, 13.

Pytkowicz, R. M., E. Atlas, and C. H. Culberson (1978). In *Oceanography Marine Biology Annual Reviews*, Vol. 15, H. Barnes, Ed., Allen and Unwin, p. 11.

Pytkowicz, R. M., and K. Johnson (1979). In *Activity Coefficients in Electrolyte Solutions*, Vol. 1, R. M. Pytkowicz, Ed., CRC Press, Boca Raton, p. 209.

Ramanathan, P. S., and H. L. Friedman (1971). *J. Chem. Phys.* **54**, 1086.

Reardon, E. J., and D. Langmuir (1976). *Geochim, Cosmochim. Acta* **40**, 549.

Reilly, P. J., R. H. Wood, and R. A. Robinson (1971). *J. Phys. Chem.* **75**, 1305.

Robinson, R. A. (1954). *J. Mar. Biol. Assoc. U.K.* **33**, 449.

Robinson, R. A., and H. S. Harned (1941). *Chem. Rev.* **28**, 452.

Robinson, R. A., and V. E. Bower (1966). *J. Res. Natl. Bureau Stds* **70A**, 305.

Robinson, R. A., and R. H. Stokes (1968). *Electrolyte Solutions*, Butterworths, London.

Robinson, R. A., and A. K. Covington (1968). *J. Res. Natl. Bureau Stds.* **72A**, 239.

Robinson, R. A., and R. H. Wood (1972). *J. Soln. Chem.* **1**, 481.

Robinson, R. A., R. N. Roy, and R. G. Bates (1974). *J. Soln. Chem.* **3**, 837.

Robinson, R. A., and R. G. Bates (1978). *Mar. Chem.* **6**, 327.

Ruby, C., and J. Kawai (1926). *J. Am. Chem. Soc.* **48**, 1119.

Setchenov, J. (1892). *Ann. Chim. Phys.* 6 **25**, 226.

Shedlovsky, T., and D. A. MacInnes (1936). *J. Am. Chem. Soc.* **58**, 1970.

Smith, R. M., and A. E. Martell (1976). *Critical Stability Constants*, Vol. 4, Plenum Press, New York.

Stokes, R. (1979). In *Activity Coefficients in Electrolyte Solutions*, R. M. Pytkowicz, Ed., CRC Press, Boca Raton, p. 1.

Stokes, R. H., and R. A. Robinson (1948). *J. Am. Chem. Soc.* **70**, 1870.

Storokin, A. V., M. D. Lagunov, and V. I. Belokoskov (1956). *Russ. J. Phys. Chem.* **40**, 1501.

Struck, B. D., and O. Schneider (1972). *J. Electronanal. Chem.* **36**, 31.

Stumm, W., and J. J. Morgan (1970). *Aquatic Chemistry: An Introduction Emphasizing Chemical Equilibria in Natural Waters*, Interscience, New York.

Thiesen, M. (1904). *Wiss. Abh. Physikalisch.-Technischen Reischsanstalt* **4**, No. 1.

Thompson, M. E. (1966). *Science* **153**, 866.

Thompson, M. E., and J. W. Ross, Jr. (1966). *Science* **154**, 1643.

Toschev, S. (1973). In *Crystal Growth*, P. Hartman, Ed., Elsevier, New York, p. 1.

Truesdell, A. H., and B. F. Jones (1969). *Chem. Geol.* **4**, 51.

van Breemen, N. (1973). *Geochim. Cosmochim. Acta* **37**, 101.

Vasil'ev, V. P., and S. R. Glavina (1976). *Soviet Electrochem.* **12**, 705.

Walker, A. C., V. B. Bray, and J. Johnston (1927). *J. Am. Chem. Soc.* **49**, 1235.

Whitfield, M. (1973). *Mar. Chem.* **1**, 251.

Whitfield, M. (1975). *Mar. Chem.* **3**, 197.

Whitfield, M. (1979). In *Activity Coefficients in Electrolyte Solutions*, R. M. Pytkowicz, Ed., CRC Press, Boca Raton, p. 153.

Young, T. F., C. R. Singleterry, and I. M. Klotz (1978). *J. Phys. Chem.* **82**, 671.

SUGGESTED READINGS

Anderson, H. L., and L. A. Petree, *J. Phys. Chem.* **74**, 1455 (1970).
Heats of mixing.

Arons, A. B., and C. F. Kientzler, *Trans. Am. Geophys. Union* **35**, 722 (1954).
Vapor pressure of sea salt solutions.

Bahe, L. W., *J. Phys. Chem.* **76**, 1062 (1972).
Structure in electrolyte solutions.

Bethe, H. A., *Proc. Royal Soc. Ser. A.* **150**, 552 (1935).
Superlattices.

Childs, C. W., C. J. Downes, and R. F. Platford, *Austr. J. Chem.* **26**, 863 (1973).
Thermodynamics of NaH_2, PO_4, and KH_2PO_4.

Dickson, A. G., and M. Whitfield, *Geochim. Cosmochim. Acta* **10**, 315 (1981).
Use of ion association models to estimate dissociation constants.

Duedall, I. W., and P. K. Weyl, *Limnol. Oceanogr.* **12**, 52 (1967).
Partial equivalent volumes in seawater.

Eisenberg, D., and W. Kanzmann, *The Structure and Properties of Water*, Oxford University Press, Oxford, 1969.

Guggenheim. E. A., *Phils. Mag.* **19**, 588 (1935).
Thermodynamics of aqueous electrolytes.

Franks, F., *Water: A Comprehensive Treatise*, Vol. 1, Plenum Press, New York, 1972.

Friedman, H. L., and C. V. Krishnan, in *Water: A Comprehensive Treatise*, Vol. 1, F. Franks, Ed., Plenum Press, New York, 1973, p. 16.
Thermodynamics of ionic hydration.

Hood, D. W., Ed., Symposium on Organic Matter in Natural

Waters, Fairbanks, *Inst. Mar. Sci., Univ. Alaska Occas. Publ.* **1**, 625 (1970).

Horne, R. A., Ed., *Water and Aqueous Solutions,* Interscience, New York, 1972.

Horne, R. A., R. A. Courant, B. R. Myers, and J. H. B. George, *J. Phys. Chem.* **68**, 2578 (1964).
Ion-exchange equilibrium at pressure.

Horne, R. A., and D. S. Johnson, *J. Chem. Phys.* **44**, 2946 (1966).
Electrical conductivity and the structure of aqueous $MgSO_4$.

Horne, R. A., and J. D. Birkett, *Electrochim. Acta* **12**, 1153 (1967).
Hydration atmosphere of aqueous alkali-metals.

Hunt, John P., *Metal Ions in Aqueous Solutions,* Benjamin, New York, 1965.

Johnson, K. S., and R. M. Pytkowicz, *Mar. Chem.* **10**, 85 (1981).
Activity coefficient of NaCl in seawater.

Justice, M.-C., and J.-C. Justice, *J. Soln. Chem.* **5**, 543 (1976).
Application of the MacMillan–Meyer theory.

Katz, A., and S. Ben-Yaakov, *Mar. Chem.* **8**, 263 (1980).
Diffusion of seawater ions and ion pairs.

Kavanan, J. L., *Water and Solute–Water Interactions,* Holden-Day, San Francisco, 1964.

Kester, D. R., *J. Geophys. Res.* **79**, 455 (1974).
Freezing point of seawater.

Kortum, G., *Treatise on Electro-Chemistry,* 2nd Ed., Elsevier, Amsterdam, 1965, Chap. III.

Koryta, J., J. Dvorak, and V. Bohackova, *Electrochemistry,* Science Paperbacks, London, 1973, p. 22.

Leyendekkers, J., *Mar. Chem.* **1**, 75 (1973).
Chemical potentials of seawater components.

Leyendekkers, J., *Mar. Chem.* **8**, 89 (1974).
Partial molal volumes of electrolytes in seawater.

Leyendekkers, J. V., *Mar. Chem.* **3**, 23 (1975)
Pressure and chemical potentials in seawater.

Madelung, E., *Physik. Z.* **19**, 524 (1948).
Madelung constant.

Mayer, J. E., *J. Chem. Phys.* **18**, 1426 (1950).
Theory of ionic solutions.

Millero, F. J., *Ann. Rev. Earth Planet. Sci.* **2**, 101 (1974).
Physical chemistry of seawater.

Pitzer, K. S., *J. Soln. Chem.* **4**, 249 (1975).
Effect of high-order terms on specific interactions.

Pitzer, K. S., and L. F. S. Silvester, *J. Soln. Chem.* **5**, 269 (1970).
Specific interactions for weak electrolytes.

Pitzer, K. S., and L. F. S. Silvester, *J. Phys. Chem.* **82**, 1239 (1978).
Specific interactions for 3-2, 4-2, and other high-valence electrolytes.

Pytkowicz, R. M., I. W. Duedall, and D. N. Connors, *Science* **152**, 640 (1966).
Activity of magnesium in seawater.

Pytkowicz, R. M., and D. R. Kester, in *Oceanography Marine Biology Annual Review,* H. Barnes, Ed., Allen and Unwin, London, 1971, p. 11.
Physical chemistry of seawater.

Reeburgh, W. S., *J. Mar. Res.* **23**, 187 (1965).
Electric conductivity of seawater.

Reilly, P. J., R. H. Wood, and R. A. Robinson, *J. Phys. Chem.* **75**, 1305 (1971).
Osmotic and activity coefficients.

Sipos, L., B. Raspor, H. W. Nürnberg, and R. M. Pytkowicz. *Mar. Chem.* **9**, 37 (1980).

van Breemen, N., *Geochim. Cosmochim. Acta* **37**, 101 (1973).
Activities of ions in natural waters.

INDEX